A TOUCH *of* BLOSSOM

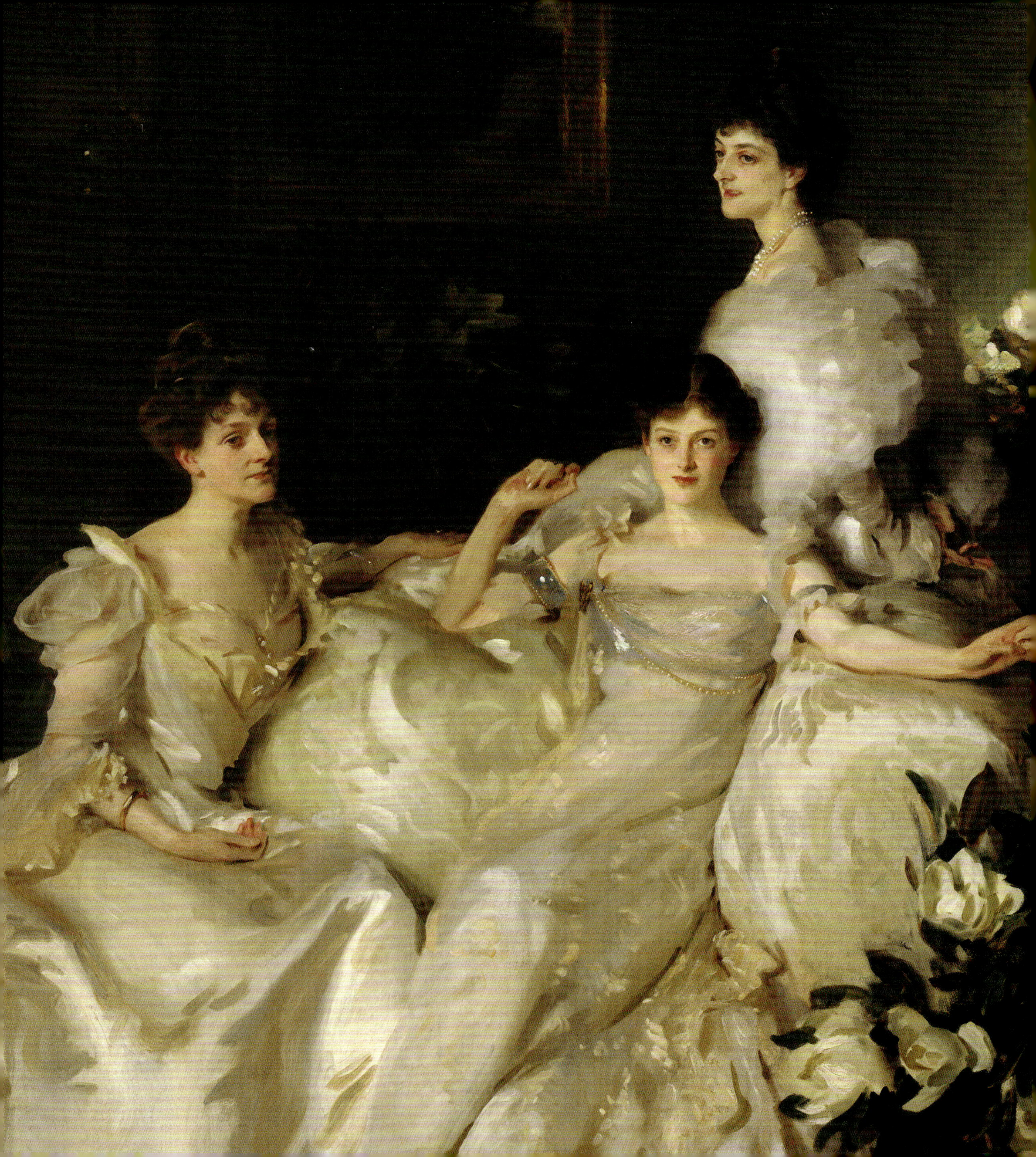

A TOUCH *of* BLOSSOM

JOHN SINGER SARGENT AND THE QUEER FLORA OF FIN-DE-SIÈCLE ART

ALISON SYME

The Pennsylvania State University Press
University Park, Pennsylvania

Publication of this book has been aided by a grant from the Wyeth Foundation for American Art Publication Fund of the College Art Association.

Frontispiece: John Singer Sargent, *The Wyndham Sisters: Lady Elcho, Mrs. Adeane, and Mrs. Tennant*, 1899 (detail). The Metropolitan Museum of Art, Catharine Lorillard Wolfe Collection, Wolfe Fund, 1927 (27.67). © The Metropolitan Museum of Art.

Page vi: František Kupka, *Self-Portrait Among Roses*, 1894–95 (detail). Museum Kampa—Jan and Meda Mládek Foundation Collection, Prague. © 2008 Artists Rights Society (ARS), New York/ ADAGP, Paris.

Library of Congress Cataloging-in-Publication Data

Syme, Alison (Alison Mairi)

A touch of blossom : John Singer Sargent and the queer flora of fin-de-siècle art / Alison Syme.

p. cm.

Includes bibliographical references and index.

Summary: "Explores the art of John Singer Sargent in the context of nineteenth-century botany, gynecology, literature, and visual culture. Argues that the artist was elaborating both a period poetics of homosexuality and a new sense of subjectivity, anticipating certain aspects of artistic modernism."—Provided by publisher.

ISBN 978-0-271-03622-9 (cloth : alk. paper)

1. Sargent, John Singer, 1856-1925—Criticism and interpretation. 2. Plants in art. 3. Plants, Sex in. 4. Art and science—History—19th century. 5. Arts and society—History—19th century. I. Sargent, John Singer, 1856-1925. II. Title. III. Title: John Singer Sargent and the queer flora of fin-de-siècle art.

N6537.S32S96 2010

759.13–dc22

2009043575

Printed in Hong Kong by Asia Pacific Offset

Published by The Pennsylvania State University Press,

University Park, PA 16802-1003

The Pennsylvania State University Press is a member of the Association of American University Presses.

It is the policy of The Pennsylvania State University Press to use acid-free paper. Publications on uncoated stock satisfy the minimum requirements of American National Standard for Information Sciences—Permanence of Paper for Printed Library Material, ANSI Z39.48–1992.

to my sweet pea

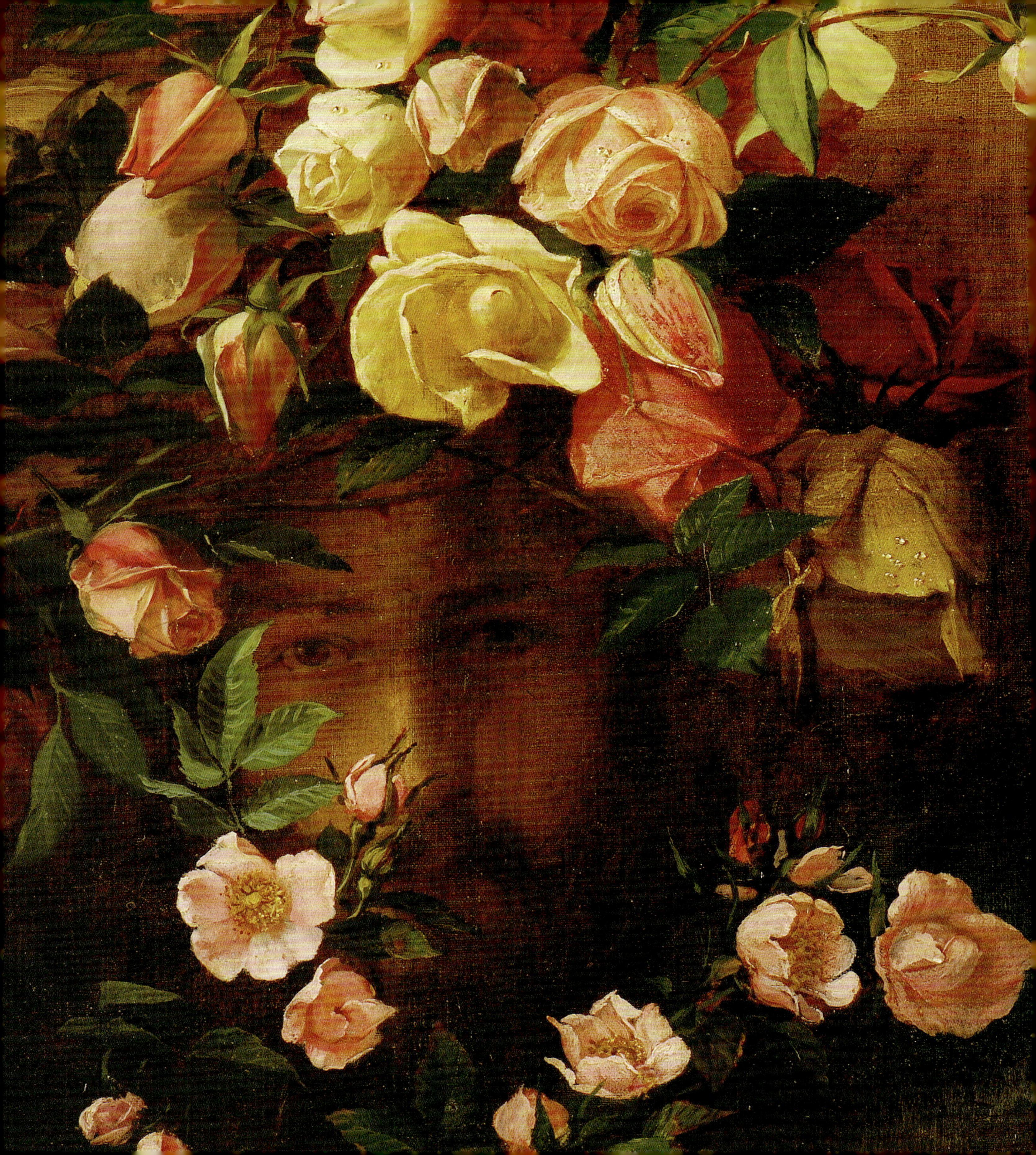

CONTENTS

viii Acknowledgments

x List of Illustrations

1 INTRODUCTION: THE BIRDS AND THE BEES

22 1 LA VIE EN ROSE

70 2 LASCIVIOUS DIGITATION, OR THE IMPORTANCE OF MANUAL STIMULATION TO THE INVERT ARTIST

108 3 DR. OCTOGYNECOLOGIST

144 4 PORTRAIT OF THE ARTIST AS A YOUNG LADYBIRD

194 5 THE STING

229 CONCLUSION: LEAF TAKING

238 Notes

288 Bibliography

308 Index

ACKNOWLEDGMENTS

I've spent a lot of time fumbling in the dark. As one might imagine, this occupation has had its pleasures as well as its frustrations. But now this book is finally ready to see the light of day, and I must bid my leafy progeny go forth and prosper. While this moment of parting is a melancholy one, it is nevertheless a delight and consolation to be able to acknowledge the many personal and professional debts I have accumulated over its long gestation period.

A Touch of Blossom grew out of a dissertation I wrote at Harvard under the supervision of Professors Ewa Lajer-Burcharth, Yve-Alain Bois, and Elaine Scarry. Each brilliant and inimitable member of this trio brought invaluable expertise to my project (then exuberantly entitled "Hedgewhores, Wagtails, Cockatrices, Whipsters: John Singer Sargent and His Coterie of Nature's Artful Dodgers"). I will always be grateful for their guidance, support, humor, and criticism. The inspiring scholarship and teaching of Norman Bryson, Harry Cooper, Barbara Johnson, Joseph Koerner, Margaretta Lovell, and Rabun Taylor were also essential to my development as an academic, and I feel privileged to have worked and studied with them. Dialogue with my classmates was equally challenging and formative: love and thanks go to Debbie Gettelman, Jordan Kantor, Anne Leonard, Meredith Martin, Benjamin Paul, Amy Powell, Zahr Said, Holger Schott, and Alicia Walker.

While a graduate student I was fortunate enough to become a Wyeth Fellow at the Center for Advanced Study in the Visual Arts. The adorable and formidable Elizabeth Cropper kept us all on our toes and was an endearingly sharp reader and critic. C. Jean Campbell, Caroline Elam, Meredith Hale, Carmenita Higginbotham, Peter Lukehart, and Therese O'Malley were all constructive interlocutors. Sally Promey and Virginia Spate deserve special mention: both specialists in what I might with hubris call my areas of expertise, they generously invested time and energy in my work, threw new light on many issues, and stand out in my mind as superb scholars and emulable human beings.

Along the way all sorts of people have helped me unstintingly. Juliet Bellow, Whitney Davis, Anne Helmreich, Lawrence Joseph, and Joanna Luft graciously shared papers with me. Many offered references and resources; Terri Hensick and Miriam Stewart allowed me access to key holdings at the Fogg Art Museum; Clive Getty let me study his copy of *Les fleurs animées*. Anne Higgonet kindly included me in her panel "New Directions in Nineteenth-Century Art" at CAA. Without University of Toronto and UTM grants this project would have been much delayed (thank you, Gage Averill and Ulli Krull!). And my colleagues at the University of Toronto, some of them former teachers—Matt Brower, Mark Cheetham, Deepali Dewan, Marc Gotlieb, Betsy Harney, Kajri Jain, Louis Kaplan, Matt Kavaler, Elizabeth Legge, Evonne Levy, Alex Nagel, Julian Patrick, Jenny Purtle, and John Ricco—have been amazing. I must also thank innumerable helpful people at libraries, archives, galleries, and museums. In particular I profess my debt and gratitude to Richard Ormond, Elizabeth Oustinoff, and Adelson Galleries for their patient fielding of inquiries and image requests. And I am grateful to my resourceful and enthusiastic research assistants Julie Boivin, Mallory Diaczun, Rose Logie, Lindsay May, and especially Richard Nalli-Petta.

My brilliant friends Jessica Humphreys, Emily Menezes, and especially Miranda Purves have nurtured this book along the way, as have my siblings, Alasdair and Catriona, and my parents, Morag and Colin Syme. I thank Eleanor Goodman at Penn State for liking this project and for her hard work, and the anonymous readers of the manuscript for their engagement with the text and their constructive criticisms. Suzanne Wolk did a marvelous job copyediting the manuscript, and the team at Penn State have made this book more beautiful than I could have imagined. Finally, I acknowledge my greatest debt, to Holger: fiendish enchanter, my partner.

ILLUSTRATIONS

1. Gustave Courbet, *The Trellis,* 1862. Oil on canvas, 109.8 x 135.2 cm. Toledo Museum of Art, purchased with funds from the Libbey Endowment, gift of Edward Drummond Libbey, 1950.309.

2. J. J. Grandville, "Le Louvre des marionnettes," in *Un autre monde* (Paris: H. Fournier, 1844), 86. Courtesy of the Thomas Fisher Rare Book Library, University of Toronto.

3. Claude Monet, *Caricature of Jules Didier,* c. 1860. Charcoal, heightened with white chalk, on blue laid paper (discolored to grey), 61.6 x 43.6 cm. Mr. and Mrs. Carter H. Harrison Collection, 1933.889, The Art Institute of Chicago. Photography © The Art Institute of Chicago.

4. Odilon Redon, *There Was Perhaps a First Vision Attempted in the Flower,* plate 2 of 8 from *Les Origines,* 1883. Lithograph in black on light grey chine laid down on white wove paper, 22.5 x 17.7 cm (image/chine), 51.1 x 35.6 cm (sheet). The Stickney Collection, 1920.1579, The Art Institute of Chicago. Photography © The Art Institute of Chicago.

5. Claude Monet, *The Playwright François Nicolaïe, Called "Clairville,"* 1858. Pencil on beige paper, 32 x 24 cm. Musée Marmottan, Paris. Photo © Bridgeman-Giraudon/Art Resource, New York.

6. Albert Moore, *William Connal, Esq. Jr., of Solsgirth,* 1883. Oil on canvas, 33 x 25.4 cm. York Museums Trust (York Art Gallery).

7. Vilhelm Pedersen, "Tommelise," 1843, in Hans Christian Andersen, *Eventyr* (Copenhagen: C. A. Reitzel, 1850), 137. Courtesy of the Thomas Fisher Rare Book Library, University of Toronto.

8. Odoardo Fialetti, *Petal Venus,* engraved by Francesco Valesio, in Adriaan van de Spiegel, *De formato foetu* (Frankfurt: Mattheus Merian, 1631), 37, plate 4. Courtesy of the Thomas Fisher Rare Book Library, University of Toronto.

9. Valentine card with cherub, c. 1850s. Hand-colored engraving and paper lace, 22 x 26 cm. Museum of London. Photo © Museum of London.

10. English postcard and American greeting card with baby-in-flower motif, both 13.7 x 8.6 cm, 1880s.

11. James McNeill Whistler, *Loie Fuller Dancing,* 1892. Pen and black ink on off-white wove paper, 20.9 x 31.9 cm. Hunterian Museum and Art Gallery, Glasgow, GLAHA 46173. © Hunterian Museum and Art Gallery, University of Glasgow.

12. James McNeill Whistler, *Three Butterflies,* c. 1890. Graphite on ivory wove card, 24.1 x 20.6 cm. Gift of Mr. Walter S. Brewster, 1933.284a, The Art Institute of Chicago. Photography © The Art Institute of Chicago.

13. Charles Ricketts, cover of Oscar Wilde, *The Picture of Dorian Gray* (London: Ward, Lock, and Co., 1891). Courtesy of Special Collections and Archives, Sydney Jones Library, University of Liverpool.

14. J. J. Grandville, "Tubéreuse et Jonquille," in *Les fleurs animées* (Paris: Garnier Frères, 1867), vol. 2.

15. Liebig card with animated flowers, 1897. Chromolithograph, 7 x 10.5 cm.

16. Thomas de La Rue, valentine, series 78, c. 1880, 12.7 x 8.7 cm.

17. Giambattista della Porta, "Orchid," in *Phytognomonica* (Rothomagi: Berthelin, 1650), 244. Courtesy of the Thomas Fisher Rare Book Library, University of Toronto.

18. Gabriel de Saint-Aubin, *Comparaison du bouton de rose,* 1781, engraved by R. Dennel. Reproduced in Émile Dacier, *Gabriel de Saint-Aubin: Peintre, dessinateur et graveur,* vol. 1 (Paris: Les Éditions G. van Oest, 1929), plate 12. Courtesy of Robarts Library, University of Toronto.

19. Dr. Warner's Coraline Corsets trade cards, 1880s, (*left*) 12.7 x 7.6 cm, (*right*) 15 x 8 cm.

20. M. Boulanger, "L'oeillet," c. 1900, postcard, 13.5 x 8.5 cm.

21. Photograph of Vita Sackville-West as a basket of wisteria, c. 1900 (?). Collection of Adam Nicholson.

22. "Last Sweet Thing in Toilettes," *Punch, or the London Charivari,* 5 July 1879, 304. Courtesy of Robarts Library, University of Toronto.

23. John Singer Sargent, *The Wyndham Sisters: Lady Elcho, Mrs. Adeane, and Mrs. Tennant,* 1899. Oil on canvas, 292.1 x 213.7 cm. The Metropolitan Museum of Art, Catharine Lorillard Wolfe Collection, Wolfe Fund, 1927 (27.67). Image © The Metropolitan Museum of Art.

24. John Singer Sargent, *Miss Beatrice Goelet,* 1890. Oil on canvas, 162.6 x 86.3 cm. Private collection. Reproduced in *The Century* 43, no. 5 (March 1892): 773.

25. Georges Rochegrosse, *Knight Among the Flower-Maidens, from Wagner's "Parsifal,"* 1894. Oil on canvas, 235 x 376 cm. Musée d'Orsay, Paris. Photo © Réunion des Musées Nationaux/Art Resource, New York.

26. Walter Crane, "Tulips" and "Convolvulus and Honeysuckle," in *Flora's Feast: A Masque of Flowers* (London: Cassell and Co., 1890), 14, 34. Courtesy of the Robertson Davies Library, Massey College, University of Toronto.

27. J. J. Grandville, "Pervenche desséchée," in *Les fleurs animées* (Paris: Gabriel de Gonet, 1847), vol. 2.

28. Alfred Le Petit, "M. Littré," in *Fleurs, fruits et légumes du jour* (Paris: Au Bureau de l'Éclipse, 1871), 11.

29. Victorian seed and fertilizer trade cards, c. 1885–90, all between 12.8 and 13.5 x 7.5 and 8 cm.

30. Le Bon Marché trade cards from the Noce de légumes and Fruits animés series, c. 1890s, each 18 x 12 cm.

31. J. J. Grandville, "Giroflée," in *Les fleurs animées* (Paris: Gabriel de Gonet, 1847), vol. 2.

32. Edward Lear, "There was an old person so silly," sketch formerly in the possession of Mrs. C. H. A. Anstruther-Duncan, in Lear, *Bosh and Nonsense* (London: Allen Lane, 1982), 44.

33. Vick's Seeds catalogue cover depicting pansy sailors, 1896, 26 x 19.6 cm.

34. Charles Ricketts, detail of the cover of Oscar Wilde, *Intentions* (London: James R. Osgood, McIlvaine, and Co., 1891). Courtesy of the Thomas Fisher Rare Book Library, University of Toronto.

35. Jacques-Émile Blanche, *Marcel Proust,* 1892. Oil on canvas, 73.5 x 60.5 cm. Musée d'Orsay, Paris. Photo © Erich Lessing/Art Resource, New York. © 2008 Artists Rights Society (ARS), New York/ADAGP, Paris.

36. Charles-Germain de Saint-Aubin, *La toilette,* c. 1750s. Etching, 34.1 x 24.7 cm. Edmond Rothschild Collection. Photo © Réunion des Musées Nationaux/Art Resource, New York.

37. Maison du Pont-Neuf trade cards, late 1880s or early 1890s, each 7.7 x 11.3 cm.

38. J. J. Grandville, "Un vieux papillon," in P.-J. Stahl, *Scènes de la vie privée et publique des animaux,* 2 vols. (Paris: J. Hetzel et Paulin, 1842), vol. 1, between 136 and 137. Courtesy of the Thomas Fisher Rare Book Library, University of Toronto.

39. J. J. Grandville, illustration for "Les aventures d'un papillon," in Stahl, *Scènes de la vie privée et publique des animaux,* 1:121. Courtesy of the Thomas Fisher Rare Book Library, University of Toronto.

40. H. J. Ford, "The Princess Carried Off by the Bees," in Andrew Lang, *The Green Fairy Book* (London: Longmans, Green, 1892), 53. Courtesy of Widener Library, Harvard College Library, 25227.31.5.

41. Chicorée trade card, 1890s, 10.4 x 6.5 cm.

42. "Spring Flowers," *Illustrated Sporting and Dramatic News,* 10 April 1879, 108. Courtesy of the Toronto Reference Library.

43. Alfred Le Petit, "La rose impériale," cover of *La Charge* 4 (1875). Courtesy of University of Glasgow Library, Department of Special Collections (HX133/1/123).

44. Théophile Thomas, maquettes of costumes for *Le roi Carotte,* 1872. Watercolor on paper. Bibliothèque musée de l'Opéra, D-216 (Z1), plate 14. Image courtesy Bibliothèque nationale de France.

45. Alfred Albert, Emma Livry's second costume for *Le papillon,* 1860. Bibliothèque musée de l'Opéra, D-216 (21), plate 61. Image courtesy Bibliothèque nationale de France.

46. Alfred Albert, "Wasp and Large Moth," maquettes of costumes for the ballet *La source,* 1866. Colored pencil on paper. Bibliothèque musée de l'Opéra, D-216 (23bis), plate 21. Image courtesy Bibliothèque nationale de France.

47. French trade cards, 1880s and 1890s, all between 10.5 and 11.5 x 6.5 and 7.5 cm; (*center*) postcard (cropped), 1900s, 14 x 9 cm.

48. Advertisement for stage version of Ouida's *Moths,* 1886, 15.7 x 20.3 cm.

49. J. J. Grandville, "Guêpe," in Stahl, *Scènes de la vie privée et publique des animaux,* vol. 2, between 116 and 117. Courtesy of the Thomas Fisher Rare Book Library, University of Toronto.

50. John Singer Sargent, *Lady Sassoon,* 1907. Oil on canvas, 157.5 x 104.1 cm. Private collection. Photo © Private Collection/The Bridgeman Art Library.

51. John Singer Sargent, *Two Men Smoking and Writing* (?), c. 1874–77. Pencil on beige paper, 10.16 x 16.83 cm. Gift of Miss Emily Sargent and Mrs. Francis Ormond, sisters of the artist, 49.148d. Collection of The Corcoran Gallery of Art, Washington, D.C.

52. John Singer Sargent, *Portrait of Carolus-Duran,* 1879. Oil on canvas, 116.8 x 96 cm. Sterling and Francine Clark Art Institute, Williamstown, Massachusetts, 1955.14.

53. Stop [L. P. G. B. Morel-Retz], "Portrait of M. Carolus Duran, in the Role of the Marquis de Méphisto," *Journal Amusant,* 5 July 1879, 5. Courtesy of the Special Collections and Archives Department, Delmar T. Oviatt Library, California State University, Northridge, California.

54. "Edison's Anti-Gravitation Under-Clothing," *Punch's Almanack for 1879,* 9 December 1878. Courtesy of Robarts Library, University of Toronto.

55. James McNeill Whistler, *Butterfly,* c. 1899. Pen and black ink on cream wove card, 12 x 15.8 cm. Gift of Mr. Walter S. Brewster, 1933.291, The Art Institute of Chicago. Photography © The Art Institute of Chicago.

56. Giovanni Morelli [Ivan Lermolieff], "Typical Hands," in *Italian Painters: Critical Studies of their Works,* vol. 1, *The Borghese and Doria Pamfili Galleries in Rome* (London: John Murray, 1892), 77. Courtesy of the Toronto Reference Library.

57. John Singer Sargent, *The Misses Vickers,* 1884. Oil on canvas, 137.8 x 182.9 cm. Graves Gallery, Museums Sheffield. Photo © Sheffield Galleries and Museums Trust, UK/The Bridgeman Art Library.

58. John Singer Sargent, *Venetian Onion Seller,* 1880–82. Oil on canvas, 95 x 70 cm. Museo Thyssen-Bornemisza, Madrid. Photo © Museo Thyssen-Bornemisza, Madrid.

59. John Singer Sargent, *Mrs. Albert Vickers,* 1884. Oil on canvas, 208.4 x 101.5 cm. Virginia Museum of Fine Arts, Richmond, the Adolph D. and Wilkins C. Williams Fund. Photo: Katherine Wetzel. © Virginia Museum of Fine Arts.

60. John Singer Sargent, *Study for Mrs. Albert Vickers* (Edith Foster), c. 1884. Charcoal on off-white wove paper, 33.3 x 23 cm. Harvard University Art Museums, Fogg Art Museum, gift of Miss Emily Sargent and Mrs. Francis Ormond in memory of their brother, John Singer Sargent, 1931.92. Photo: Imaging Department. © President and Fellows of Harvard College.

61. Detail of *Study for Mrs. Albert Vickers.*

62. John Singer Sargent, *Léon Delafosse,* c. 1895–98. Oil on canvas, 101 x 59.4 cm. Seattle Art Museum, 2001.17. Given in honor of Trevor Fairbrother by Mr. and Mrs. Prentice Bloedel by exchange, and by Robert M. Arnold, Tom and Ann Barwick, Frank Bayley, Jeffrey and Susan Brotman, Contemporary Art Council, Council of American Art, Jane and David R. Davis, Decorative Arts and Paintings Council, Robert B. Dootson, Mr. and Mrs. Barney Ebsworth, P. Raaze Garrison, Lyn and Gerald Grinstein, Helen and Max Gurvich, Marshall Hatch, John and Ann Hauberg, Richard and Betty Hedreen, Marty Ann and Henry James, Mrs. Janet W. Ketcham, Allan and Mary Kollar, Greg Kucera and Larry Yocom, Rufus and Pat Lumry, Byron R. Meyer, Ruth J. Nutt, Scotty Ray, Gladys and Sam Rubenstein, Mr. and Mrs. Allen Vance Salsbury, Herman and Faye Sarkowsky, Mr. and Mrs. Douglas Scheumann, Seattle Art Museum Supporters, Jon and Mary Shirley, Joan and Harry Stonecipher, Dean and Mary Thornton, William and Ruth True, Volunteers Association, Ms. Susan Winokur and Mr. Paul Leach, The Virginia Wright Fund, Charlie and Barbara Wright, Howard Wright and Kate Janeway, Merrill Wright, and Mrs. T. Evans Wyckoff. Photo: Paul Macapia.

63. John Singer Sargent, *Helen Dunham,* 1892. Oil on canvas, 121.5 x 81.3 cm. Private collection. Photo: Peter A. Juley and Son Collection, Smithsonian American Art Museum, Washington, D.C.

64. John Singer Sargent, *Beatrice Townsend,* c. 1882. Oil on canvas, 81.9 x 58.4 cm. Private collection. Photo © Geoffrey Clements/CORBIS.

65. Abraham Kent (?), sculpted hands of royal child(ren), possibly Princess Victoria's, c. 1840s. Marble. RCIN 34358, The Royal Collection. © 2008 Her Majesty Queen Elizabeth II.

66. Visiting and greeting cards and scrap depicting floral hands, later nineteenth century.

67. John Singer Sargent, *Mrs. Charles Thursby,* c. 1897–98. Oil on canvas, 198 x 101 cm. Purchase by exchange, 1985. Gift of Mr. and Mrs. J. Duncan Pitney, Emilie Coles (from the J. Ackerman Coles Collection), Mrs. Lewis B. Ballantyne, Mrs. Owen Winston, and the Bequest of Louis Bamberger. Collection of The Newark Museum, Newark, New Jersey.

68. John Singer Sargent, *Brigadier Archibald Campbell Douglas Dick,* 1886. Oil on canvas, 160 x 90.8 cm. Private collection. Photo courtesy of Spanierman Gallery, LLC, New York.

69. John Singer Sargent, *Mrs. Thomas Wodehouse Legh,* 1884. Oil on canvas, 119.4 x 94 cm. Private collection. Photo courtesy of Adelson Galleries, New York.

70. Detail of *Mrs. Thomas Wodehouse Legh.*

71. John Singer Sargent, *Lady Agnew of Lochnaw,* 1892. Oil on canvas, 127 x 101 cm. The National Gallery of Scotland, Edinburgh, purchased with the aid of the Cowan Smith Bequest Fund, 1925.

72. Giambattista della Porta, "Orchid," in *Phytognomonica* (Rothomagi: Berthelin, 1650), 240. Courtesy of the Thomas Fisher Rare Book Library, University of Toronto.

73. Max Beerbohm, "London in November, and Mr. Henry James in London," in *A Book of Caricatures* (London: Methuen, 1907), plate 11. Houghton Library, fEC9. B3927.907b (B). By permission of the Houghton Library, Harvard University.

74. James Hadley, *Aesthetic Teapot,* 1882. Glazed parian, 15.7 x 17.5 x 8.5 cm. Worcester Porcelain Museum. Photo by John Everett.

75. "L'anguille," c. 1910, postcard, 14 x 9 cm.

76. "Queer Fish at the Westminster Aquarium," *Illustrated Sporting and Dramatic News,* 1 July 1876, 333. Courtesy of the Toronto Reference Library.

77. John Singer Sargent, *Two Octopi,* c. 1875. Oil on canvas, 40.6 x 32.1 cm. Private collection. Photo: Peter A. Juley and Son Collection, Smithsonian American Art Museum, Washington, D.C.

78. Detail of "Orchis mascula," in Charles Darwin, *The Various Contrivances by which Orchids are Fertilised by Insects,* 2nd ed. (London: John Murray, 1877), 8, fig. 1. Courtesy of the Thomas Fisher Rare Book Library, University of Toronto.

79. Advertisement for sphincter grip armoured hose, in *Garden-Work for Villa, Suburban, Town, and Cottage Gardens,* 29 June 1889, after 307. NLS shelfmark Q.55. Courtesy of the Trustees of the National Library of Scotland.

80. John Singer Sargent, *Hercules and the Hydra,* 1922–25. Oil on canvas, 347.98 x 317.5 cm. Museum of Fine Arts, Boston. Francis Bartlett Donation of 1912 and Picture Fund, 25.647. Photograph © 2010 Museum of Fine Arts, Boston.

81. Émile Gallé, *Hand with Seaweed and Shells,* 1904. Glass modeled under heat with inclusions of metallic oxides, veins, applications in low and high relief, and wheel engraving, 33.4 x 13.4 cm. Musée d'Orsay, Paris. Photo © Réunion des Musées Nationaux/Art Resource, New York.

82. John Singer Sargent, *Jacques-Émile Blanche,* c. 1886. Oil on canvas, 81.9 x 48.9 cm. Musée des Beaux-Arts, Rouen. Photo © Bridgeman-Giraudon/Art Resource, New York.

83. Jacques-Émile Blanche, *Self-Portrait with Raphael de Ochoa,* 1890. Oil on fabric, 99.2 x 71 cm. © The Cleveland Museum of Art, bequest of Noah L. Butkin, 1980.230. © 2008 Artists Rights Society (ARS), New York/ADAGP, Paris.

84. John Singer Sargent, *Paul Helleu Lying in a Field,* c. 1889. Pastel on paper, 26 x 37.7 cm. British Museum, London. Presented by Mrs. Ormond and Miss Sargent, 1959, 0102.4. © The Trustees of the British Museum.

85. John Singer Sargent, *Robert Louis Stevenson,* 1887. Oil on canvas, 51 x 61.8 cm. Bequest of Charles Phelps and Anna Sinton Taft, 1931.472. The Taft Museum of Art. Photo: Tony Walsh, Cincinnati, Ohio. Courtesy of the Taft Museum of Art, Cincinnati, Ohio.

86. John Singer Sargent, *Paul Helleu,* c. 1880. Pastel on thin brown wove paper, 49 x 43.9 cm. Harvard University Art Museums, Fogg Art Museum, bequest of Mrs. Annie Swan Coburn, 1933.18. Photo: Allan Macintyre. © President and Fellows of Harvard College.

87. Detail of *Paul Helleu.*

88. John Singer Sargent, *Dr. Pozzi at Home,* 1881. Oil on canvas, 202.9 x 102.2 cm. The Armand Hammer Collection. Gift of the Armand Hammer Foundation, Hammer Museum, Los Angeles, California.

89. Charles Nicolet, "Bimanual palpation," in Samuel Pozzi, *Traité de gynécologie clinique et opératoire* (Paris: G. Masson, 1890), 101, fig. 70.

90. "Myosarcoma of the cervix in the form of a cluster of grapes," reprinted from Pernice, and "Fibrous adenoma of the uterus," reprinted from Thomas, in Pozzi, *Traité de gynécologie* (1890), 376, figs. 199 and 200.

91. "Cluster of grapes imperfectly fertilized," in *Vick's Illustrated Monthly Magazine,* December 1883, 381. Courtesy of Widener Library, Harvard College Library, Sci 1621.45.4(6).

92. "Ovarian dermoid containing 3,930 epithelial balls," in J. Bland-Sutton, *Surgical Diseases of the Ovaries and Fallopian Tubes, including Tubal Pregnancy* (London: Cassell and Co., 1896), 65, fig. 38. Courtesy of the Gerstein Science Information Centre, University of Toronto.

93. "Fibro-myxoma of the tubal fimbriae," reprinted from Martin, in Bland-Sutton, *Surgical Diseases of the Ovaries and Fallopian Tubes,* 230, fig. 103. Courtesy of the Gerstein Science Information Centre, University of Toronto.

94. "Enormous ovarian cyst in a girl seventeen years of age," reprinted from Dayot, in Bland-Sutton, *Surgical Diseases of the Ovaries and Fallopian Tubes,* 82, fig. 50. Courtesy of the Gerstein Science Information Centre, University of Toronto.

95. "Polycystic ovarian tumor," in Louis A. Rodenstein, "A Case of Enormous Polycystic Ovarian Tumor," *American Journal of Obstetrics and Diseases of Women and Children* 12 (April 1879): 304. Courtesy of the Gerstein Science Information Centre, University of Toronto.

96. "Ovarian cyst complicated by pregnancy," reprinted in Pozzi, *Traité de gynécologie* (1890), 746, fig. 365.

97. Charles Nicolet, "Colpo-perineoplasty by flap-splitting," in Pozzi, *Traité de gynécologie* (1890), 541, fig. 305.

98. Charles Nicolet, "Suture of the abdominal walls about the pedicle," in Pozzi, *Traité de gynécologie* (1890), 310, fig. 161.

99. Detail of "Cypripedium sedeni," in Benjamin Samuel Williams, *The Orchid-Grower's Manual,* 6th ed. (London: Victoria and Paradise Nurseries, 1885), after 256.

100. "Actinia mesembryanthemum," in Philip Henry Gosse, *Actinologia Britannica: A History of the British Sea-Anemones and Corals* (London: Van Voorst, 1860), plate 6, between 216 and 217. Facsimiles of Gosse's drawings made by W. Dickes. Courtesy of the Thomas Fisher Rare Book Library, University of Toronto.

101. "Actiniae," in Ernst Haeckel, *Kunstformen der Natur* (Leipzig: Verlag des Bibliographischen Instituts, 1904), plate 49.

102. "Transverse section," in Robert Barnes, *Traité clinique des maladies des femmes,* trans. A. Cordes (Paris: G. Masson, 1876), 3, fig. 1. Courtesy of the Gerstein Science Information Centre, University of Toronto.

103. Charles Nicolet, "Prolapse of the uterus," in Pozzi, *Traité de gynécologie* (1890), 518, fig. 282.

104. "La perle," c. 1900, postcard, 14 x 9 cm.

105. Photograph of Pozzi's ward, Hôpital Broca, c. 1900. Archives of the Musée de l'Assistance Publique–Hôpitaux de Paris.

106. Georges Clairin, *Health Restored to the Sick,* 1898. Oil on canvas, 275 x 440 cm. Musée de l'Assistance Publique–Hôpitaux de Paris.

107. "Simple vegetations of the vulva," reprinted from Tarnier, in Pozzi, *Traité de gynécologie* (1890), 1014, fig. 456.

108. "Bullous syphilide rupia of the face and chest; bullous syphilide rupia, with cicatrices of tubercles," in Auguste Cullerier, *Atlas of Venereal Diseases,* trans. Freeman J. Bumstead (Philadelphia: Henry C. Lea, 1868), plate 23.

109. Sandro Botticelli, detail of Flora in *The Primavera,* 1477. Tempera on panel, whole 203 x 314 cm. Uffizi, Florence. Photo © Scala/Art Resource, New York.

110. Detail of "Cypripedium," in Darwin, *The Various Contrivances by which Orchids are Fertilised by Insects,* 2nd ed., 227, fig. 35. Courtesy of the Thomas Fisher Rare Book Library, University of Toronto.

111. "How far is it from this to this?" *Washington Post,* 22 January 1882, 4.

112. Charles Nicolet, "Partial pseudo-hermaphroditism in a male (androgyny)," in Pozzi, *Traité de gynécologie* (1890), 1070, fig. 469.

113. Charles Nicolet, "Anomaly of the hymen," in Pozzi, *Traité de gynécologie* (1890), 1064, fig. 466.

114. Gustave Courbet, *The Origin of the World,* 1866. Oil on canvas, 46 x 55 cm. Musée d'Orsay, Paris. Photo © Réunion des Musées Nationaux/Art Resource, New York.

115. Gustave Courbet, *The Source of the Loue,* c. 1864. Oil on canvas, 107.3 x 137.5 cm. Albright-Knox Art Gallery, Buffalo, New York. George B. and Jenny R. Mathews Fund, 1959.

116. Gustave Courbet, *La dame verte,* early 1840s. Pencil on paper, 14 x 21.7 cm. In album "paysage rocheux," RF29234, folio 10, recto. Musée du Louvre, Paris. Photo © Réunion des Musées Nationaux/Art Resource, New York.

117. René Lalique, *Gynécologie,* c. 1898. Ivory, 17 x 11.5 cm. Robert Zehil Collection. Photo © Robert Zehil Collection, Monte Carlo. © 2008 Artists Rights Society (ARS), New York/ADAGP, Paris.

118. Charles Nicolet, "Absence of vagina and uterus," in Pozzi, *Traité de gynécologie* (1890), 1090, fig. 475.

119. "Mikschik's instrument for excision of uterine polyps," in F. W. von Scanzoni, *Lehrbuch der Krankheiten der weiblichen Sexualorgane* (Vienna: Wilhelm Braumüller, 1857), 220. Courtesy of the Gerstein Science Information Centre, University of Toronto.

120. "Reduction of a retroflexion of the uterus by the sound," reprinted from Barnes, in Pozzi, *Traité de gynécologie* (1890), 474, fig. 253.

121. "Neugebauer's speculum and retracting hooks in place," reprinted in Pozzi, *Traité de gynécologie* (1890), 912, fig. 395.

122. John Singer Sargent, *Madame X,* 1883–84. Oil on canvas, 208.6 x 109.9 cm. The Metropolitan Museum of Art, Arthur Hoppock Hearn Fund, 1916 (16.53). Image © The Metropolitan Museum of Art.

123. J. J. Grandville, "Pensée," *Les fleurs animées* (Paris: Garnier Frères, 1867), vol. 1.

124. John Singer Sargent, *Carnation, Lily, Lily, Rose,* 1885–86. Oil on canvas, 174 x 153.7 cm. Tate Britain. © Tate, London, 2008.

125. William Dyce, *Titian Preparing to Make His First Essay in Colouring,* 1856–57. Oil on canvas, 100.5 x 79 cm. Aberdeen Art Gallery and Museums Collections.

126. Martin Johnson Heade, *Roses on a Palette,* c. 1883–1900. Oil on canvas, 35.7 x 56.2 cm. Questroyal Fine Art, LLC.

127. Trade cards and assorted scrap with palettes, 1870–1910.

128. Cover of *Maule's Seed Catalog,* 1896, 27 x 19.7 cm. Courtesy of Smithsonian Institution Libraries, Washington, D.C.

129. "The Man Whistler at the Telephone," *Life* magazine, 8 November 1894, 293.

130. John Singer Sargent, *William M. Chase,* 1902. Oil on canvas, 158.8 x 105.1 cm. The Metropolitan Museum of Art, gift of the Pupils of Mr. Chase, 1905 (05.33). Image © The Metropolitan Museum of Art.

131. Detail of *William M. Chase.*

132. John Singer Sargent, *Claude Monet Painting by the Edge of a Wood,* 1885. Oil on canvas, 54 x 64.8 cm. Tate Britain. © Tate, London, 2008.

133. Detail of *Claude Monet Painting by the Edge of a Wood.*

134. John Singer Sargent, *An Artist in His Studio,* 1904. Oil on canvas, 56.2 x 72.07 cm. Museum of Fine Arts, Boston. The Hayden Collection—Charles Henry Hayden Fund, 05.56. Photograph © 2010 Museum of Fine Arts, Boston.

135. John Singer Sargent, *Padre Sebastiano,* c. 1904–6. Oil on canvas, 56.5 x 71.1 cm. The Metropolitan Museum of Art. Rogers Fund, 1910 (11.30). Image © The Metropolitan Museum of Art.

136. Photograph of Sargent painting *Carnation, Lily, Lily, Rose* at Broadway in Worcestershire, 1885–86. Gelatin silver print on paper, 11.2 x 9.4 cm. Harvard University Art Museums, Fogg Art Museum, gift of Mrs. Francis Ormond, 1937.7.27.1.A. Photo: Imaging Department. © President and Fellows of Harvard College.

137. Saxoléine trade card, 1890s. Chromolithograph, 14.1 x 9.3 cm.

138. John Singer Sargent, *Study for Carnation, Lily, Lily, Rose,* c. 1885. Graphite on darkened off-white wove paper, whole page 24.7 x 34.6 cm. Harvard University Art Museums, Fogg Art Museum, gift of Mrs. Francis Ormond, 1937.7.21.11, verso. Photo: Allan Macintyre. © President and Fellows of Harvard College.

139. John Singer Sargent, *Study of Polly Barnard for Carnation, Lily, Lily, Rose,* c. 1885. Oil on canvas, 49.5 x 39.4 cm. Private collection. Photo courtesy of Adelson Galleries, New York.

140. Detail of *Carnation, Lily, Lily, Rose.*

141. Claude Monet, detail of *Impression, Sunrise,* 1872. Oil on canvas, whole 48 x 63 cm. Musée Marmottan, Paris. Photo © Erich Lessing/Art Resource, New York.

142. Francis Frith, photograph of Pangbourne Weir, 1890. © The Francis Frith Collection.

143. John Singer Sargent, *Study for Carnation, Lily, Lily, Rose,* 1885. Oil on canvas, 72.4 x 49.5 cm. Private collection. Photo courtesy of Adelson Galleries, New York.

144. Alfred Le Petit, "Louis Blanc," in *Fleurs, fruits et légumes du jour* (Paris: Au Bureau de l'Éclipse, 1871), 13.

145. Paul Gauguin, *Still Life with Japanese Print,* 1889. Oil on canvas, 85 x 115 cm. Private Collection. Photo © Private Collection/Lauros/Giraudon/The Bridgeman Art Library.

146. Edward Burne-Jones, *Romaunt of the Rose: The Heart of the Rose,* 1901. Wool and silk tapestry on cotton warp, 150 x 201 cm. Executed by Morris and Company. Badisches Landesmuseum, Karlsruhe.

147. Caricature of John Singer Sargent painting *Carnation, Lily, Lily, Rose* (possibly by Sargent), c. 1885–86. Ink on paper, 8 x 8 cm. Courtesy of the Francis David Millet and Millet Family Papers, 1858–94. Archives of American Art, Smithsonian Institution, Washington, D.C.

148. John Singer Sargent, *Rose Branch, Study for Carnation, Lily, Lily, Rose,* 1885–86. Graphite on grey-green wove paper, 34.8 x 24.6 cm. The Metropolitan Museum of Art, gift of Mrs. Francis Ormond, 1950 (50.130.142g). Image © The Metropolitan Museum of Art.

149. John Singer Sargent, *Self-Portrait,* 1906. Oil on canvas, 70 x 53 cm. Uffizi, Florence. Photo © Scala/Art Resource, New York.

150. John Singer Sargent, *Home Fields,* c. 1885. Oil on canvas, 73 x 96.5 cm. Detroit Institute of Arts. City of Detroit Purchase. Photo © 1985, Detroit Institute of Arts.

151. John Singer Sargent, *Garden Study of the Vickers Children,* 1884. Oil on canvas, 137.6 x 91.1 cm. Collection of the Flint Institute of Arts, Flint, Michigan, 1972.47.

152. Salomon Trismosin, "Hermaphrodite with Egg," in *Splendor Solis,* 1582, Harley 3469 folio 19, verso. The British Library, London. © The British Library Board, All Rights Reserved, 2009.

153. John Singer Sargent, *Two Girls on a Lawn,* c. 1889. Oil on canvas, 53.7 x 64.1 cm. The Metropolitan Museum of Art, gift of Mrs. Francis Ormond, 1950 (50.130.20). Image © The Metropolitan Museum of Art.

154. Photograph of John and Emily Sargent, photographer unknown, c. 1867. Private collection. Photo courtesy of Adelson Galleries, New York.

155. John Singer Sargent, *Nude Oriental Youth,* c. 1878–79. Oil on canvas, 90.3 x 64.8 cm. The Western Reserve Historical Society, Cleveland, Ohio, 76.46.5.

156. John Singer Sargent, *Gathering Flowers at Twilight,* c. 1883. Oil on canvas. Private collection. Photo © Christie's Images/The Bridgeman Art Library.

157. John Singer Sargent, *Helen Sears,* 1895. Oil on canvas, 167.3 cm x 91.4 cm. Museum of Fine Arts, Boston, gift of Mrs. J. D. Cameron Bradley, 55.1116. Photograph © 2010 Museum of Fine Arts, Boston.

158. Detail of *Helen Sears.*

159. Linley Sambourne, "Flowers of 'Culture,' or A Swinburne-Jones Cutting," in *Punch, or the London Charivari,* 15 January 1881, 22. Courtesy of Robarts Library, University of Toronto.

160. Linley Sambourne, "Our Guide to the Grosvenor Gallery," in *Punch, or the London Charivari,* 22 June 1878, 285. Courtesy of Robarts Library, University of Toronto.

161. Linley Sambourne, "O.W.," in *Punch, or the London Charivari,* 25 June 1881, 298. Courtesy of Robarts Library, University of Toronto.

162. Edward Lear, "Queeriflora Babyoïdes," in *Laughable Lyrics: A Fourth Book of Nonsense Poems, Songs, Botany, Music, &c.* (London: Robert John Bush, 1876). Houghton Library, Typ 805LN4.76 (A). By permission of the Houghton Library, Harvard University.

163. Paul Helleu, *Hydrangeas and Bats,* n.d. Pastel on paper laid on canvas, 69 x 99.5 cm. Private collection. Image courtesy of Nevill Keating Pictures, Ltd.

164. "Robert de Montesquiou en chasseur de chauves-souris, château de Charnizay." Manuscrits occidentaux, NAF 15039, folio 101. Bibliothèque nationale de France.

165. John Singer Sargent, *Portrait of Ernest-Ange Duez,* 1884–86. Oil on canvas, 73.7 x 60.3 cm. Montclair Art Museum, Montclair, New Jersey, gift of Dr. Arthur Hunter in memory of Ethel Parsons Hunter, 1962.32. Courtesy of Montclair Art Museum.

166. John Singer Sargent, *Portrait of Katharine Chase Pratt,* 1890. Oil on canvas, 101.6 x 76.5 cm. Worcester Art Museum. Photo © Worcester Art Museum, Worcester, Massachusetts/The Bridgeman Art Library.

167. Jacques-Émile Blanche, *Little Girl with Hydrangeas,* 1887. Oil on canvas, 92 x 64.8 cm. Private collection. Photograph courtesy of Sotheby's, Inc., © 1996. © 2008 Artists Rights Society (ARS), New York/ADAGP, Paris.

168. Sarah Choate Sears, "Helen Sears with Japanese Lantern," 1895. Platinum print, 23.7 x 13.2 cm. Descendants of Sarah Choate Sears. Photograph courtesy of the Museum of Fine Arts, Boston.

169. John Singer Sargent, *Frederick Law Olmsted,* 1895. Oil on canvas, 254 x 139.7 cm. Biltmore Estate. Used with permission from The Biltmore Company, Asheville, North Carolina.

170. John Singer Sargent, *The Hermit (Il Solitario),* 1908. Oil on canvas, 95.9 x 96.5 cm. The Metropolitan Museum of Art. Rogers Fund, 1911 (11.31). Image © The Metropolitan Museum of Art.

171. Detail of *Il Solitario.*

172. James McNeill Whistler, *Arrangement in Black and Gold: Comte Robert de Montesquiou-Fezensac,* 1891–92. Oil on canvas, 208.6 x 91.76 cm. © The Frick Collection, New York.

173. James McNeill Whistler, *Large Bowl and Cover,* c. 1876–78. Pencil, brown ink, and wash on cream wove paper, 22.2 x 17.2 cm. Glasgow Art Gallery and Museum. © Culture and Sport Glasgow (Museums).

174. John Singer Sargent, *Pressing the Grapes: Florentine Wine Cellar,* c. 1882. Oil on canvas, 83.5 x 49.9 cm. The Beaverbrook Art Gallery/The Beaverbrook Foundation (in dispute, 2004), Fredericton, New Brunswick.

175. John Singer Sargent, *Sir Frank Swettenham,* 1904. Oil on canvas, 170.8 x 110.5 cm. National Portrait Gallery, London. Photo © National Portrait Gallery, London.

176. John Singer Sargent, *Lord Ribblesdale,* 1902. Oil on canvas, 258.5 x 143.5 cm. The National Gallery, London. Presented by Lord Ribblesdale in memory of Lady Ribblesdale and his sons, Captain the Hon. Thomas Lister and Lieutenant the Hon. Charles Lister, 1916. © The National Gallery, London.

177. John Singer Sargent, sketch for *Lord Ribblesdale,* 1902. Oil and crayon on canvas, 87.6 x 61 cm. Private collection. Photo courtesy of Adelson Galleries, New York.

178. John Singer Sargent, *The Honourable Victoria Stanley,* 1899. Oil on canvas, 196.8 x 105.4 cm. Private collection. Photo courtesy of Adelson Galleries, New York.

179. John Singer Sargent, *Portrait of Mrs. Robert Harrison,* 1886. Oil on canvas, 157.8 x 80.3 cm. Tate Britain. © Tate, London, 2008.

180. John Singer Sargent, *Elsie Palmer,* 1889–90. Oil on canvas, 190.5 x 114.3 cm. Colorado Springs Fine Arts Center, FA1969.3. Funds acquired through public subscription and Debutante Ball Purchase Fund.

181. "An Aesthetic lady clutching her mahl stick," *Illustrated Sporting and Dramatic News* (?), c. 1880.

182. John Singer Sargent, *Flora Priestley,* c. 1889. Oil on canvas, 91.4 x 63.5 cm. Tate Britain. © Tate, London, 2008.

183. John Singer Sargent, *Robert Louis Stevenson and His Wife,* 1885. Oil on canvas, 52.1 x 62.2 cm. Private collection. Photo © Private Collection/The Bridgeman Art Library.

184. John Singer Sargent, *Two Studies for a Portrait of Robert Louis Stevenson,* 1880–85. Charcoal on darkened off-white wove paper, 24.7 x 34.6 cm. Harvard University Art Museums, Fogg Art Museum, gift of Mrs. Francis Ormond, 1937.7.21.7. Photo: Allan Macintyre. © President and Fellows of Harvard College.

185. John Singer Sargent, *W. Graham Robertson,* 1894. Oil on canvas, 230.5 x 118.7 cm. Tate Britain. © Tate, London, 2008.

186. John Singer Sargent, *Standing Male Nude with Arms Akimbo,* c. 1890–1915. Charcoal on off-white laid paper, 63.5 x 48.7 cm. Harvard University Art Museums, Fogg Art Museum, gift of Mrs. Francis Ormond, 1937.9.9. Photo: Katya Kallsen. © President and Fellows of Harvard College.

187. John Singer Sargent, *Study for W. Graham Robertson,* 1894. Oil on canvas, 91.5 x 71.2 cm. Private collection. Photo courtesy of Spanierman Gallery, LLC, New York.

188. Raphael Tuck and Sons, "May all your hopes blossom," greeting card, 1880s, 11 x 15.4 cm.

189. John Singer Sargent, *Portrait of the Artist Alberto Falchetti,* c. 1905. Oil on canvas, 74.9 x 54.6 cm. Private collection. Photo © Private Collection/The Bridgeman Art Library.

190. John Singer Sargent, *Edward Robinson,* 1903. Oil on canvas, 143.5 x 92.1 cm. The Metropolitan Museum of Art, gift of Mrs. Edward Robinson, 1931 (31.60). Image © The Metropolitan Museum of Art.

191. John Singer Sargent, *Alfred, Son of Asher Wertheimer,* c. 1901. Oil on canvas, 163 x 115 cm. Tate Britain. © Tate, London, 2008.

192. John Singer Sargent, *Hell,* installed 1916. Oil and gilded or painted Lincrusta-Walton on canvas, 254.6 x 507.4 cm. West wall lunette, north end, Special Collections Hall, Boston Public Library. Courtesy of the Trustees of the Boston Public Library.

193. John Singer Sargent, *Man Screaming, Study for "Hell"* (study for mural paintings installed in three phases [1895, 1905, 1916] at the Boston Public Library, and related drawings), 1890–1916. Charcoal and stump with pencil on beige paper, 61.9 x 47.6 cm. Gift of Miss Emily Sargent and Mrs. Francis Ormond, sisters of the artist, 49.99. Collection of The Corcoran Gallery of Art, Washington, D.C.

194. John Singer Sargent, *Tommies Bathing,* 1918. Watercolor and graphite on white wove paper, 38.9 x 52.7 cm. The Metropolitan Museum of Art, gift of Mrs. Francis Ormond, 1950 (50.130.48). Image © The Metropolitan Museum of Art.

195. John Singer Sargent, *Group with Parasols,* c. 1905. Oil on canvas, 55.3 x 70.8 cm. Private collection. Photo courtesy of Adelson Galleries, New York.

196. František Kupka, *Self-Portrait Among Roses,* 1894–95. Oil on wood, 65 x 42 cm. Museum Kampa—Jan and Meda Mládek Foundation Collection, Prague. © 2008 Artists Rights Society (ARS), New York/ ADAGP, Paris.

197. František Kupka, *Study for Roman du rose,* 1911–24. Oil on canvas, 80 x 67 cm. Museum Kampa—Jan and Meda Mládek Foundation Collection, Prague. © 2008 Artists Rights Society (ARS), New York/ ADAGP, Paris.

198. Gilbert and George, *A Touch of Blossom,* 1971. Offset lithograph, 30 x 40 cm. Postal sculpture. © Gilbert and George, courtesy of the artists, Jay Jopling/ White Cube (London).

199. Gilbert and George, *George the Cunt and Gilbert the Shit,* 1969. Magazine sculpture. © Gilbert and George, courtesy of the artists, Jay Jopling/White Cube (London).

200. Photograph of John Singer Sargent, c. 1884. Reproduced in Evan Charteris, *John Sargent* (New York: Charles Scribner's Sons, 1927), facing page 78.

INTRODUCTION

THE BIRDS AND THE BEES

The work of art always tells the story of the cross-fertilization of the mind by the outside world and of the outside world by the mind.

—CLAUDETTE SARTILIOT, *HERBARIUM VERBARIUM: THE DISCOURSE OF FLOWERS*, 63

The story of art has ever been a melliferous tale of the birds and the bees and the flowers. Plato described the poet as "a light and winged thing," flitting "like the bees" from flower to flower in the "gardens of the Muses,"[1] and from antiquity through the Renaissance—following the injunctions of Seneca, Virgil, Petrarch, and others—writers wrote "as the bees make honey,"[2] and painters searched "the pictorial gardens" for "their sweet juices."[3] The ideal artist, according to Carlo Malvasia, "like a bee amongst the flowers," passed "amongst all things, sampling the most exquisite and most perfect."[4] Savoring a beautiful arrangement here, an enchanting shade there, the apian painter-poet distilled from diverse nectars a "delicious compound," a creation that evinced its origins yet was "clearly a different thing from that whence it came."[5] The true artist, in other words, manufactured a distinctive honey, sprung from imitation yet freshly formulated, which could be identified by those with discerning taste.

In the eighteenth century, Winckelmann's ideal artist was still beelike,[6] but new models of the artist cropped up with increasing frequency over the next century and a half: Romantic monk, Baudelairean *flâneur,* Victorian anthropologist, willful primitive, and so on. In the Darwin-informed milieu of the *fin de siècle,* the analogy of artist and bee regained its potency. Proust at school, for example, was already marked by "the ominous signs of the pure artist," with his "antennae, his multiply-faceted bee eyes."[7] The allegory had been modernized, though: a broad cultural awareness of co-evolutionary relationships and the mechanics of pollination occasioned a shift in focus for this metaphorics, from mellification to pollen brushing, fertilization, and floral reproduction.

A poetics of vegetal creation became central to literature and the visual arts. As Philip Knight has demonstrated, for example, the history of nineteenth-century French poetry *is* the "history of its flower poetics" and "floral tropology."[8] Baudelaire's *Fleurs du mal* was a literal *anthology* (from the Greek, a collection of flowers): "I've made a new flower," the poet would write to describe his progress.[9] After Baudelaire, Rimbaud "tried to invent new flowers" along with "new tongues"[10] (new poetic organs as well as languages), and Mallarmé's performative poetics gave rise to ideal lingual blossomings: "I say: a flower!"[11]

In the field of painting, the representation of the floral environment and its phantasmatically assumed interspecies dynamics effected a radical rejuvenation of the medium across movements and countries over the course of the century.[12] Whistler asserted that "the work of art worthy of the name" had to be made "the way nature produces plants,"[13] and Courbet famously painted "as an apple-tree produces apples."[14] Des Esseintes, Huysmans's Decadent paragon in *À rebours,*

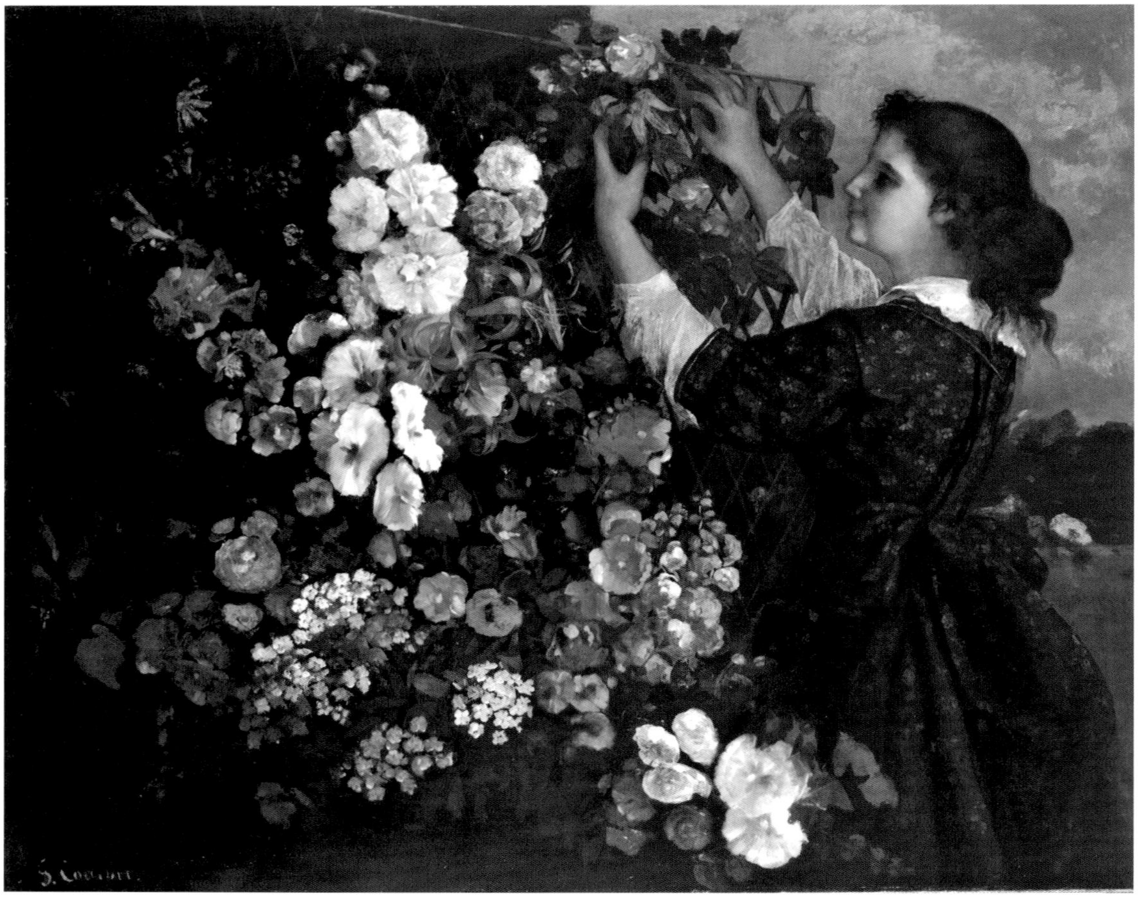

I
Gustave Courbet, *The Trellis,* 1862. Oil on canvas, 109.8 x 135.2 cm. Toledo Museum of Art, purchased with funds from the Libbey Endowment, gift of Edward Drummond Libbey, 1950.309.

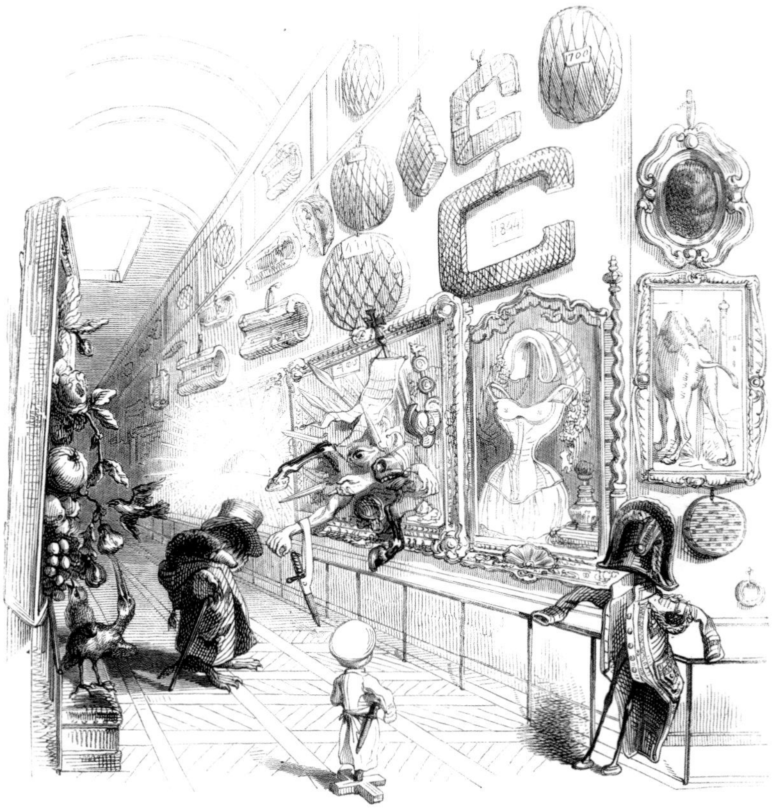

2
J. J. Grandville, "Le Louvre des marionnettes," in *Un autre monde* (Paris: H. Fournier, 1844), 86.

declared with characteristic extremism that the *only* true artists "are the horticulturalists,"[15] and Monet, who considered his garden his "most beautiful masterpiece," would have concurred.[16] John Singer Sargent's paintings will be central examples in this book; they seemed, to his contemporaries, to "burst into bloom."[17]

Courbet's art was an important precedent for later floriferous aesthetics. *The Trellis* from 1862 in particular (fig. 1), as Michael Fried has suggested, envisions a fantastically fresh and natural form of painting by the gardener-artist.[18] The shape of the canvas is echoed within the painting by the lattice, across which flowers of the most astonishing variety—poppies, scarlet trumpet vine, anemones, passion flowers, lilies, clematis, rose of Sharon, hollyhocks, and hydrangeas—burst into bloom and out of the frame. Almost twenty years earlier, J. J. Grandville had identified the living vegetable work as one of the modern genres of painting (fig. 2). In the section of *Un autre monde* titled "Le Louvre des marionnettes," an illustration of provocative pictures includes a blinding sunrise; a battle scene in which a standard, spears, and limbs equine and human thrust furiously out of the frame; a "womannequin" in a state of semiundress; an Orientalist view of the backside of a urinating camel; and, at the left, parodying Zeuxis's grapes, a painting of roses, pears, apples, and grapes that grow out of the frame and are pecked at by birds.[19] One of several trends in midcentury art, the fantasy of art as *nature vivante,* rather than *nature morte,* became increasingly important as the century progressed. The Zeuxis myth was reworked to emphasize the lure of flora: while John Everett Millais was working on his painting *Apple Blossoms* in 1858, "the bees used often to settle on the bunches of blossom, thinking them real flowers from which they might make their honey."[20]

"The way nature produces plants," to return to Whistler's formulation, is through cross- or self-fertilization, and as the scientific world came to understand the intricacies of plant reproduction through pollination, many individuals in the artistic world—creators and critics—embraced the metaphorical equation of painters as pollinators, though not necessarily bees. Whistler most famously identified with and signed himself as a butterfly, but he was certainly not alone in this eccentricity: Monet drew the artist and illustrator Jules Didier as an "homme-papillon" with appropriately elongated proboscis (fig. 3), and various writers contended that the butterfly was "the artist's true symbol." The identification of the two creatures was not predicated on a fantasy of sublimation—of the artist reborn after struggling in the carnal mire to a state of genius, of increased clarity and scope of vision. Rather, pollinating habits were the shared trait. The butterfly's flights "from one near object to another" prompted one writer in the *Monthly Illustrator* to exclaim, "How similar is [the artist] to the insect! What an unparalleled voracity and what a quickness with which he digests impressions!"[21] Other artists were moths, dragonflies, birds, and bats; some even flitted between different pollinator personae. Sargent was at times a wagtail, at times a bee. As we shall see, the concepts of pollination and artistic modernism became intertwined for some of these creative creatures; thus Didier, who had won the conservative Prix de Rome shortly before Monet's caricature, is restrained on a leash by a centaur—the classical tradition—and not at liberty to seek out his own impressions.[22]

While in the ancient model the apian artist was a divinely inspired gatherer and alchemist whose interior distillation of ideas gave rise to more perfect imitations, the modern painter-pollinator had quite another orientation. A carrier of pigment and pollen between surfaces, the artist's activity came to be figured as a physical and erotic interaction between creature and environment. The creative organ was no longer the mind, but the hand—a hand conceived not as the instrument of instruments but as an organ co-opted by a drive. As butterflies uncoil and dip their proboscides, and birds their beaks into nectaries, so the artist, naturally, dipped his or her brush. The hand-pollination of plants, in fact, was usually effected by a paintbrush or another artistic implement, such as a pencil or a needle. Artists' and insects' instruments were easily conflated; Darwin even tells us that "bees will act like a camel-hair pencil,"[23] intriguingly suggesting, thirty years earlier than Oscar Wilde, that life imitates art rather than the other way around.[24]

In addition to the artist-as-pollinator, the artist-as-plant was a flourishing paradigm in the nineteenth century. The "precious floweret with which certain of the old masters used to sign their pictures"[25] throve again in the second half of the century, for instance with Albert Moore's anthemion.[26] Degas wanted the Impressionist group to be called "La Capucine" and to adopt the nasturtium as its emblem,[27] and Oscar Wilde was identified with various flowers: lilies, daisies, daffodils, sunflowers, and green carnations.[28] Nominally vegetal artists abounded, from the fictional Basil Hallward to the living Nathaniel Hawthorne, Pierre Loti,[29] George de Forest Brush, Flora Priestley, and Violet Paget, to name a very few.

Odilon Redon, an artist directly inspired by Darwin, took up the latter's suggestion that "the petal's sensitivity to light" was "precocious of our own perception,"[30] drawing the blossom as a receptive organ in *There Was Perhaps a First Vision Attempted in the Flower* (fig. 4), one of eight lithographs in the *Origines* series from 1883. In the print a solitary flower unfurls lid after petal-lashed lid, which open onto a central orb in which the iris

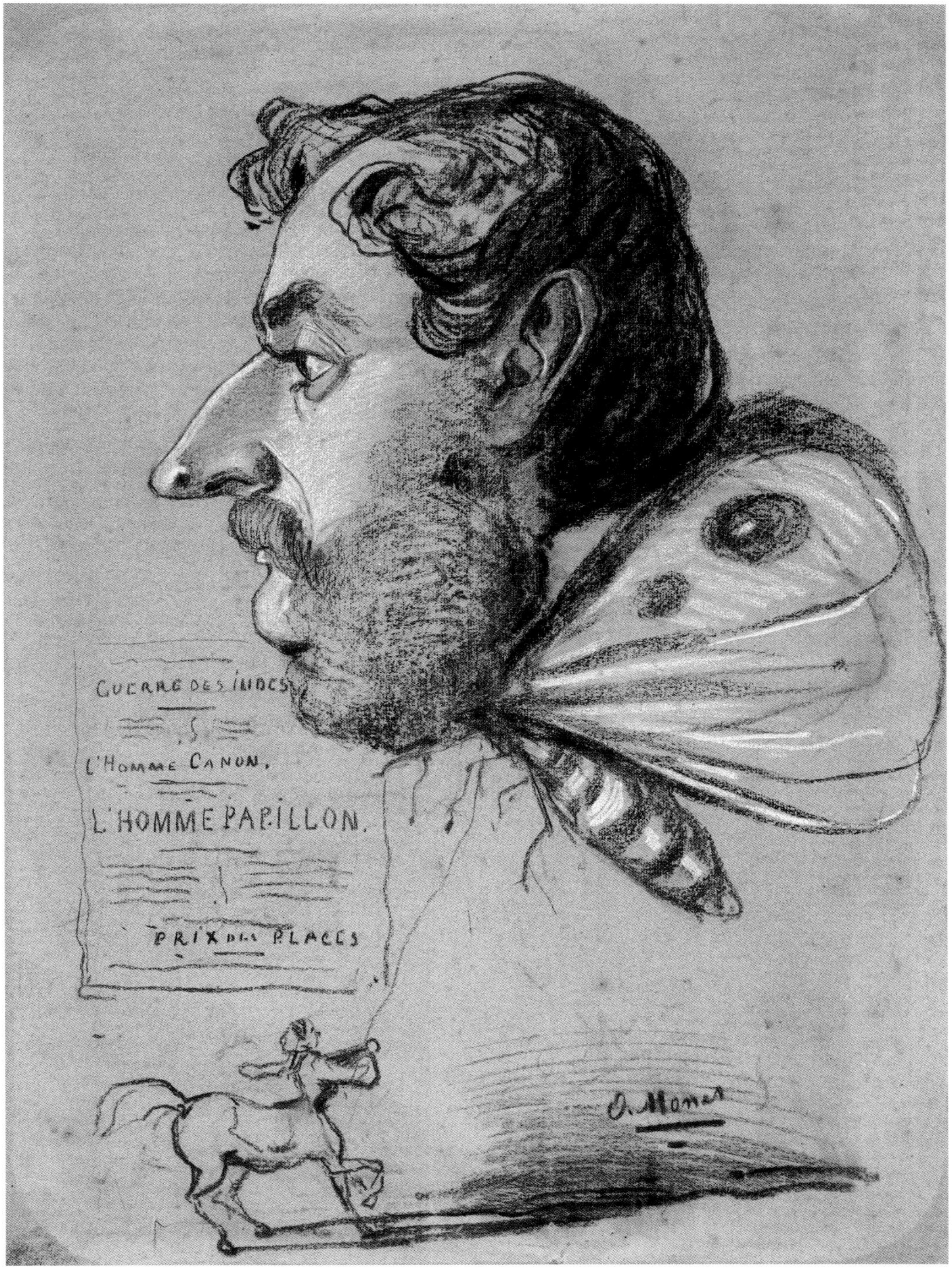

3
Claude Monet, *Caricature of Jules Didier*, c. 1860. Charcoal, heightened with white chalk, on blue laid paper (discolored to grey), 61.6 x 43.6 cm. Mr. and Mrs. Carter H. Harrison Collection, 1933.889, The Art Institute of Chicago.

4

Odilon Redon, ***There Was Perhaps a First Vision Attempted in the Flower,*** plate 2 of 8 from *Les Origines,* 1883. Lithograph in black on light grey chine laid down on white wove paper, 22.5 x 17.7 cm (image/chine), 51.1 x 35.6 cm (sheet). The Stickney Collection, 1920.1579, The Art Institute of Chicago.

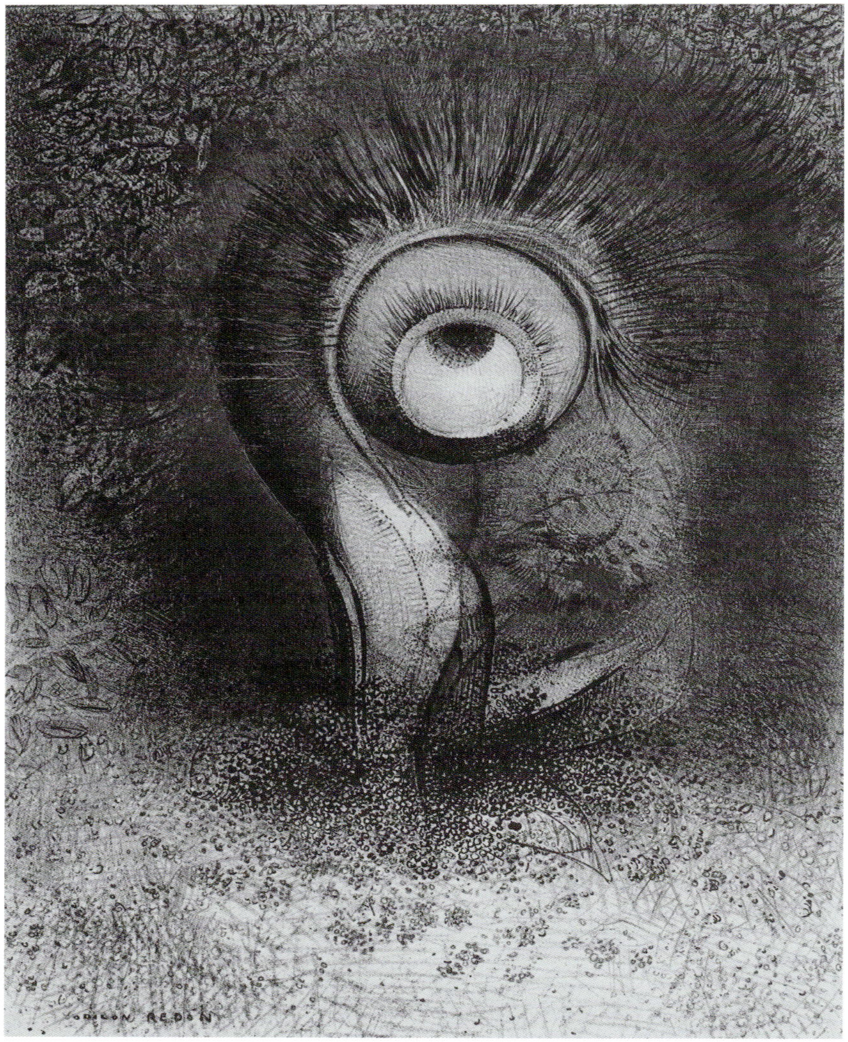

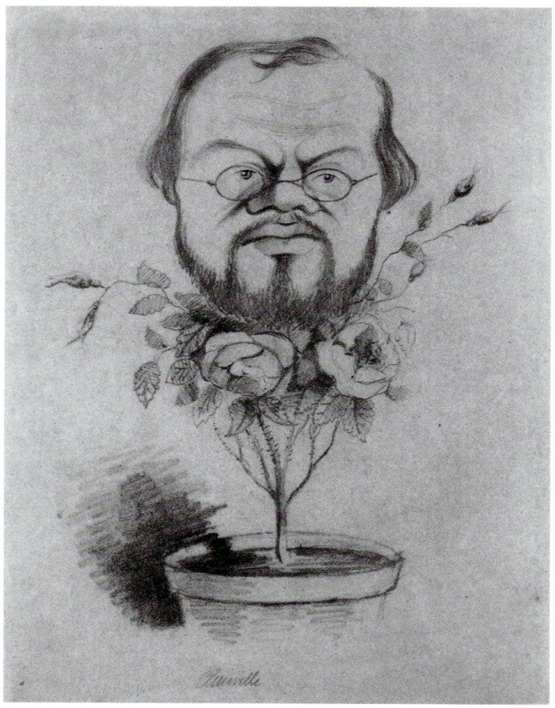

5

Claude Monet, ***The Playwright François Nicolaïe, Called "Clairville,"*** 1858. Pencil on beige paper, 32 x 24 cm. Musée Marmottan, Paris.

6
Albert Moore, *William Connal, Esq. Jr., of Solsgirth*, 1883. Oil on canvas, 33 x 25.4 cm. York Museums Trust (York Art Gallery).

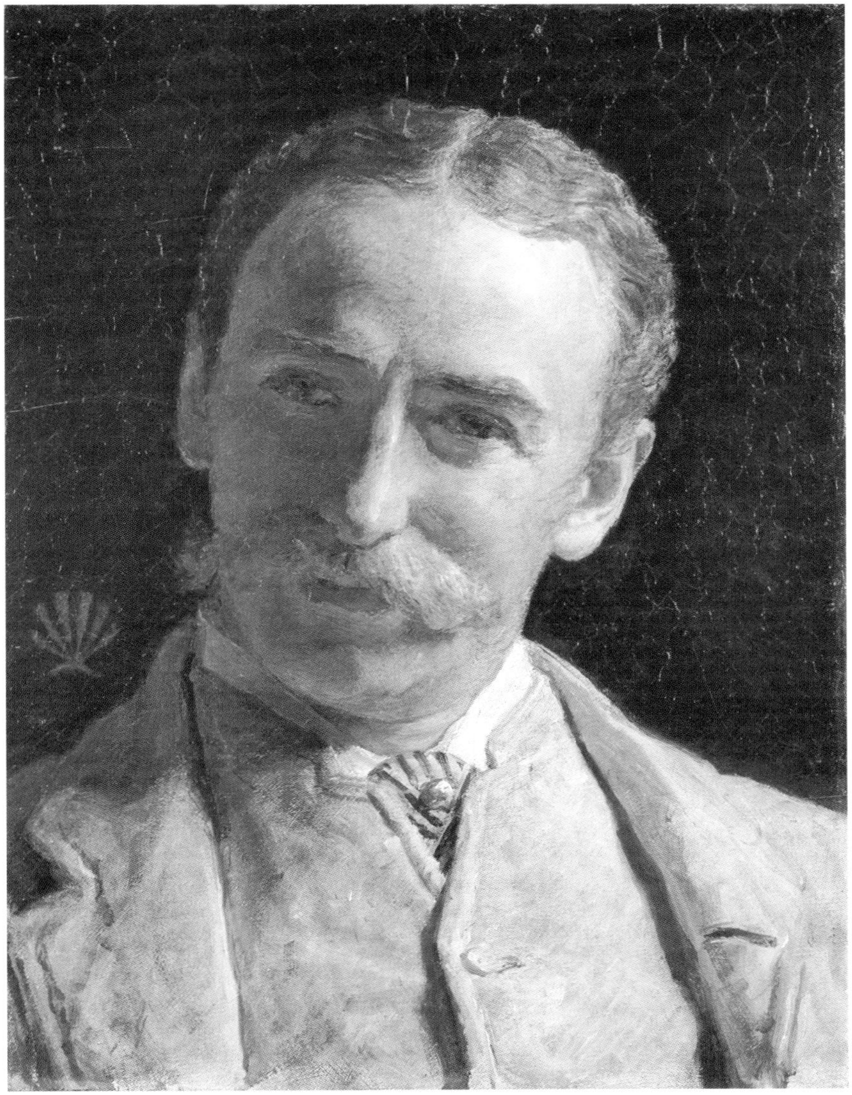

and pupil appear as a blossom; seeds like tears have fallen from it onto the veined ground. This image of budding vision is simultaneously a model of the developing artist; Redon advised pupils "to cultivate only the flower of their own garden"[31]—a unique plant synonymous with a particular vision. Leonardo was a paradigmatic example of this singular horticultural artistry: "out of the secret places of a unique temperament," Pater wrote, "he brought strange blossoms and fruits hitherto unknown."[32] The original flower of art was rare. Max Nordau pathologized those who climbed on the success of others: while the healthy poet was a "chlorophyllic plant," the degenerate one was a parasitic growth.[33] The issue of rarity, though, was not so key in the popular imagination, where to be prolific was to be floriferous, as another Monet caricature of the playwright Clairville demonstrates (fig. 5).

The floral artist required pollination in order to bring forth flowers of art. Henry James, for whom the artist was often a flower,[34] described his creative process as beginning with "a single small seed," "minute and wind-blown," which provided the "germ of a 'story.'" This "stray suggestion" was a "vital particle," a pollen-like "grain of gold" possessed of "the power to penetrate as finely as possible." With a "prick," "a single touch," the artist was fertilized; eventually "the flower of conception . . . bloomed."[35] James recounts the process in microscopic detail, but in other works the advertisement of cross-fertilization is less overt. In Moore's portrait of William Connal (fig. 6), the sitter's conspicuously empty buttonhole declares that he is not the flowery sort. His yellow and black bee-striped tie, on the other hand, signals his apian identification, as does the overall color scheme of black, yellow, and gold. Connal employed the honeybee "as his personal emblem" and used it on his stationery.[36] In the picture, his head is turning toward the anthemion above his shoulder, his golden moustache tantalizingly close to brushing the flower. The nature of the reproductive connection between pollinating model-bee and artist-flower is thus delicately invoked.[37]

Linked to contemporary theories of cross-fertilization, the floral birth of the work of art was also a modern version of the enduring fantasy of artistic parturition. During the Renaissance the conception and gestation of children and ideas were considered analogous acts: "the works I leave behind shall be my children," Michelangelo declared.[38] Sir Philip Sidney figured his poetic self as "great with child to speak," and when he could not express himself he felt "helpless" in his "throes."[39] In the nineteenth century the trope of the artist as fecund, multiparous mother was still very much in circulation. The painters Claude Lantier and Bongrand in Zola's *L'oeuvre,* for example, both suffer from morning sickness and birth pangs.[40] The fiction of art's botanical reproduction, however, had the benefit of diverting attention from the collision of ground and vehicle in the creation-as-procreation metaphor: the bother that men cannot bear children.[41] In fact, the fantasy of vegetable reproduction could allay anxieties stemming from physiological incapacity of whatever sort, and was an alternative form of imaginary reproduction for both men and women.

In Hans Christian Andersen's fairy tale "Thumbelina," an old woman who desires a child resorts to this vicarious form of procreation. A witch gives her a grain of barley to plant in a flowerpot, a large tulip grows, and when it opens, "a tiny little girl" is found sitting "in the center of the flower on the green stigma" (fig. 7).[42] This substitution of blossom for uterus was facilitated by the long-standing association of the female reproductive organs with flowers; Odoardo Fialetti's seventeenth-century *Petal Venus* (fig. 8), which depicts the dissected pregnant womb as a gorgeously unfurled bloom, is an exemplary early visualization of this conception. In the nineteenth century the blossom-womb became a common motif in visual culture. A midcentury Valentine (fig. 9) uses a blossom-womb to suggest the birth of Love—the golden-haired, blue-eyed, winged cupid cupped in the pale pink rose's corolla—as well as the births to which love might lead.[43] Two later, mass-produced examples illustrate the continuing appeal of the theme (fig. 10): on the left, a "congratulations" card celebrates the birth of a child, pictured as a rosy-cheeked baby calmly nestled in the pink and red uterine folds of a flower; on the right, a tulip opens to reveal a thoughtful infant.

Terry Castle has traced the vicissitudes of the birth topos from Plato onward, focusing specifically on the shift from its repudiation in the neoclassical period to its embrace by the Romantics. Alexander Pope had argued that "artistic activity should not be linked, even metaphorically, to physiological processes"; to embrace such a connection would be "to abdicate from intellectual responsibility, from reason, from creative independence."[44] Works of poet *born* were therefore troped as deformed, aborted, or otherwise monstrous offspring in the eighteenth century.[45] Early nineteenth-century

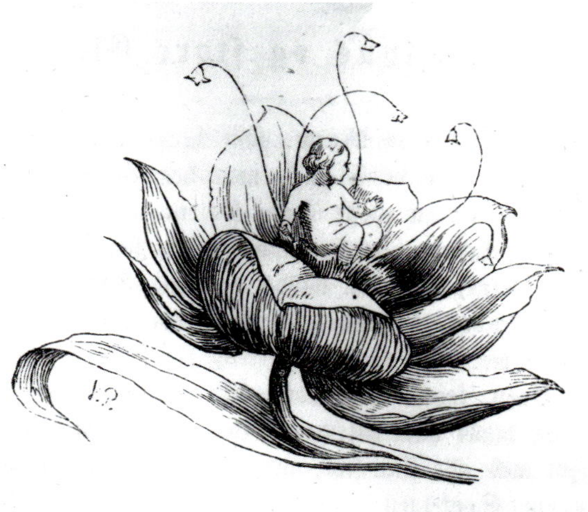

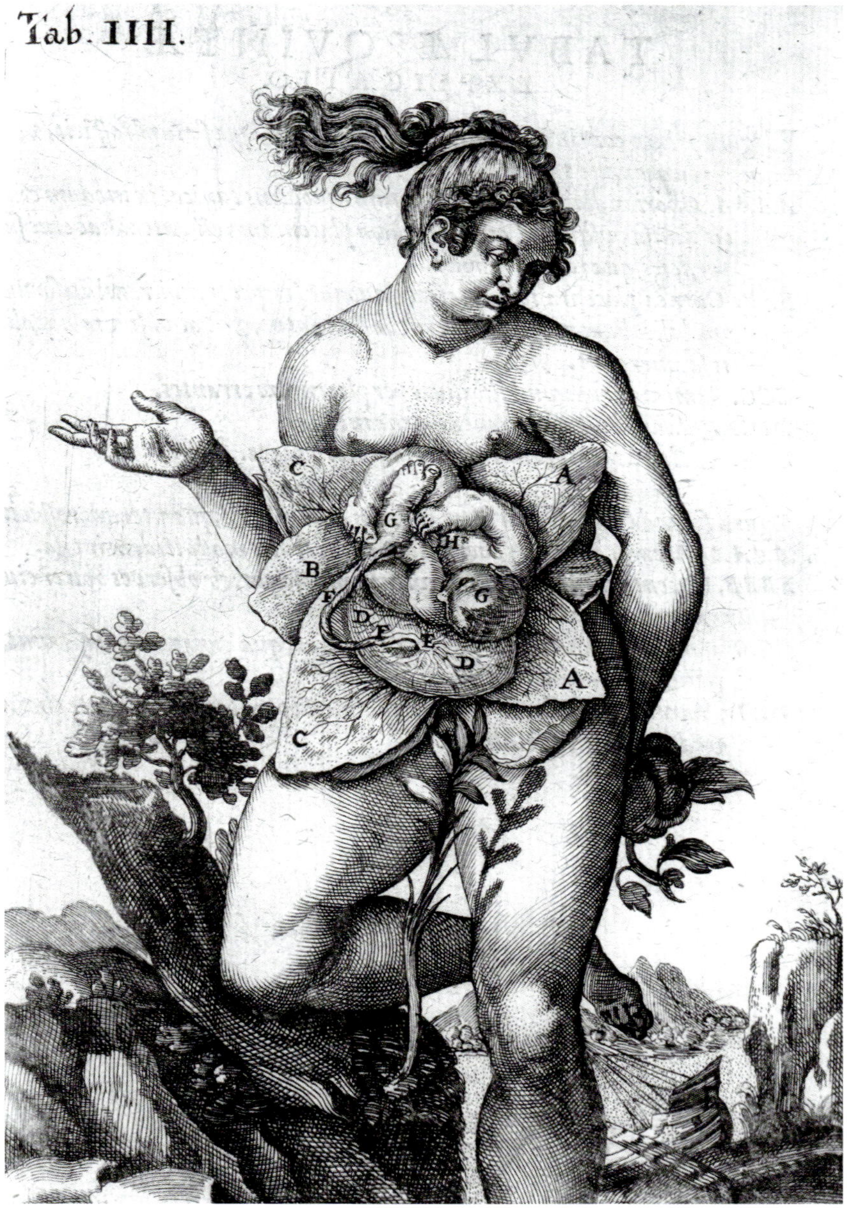

7
Vilhelm Pedersen, "Tommelise," 1843, in Hans Christian Andersen, *Eventyr* (Copenhagen: C. A. Reitzel, 1850), 137.

8
Odoardo Fialetti, *Petal Venus,* engraved by Francesco Valesio, in Adriaan van de Spiegel, *De formato foetu* (Frankfurt: Mattheus Merian, 1631), 37, plate 4.

authors and artists rejected this model, conceiving of artistic creation as an organic, physiological process and exalting the metaphor of the work of art as human birth. This organic fantasy was predicated on the imaginative androgyny considered the mark of the true Romantic (and later the Decadent) artist.[46]

Coleridge introduced the botanical variation of this Romantic birth topos by figuring his poems as vegetal productions.[47] In his criticism, "authors, characters, poetic genres, poetic passages, words, meter, logic become seeds, trees, flowers, blossoms, fruit, bark, and sap."[48] The shift from the figure of the artist as androgyne to the artist as plant was enabled by the latter's bisexuality: the German Romantics saw the plant "as a marker of primal undividedness,"[49] its capacity for self-fertilization enabling extraordinary originality. As Novalis put it in "Pollen," published in the Schlegel brothers' *Athenäum* in 1798, "the first genius that penetrated into itself discovered thereby the typological seed of an infinite world."[50] He

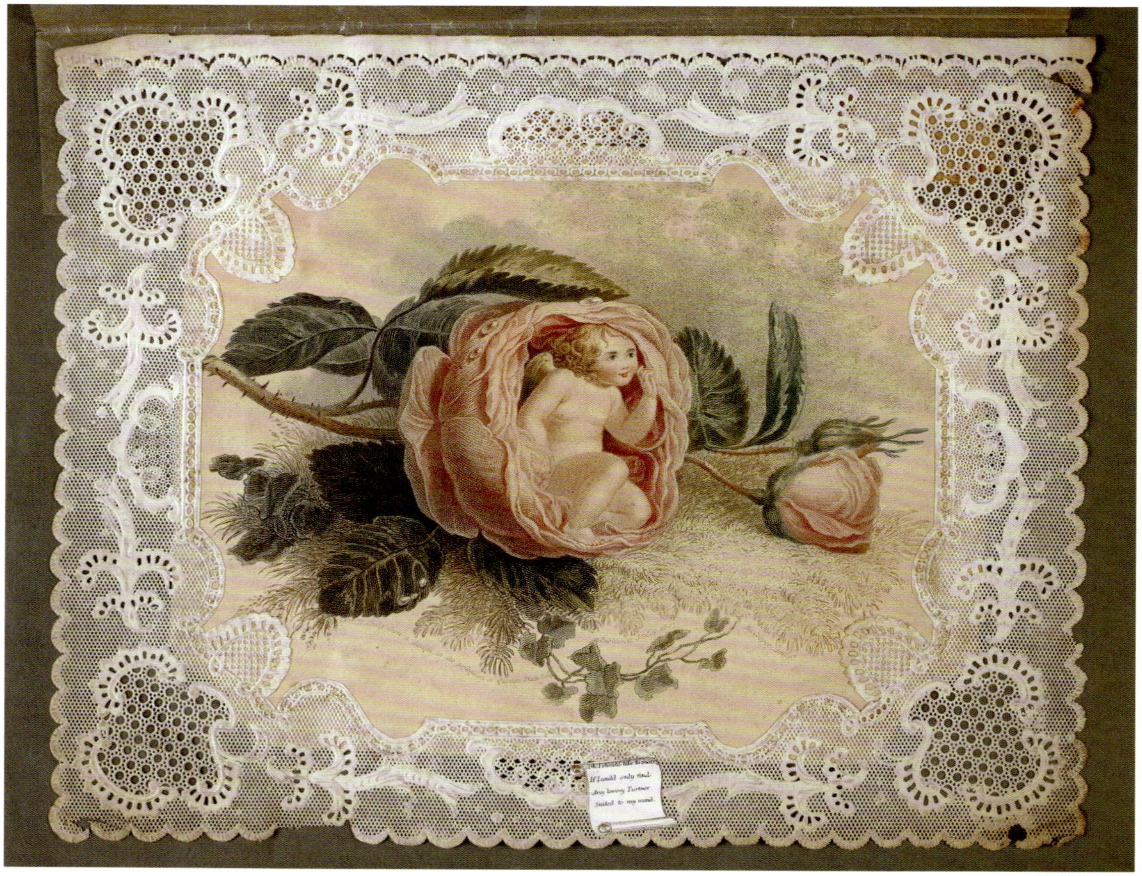

9
Valentine card with cherub, c. 1850s. Hand-colored engraving and paper lace, 22 x 26 cm. Museum of London.

10

English postcard and American greeting card with baby-in-flower motif, both 13.7 x 8.6 cm, 1880s.

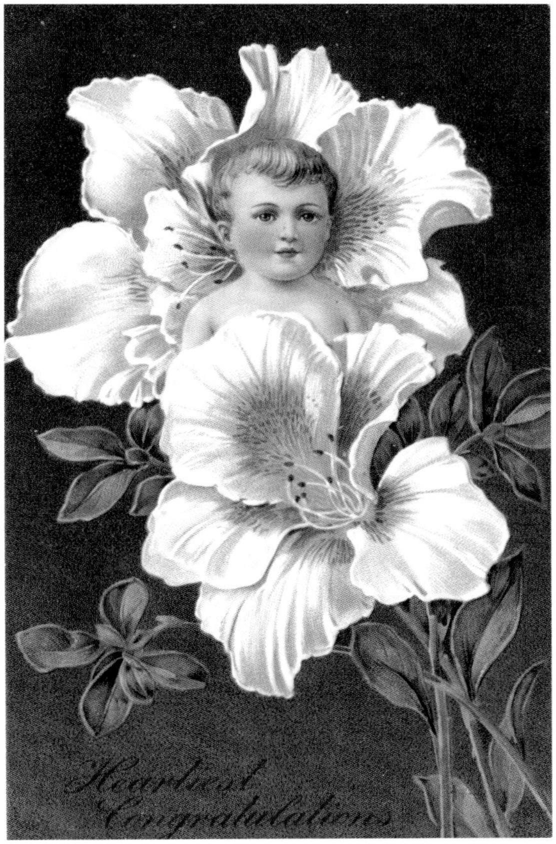

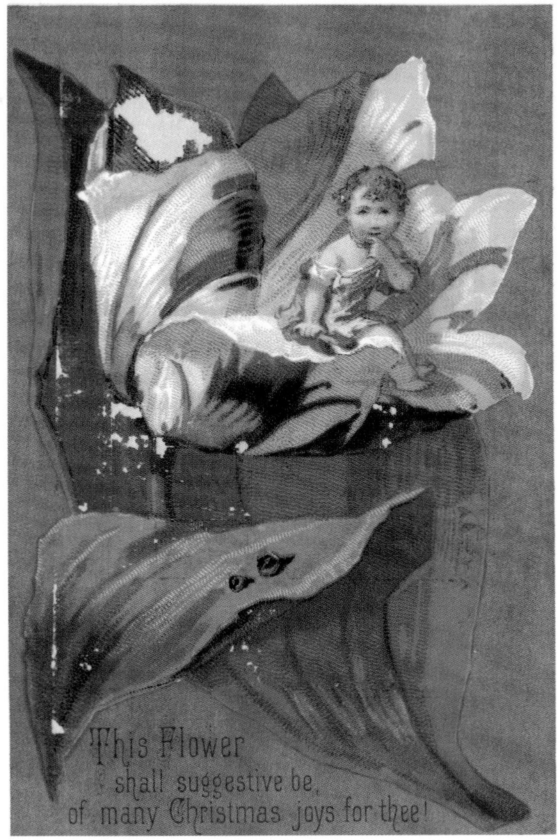

identified both genius and its works botanically—his fragments were pollen, "literary seeds" (some of which would "hopefully sprout"), and flowers: "Even philosophy has its blossoms. That is, its thoughts."[51] At midcentury the Pre-Raphaelites conceived of themselves as such seminal artists and gave voice to their conviction in their journal *The Germ*. Coleridge's and these other early examples of vegetal artists and works of art emphasized the plants' growth, nourishment, and relationship to the environment only insofar as they facilitated the development of an interior world.[52] In contrast, their successors, steeped, as peers of Darwin, in the details of pollinator-plant relationships, pushed the allegory into botany's expanded field—plants' contexts and creatural connections—and floral births became the result of cross-pollination.

Critics and historians adapted to this model of art production. In *Philosophie de l'art,* Hippolyte Taine described his method as "a sort of botany, applied not to plants, but to human works." He vowed not to neglect the broader milieu in which works of art were created, for "the productions of the human mind, like those of living nature, are only understandable through their environment," and every region has "its own vegetation and its own culture."[53] John Addington Symonds, who penned studies of Michelangelo, Whitman, Cellini, Marlowe, and other artists, outlined a similar interpretive model in which the critic would function as a "natural historian of art and literature."[54]

The cogency of a garden metaphorics imaging artists as pollinators and plants, and artistic creation as cross-fertilization, was predicated on more than a fashion for *plein-air* painting. Also at stake was something more difficult to attain and too risky or risqué to formulate directly: the naturalization of alternative sexualities and identities. In the last few decades of the nineteenth century, sexologists frequently used botanical and zoological examples of hermaphroditism to explain both the "latent organic bi-sexuality of each sex" and the varieties of sexual behavior "inverts" exhibited.[55] At the time, the terms "sexual inversion" and "invert" were used loosely to indicate a range of nonnormative behaviors and desires,[56] most of which involved multiple sex and/or gender identifications.[57] The hermaphrodite and the invert seem to have entered *fin-de-siècle* medical discourse "side by side." Inverts were seen as would-be, "incomplete" hermaphrodites,[58] "psychosexual hermaphrodites," and evolutionary throwbacks to originary bisexuality. Sexologists catalogued sexual types as Linnean botanists classified flowers, based on the arrangements of their physical and mental sexual characteristics, their combination of pistils and stamens, as it were. If in earlier periods hermaphrodites were considered monstrosities, in the nineteenth century they came to be seen as anomalies.[59]

Inverts were not only intimately identified with hermaphroditic flowers, however; they were also imagined as pollinators. Arthur Schopenhauer, in the 1859 edition of *The World as Will and Idea,* likened the pederast to a fly: "The sense of beauty which instinctively guides the selection for the satisfaction of sexual passion is led astray when it degenerates into the tendency to pederasty; analogous to the fact that the blue-bottle (*Musca vomitoria*), instead of depositing its eggs, according to instinct, in putrefying flesh, lays them in the blossom of the *Arum dracunculus,* deceived by the cadaverous smell of this plant."[60] Beyond the fact that both heterosexual male and pederast are imaged as female blue flies here, Schopenhauer's analogy is anything but straightforward: it links sexual deviance with a variant aesthetics (an errant "sense of beauty") and alternate form of procreation, both of which are natural, if not the norm, and even perhaps enviable (some might, after all, prefer a flower to rotting flesh).

Whether identifying as pollinators or plants, invert artists working in diverse disciplines mobilized the ideas of cross-fertilization and the hermaphroditic sexuality of flowers to "naturalize" sexual inversion—a strategy most famously employed by Proust in the orchid and bee scene of *À la recherche du temps perdu.* In Proust's novel, the narrator, playing botanist, watches for the arrival of an insect to fertilize the rare orchid of his neighbor, the duchesse de Guermantes, which has been put out into the courtyard of their Paris hotel in the hope that an insect carrying the particular pollen required will make an appearance. While he waits, concealed behind the shutters of an open window, a chance encounter of the baron de Charlus and the tailor Jupien outside gives rise to a natural historical *tableau vivant*. Jupien is mesmerized by Charlus; at first rooted to the spot "like a plant,"[61] after a moment he begins to strike "poses with the coquetry that the orchid might have adopted on the providential arrival of the [bumble]bee."[62] Charlus, in turn, "humming like a great bumble-bee,"[63] is lured by the tailor's flowery wiles into his abode. At this moment a bee, perhaps the very one the orchid and the narrator have been awaiting, flies into the courtyard. The young voyeur, however, chooses to follow the Jupien-Charlus courtship; the "beauty" of its "naturalness" steadily increases,[64] and it results in a "fertilization."[65]

E. M. Forster employed similar imagery to naturalize the male lovers in *Maurice*. In the gardens at Penge, the batlike Maurice, who "liked being out of doors, among the robins and bats, stealing hither and thither bare-headed," finds himself stirred by the evening primroses "expanding in the shrubbery." While wandering in the park at night he twice bumps into Alec Scudder, a gardener and "haunter of shrubberies," who is not recognized at first by sight, but rather by touch in the darkness: "he struck against corduroys, and was held for a moment by both elbows" (later he finds himself "strolling on and again colliding with corduroys").[66] After two brushes with the floral Alec, Maurice returns inside to discover that his head is "all yellow with evening primrose pollen."[67] Later that night Alec climbs into Maurice's room and their fertilizing contact resumes.

Such twentieth-century examples of homoerotic pollination have nineteenth-century precedents. Like the narrator in *À la recherche du temps perdu*, Dorian Gray is initiated into the world of sexual perversion when he witnesses a scene of pollination. It occurs outside the studio of Basil Hallward, creator of the portrait that will be endowed with life. His mind having been prepared by the vegetable painter's admiration of his beauty and Lord Henry Wotton's probing comments, which set him "vibrating and throbbing," Dorian sees a furry bee "creeping into the stained trumpet of a Tyrian convolvulus. The flower seemed to quiver, and then swayed gently to and fro."[68] Fifty years earlier, in Balzac's *Illusions perdues*, the flower-poet Lucien de Rubempré, né Chardon (thistle), author of *Les marguerites* (an anthology), is contemplating suicide and carrying a large bunch of sedum (yellow flowers), when he meets his black and golden bee in the form of the Abbé (*abeille*) Carlos Herrera.[69] George Bauer has demonstrated that the meeting of Lucien and Carlos in the novel became "a model for gay horticulture" and inspired Proust.[70] As we shall see, however, the chance outdoor meeting is but one of many botanical trysts and tropes used to describe variant sexual behaviors.

The naturalization of queer encounters in these episodes is not merely a matter of defensive mimicry. Proust also invokes the cross-fertilization analogy to naturalize the offspring of the invert artist's creativity. Before the Jupien-Charlus scene begins to unfold, the narrator discloses that he had "already drawn from the visible stratagems of flowers a conclusion that bore upon a whole unconscious element of literary production."[71] This "unconscious element" of creation in the later nineteenth and early twentieth centuries was more complex than the Romantic vegetable's abdication from reason. The earlier period's conception of the creative process as a kind of vegetable growth had its roots in German philosophy—specifically in what Meyer Abrams labeled "German theories of vegetable genius"[72]—and as we have seen, reflected a larger shift in the understanding of art, from a "supremely purposeful activity" to one that is "a spontaneous process independent of intention, precept, or even consciousness."[73] While the *fin-de-siècle* pollination model of art-making continued to emphasize the activity's unpurposive character, it stressed art's roots in instinct, allowing its simultaneously deeply sexual, communal, atavistic, and transspecies qualities to surface. These elements are emphasized in another late nineteenth-century horticultural model of painterly creation deriving from evolutionary theory, which posited that styles of art spring up periodically, as do hereditary traits. "In art as in gardens," its anonymous author argued in *L'Art Moderne,* one occasionally sees long ago uprooted or dead flowers appear. Sargent and Whistler are cited as key examples of this "sort of phenomenon of atavism," the sexual charge of which is suggested by the author's gloss: "it suffices that a germ, buried no matter how deeply, climbs little by little to the surface and sprays its shoot at the gardener's beard."[74]

But to return to the cross-fertilizing encounter, Julia Kristeva has shown in her analysis of Proust's Darwinian tableau that in the act of cross-fertilization we realize our innate capacity to multiply and mingle subject positions, to become "at once a space (a bisexual plant) and a force (an ephemeral insect),"[75] or to experience "a becoming-wasp of the orchid and a becoming-orchid of the wasp,"[76] as Gilles Deleuze and Félix Guattari describe it. We need not look to twentieth-century theorists for this insight, however. Michelet describes botany as "nothing other than the science of Love. A science barely distinct from Zoology. They translate each other. Love is the hazy terrain where the animal is flower, the flower is animated."[77] Love, cross-fertilization, and the creation of art became virtually synonymous in the "pictorial gardens" of the *fin de siècle.*

That any of these three experiences could induce metamorphosis had been widely acknowledged for some time. Charles Didelot's 1796 ballet *Flore et Zéphire,* for instance, is simultaneously the story of a romance, a pollination, and the origins of painting. Zephyr, the wind god, falls in love with the nymph Flora and asks Cupid to help him win her affections. The two immortals journey to Flora's rose garden. She is momentarily absent, but another attractive young nymph who lies sleeping catches Zephyr's eye, and the inconstant wind starts stroking her with his wings. She awakes, and Zephyr continues his flirtation. They dance, but when the nymph spies Zephyr's shadow on a temple wall she decides to trace it, and thus discovers the art of painting. Flora at last appears and Zephyr's attention reverts to his first love. Cupid, dismayed by the wind god's fickleness, agrees to unite him with the anemophilous (wind-pollinated) flower of his affections, but only after he has transformed the pair. He removes Zephyr's wings and gives them to his bride-to-be, grounding the groom. Love, then, literally gives flowers wings and the wind (the pollinator in this story), roots.[78]

In the later nineteenth century, botany and zoology largely took the place of mythology as explanations of such reciprocal metamorphoses, though the idea of eros as agent remained. Interspecies transformations accelerated; in Arthur Symons's poem "Violet" the femme-fleur's kisses are floral and birdlike by turns: "How this one clings and how that uncloses / From bud to flower in the way of roses," while another is "fluttering swift as a bird's kiss," a "bird that taps at a leafy lattice."[79] Subjects became protean, significantly, through their own various and interchangeable acts of art-making, lovemaking, and pollinating. Dance captured this embodied, performative poetics particularly well. Jean Lorrain's 1882 poem "Bathyle," about a male dancer in a sailors' dive, traces the becoming-insect and becoming-flower effected by the performance of pollination: "While he strips rose petals flying from blade to blade, his tunic parts over the roundness of his loins like a lily opening its petals."[80]

As the century drew to a close, such nonheteronormative manifestations of cross-fertilization became more common. Whistler, Mallarmé, Lorrain, Rodin, and others repeatedly sketched the "woman-flower, woman-bird, woman-dragonfly, woman-butterfly" Loïe Fuller (fig. 11).[81] Considered the first modernist dancer, Fuller debuted at the Folies-Bergère in 1892 and became an instant sensation. She was notably queer;[82] she

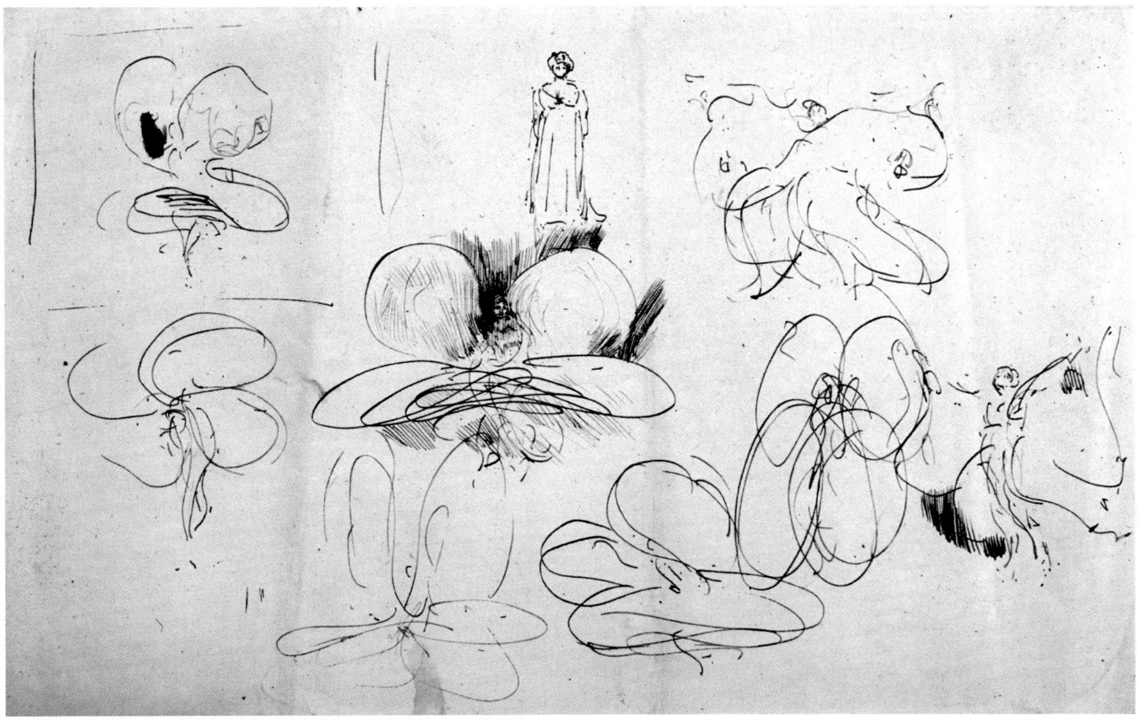

11
James McNeill Whistler, *Loie Fuller Dancing*, 1892. Pen and black ink on off-white wove paper, 20.9 x 31.9 cm. Hunterian Museum and Art Gallery, Glasgow, GLAHA 46173.

surrounded herself with a bevy of beautiful young students—"butterflies"—who were always "alternately stroking her hands and kissing her." In this harem her collaborator and romantic partner of thirty years, Gabrielle Bloch, was an insect, a strange figure "in a black tailor-made" suit who "circulated around the bevy of brightly colored butterflies like some scarab of ancient Egypt."[83] In her performances Fuller used bamboo rods to extend her arms; with these prostheses she manipulated hundreds of yards of white silk to create her metamorphic dances: *Butterfly, Violet, Lily, Orchid, Black Moth*.[84] Isadora Duncan emphasized her floral metamorphoses: "she turned to many coloured, shining orchids, to a wavering, flowing sea flower, and at length to a spiral-like lily."[85] Conversely, Roger Marx imaged her as a pollinator: she "skims the ground with the lightness of a dragonfly, hops and barely settles like a bird, glides, while shaking her trembling wings, like a bat."[86]

The complete artist, then, capitalized on the polyvalence the act of floral fertilization allowed, becoming both a pollinator *and* a flower. The poet Robert de Montesquiou identified himself as a bat and a blue hydrangea, and, most eloquently, Whistler signed himself not as a butterfly alone but an ecstatic one upon a pansy, copulating with it (fig. 12). Charles Ricketts's design for the cover of the 1891 edition of *Dorian Gray* (fig. 13) includes a repeated motif variously identified as a butterfly, marigolds, asphodels, or an abstraction. While Paul van Capelleveen claims to identify it definitively as the hearts of plucked daisies linked by a chain of petals,[87] the simultaneously floral and alate qualities of the motif (the V-shaped pattern appears against the grey ground like a flock of bats in foggy twilight) situate its meaning in the hybrid realm of cross-fertilization. Similarly, Walter Pater's claim that Michelangelo's sculpture brings "into one's thoughts a swarm of birds and flowers and insects"[88] is directly connected to the contemporary invert erotics of artistic pollination, for this model held out an ambrosial promise: the work of art in the age of botanical reproduction allowed the invert artist, as a *self*-cross-fertilizing creature, to be both a father and a mother, a fertile hermaphrodite, liberated from the typical nineteenth-century invert's fear of sterility, "the fangs of fruitless longing."[89]

In this study, John Singer Sargent is my primary example of botanical artistry. He has been called "a painter of natural history of the first rank," perhaps a surprising accolade for a society portraitist.[90] Interest in the natural world, however, was instilled in him as a child, providing a constant occupation throughout his nomadic upbringing and irregular education. In a letter of 1863 or 1864, FitzWilliam Sargent, the artist's father, wrote to his parents, "I keep [John] well supplied with interesting books of natural history, of which he is very fond, containing well-drawn pictures of birds and animals, with a sufficient amount of text to interest him in their habits, etc. He is quite a close observer of animated nature, so much so that, by carefully comparing what he sees with what he reads in his books, he is enabled to distinguish the birds which he sees about where he happens to be."[91] Such observations inspired the painter's first artistic forays; some of his earliest sketches are of birds among flowers and berries.[92] In later life he

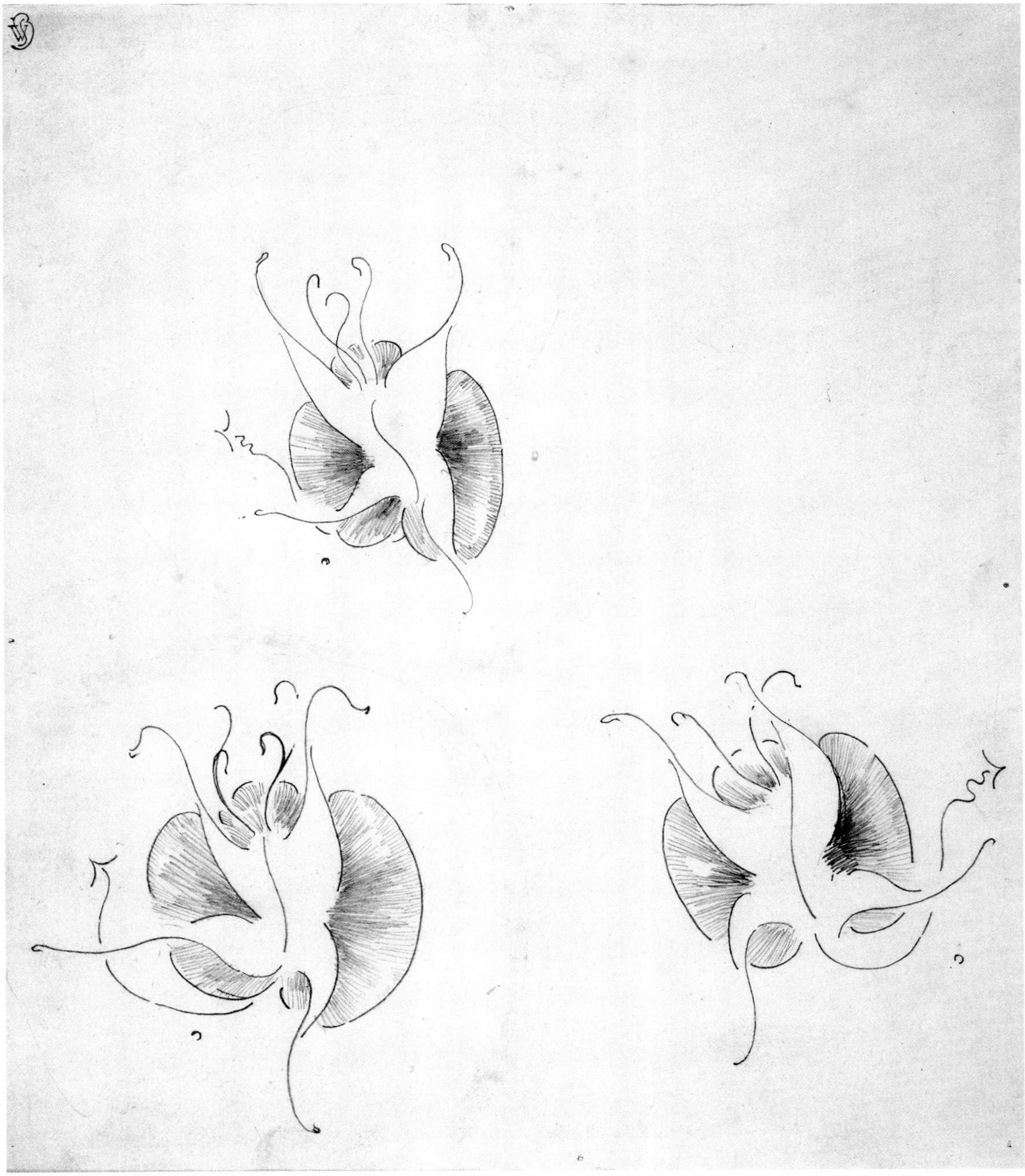

12
James McNeill Whistler, *Three Butterflies*, c. 1890. Graphite on ivory wove card, 24.1 x 20.6 cm. Gift of Mr. Walter S. Brewster, 1933.284a, The Art Institute of Chicago.

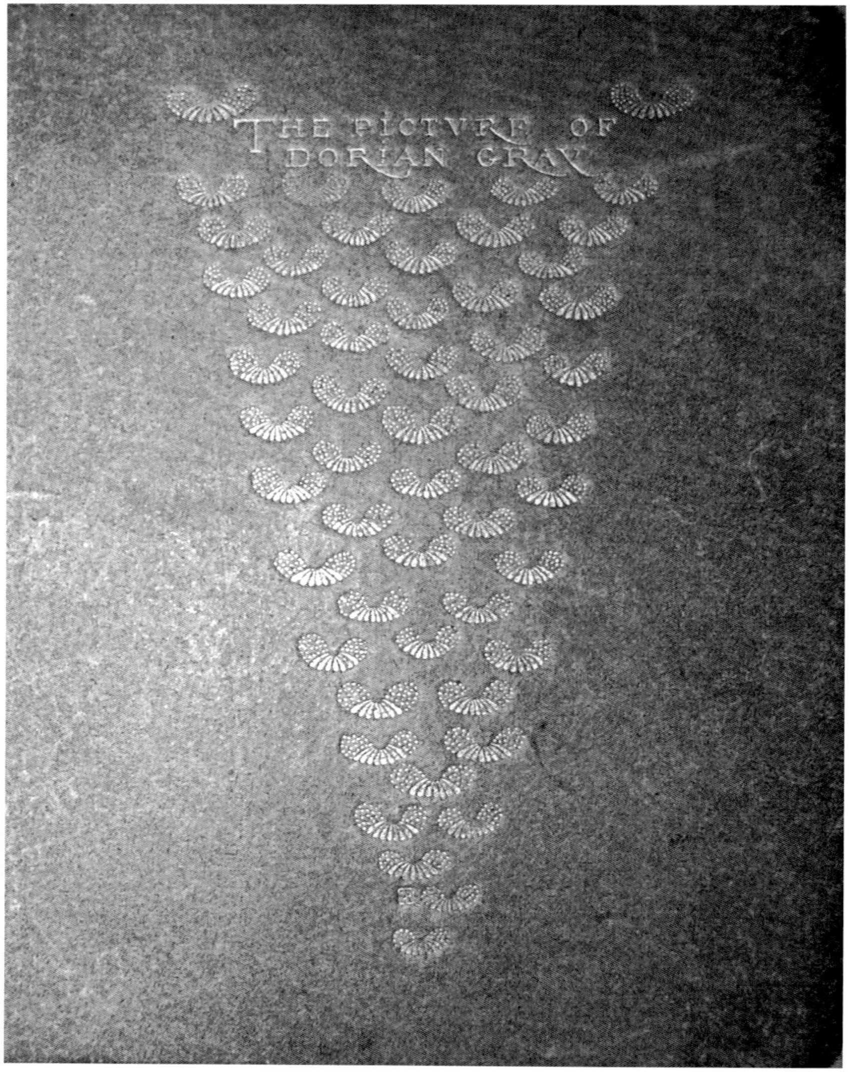

13
Charles Ricketts, cover of Oscar Wilde, *The Picture of Dorian Gray* (London: Ward, Lock, and Co., 1891).

remained "ever alert" to the natural world, his "joy over the petal of a flower, over a feather of a small bird, the mystery of the propelling power of a little snake in the grass" noted by his peers. Sargent "knew a great deal about natural history,"[93] and even while studying and working in Paris he continued to add to his butterfly collection.[94]

Unsurprisingly, Sargent found the correspondence between art and vegetal creation compelling. He was a successful horticultural artist, unlike some of his peers. Faced with Sargent's vegetal fecundity while painting at Calcot in 1888, Dennis Miller Bunker described himself sitting dejectedly "among the ruins" of his own sad "vegetable works,"[95] which were, significantly, rather reactionary efforts.[96] Like Whistler, Sargent considered the genesis of the modernist work of art an act of pollination, and he shared the period's notion of the artist as both pollinator and flowering plant. Starting in the 1880s, many of his canvases, particularly his noncommissioned paintings, allegorize artistic creation in general as cross-fertilization, and painting more specifically as hand-pollination.

The botanical model's capacity to naturalize inversion was an essential part of its allure for the painter. An invert himself,[97] Sargent was a friend of many famous inverts and investigators of inversion; his circle included Henry James, Edmund Gosse, Violet Paget (a.k.a. Vernon Lee, a nominal transvestite as well as a floral artist), Oscar Wilde, Robert de Montesquiou (the model for Proust's baron de Charlus), Dr. Samuel-Jean Pozzi, W. Graham Robertson, Walter Pater, Robert Louis Stevenson, Marc-André Raffalovich, John Addington Symonds, and James McNeill Whistler.

Discussion of the vegetable poetics of these artists and writers—Sargent's polyglot, cosmopolitan coterie—will spring up throughout these pages, for their work illuminates aspects of his and vice versa.[98] As we shall see, for Sargent the "pictorial garden" was a locus wherein modern conceptions of both art and sexuality, ineluctably intertwined for him, could be explored in a material way, through the touch of paint and pollen, and where the heteronormative rhetoric of production dominant in the cultures of Europe and North America could be eluded. Naturalizing both his inversion and his artistic production, Sargent's oeuvre visually elaborates a period poetics of queer sexuality and implicitly formulates a new sense of artistic subjectivity. An alternative cultural trajectory emerges through close investigation of this fertile period, authorizing alternative modernist genealogies and identities.

Beginning with a broad sweep, my first chapter, "La Vie en Rose," investigates the nineteenth century's fascination with the sex lives of flowers by examining evidence from diverse realms: visual culture (cartoons, advertisements, greeting cards, seed catalogues, costume design), the world of botany (treatises and their illustrations), literature, letters, lexicography, and the visual arts. Cataloguing the perversions to which plants were prone and charting the rise of pollinator erotics, I bring to light some little-known queer readings of botanical reproduction, from which the artists in this study took in part their inspiration even as they contributed to this interpretive legacy.

In chapter 2, "Lascivious Digitation, or The Importance of Manual Stimulation to the Invert Artist," the focus shifts to Sargent's work, in particular to his often very floral and gender-ambiguous painted hands, which can be situated within an iconographic tradition operating in the work of Gautier, Huysmans, Wilde, Proust, and others that capitalizes on the hermaphroditism of flowers. By grafting onto the hands of his subjects a floral model of hermaphroditic sexuality and exposed "interiority" (flowers' reproductive organs are on full display), Sargent endowed hands with their own potentially dangerous semantic and narrative life in his pictures, one where the boundaries between species and sexes are blurred, and identity's very "nature" is queried. Floral hands inevitably give rise to pollinator erotics, and the dramatically brushed appendages in these works intimate interspecies contact.

Chapter 3, "Dr. Octogynecologist," branches out into Sargent's coterie to elucidate what the touch of floral hands implies through close analysis of Dr. Samuel-Jean Pozzi's gynecological treatise and two portraits by Sargent, of the doctor himself and one of his most famous patients, Virginie Gautreau. Pozzi modernized French gynecology and notably popularized the bimanual technique of examination (one hand on the inside, one hand on the outside). Proust describes the creative work of the artist as consisting of a sort of gynecological probing of the outside world, and I argue that just as Pozzi was probing and palpating the space of biological creation, Sargent, whose habitual response to paintings was to caress their surfaces, conceived of his hands as probing the space of artistic creation. Pozzi's treatise also provides further insight into the painter's efforts to naturalize sexual inversion through a garden metaphorics. The gynecologist's descriptions of the "vegetations" of the womb and other anatomical growths conjure up a pastoral idyll, and these nonprocreative but natural bodily productions exemplify a model of creativity able to reconcile human and vegetable modes of reproduction.

In my fourth chapter, "Portrait of the Artist as a Young Ladybird," I read Sargent's paintings of female children in gardens and conservatories as veiled self-portraits and allegories of the process of painting—the hand-pollination the artist accomplishes with his brush and "pure tact of vision," as Henry James put it.[99] Two canvases in particular, *Carnation, Lily, Lily, Rose* and *Helen Sears,* picture the very act of hand-pollination. While avowing the naturalness and fruitfulness of the artist's and invert's creation, however, these superficially innocent paintings nevertheless register to different degrees Sargent's anxieties as he negotiated his artistic, sexual, and national identities through them.

Chapter 5, "The Sting," describes the less nostalgic and often more sadomasochistic practices of other pollinators, such as Whistler (the Butterfly) and Montesquiou (the Bat), and examines Sargent's portraits of fellow artists, including W. Graham Robertson and Robert Louis Stevenson. These paintings unleash the metamorphic potential of their subjects, who are transmuted into flowers and trees, moths and bees, that partake in scenes of pollination. The anatomy of the architecture in which the sitters are depicted also contributes to another, shadier and more aggressive kind of botanical narrative, one of back-door pollination (some pollinators drill or bite though the backside of the nectary rather than going through the "normal" channel). Both readings are essential to understanding the erotic charge of these works, in which the dangers and pleasures of invert sexuality are addressed.

The Conclusion, "Leaf Taking," situates Sargent's model of art in *fin-de-siècle* poetics and the development of artistic modernism and briefly traces the legacy of this poetics of vegetable artistic production in the twentieth century down two paths. One runs through the field of abstract art, and I follow its metaphorics from Kupka's paintings of pistils and stamens up to Laib's pollen planes. The other leads from photography to new media, from the botanical photogenic drawings of Henry Fox Talbot to the homoerotic work of von Gloeden, Gilbert and George, and others.

1

LA VIE EN ROSE

All the letters I can write
Are not fair as this—
Syllables of Velvet—
Sentences of Plush,
Depths of Ruby, undrained,
Hid, Lip, for Thee—
Play it were a Humming Bird—
And just sipped—me—

—EMILY DICKINSON, POEM 334 (1862)

Baudelaire believed that "a childhood awareness of flowers" counted "among the determinants of an artist's cast of mind."[1] Perhaps unsurprisingly, such sensitivity to the vegetal world was also considered a determinant of sexual deviance. From her earliest years, the Decadent novelist and playwright Rachilde averred, she had been prey to the perverting effect of "too many periwinkles, too much ivy, too much Virginia creeper, too many weeping willows, and too many truffles."[2] Proust's boyhood passion for hawthorns, like those of André Gide and Edmund Gosse for polyps and anemones (those living flowers of the sea), had a similar corollary.[3] Whence this affinity of pervert and plant, noted but unexplained by contemporary sexologists?[4] Over the next pages, dear reader, we shall discover that a broad cultural fascination with the social and sexual lives of flowers in the nineteenth century, explored in novels, poetry, ballets, pantomimes, paintings, prints, and visual ephemera as well as scientific treatises, classified and encouraged these cathexes.

"Flowers live, and feel, and desire," declared Michelet.[5] Ruskin too asserted the emotional susceptibility of the vegetable species, describing the time of blossoms as the most intimate and intense stage in the relations of plants, "correspondent to the joy of love in human creatures, and having the same object in the continuance of the race," efflorescence being the plant's expression of its "inner rapture."[6] J. J. Grandville's 1847 *Les fleurs animées* was a particularly popular manifestation of the predilection for floral romance. In the text and images, inner states animate the personified flowers more than external action; the illustration of "Tubéreuse et Jonquille" (Tuberose and Jonquil) (fig. 14), for example, pictures an imminent yet discreetly withheld embrace at a conservatory window, while the two flowers debate the relative merits of desire and pleasure.[7] In the latter half of the century, trade and greeting cards grew rife with such delicate floral courtships. A typical example from one of the dozens of Liebig card series devoted to animated flowers depicts a scene of blossoming love in a classical landscape (fig. 15).[8] A red dahlia—his magnificent petals unfolded to reveal his tender heart—woos an anemone, who smiles coyly at her would-be lover from behind her fan. Cupid was a busy bee in every garden; in an 1880s de La Rue valentine card (fig. 16), he literally

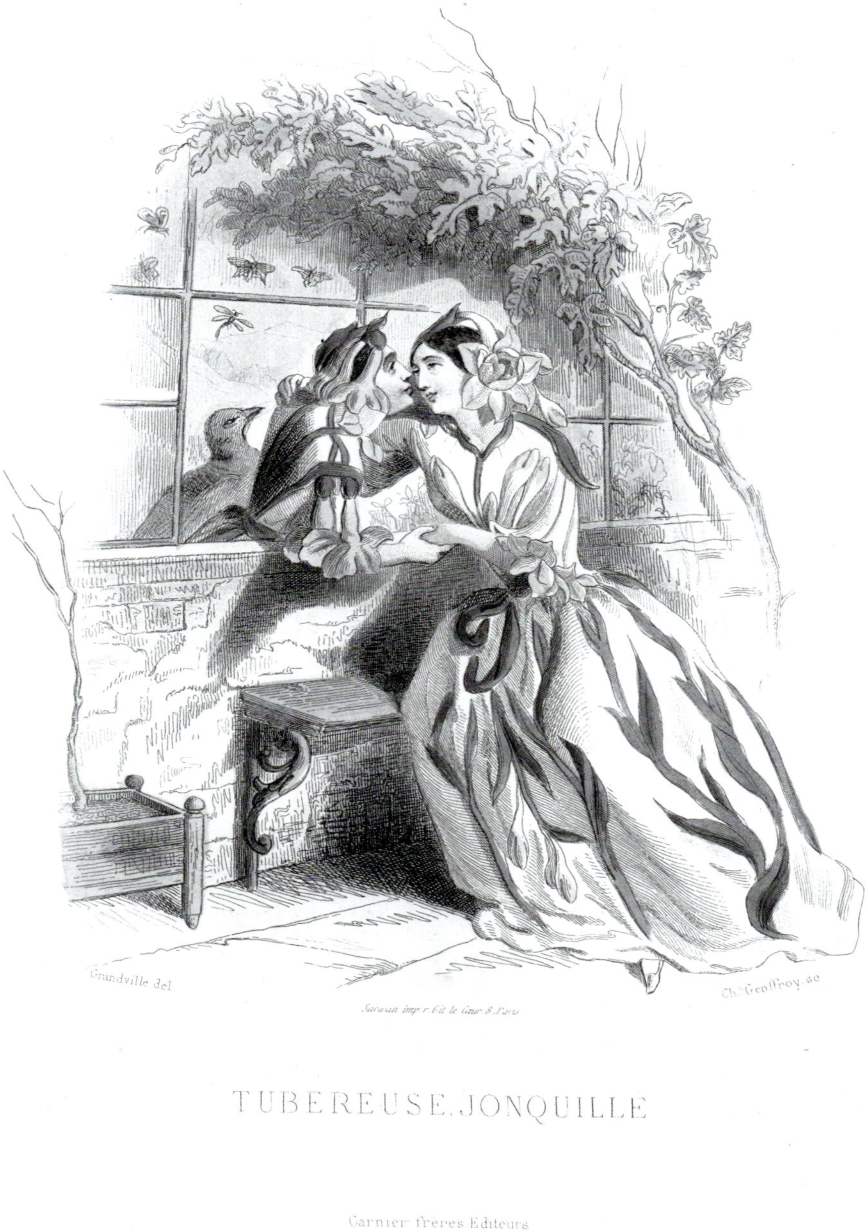

14
J. J. Grandville, "Tubéreuse et Jonquille," in *Les fleurs animées* (Paris: Garnier Frères, 1867), vol. 2.

15

Liebig card with animated flowers, 1897. Chromolithograph, 7 x 10.5 cm.

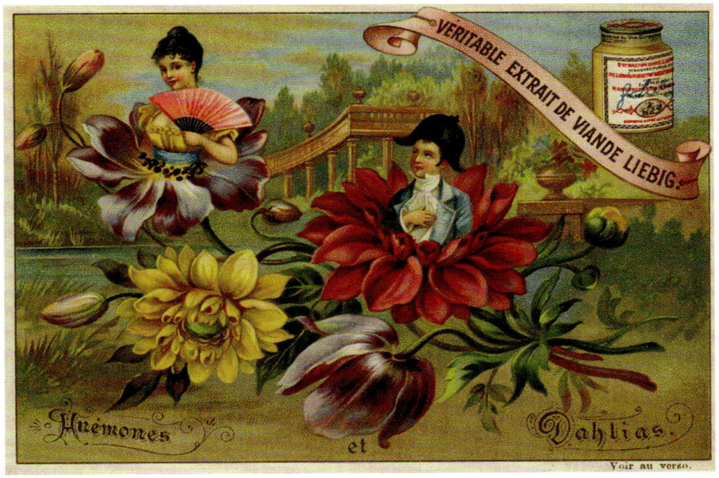

16

Thomas de La Rue, valentine, series 78, c. 1880, 12.7 x 8.7 cm.

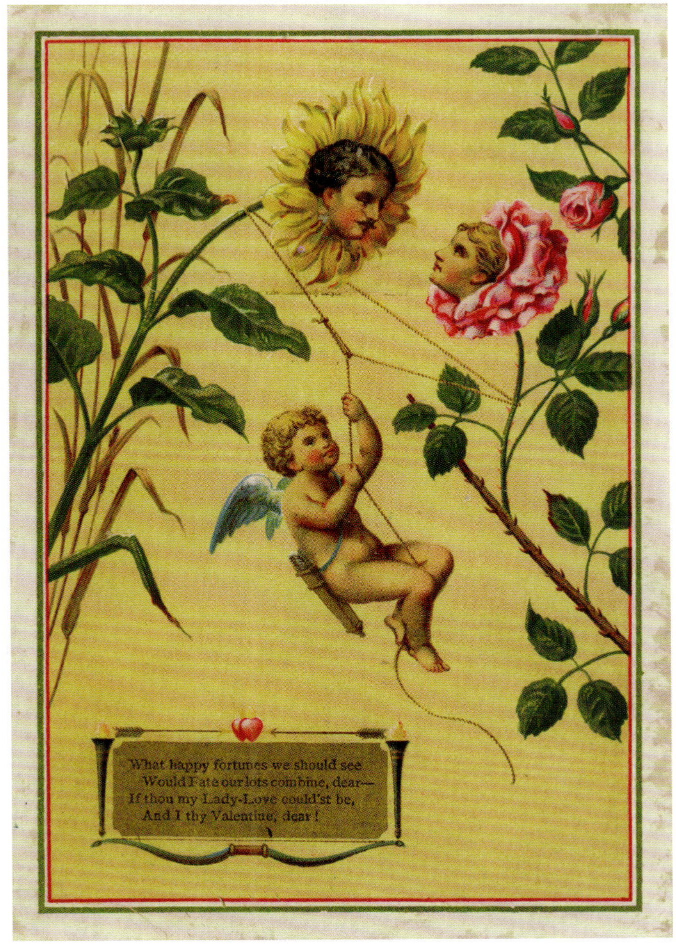

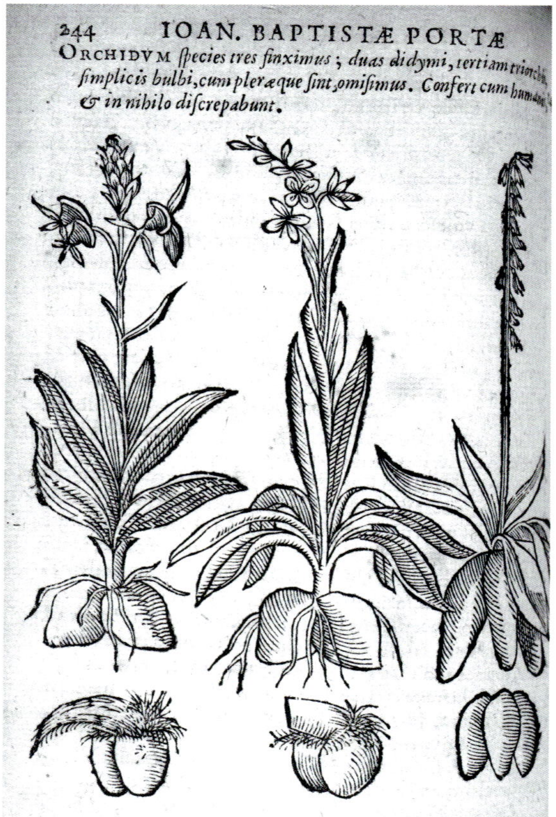

17
Giambattista della Porta, "Orchid," in *Phytognomonica* (Rothomagi: Berthelin, 1650), 244.

draws a moustached yellow daisy and a pink rose together.

The popularization of Linnaean botany in the eighteenth century had awoken a widespread empathy with the vegetable world. Linnaeus saw every flower as a bridal bed awaiting the consummation of marriages between pistils and stamens, and his work not only called attention to the romantic inclinations of these humanized plants but also provided a vocabulary with which to discuss them.[9] Erasmus Darwin, Charles's grandfather, could thus describe vegetable amours with piquant precision in his popular poems *The Temple of Nature* and *The Loves of the Plants,* in which "the wakeful Anther in his silken bed / O'er the pleas'd Stigma bows his waxen head" again and again.[10] A voyeuristic pleasure awaited readers, who learned that "on green leaves the sexual pleasures dwell,"[11] and could revel in visions of requited love: "With honey'd lips enamour'd Woodbines meet, / Clasp with fond arms, and mix their kisses sweet."[12]

Floral sexuality had been anthropomorphized long before Linnaeus, of course. The names of flowers, such as orchid, from the Greek *orkhis* (testicle), or vanilla, from the Latin *vagina,*[13] demonstrate the antiquity of these associations, and earlier texts, both literary and scientific, had made the same morphological arguments as the Swedish sage, sometimes quite graphically (fig. 17).[14] In the eighteenth century, however, these parallels attained an unprecedented imaginary resonance. The 1730s even saw the emergence of a new variety of erotica inspired by botany, consisting of titillating "natural historical" descriptions of the tree of life, or *arbor vitae,* and the *Frutex Vulvaria*—"a flat low Shrub, which always grows in a moist warm valley, at the foot of a little hill."[15] Such relatively straightforward exploitation of the suggestiveness of plant morphology and habitat persisted for more than a century, as a pseudoscientific report from the erotic magazine *The Exquisite* (1842) attests:

> The tree of life, then, is a succulent plant, consisting of one only straight stem, on the top of which is a *pistillim* or *apex,* sometimes of a grandiform appearance, and not unlike a

> May-cherry, though, at other seasons more resembling the Avellana or filbeard tree. Its fruits, contrary to most others, grow near the root; they are usually two in number; in size somewhat exceeding that of an ordinary nutmeg, and are both contained in one Siliqua, or purse, which, together with the whole root of the plant, is commonly beset with innumerable fibrilla, or capillary, tendrils.[16]

Julie Peakman has argued that the mobilization of such botanical analogies facilitated the "expression of alternative attitudes on sex," and we shall examine both how this was accomplished and why such analogies became relevant to the creative production of invert artists.[17] First, however, we shall explore the scope of this botanical erotica.

If floral sexuality was described in terms of the human, the reverse was also the case. Linnaeus called his wife his "monandrian lily"—a lily with only one stamen (i.e., man) to satisfy its needs—and looser analogies abounded in popular culture.[18] Widespread botanical knowledge and the Rococo cultivation of witty analogies, floral decoration, and garden activities probably account for the prevalence of floral double entendres, symbols, and metaphors, both visual and verbal, in the eighteenth century. The rose as a metaphor for virginity, for example, inspired scenes of literal and symbolic defloration like Fragonard's *Sacrifice of the Rose;*[19] secret love societies like the Order of the Knights and Nymphs of the Rose in France were devoted to it.[20] Gabriel de Saint-Aubin's 1781 print *Comparaison du bouton de rose* (Comparison of the Rosebud) (fig. 18) depicts an incident of spontaneously arising and simultaneously arousing curiosity about floral-human equivalences. A woman in a state of semi-undress examines the juxtaposition of her nipple and a rosebud in a butterfly-shaped mirror, her spread legs suggesting the possibility of further avenues of intimate exploration. Although the title puns on the polyvalence of the word "bouton," which means both nipple and bud, it does not limit the number of comparisons the figure or the viewer might make.

In the nineteenth century female maturation was habitually described in botanical terms. A healthy "bloom," as Amy King has shown, was a sign of sexual readiness and/or marriageability, and the telltale lily-white and rose-red hues of the perfect complexion were exalted.[21] *Jeunes filles en fleur* received a great deal of literary attention, and every sign of ripening sexuality, from the blush to the burgeoning bosom, was scrutinized. Edmund Gosse described the floral unfolding of pubescent girls in his poem "Moorland":

> . . . the sweet child-maiden
> We admired so well, fruit-laden,
>
> Now, like an expanded bud,
> Must be blown to womanhood,
> And the fuller lips and bosom
> Must proclaim the perfect blossom.[22]

The budlike breast was a signifier of ideal nubility that corsets could help achieve, though some preferred full-blooming bosoms, like that of Balzac's mature heroine, a lily of the valley "in flower already,"[23] or the even headier charms of a bust "swollen" like "a seed-pod of a flower surfeited with pollen."[24] Nipples themselves, as Saint-

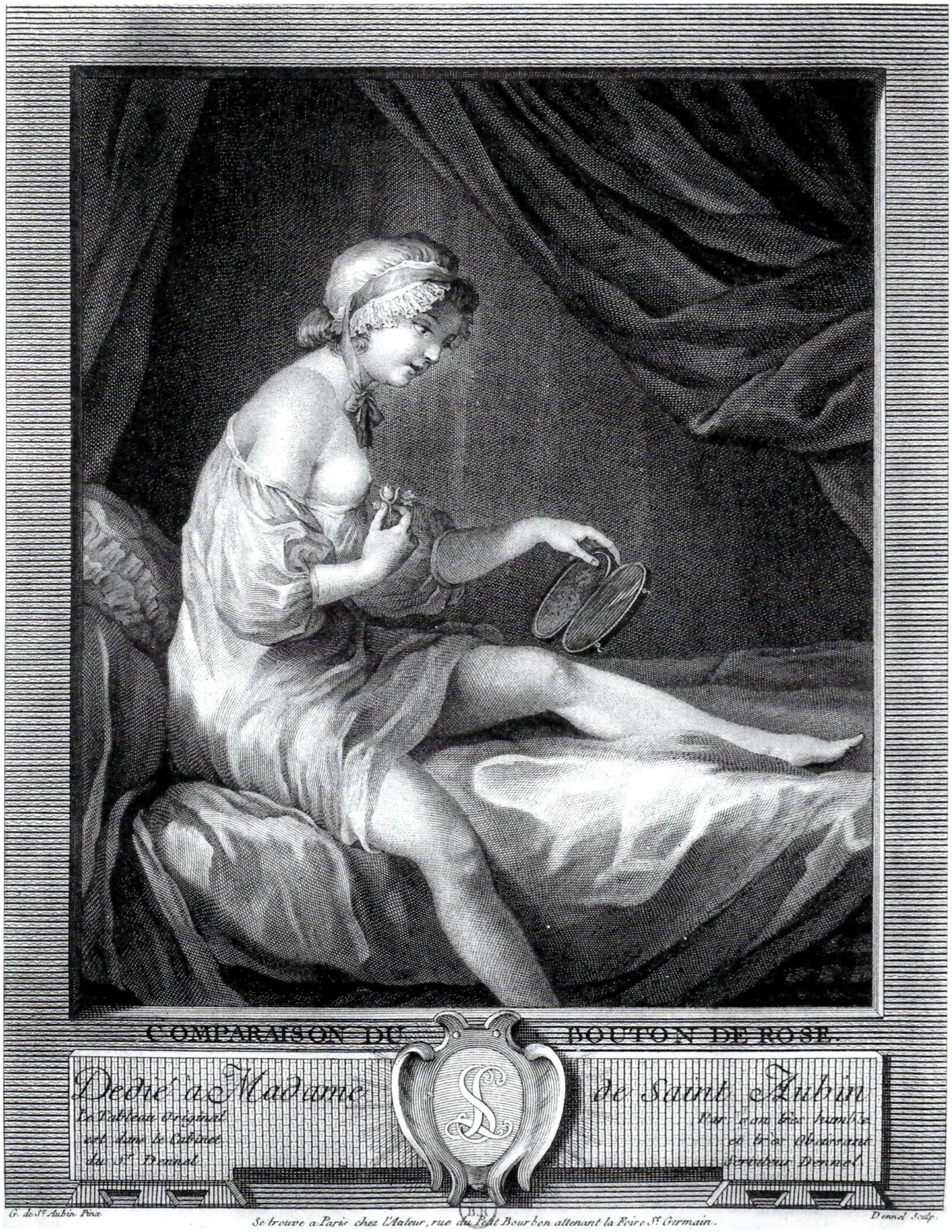

18
Gabriel de Saint-Aubin, *Comparaison du bouton de rose,* 1781, engraved by R. Dennel. Reproduced in Émile Dacier, *Gabriel de Saint-Aubin: Peintre, dessinateur et graveur,* vol. 1 (Paris: Les Éditions G. van Oest, 1929), plate 12.

Aubin's print had already suggested, could be buds "about to blossom."[25] Corset trade cards, which frequently depicted models in floral worlds, occasionally made the female anatomy explicitly vegetable. Dr. Warner's Coraline Corsets, for example, restrain bodies bursting with luxurious vegetal life (fig. 19). *Fin-de-siècle* painting confirmed the equation of bloom and sexual ripeness in the most starkly literal terms: in Charles Courtney Curran's 1890 painting *Scent of the Rose* the nude upper half of a female body "emerges pistil-like"[26] from the corolla of a creamy-white rose. A French turn-of-the-century postcard entitled "L'oeillet" (The Carnation) (fig. 20) makes use of the same motif, a nude female torso emerging from the petals of a carnation.[27]

The more intimate botany of the female body—its rosy folds and mossy charms—was not neglected. Slang words for pubic hair included "grass," "humid forest," "green-grove," and "sweet-briar."[28] The vagina was a garden, a corolla, or a calyx:[29] Des Esseintes sees "the wild Nidularium" blossoming between the "upraised thighs"[30] of his dream woman (the legs leading up to such blooms

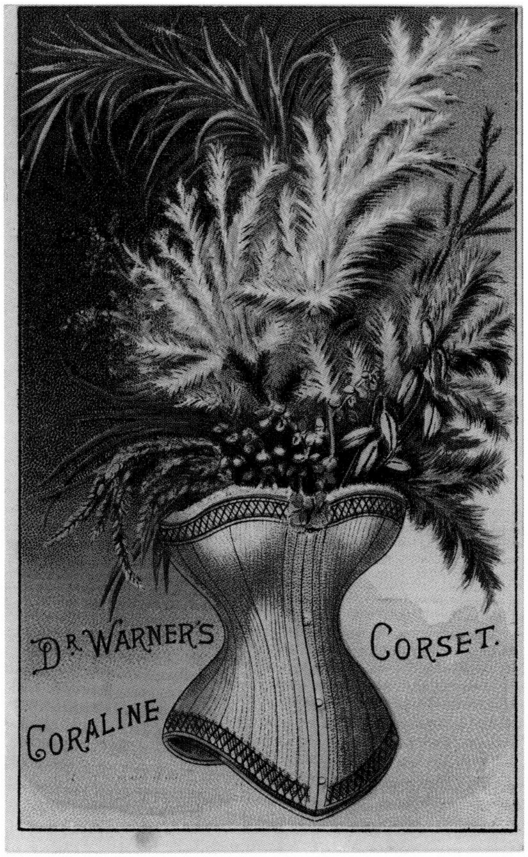

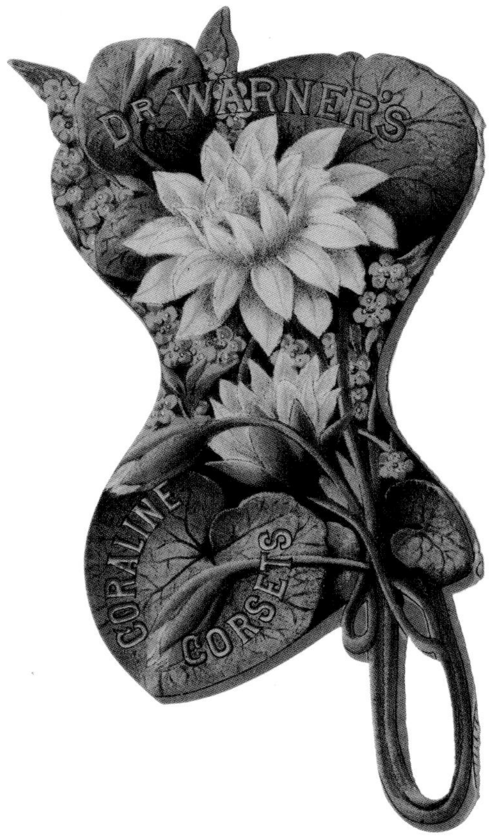

19
Dr. Warner's Coraline Corsets trade cards, 1880s, (*left*) 12.7 x 7.6 cm, (*right*) 15 x 8 cm.

20

M. Boulanger, "L'oeillet," c. 1900, postcard, 13.5 x 8.5 cm.

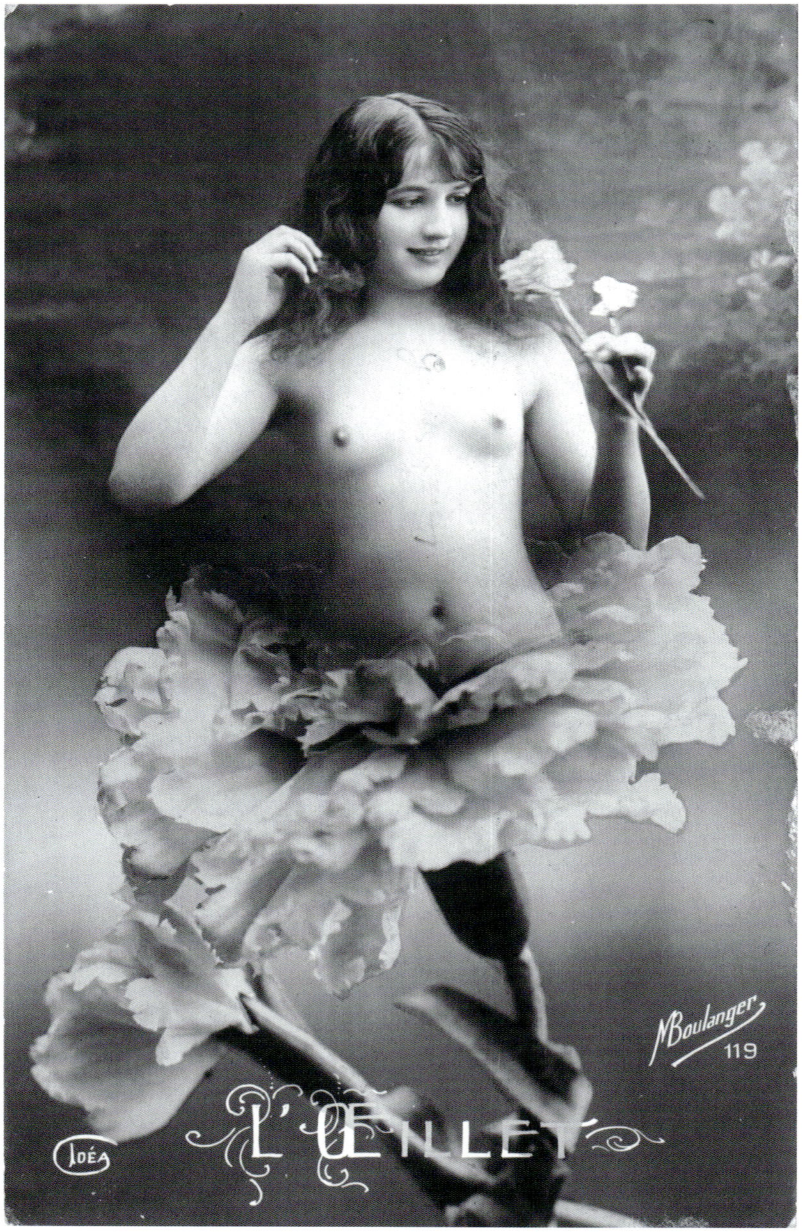

were, of course, "stems").[31] Labia, which relate to floral labella both linguistically and functionally, were termed "petals" or "violets."[32] Wilde described the clitoris as "a tiny bud—a living flower of flesh and blood," which, when excited, begins "to expand its petals and shed forth its ambrosial dew."[33] We need not turn to Decadent or pornographic texts for examples of this analogy, however. The poems of Emily Dickinson and other women writers of the nineteenth century are sprinkled with vegetable clitorises in the form of "buds, berries, seeds," peas, nuts, and "compact flowers such as snowdrops and crocuses," as Paula Bennett has shown.[34] To give but one example, from Dickinson:

> Forbidden Fruit a flavor has
> That lawful Orchards mocks—
> How luscious lies within the Pod
> The Pea that Duty locks—[35]

French slang for the clitoris also relied on its similarity to small nuts and legumes; "coffee bean," "lentil," "bean," and "hazelnut" were all synonyms.[36] The erotic imagination, however, was not limited to vegetal minutiae: in their journal the Goncourt brothers muse that "in eating a banana" one is consuming "the clitoris of a fruit."[37] Other erogenous zones were also subject to vegetable blazoning. The mouth was frequently a flower, its lips "firm buds ready to blossom";[38] when kissed its "odorous rose-red petals part";[39] Du Maurier's Trilby's "little tongue is scooped out like the leaf of a pink peony."[40]

Comparisons of human reproductive organs and vegetable features did not begin and end with morphology, though: a number of physiological processes were shared. Writing on menstruation, the American doctor Mary Jacobi noted that "the woman buds as surely and as incessantly as the plant, continually generating not only the reproductive cell, but the nutritive material without which this would be useless,"[41] and for centuries menstruation had been referred to as "flowers" in English and "fleurs" in French.[42] In *La dame aux camélias,* the floral heroine (Marguerite, a daisy) wears red camellias to signal to her lovers that she is menstruating and white ones the rest of the time.[43] Women's sexual activity of all sorts, moreover, was floral. To lose one's virginity was, of course, to be deflowered, but other acts of vegetable cultivation or maturation abounded: copulation with a woman entailed picking or gathering strawberries, nuts, flowers, buds, and so on, and female masturbation was a matter of blooming or shedding petals.[44]

In the face of all this teeming erotic plant life, it is astonishing to consider the popularity the idea of the *femme-fleur* enjoyed. It flourished nonetheless, along with the fiction of the figuration's innocence and ethereality. First names like Rose, Lily, Iris, Olive, Pansy, Ivy, Flora, Violet, and Daisy became popular, as did epithets like "bud" and "bloss."[45] A healthy bloom, as previously mentioned, was vital. Soap brands like "Laird's Bloom of Youth" and "White Lilac" helped maintain it. And for those whose pellicles weren't so petaline, the right shades were attainable through mass-marketed "liquid blooms" and "Lillie Powders."[46] Other vegetal enticements were also cultivated: delicate floral scents replaced the more potent animal-based perfumes (civet, musk, and ambergris) of earlier periods,[47] and young women's sleeves, skirts, bodices, and hair were decorated with real or artificial flowers.[48] A flower-costuming industry sprang up around flower-girl spectacles at the theater and ballet, Victorian May Day festivals, and the like.[49] The young Vita Sackville-West was photographed, trellised and blooming wisteria, in one of the forms such attire could assume (fig. 21). Grandville's *Fleurs animées* (fig. 14) provided another influential model: on plain white dresses leaves and petals often appear as slashings or embroidery, though morphological similarities, for example between the trumpeting of sleeves and corollas, are also exploited. Ladies' everyday fashions even reflected the floral craze, as cartoons in *Punch* like "Last Sweet Thing in Toilettes" (fig. 22) and literary descriptions attest. In Ouida's *Moths* the courtesan Noisette is first encountered as "a handsome woman in a black dress, with knots of pink; she had a dome-like pile of glistening hair, gorgeous beauty, a splendid bust; she looked like a rose-hued rhododendron made human."[50]

In painting, the *femme-fleur* was a popular motif. Annette Stott, looking primarily at American art, includes Sargent in her list of its devotees.[51] In his portrait of the Wyndham sisters (fig. 23), for example, the sitters, all in petally white among green cushions and against a green background, are "huge white flowers on a sumptuous sofa."[52] The three floral arrangements visible in the room echo the placement of the figures, as do three of the large, creamy magnolia blossoms at the bottom right.[53] A quick sampling of contemporary reactions to Sargent's art yields a posy of blossoming subjects: Rodin famously deemed the Hunter sisters "a bouquet of flowers";[54] Joséphin Péladan described the Boit daughters as

21

Photograph of Vita Sackville-West as a basket of wisteria, c. 1900 (?). Collection of Adam Nicholson.

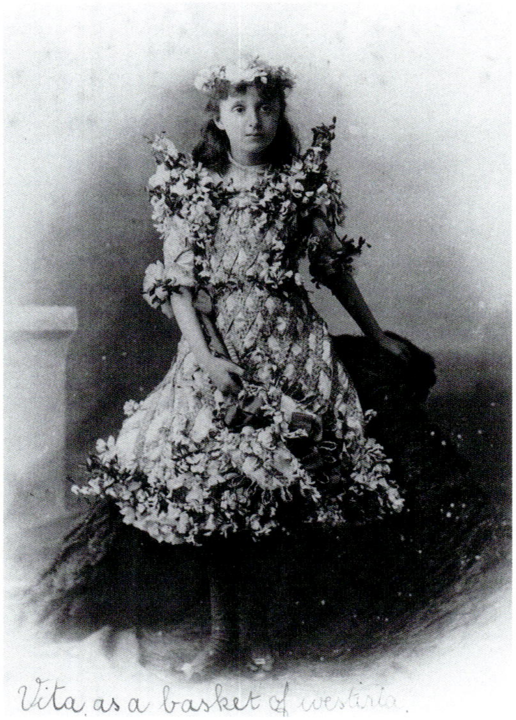

22

"Last Sweet Thing in Toilettes," ***Punch, or the London Charivari,*** **5 July 1879, 304.**

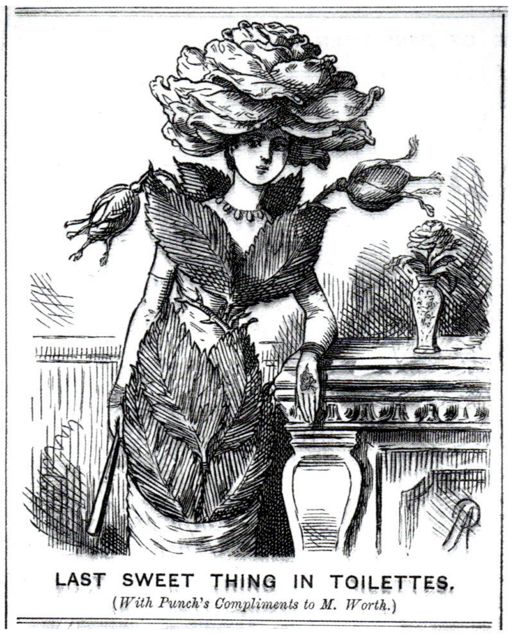

"aristocratic flowers";[55] and the painting of *Mrs. Carl Meyer and Her Children* was perceived, when first exhibited, as having "the fragility and the complexion of a flower."[56] Occasionally writers named the species of Sargent's sitters. An artist peer saw Beatrice Goelet (fig. 24), clad in a strange red, gold, and green dress and emerging from an earthy brown background, as belonging to "the streaked variety of tulips."[57] Paul Bourget declared Isabella Stewart Gardner "a living orchid"[58] in her portrait, and the critic Armand Sylvestre found himself "infinitely charmed" by the affinity between human and plant expressed in *Mme. Pailleron,* where "a steep slope of tulips stretching their curious corollas toward her" brought "to mind the poetry of Charles Baudelaire."[59] *La Carmencita,* according to Claude Phillips, was "a veritable 'Fleur du Mal' such as would have delighted Baudelaire himself," a creation "sprung from the artificial soil of the expiring century."[60]

Stott argues that "nineteenth-century artists used the flower as a versatile sign of female sexuality, to represent anything from moral laxity to innocent chastity,"[61] and Sargent's *femmes-fleurs* appear to have run the gamut of significations for his contemporaries, as the allusions to Baudelaire and orchids suggest. Georges Rochegrosse's ravishing and enormously popular 1894 painting *Le chevalier aux fleurs,* based on the flower-maidens' scene in Wagner's opera *Parsifal* (fig. 25), represents a female sexuality that is decidedly at the salacious end of the spectrum.[62] The flower maidens in the opera, which was first performed in 1882, may have been inspired by Grandville's *Fleurs animées,*[63] but their brazen behavior verged on the indecent. An 1879 English translation of

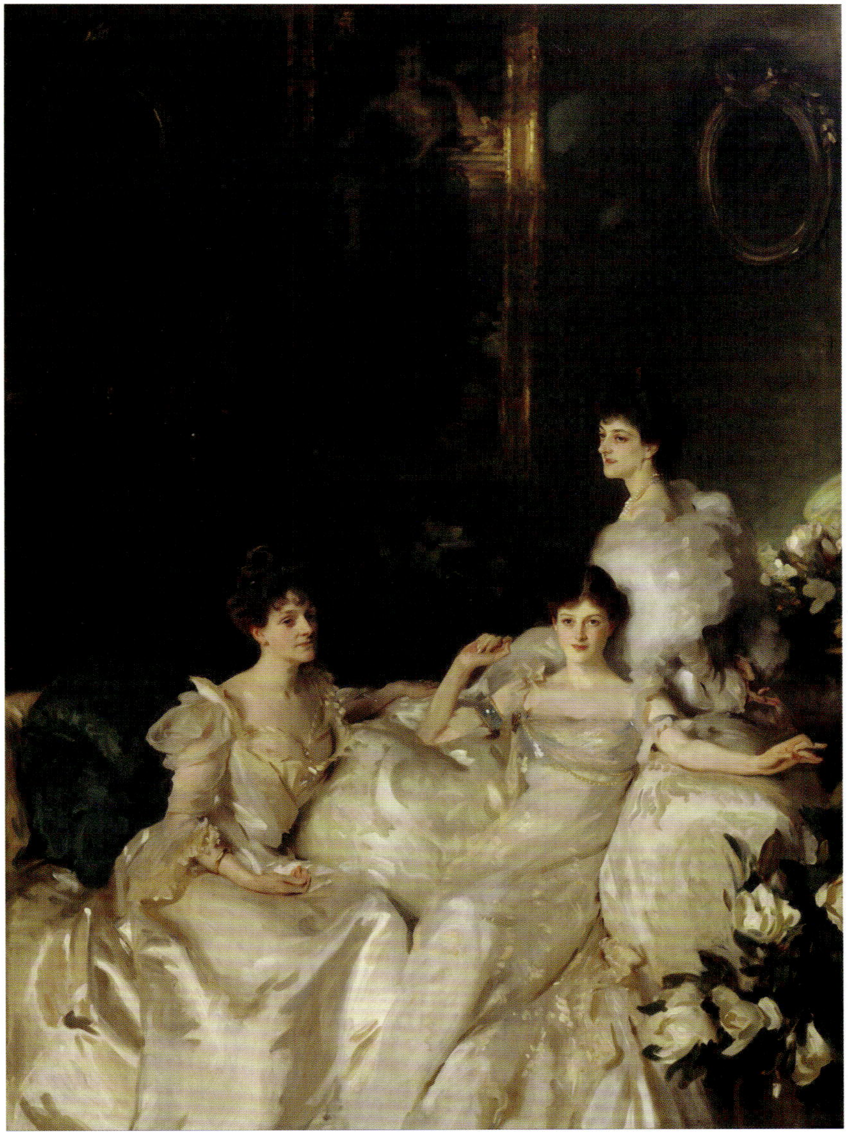

23
John Singer Sargent, *The Wyndham Sisters: Lady Elcho, Mrs. Adeane, and Mrs. Tennant,* 1899. Oil on canvas, 292.1 x 213.7 cm. The Metropolitan Museum of Art, Catharine Lorillard Wolfe Collection, Wolfe Fund, 1927 (27.67).

24
John Singer Sargent, *Miss Beatrice Goelet,* 1890. Oil on canvas, 162.6 x 86.3 cm. Private collection. Reproduced in *The Century* 43, no. 5 (March 1892): 773.

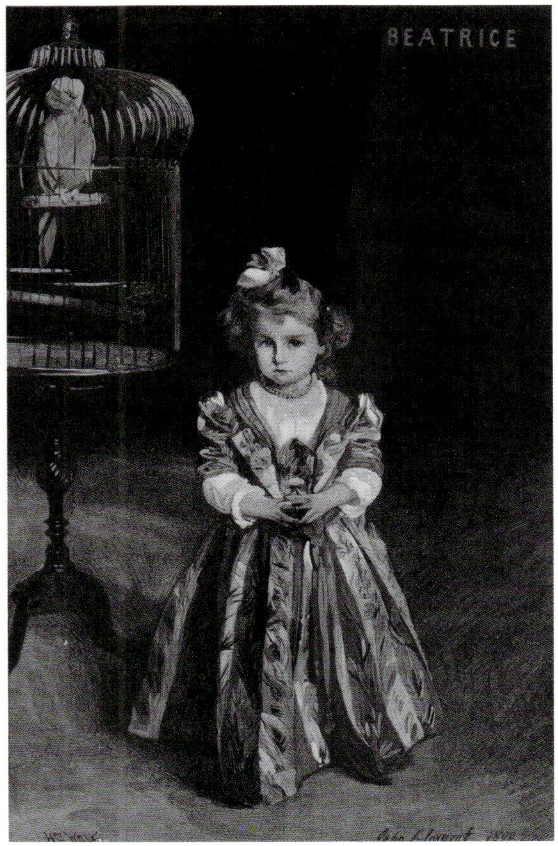

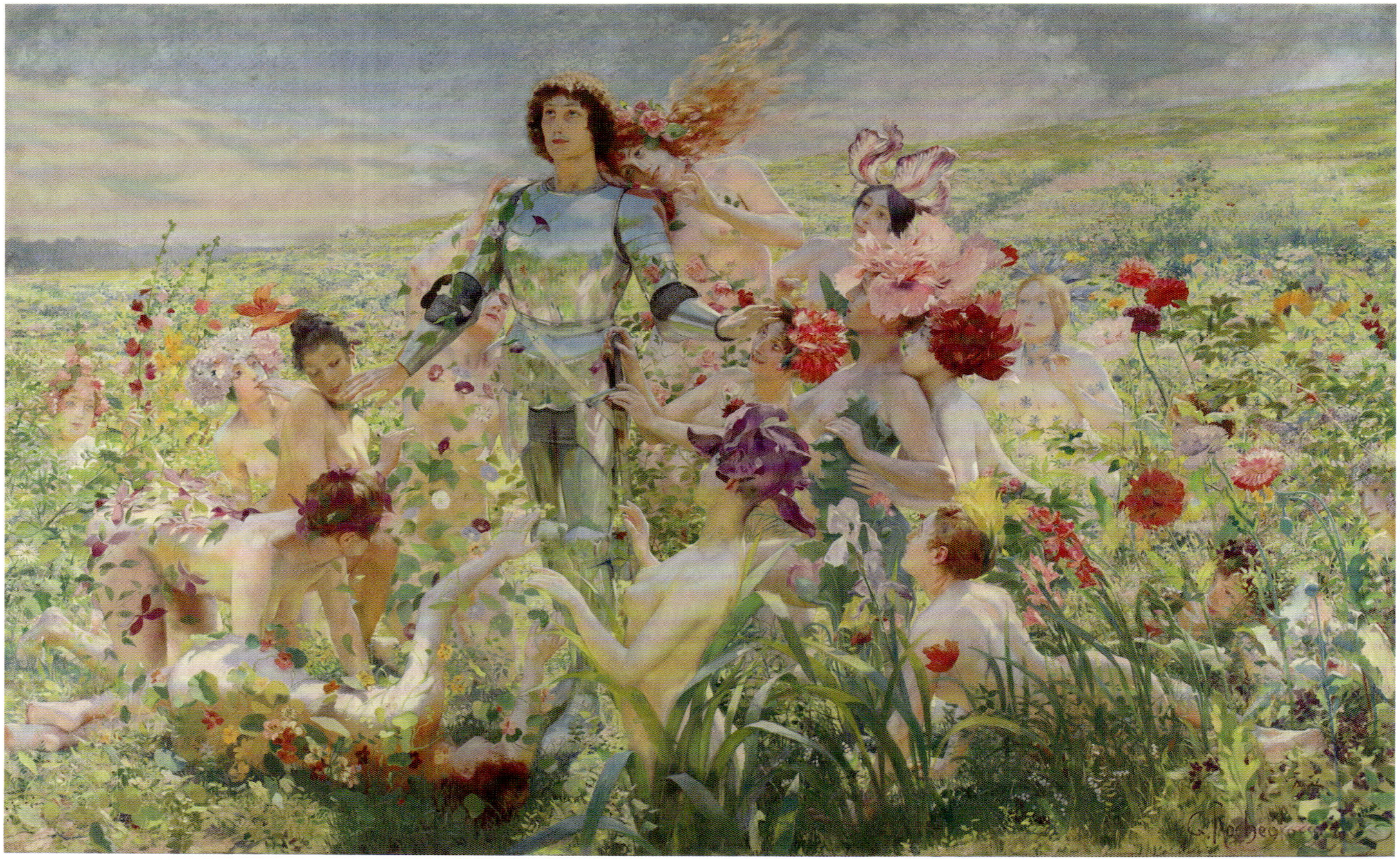

25
Georges Rochegrosse, *Knight Among the Flower-Maidens, from Wagner's "Parsifal,"* 1894. Oil on canvas, 235 x 376 cm. Musée d'Orsay, Paris.

Wagner's lyrics retains the sexual innuendo of the following exchange, particularly in the last line. The flower maidens sing,

> Come! Come!
> Handsome stripling,
> I'll be thy flower!
> Sweetly dancing and rippling
> Bliss unshadowed I'll shower

Parsifal responds, "Ye wild crowd of beautiful flowers, / If I am to play, ye must widen your bowers."[64] In Rochegrosse's painting the naked, stroking, twining *femmes-fleurs* prove themselves willing to assume any position to fulfill male fantasy: convolvulus clings while carnation fingers the hilt of the knight's sword, hydrangea sucks her own digits, nasturtium writhes on the ground, iris kneels, and clematis readies herself on all fours. If modesty and innocence are no longer part of the picture here, Albert Flament's account of the Mardi Gras carnival in Paris in 1896, where "the float of the *femmes-fleurs,* during a quarter of an hour's delay, served as a urinal for the antique warriors, while the *femmes-fleurs* ate hard-boiled eggs,"[65] is certainly a nadir in the history of the trope. The ideal was also undermined by the prostitution of "the night-blooming plants of the town,"[66] including "garden-whores," "columbines," elderly "artichokes," and "hedge-whores,"[67] in addition to the more common "violets" and "camellias."[68] Even an emblem of purity like the *Victoria Regia* (the giant water lily named after Queen Victoria) could fall from grace. Thomas Wentworth Higginson described its "beautiful disrobing"—the bud unfolding petal by petal—with a voyeur's "sense of guilt." He found the floral body laid bare after this unconscious striptease, however, shocking: "Can this be the virgin Victoria,—this thing of crimson passion, this pile of pink and yellow, relaxed, expanded, voluptuous, lolling languidly upon the water, never to rise again?" It is "'Hebe turned to Magdalen,'" he concluded.[69]

The *fin-de-siècle* increase in floral-female imagery, which often rendered women "both passive and consumable," has been linked to the threat of the New Woman.[70] The fetishistic and dehumanizing aspect of the conflation of women and flowers was criticized by Iota in *A Yellow Aster,* a popular New Woman novel. The heroine dismisses her scores of suitors: "My lovers? They weren't lovers at all, they were explorers, experimental philosophers. They had the same strong yearning for me that a botanist has for a blue chrysanthemum, or a yellow aster."[71] Perhaps such critiques of the female rarefaction, vegetation, and collection demanded by the *femme-fleur* model inspired the suffragette attack on the orchid houses at the Royal Botanic Gardens at Kew in 1913, which targeted the rarest and most expensive plants.[72]

Certainly the floral model of femininity, with its eroticization of innocence and growth, had potentially disturbing implications. Henry James explored these in *Watch and Ward,* where the pedophilic protagonist is described as "a wandering botanist" watching "for the first woodland violets for the year, for the shy field-flower of spontaneous affection" in his young charge. When these are not forthcoming he wonders whether, "at the worst, a little precursory love-making would do any harm. The ground might be gently tickled to receive his own sowing; the petals of the young girl's nature, playfully forced apart, would leave the golden heart of the flower but the more accessible to his own vertical rays."[73] The charms of the tightly closed bud and the act of "forcing," as we shall see, dominated other areas of erotic life as well.

While never as pervasive an image as the *femme-fleur,* over the course of the nineteenth century the image of the *fleur du mâle* became increasingly common. From Grandville's *Fleurs animées* to Walter Crane's 1889 *Flora's Feast: A Masque of Flowers,* just over forty years apart, the number of male flowers represented increased from a tenth to half of the figures. Crane's man-flowers (fig. 26) range from the burly tulip type—proud if flaming warriors keeping a firm grip on their thrusting stems, their torn leaves suggesting battle scars—to the ephebish and narcissus-like honeysuckle, who coldly ignores the advances of the cloying convolvulus. The stamens of these man-plants are wrapped in petals. Even as Sargent's women and children have been seen as typically "spread out like petals of a great Victorian rose in poses of affected lassitude,"[74] his *fleurs du mâle,* too, are often languorously spread open, as we shall see.

For an illustration of how the male body becomes subject to floral troping, we can turn to Zola's description of a *tableau vivant* in *La curée,* which follows a standard metamorphic formula linking head and blossom. The character Maxime, who plays Narcissus, does not move and yet appears to be turning into a flower:

> His limbs became verdant, elongated, in his tight-fitting dress of green satin; the flexible stalk, formed by his legs slightly bent, was on the point of sinking into the ground and taking root, while his chest, adorned with broad lappets of white satin, blossomed into a wondrous corolla. Maxime's fair hair completed the illusion, and with its long curls set yellow pistils amid the whiteness of the petals. And the great nascent flower, still human, inclined its head towards the spring, its eyes moistened, its countenance smiling with voluptuous ecstasy, as though the

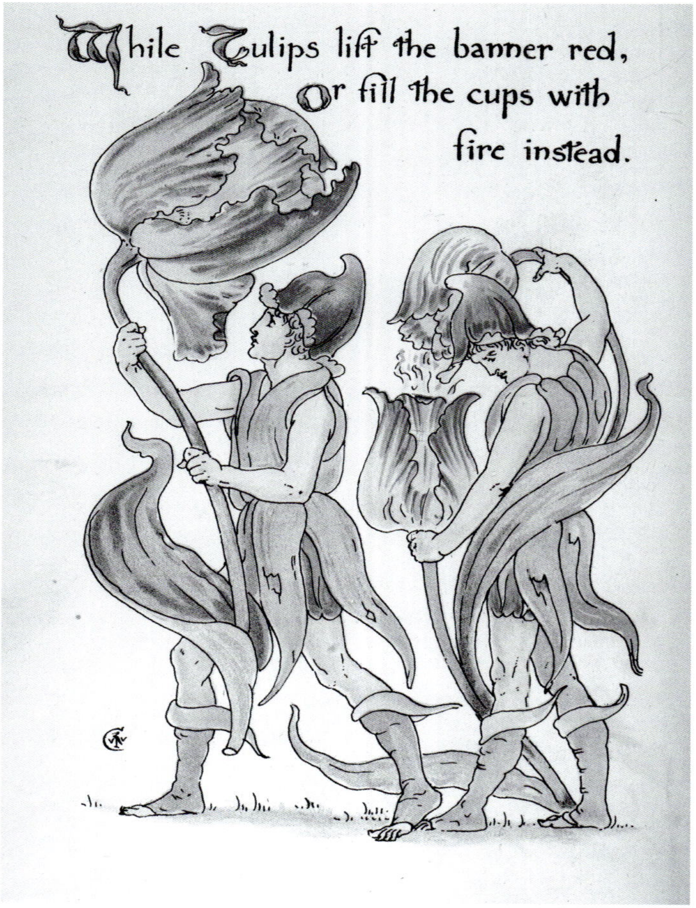

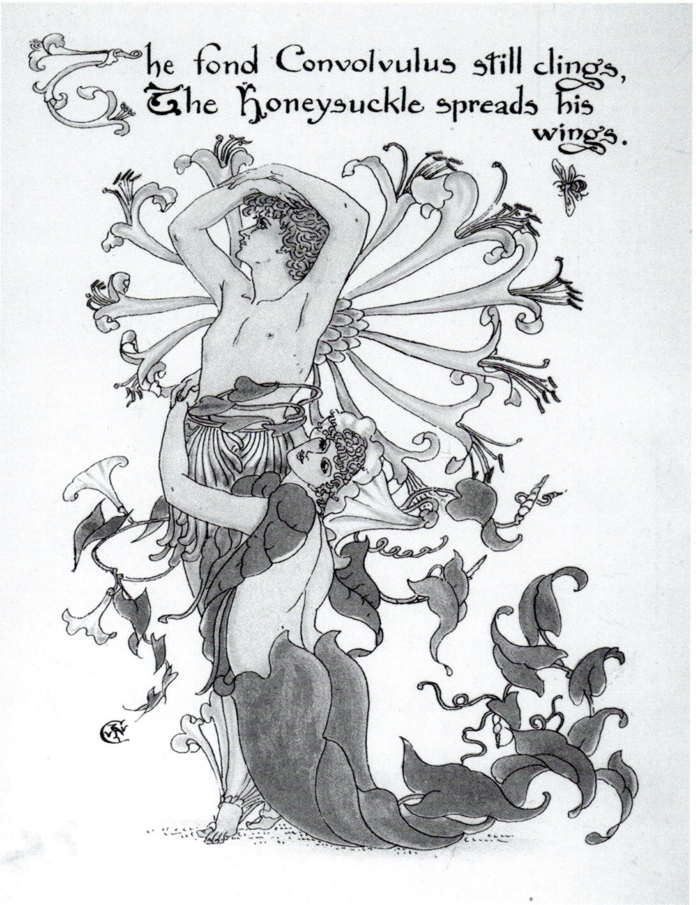

26
Walter Crane, "Tulips" and "Convolvulus and Honeysuckle," in *Flora's Feast: A Masque of Flowers* (London: Cassell and Co., 1890), 14, 34.

> beauteous Narcissus had at last in death satisfied the passion with which he had inspired himself.[75]

This imaginary metamorphosis, catalyzed by the mutual desire of writer, reader, and character, is experienced sexually. The body's transmutation under the gaze brings it through arousal to climax: it slowly lengthens, reveals its head, and finally, voluptuously, secretes some liquid.

If the whole man could flower, as it were, the finer points of male anatomy, sexuality, and physiology were also subject to vegetal comparison. Testicles were known as "olives," "chestnuts," "plums," and even "gourds."[76] "Nectar" was slang for semen,[77] which smelled, according to one American physician, like "the roots of the orchis, the iuli of chestnuts, and the antherae of many plants."[78] In Cleland's *Fanny Hill* the penis is identified as "the true, the genuine sensitive plant, which, instead of shrinking from the touch, joys to meet it, and swells and vegetates under it."[79] Other species of male member were also identified, including the "carrot," "mushroom," "asparagus," and "radish." More generic organ "types" such as "stem," "branch," "root," and "thorn" were also common.[80] Wilde, always able to squeeze the most juice out of his vegetable metaphors for the various sexual organs and functions, invigorated the old penis/stamen analogy by dramatizing the sensitivity of the anther and glans: "like a heavy laden stamen when kissed by the deflowering wind scatters its pollen on the open ovaries around it, so, hardly had the turgid and overflowing phallus touched the tiny clitoris when it jutted forth its sappy seed not only on it, but it squirted over all the surrounding parts."[81] In addition to spilling seed, ejaculation was an act of "watering" the garden or specific plants in it, depending on the anatomical locus involved.[82] "Planting" the garden or particular flowers meant copulation or male masturbation.[83]

The rampant floral becoming that characterized the later nineteenth century was not without a lighter side. Consider Grandville's "Pervenche desséchée" (Dried Periwinkle) (fig. 27) or Alfred Le Petit's caricature of the dictionarist Émile Littré (fig. 28) in *Fleurs, fruits et légumes du jour,* the pages of which are populated entirely by man-flowers.[84] Both images employ *trompe-l'oeil* conventions—curling pages, pins, a fly crawling on a surface—to reinforce the flatness of these pressed flowers. The dry humor of these pieces extends to commentary on genre: Littré's dictionary on its stand is a *herbarium verbarium,* to use Claudette Sartiliot's phrase, the author a dried specimen (Littré had in fact spent three years studying botany at the Jardin Botanique before pursuing his career as a lexicographer, shifting from the study of vegetable to linguistic roots),[85] while Grandville's forlorn periwinkle on its single sheet mocks the "innocent" flower-pressing activity of young botanists.

Social satire underlies a variety of Victorian seed advertisements featuring animate vegetables (fig. 29), which gently mock social types, from respectable promenading pairs of well-dressed pea-ple to more ribald, ruddy-cheeked radishes and the higher class of carrot. Le Bon Marché, the Parisian department store, produced numerous vegetal trade cards. The "Noce de légumes" and "Fruits animés" series (fig. 30), both eight-card

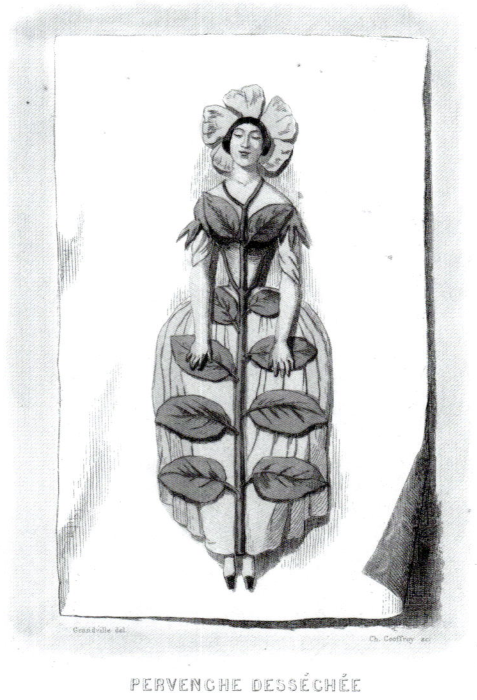

27
J. J. Grandville, "Pervenche desséchée," in *Les fleurs animées* (Paris: Gabriel de Gonet, 1847), vol. 2.

sets, likewise depict the gamut of classes, from a primly erect maid of honor and groomsman—a dandy leek and ladylike spear of asparagus—to plump apples exhibiting more rustic mores, from poor chestnut vendors to elegant pears in urban drawing rooms.[86]

The vegetal fantasies that flourished in the theater sometimes involved class struggle. In Dionysius Boucicault's *Babil and Bijou,* which opened at Covent Garden in 1872, common vegetables and fruits (cabbage, turnip, beetroot, and tomato, to name a few) revolt against "the tyranny of the Court of Flowers" and its insect allies (a noble cast of characters including Princess Rosebud, Queen Camelia, Lord Pink, and Lord Butterfly).[87] Jacques Offenbach and Victorien Sardou's operetta of the same year, *Le roi Carotte,* was a political satire, about radicals, of course, featuring a ballet of black radishes.[88] Lighter fare—romantic operas like *Ciboulette* (Chives) which recounts the courtship of the titular heroine and a man born of cabbages,[89] and pantomimes like *La reine des carottes,* by the germanely pseudonymed Champfleury[90]—was also popular.

Like the comparability of human and floral sexualities, interspecies compatibility had long been acknowledged. In the romances of Ovid's *Metamorphoses,* for example, as Elaine Scarry puts it, "even after the transformation" of one party "into bark-covered, branching laurel or into reedy grasses, the lovemaking goes on in graphically vegetable detail."[91] The lover in Guillaume de Lorris and Jean de Meun's thirteenth-century *Roman de la rose* is unmistakably in love with a plant, and delights in its species-particular and difficult-to-anthropomorphize features, such as the "four pairs of leaves" Nature had placed around the bud "with great skill, one after the other."[92] In the seventeenth century Andrew Marvell, who reasoned that "Apollo hunted Daphne so, / Only that she might laurel grow,"[93] relished not only the beauty of diverse plants but also the diversity of their erotic aptitudes. Ivy, he wrote, "with familiar trails, / Me licks, and clasps, and curls, and hales," and while he begged the brambles to "chain me

FLEURS, FRUITS & LÉGUMES DU JOUR.

FLEUR DESSÉCHÉE. 11

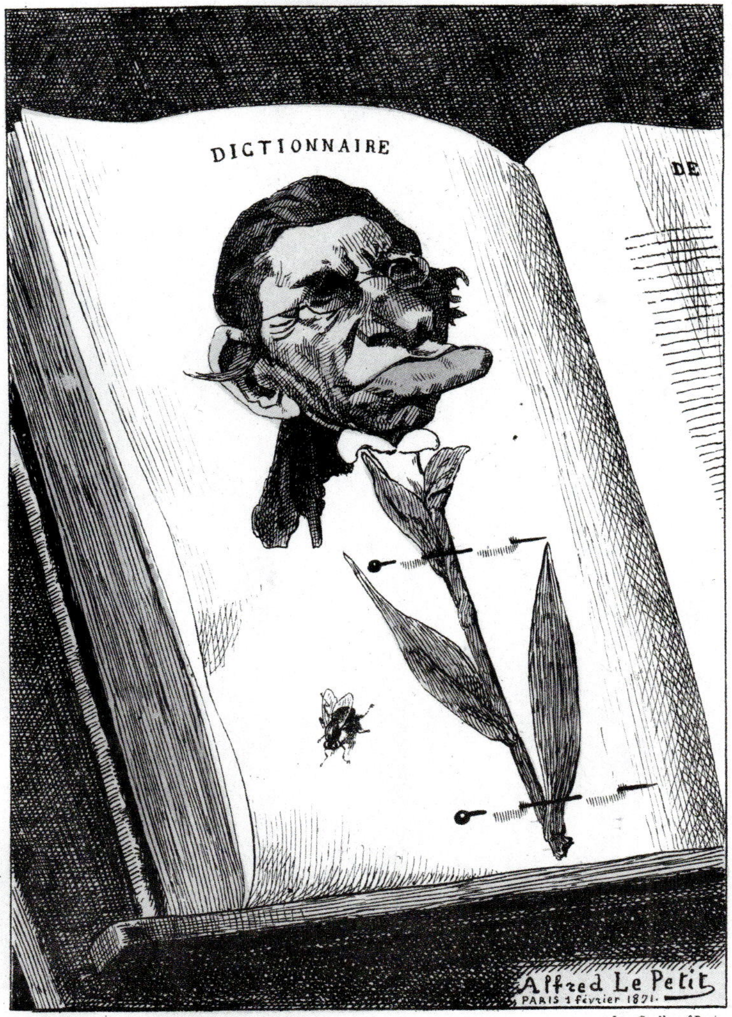

Au Bureau de L'ECLIPSE 16, Rue du Croissant PARIS — Imp. Coulbœuf, Paris.

M. LITTRÉ.

Littré, la fleur de la science,
D'un Dictionnaire accoucha;
A ce travail de patience,
La pauvre fleur se dessécha.

28
Alfred Le Petit, "M. Littré," in *Fleurs, fruits et légumes du jour* (Paris: Au Bureau de l'Éclipse, 1871), 11.

29

Victorian seed and fertilizer trade cards, c. 1885–90, all between 12.8 and 13.5 x 7.5 and 8 cm.

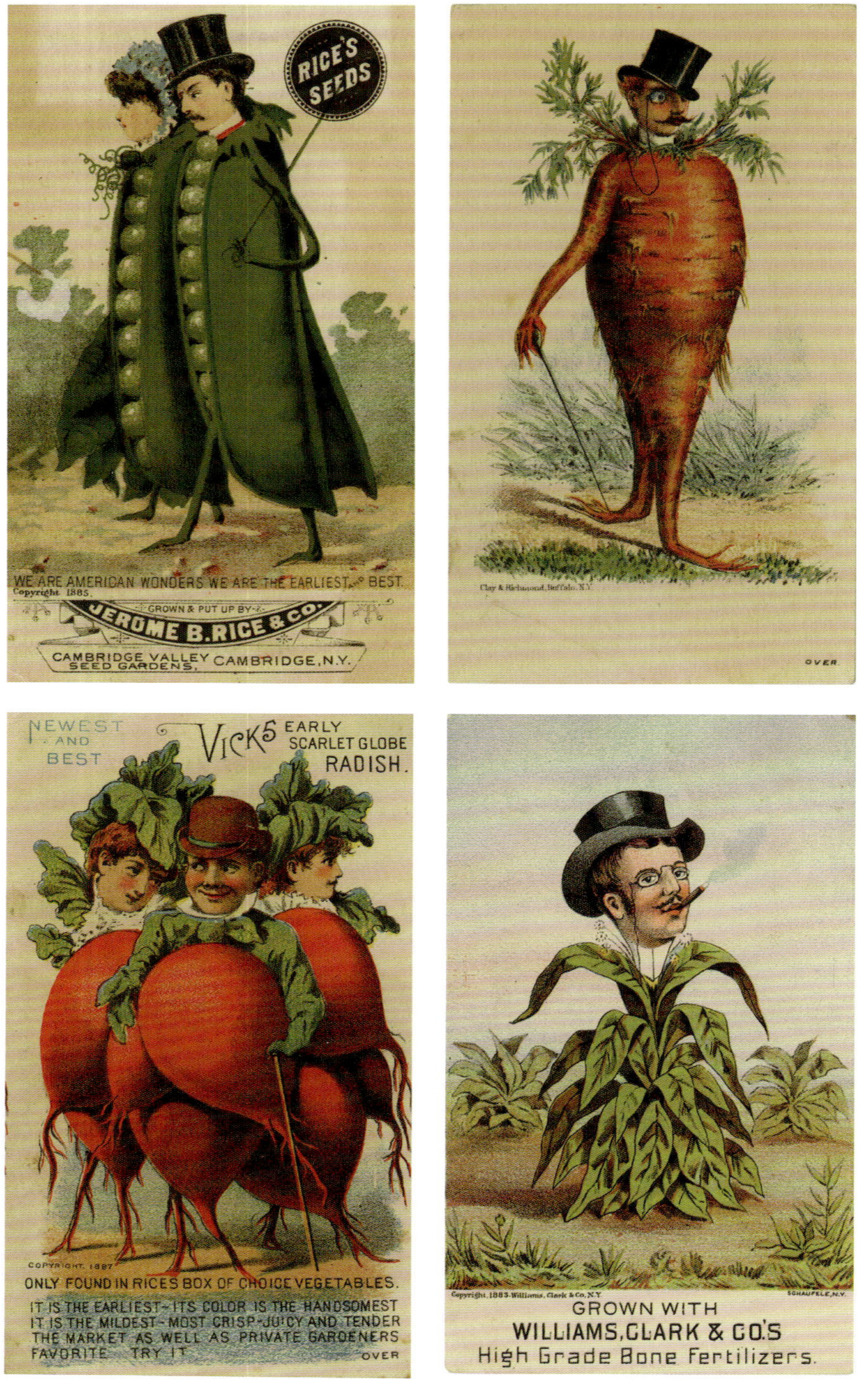

30
Le Bon Marché trade cards from the Noce de légumes and Fruits animés series, c. 1890s, each 18 x 12 cm.

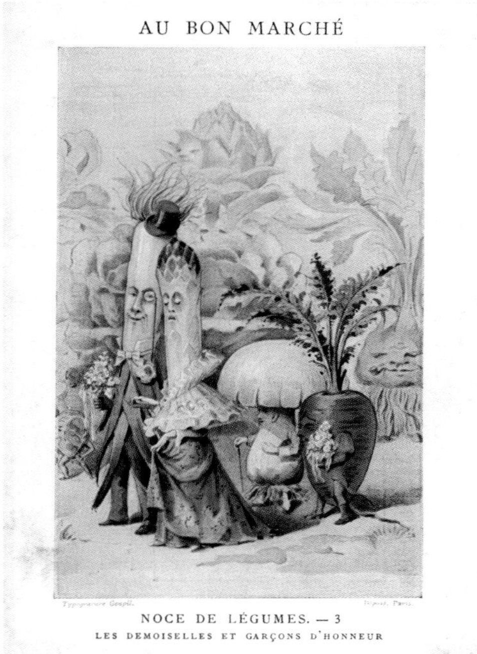

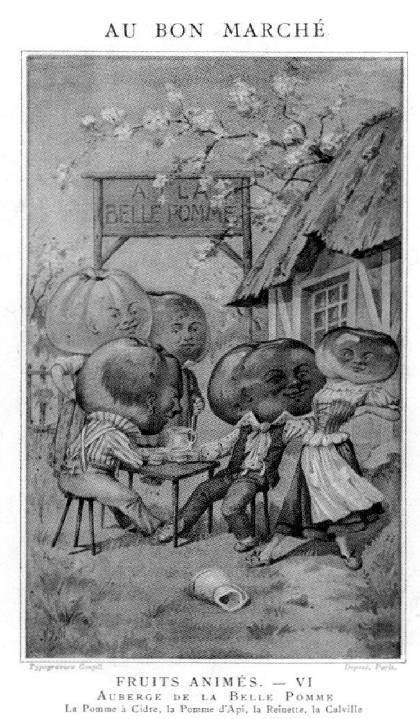

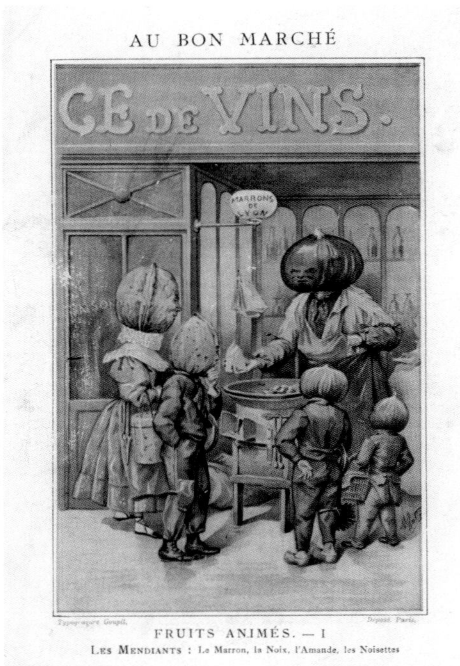

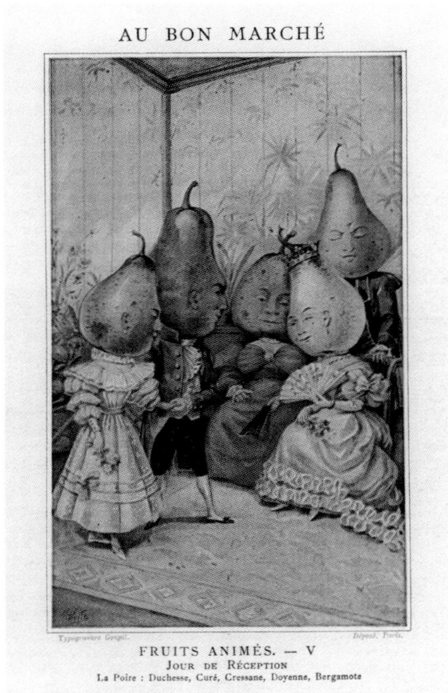

too," he urged the "courteous briars" to "nail" him "through."[94] The poet did not hesitate to declare the superiority of vegetable zeal to that of human lovers: "No white nor red was ever seen / So am'rous as this lovely green."[95] Such interspecies couplings could bear fruit; the inseminator of Fialetti's *Petal Venus* (fig. 8) is pictured as a virile shoot thrusting toward her genitals.

Nineteenth-century literature detailed every facet and phase of interspecies affairs. In Hawthorne's *Marble Faun*, for example, Donatello throws "himself at full length on the turf, and presse[s] down his lips, kissing the violets and daisies, which [kiss] him back again, though shyly, in their maiden fashion."[96] Less chastely, a "full-grown" lily in Whitman's 1881 poem "Faces" cries out to a "limber-hipp'd man near the garden pickets": "*Fill me with albescent honey, bend down to me, / Rub to me with your chafing beard, rub to my breast and shoulders.*"[97] Flowers in their beds, it seems, could be quite dirty.

In many novels the hero must choose between the love of a plant and a human, as in Saintine's *Picciola* (one of Proust's favorite childhood books),[98] or Dumas *père*'s *La tulipe noire*. In these stories the threat that the prized plant might be stolen serves primarily to heighten the tension between human and floral rivals: "I see," Rosa says in the latter, bursting into tears, "I see that you love your tulips so much that there is no more room in your heart left for another affection." The narrator confirms the justness of the charge: "We have to admit, to the disgrace of our hero and of horticulture, that of his two loves the one Cornelius felt the most inclined to regret was the love of Rosa."[99] One of Grandville's plates for *Les fleurs animées* depicts the literal rape of a blossom such protagonists fear (fig. 31). Like Saintine's *Picciola*, Grandville's wallflower is the solace of a prisoner who can see it from his window; it is "his joy, his consolation, his sole hope. He loved it as one loves a woman."[100] With naked feet and torn roots exposed, the yellow wallflower is borne off by a collector, whose not unsizeable equipment is on display, to become another ornament in his cap.

Women too were susceptible to the allure of vegetables, perhaps owing to the rise of their botanical activities since the late eighteenth century.[101] Richard Polwhele, who considered botanical education for girls mere "instructions in priapism," warned against such vegetable fiddling more or less in vain. His 1798 poem *The Unsex'd Females* envisions the scene:

> With bliss botanic as their bosoms heave,
> Still pluck the forbidden fruit, with mother Eve,
> For puberty in sighing florets pant,
> Or point the prostitution of a plant;
> Dissect its organ of unhallow'd lust,
> And fondly gaze the titillating dust.[102]

Ruskin's desire to rename orchids something less obscene betrays a similar fear for the modesty of young amateur botanists.[103] The female attraction to plants, however, was both natural and a product of culture. In his 1877 treatise *Physiological Aesthetics*, Grant Allen argued that, "amongst civilized men," the "amatory passion" of youth "manifests itself at first in the form of ideal cravings." Women expend this emotion "upon ideal heroes" or "upon plants and flowers,"[104] and can revert to such "cravings" in the absence of

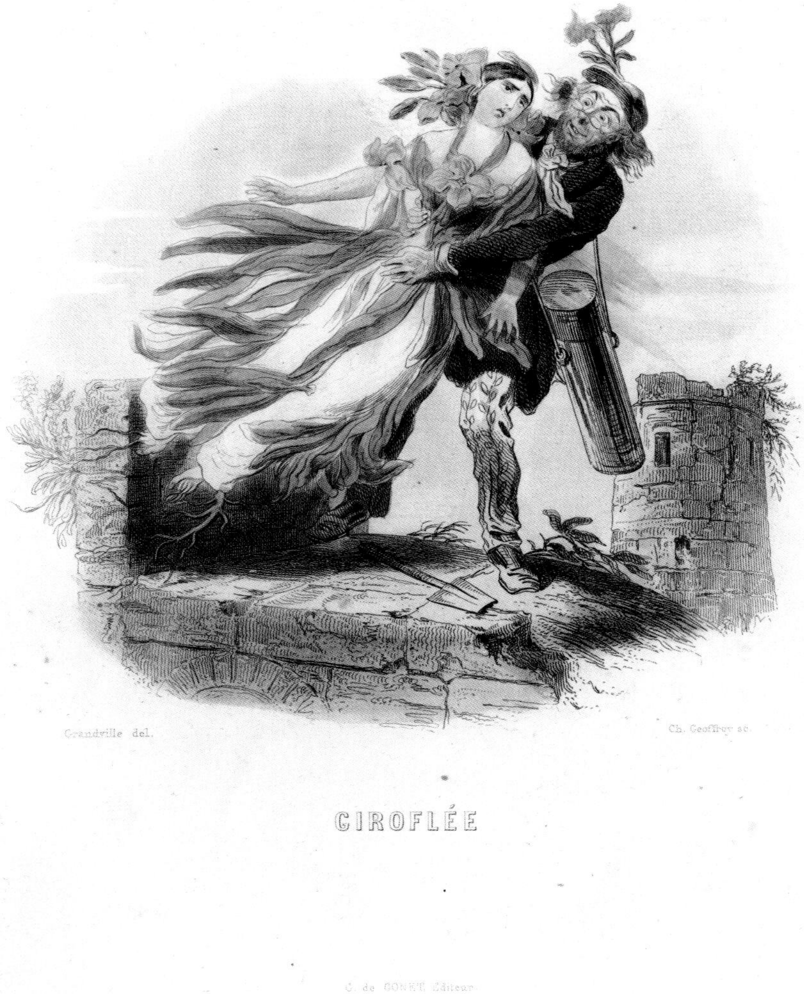

31
J. J. Grandville, "Giroflée," in *Les fleurs animées* (Paris: Gabriel de Gonet, 1847), vol. 2.

human love objects; in Balzac's *Splendeurs et misères des courtisanes,* Esther "flirt[s] with the trees" when deprived of Lucien de Rubempré.[105]

As with their male counterparts, though, the risk that females' vegetable desire should prove more powerful than the human is always lurking. In Rachilde's play *Volupté* a young girl tells her lover:

> I'll go lie down on the ground up against a big pink hyacinth which has grown at the back of the garden, near a hedge. . . . I pull my skirt up over my head and I put my arms around the flower so that all of its perfume rises up to my nose, and I breathe . . . I breathe. . . . It seems like I'm eating honey while the bees fly by and brush my eyelids with their sugary wings! (*She is almost swooning.*) You can't understand any of this! But it is so delicious that it makes me forget you![106]

Pulling her skirt over her head, the character exposes her lower half only to reject the genital pleasure associated with it for that of the synaesthetic (oral, tactile, and olfactory), orgiastic floral embrace. In this act of lovemaking she becomes, with her stems revealed and head contained in the bell of her gown, a flower herself.

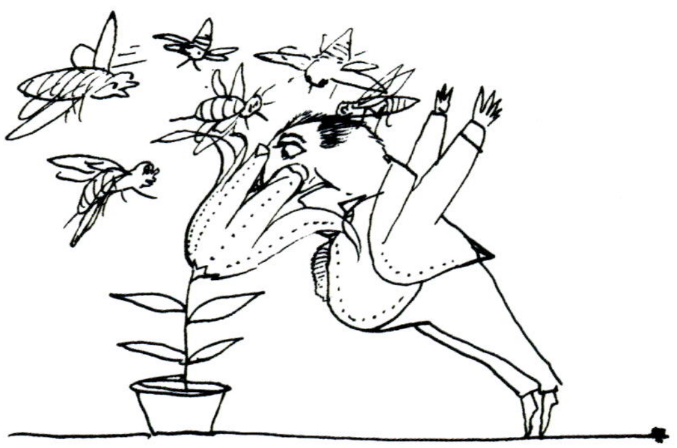

32
Edward Lear, "There was an old person so silly," sketch formerly in the possession of Mrs. C. H. A. Anstruther-Duncan, in Lear, *Bosh and Nonsense* (London: Allen Lane, 1982), 44.

Vegetal arousal came to be associated with Decadents and Aesthetes, and was satirized accordingly. "A Maudle-In Ballad to His Lily" in *Punch,* for example, pictures a languid Aesthetic poet lauding his "lank, limp," "flower-flap chilly" calla "lily-love" and wondering how to woo it: "Shall I buzz like [a] bee, with my face thrust in / Thy choice, chaste chalice"?[107] Edward Lear mocked such plant-inspired passions in the sketch accompanying the early limerick "There was an old person so silly" (fig. 32), which shows a portly old man with his face thrust in the choice chalice of a lily. The buttons dotting the waistcoat over his swollen abdomen and the flower's matching marks declare the two creatures' affinity, just as their mutual inclination is literalized in their postures. The bees swarming above the blossom, however, resent the invasion of their territory by a human. They sting him and fill the flower-lover with "despair."[108] Gilbert and Sullivan's opera of 1881, *Patience,* further ridiculed such proclivities. The poet Bunthorne, a scam artist modeled on Oscar Wilde, cunningly exploits the popular notion that "a sentimental passion of a vegetable fashion," in other words "an attachment *à la* Plato for a bashful young potato, or a not-too-French French bean," is a sign of purity. "If he's content with a vegetable love which would certainly not suit *me,*" the public is imagined as thinking to itself, "why, what a most particularly pure young man this pure young man must be!"[109]

On a more clinical note, Max Nordau concluded that vegetable desire was the perverse product of "morbidly-sensitive natures in whom the most insignificant phenomena create the most astonishing perceptions"—only degenerates "shudder at the contact of a flower."[110] Yet if derided and deemed a reliable index of effeminacy, neurasthenia, and the decay of civilization, the phenomenon was nevertheless also given a historical explanation by anthropological studies. In *The Golden Bough,* which Sargent much admired and quoted,[111] James Frazer devotes a section to "the influence of the sexes" on crops and fruit trees. "In some parts of Amboyna," he tells us, "to ensure the fruitfulness of trees" "when the state of the clove plantation indicates that the crop is likely to be scanty, the men go naked to the plantations by night, and there seek to fertilise the trees precisely as they would impregnate women, while at the same time they call out for 'More cloves!' This is supposed to make the trees bear fruit more abundantly."[112] Both the quest for fruitfulness and the aura of degenerate or atavistic desire surrounding human-vegetable amours are particularly relevant to nonheteronormative manifestations of romantic botany, to which we shall now turn.

André Gide recalled his adolescent arousal at the titular hero's vegetal metamorphosis in George Sand's *Gribouille*. The transformation takes place in a pool; in Gide's description, "as soon as he lets himself go, he floats; then he feels himself becoming tiny, light, bizarre, vegetal; he sprouts leaves all over his body; and soon the water can gently deposit on the bank the delicate branch of oak that our friend Gribouille has become." Gide acknowledges that Sand probably had no intention of writing anything "debauching" in this scene but avows that "no schoolboy was ever troubled by any page of *Aphrodite* as much as I—ignorant little boy that I was—by the vegetal metamorphosis of Gribouille."[113]

While the heterosexual varieties of botanical erotica have been explored by scholars, an equally florid history of the nonnormative sexual escapades of humans and plants from Ovid onward remains relatively unplumbed,[114] although such stories' presence in the cultural imaginary probably in large part determined the fascination Gribouille's metamorphosis held for Gide, who was well versed in both the classics and botany.[115] The *Roman de la rose,* for example, has traditionally been read as an allegory of heterosexual love, but this interpretation elides the indeterminacy of the rosebud's gendering. The bud-plucking scene can be read as hetero- or homosexual sex or rape, or as male masturbation: the lover first sees the rose in the fountain of Narcissus, and the poem's climax, as Joanna Luft notes, features only "one rosebud and one pair of hands."[116] The lover's phallic description of the botanical love object—"the stem was straight as a sapling, and the bud sat on the top, neither bent nor inclined"[117]—suggests that the romance's conclusion is either a scene of autoerotic consummation or one of manually induced pleasure for the plant: "I began, very softly, without pricking myself, to shake the bud. . . . Finally, I scattered a little seed on the bud when I shook it."[118] For Marvell the garden was a place where he could escape the female sex; he envied "that happy garden-state, / While man there walked without a mate."[119]

Eighteenth-century botanists weren't hesitant to admit that flowers were beset by "vices." Darwin *grand-père*'s *Loves of the Plants,* for example, does not limit itself to recounting the affections of tender husbands and blushing brides, but exposes the clandestine marriages, incest, miscegenation, harlotry, infanticide, suicide, necrophilia, drunkenness, and diseases to which various plants are prone.[120] The descriptions of such "unallow'd desires" and "illicit loves"[121] are allusive and vague for the most part, but they nevertheless opened the door to the naturalization of nonreproductive sexuality.[122] Writers quickly became more explicit about what alternative pleasures flora afforded them—of plucking and arranging, for example. Byron wrote to Charles Skinner Matthews in 1809, "We are surrounded by Hyacinths," as sexually available youths in Georgian England were often termed,[123] "& other flowers of the most fragrant nature, & I have some intention of culling a handsome Bouquet to compare with the exotics I hope to meet in Asia."[124] Matthews's reply is even more botanically precising, and puns on the makeup of the poet's "anthology": "As to your Botanical pursuits, I take it that the flowers you will be most desirous of culling will be of the class polyandria [with stamens] and not monogynia but *no*gynia. . . . If you should find anything remarkable in the botanical line, pray send me word of it, who take an extreme interest in your anthology."[125] The Marquis de Sade had regularly given his sodomites floral names—Saint-Florent, La Rose, Hyacinthe, and Narcisse, for example, in *Justine* and *Les 120 journées de Sodome*—and in the nineteenth century this tradition continued. "Hyacinth" retained its innuendo, and a "narcissus" was "an attractive younger man."[126] Dorian Gray is not only an exquisite plant blossoming with "lilies" and "roses" (sheltered by Leaf, his housekeeper), but also an "Adonis," a "Narcissus."[127] Inverts were also called, from at least the 1880s on, "pansies," "buttercups," "daisies," "violets," "blossoms," and even more generally, "horticultural lads."[128] A Vick's Seeds advertisement from the 1890s (fig. 33) conflates the already queer figures of the pansy and the sailor into a suggestive hybrid. The back cover of a seed catalogue, the image depicts two flowery sailors, overseen by their captain, raising anchor. The anchor's chain for the moment imprisons the blooming asters, but when the "sailor lads so bold and free / Put out again," the poem at the bottom right assures us, dissemination will occur.

Vegetable code words and signals proliferated; think of Wilde's green carnations,[129] Whitman's calamus,[130] or Claggart's knowing warning in *Billy Budd* that "a man-trap may be under the ruddy-tipped daisies" (he is referring to Billy, of course, the bud in question).[131] When Jupien offers the services of his male brothel to the narrator in *À la recherche du temps perdu,* he makes the latter's voyeuristic proclivities a matter of vegetable preference: "If ever you are curious, one evening, to see, I will not say forty but a dozen thieves, you have only to come here; to know whether I am in the house you have only to look up at that window; if I leave my little window open with a light visible it means that I am in the house and you may come in; it is my private Sesame. I say only Sesame. As for Lilies, if they are what you seek I advise you to

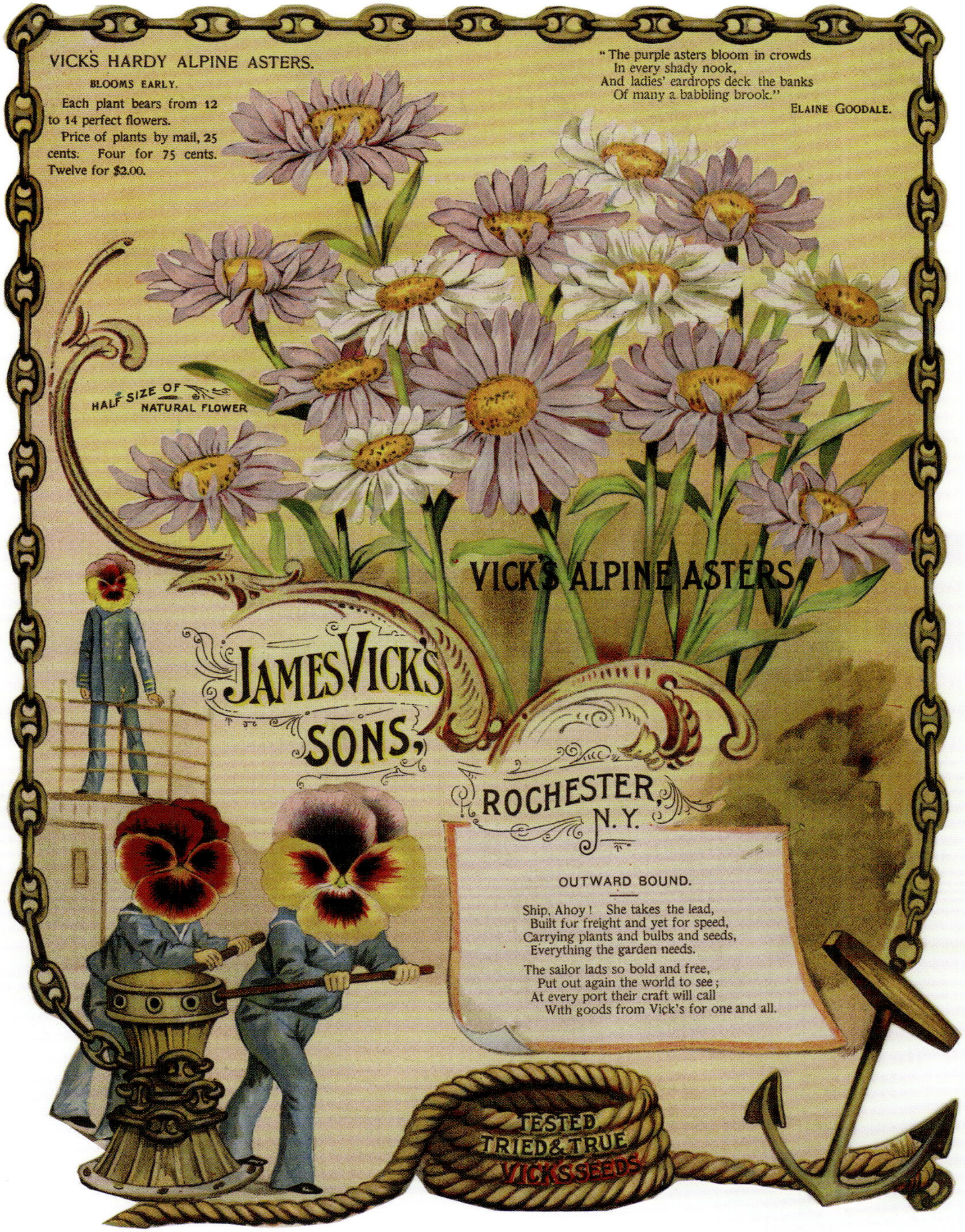

33
Vick's Seeds catalogue cover depicting pansy sailors, 1896, 26 x 19.6 cm.

go elsewhere."[132] This is a deliciously perverse co-option of Ruskin through the opening provided by the sesame of the Arabian nights (Ruskin's *Sesame and Lilies,* first published in 1865, describes the behaviors proper to each sex), but the valence of Jupien's terms is predicated as much on contemporary slang as on the logic of the novel's intertextual tapestry.

If flowers lent themselves to the discourse of male inverts, the "language of flowers" itself, which became something of a craze in the nineteenth century, had its roots in Sapphic love. The Western language of flowers developed out of *sélam,* the flower language of the harem.[133] According to the Orientalist Joseph Hammer-Purgstall, *sélam* was invented by women of the harem "in the leisure of their lonely life"; they employed it "as a code for lesbian attachments."[134] Lesbians themselves became, for the latter half of the nineteenth century, *fleurs du mal.*[135] Even Grandville courted such associations; his "Tubéreuse et Jonquille" (fig. 14), both female in the text, look like entwined lovers.

As with its heterosexual counterpart, vegetable erotica of the more exotic sort did not neglect the botany of the body. Everything was subject to vegetable troping, from Bosie's "rose-leaf lips" to "the juice of the water lily" in Monsieur Auguste's veins;[136] Marc-André Raffalovich was once "chaffed" for having "extolled someone's pollen-coloured hair."[137] The youthful Verlaine, in a Baudelaire-inspired poem called "Spring," articulated the yearnings of an experienced lesbian for one as yet chaste in a language that is by now familiar:

> Let my fingers roam in the moss
> Where the pink bud shines.
>
> Let me, among the fair grass,
> Drink the dewdrops
> From which the tender flower is watered—[138]

The anal sphincter, male or female, is frequently a strawberry or a bud in Sade's writings: "Peer into that appetizing behind, and lo! a rosebud would offer itself to your gaze, and it was in all its bloom and in the most tender pink Nature wished you to behold it."[139] On a trip to Italy in 1850 Théophile Gautier encountered "fairy-like" frescoes in a male brothel representing "bouquets of roses opening like blonds' assholes, with a touch of purple in the middle." He found it "very agreeable to crouch down, having an eye on these flowery anuses, or these anal flowers, unfolding their petals: the puckers of a sphincter ready to swallow a prick or to spew out a turd."[140]

If roses and rosebuds were the sphincter metaphor of popular choice[141]—the act of tonguing the anus, Wilde tells us, is called "une feuille de rose"[142]—other varieties also flourished. Rimbaud and Verlaine, for example, saw the orifice in question as

> Dark and puckered like a purple carnation
> It breathes, humbly nestled among the moss
> Still wet with love

in their "Sonnet from the Asshole."[143] And in *The Picture of Dorian Gray* the bee sensually explores a "Tyrian" (purple or crimson) convolvulus in the scene of Dorian's awakening. Buttocks inspired

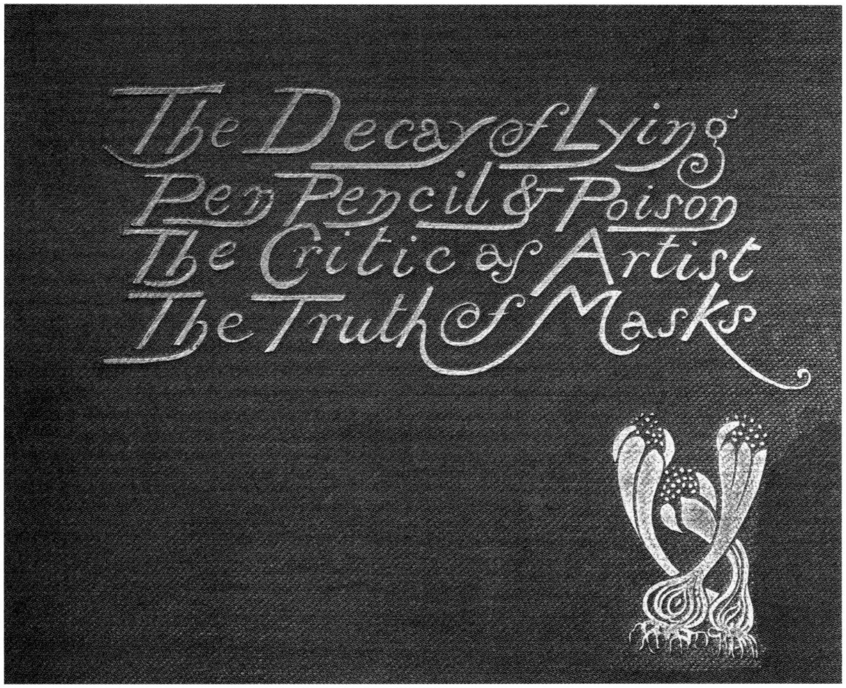

34
Charles Ricketts, detail of the cover of Oscar Wilde, *Intentions* (London: James R. Osgood, McIlvaine, and Co., 1891).

less delicate vegetal visions. In Huysmans's novel *En rade,* Jacques dreams of falling down a ladder into a gallery full of pumpkins: "They were palpitating visibly, reaching skywards feverishly, tugging at the stalks that attached them to the ground. Jacques's first impression was that he was looking at a field of Mongolian bottoms, a vegetable garden of behinds all belonging to the yellow race. He examined the deep grooves, well arched, that were sinking into the plump epidermis of these bright orange spheres. Then, a filthy curiosity seized hold of him."[144] The underside or backside of these lewdly animate, straining orbs arouses an unnamable desire.[145]

The penis retained its troping as stamen, stem, branch, or plant, the latter often, as in the *Roman de la rose,* broken down into component parts. A description penned by an early twentieth-century tourist in Venice, renowned for the glory of its gondoliers, attests to the omnipresence of floral tropes in inverts' discourse: a young stevedore strips off for him in an empty restaurant to display "closely folded thighs with (in the midst of the black blossom of exuberant robustitude) a yard like a rose-tipped lance."[146] Teleny's glans is like an "apricot" and a pink bud; des Grieux revels in its "tiny rosy lips that, bulged out with love, opened and spattered a tiny drop of sparkling dew."[147] Charles Ricketts's 1891 design for the cover of *Intentions* (fig. 34), a collection of Wilde's essays, places three stamped gold crocuses on a green cloth ground.[148] Like the paradoxical titles of the essays in the volume ("The Decay of Lying," "Pen, Pencil, and Poison," "The Critic as Artist," and "The Truth of Masks"), the crocuses demand perverse reading: the phallic plants curve up from hairy, testicular bulbs; their petals part slightly at the top to release the pollen spuming from the flowers' anthers.

Alongside such idyllic visions of natural abundance, however, darker images of crude, parasitic, and poisonous plants are to be found. John Addington Symonds, an early (and conflicted) defender of inversion, described the fateful allure of this sort of plant life in his poem "The Valley of Vain Desires," set in a sphincterous gorge rank with fungi of "fleshly shapes obscene."[149] And while the Marquis of Queensberry couldn't spell sodomite, he nevertheless understood the potentially derogatory queer connotations of vegetables: on the opening night of *The Importance*

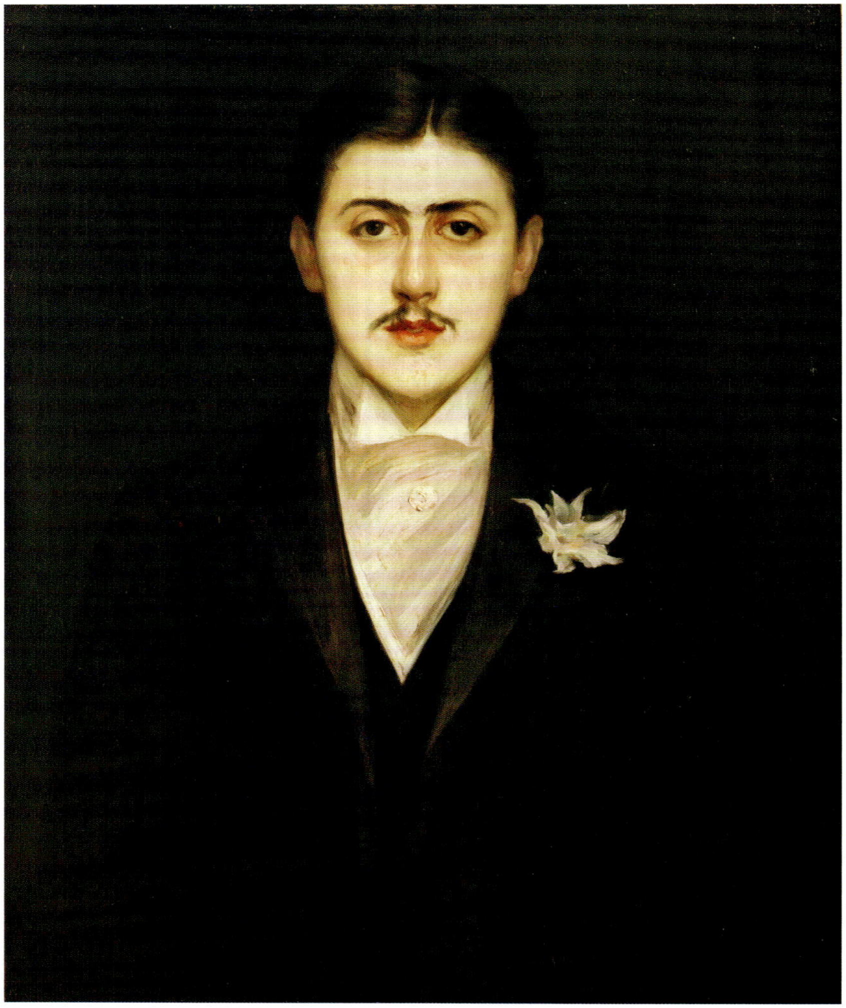

35
Jacques-Émile Blanche, *Marcel Proust*, 1892. Oil on canvas, 73.5 x 60.5 cm. Musée d'Orsay, Paris. © 2008 Artists Rights Society (ARS), New York/ADAGP, Paris.

of Being Earnest, Valentine's Day 1895, he left Wilde a bouquet of carrots and turnips at the theater.[150]

Gardening tools had their place in this poetics. As we saw earlier, ejaculation was sometimes troped as irrigation, the penis as a watering can. It could also be a stake or a rake. In the prologue to his *Roman Elegies,* Goethe equated sodomy and gardening, advising the reader to punish "the backside" of the man who turns "up his nose" at "the golden fruit of life" (i.e., procreative sex) "with one thrust of that red stake that sprouts from your loins."[151] Richard Burton and L. C. Smithers compiled a two-page list of "agricultural and horticultural terms used tropically in a venereal sense" in their 1890 collection and translation of erotic Latin poems, *Priapeia;* it includes "sickle," "stake," "bolt," "hoe," and "ploughshare" as synonyms for penis.[152] And for Proust the male invert attaches himself to a man "as the convolvulus throws out its tendrils wherever it finds a pick or a rake up which to climb."[153]

We have thus far examined floral bodies and their parts; floral adornments, however, were also important to invert erotics. Bouquets, nosegays, posies, and buttonholes could be suggestive.[154] The buttonhole in particular, the only socially acceptable form of male floral adornment, became a charged site in Wilde's and the *fin de siècle*'s homoerotic imaginary. Associated with dandies, Aesthetes, and Decadents, the exotic boutonnière was a sign of questionable masculinity. The insertion of flowers into eyelets, furthermore, functioned as a symbol of and precursor to sexual activity. In an early draft of *À la recherche du temps perdu,* Jupien is a florist; to contrive an encounter with him, Charlus sheds the rose in his lapel and asks Jupien for another; Jupien inserts it before they enter into more fleshy embraces.[155] Teleny habitually wears "at his button-hole a bunch of white heliotrope, although camellias and gardenias" are "in fashion." When he first meets Camille des Grieux, he seductively slips his posy into his future lover's hole.[156] The transfer of the white spray involves a bodily embrace: "And taking the flowers from his button-hole, he put them into mine with one hand, whilst he slipped his left arm round my waist and clasped me tightly, pressing me against his whole body for a few seconds."[157] Buttonholes in portraits often suggest something of their wearers' characters. In Jacques-Émile Blanche's 1892 portrait of Proust (fig. 35), for example, the young writer sports in the lapel of his black tuxedo a white orchid, a species whose decadent delicacy and lewd associations were widely understood. And a portrait of Marc-André Raffalovich by Sydney Starr from around 1889 depicts the writer and theorist of inversion fingering his own flower.[158]

The variety of erotic experiences afforded by the human-plant analogy becomes even more complicated when we take into account the pollinator erotics that came to compete with purely floral romances over the course of the nineteenth century. Flowers, already desiring subjects, looked beyond their corollas for love, and new players entered the scene. Blossoms were no longer seen as "simple passive victim[s] in the busy bee's sweet pillage," but rather as "conscious" beings, aspiring to, and often manipulating (as Darwin's word "contrivances" suggests) other creatures into "companionships."[159] Knowledge of pollination was disseminated through studies like Darwin's *The Various Contrivances by which Orchids are Fertilised by Insects* and books marketed to a more popular audience, such as William Hamilton Gibson's *Blossom Hosts and Insect Guests: How the Heath Family, The Bluets, The Figworts, The Orchids, and Similar Wild Flowers Welcome The Bee, The Fly, The Wasp, The Moth, and Other Faithful Insects.*[160] Companions to the flowers, with their rather familiar habits of brushing, dipping, sucking, licking, and biting, came to play starring roles in all sorts of interspecies romances. Even the most pedagogically earnest natural history texts on cross-fertilization could be highly suggestive; Philip Gosse, for example, described how bees "enter tubular flowers, and lick their bottoms."[161]

Like flowers, pollinators were popularly troped as human and vice versa. In the visual arts this identification can be traced back to at least the eighteenth century, notably to Charles-Germain de Saint-Aubin's *Essai de papillonneries humaines.* The plates of the two series depict social butterflies engaged in various artful pursuits in which the orchestration of elegant visual display—through acrobatics, dance, fireworks, theatricals,[162] or other means—is shown to be the natural inclination of these fanciful and ephemeral creatures. In *La toilette* (fig. 36) a fairy boudoir is the site of lacy-winged lepidopteran primping and preening.

Nineteenth-century pollinators were fundamentally social creatures. Proust described the reproduction of flowers as carried out by bees leaving calling cards behind on their rounds of social calls;[163] insects' other civilized mores included shopping and strolling (fig. 37). More explicitly erotic behavior was widely represented as well. Grandville's drawing of a lecherous lepidopter (fig. 38) in *Scènes de la vie privée et publique des animaux* shows the butterfly pausing to inspect, with the help of his lorgnette, a tempting bouquet. In the story a rose fends off the forward butterfly with a thorny arm (fig. 39): "'Don't come anywhere near me!' replied the rose with contempt. 'You are presumptuous and you know how to flatter; you are a deceiver, don't come closer!'"[164] The baby-bud

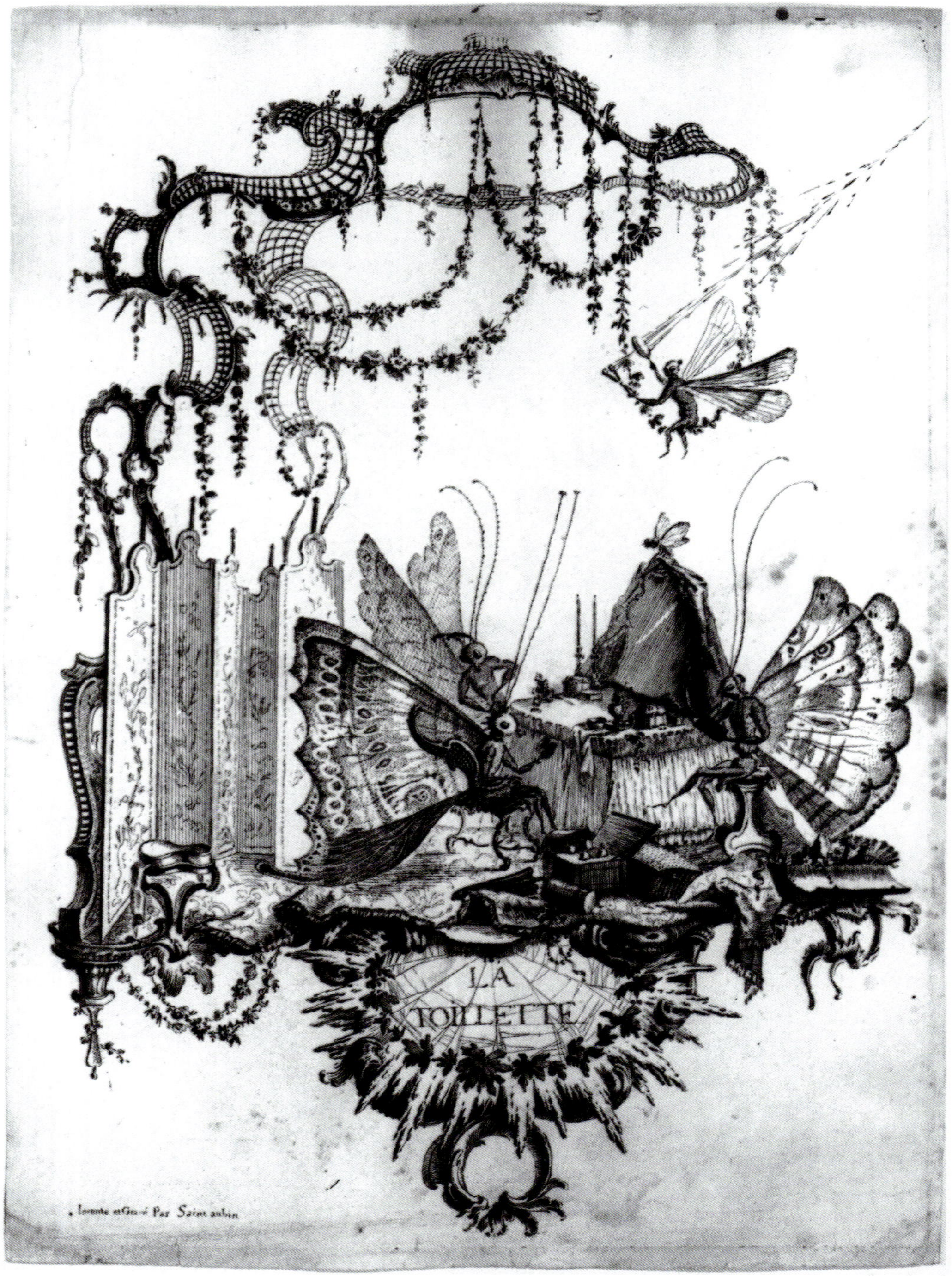

36

Charles-Germain de Saint-Aubin, *La toilette*, c. 1750s. Etching, 34.1 x 24.7 cm. Edmond Rothschild Collection.

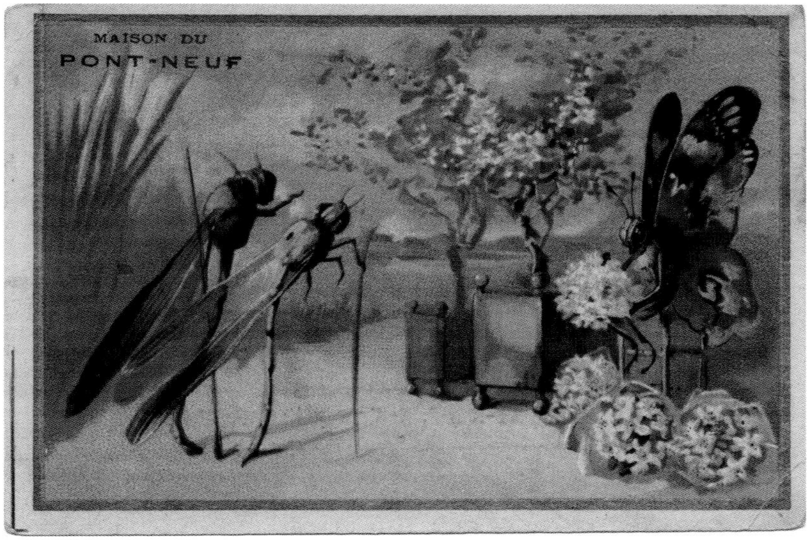

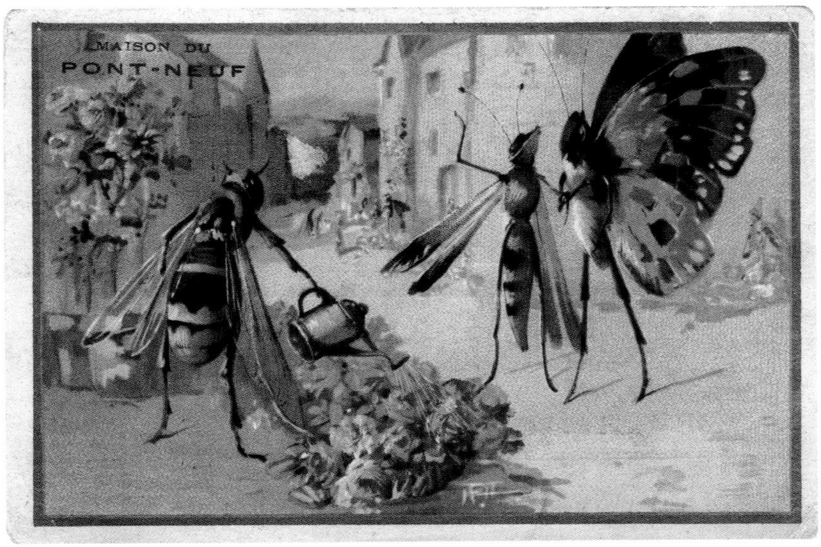

37
Maison du Pont-Neuf trade cards, late 1880s or early 1890s, each 7.7 x 11.3 cm.

behind her suggests single motherhood, and her speech reflects knowledge gained through experience. Later, and in a similar vein, *Punch* provided its readers with edifying entomological studies of various inconstant insect types such as the "Music Hall Moth or Nocturnal Cad-Fly."[165] As with *Les fleurs animées,* Grandville almost never represented moments of consummation, choosing to focus on the moments leading up to or following cross-fertilization.

Other artists were less discreet in their representation of interspecies amours. In Whitman's "Bunch Poem," the "hairy wild-bee" "gripes the full-grown lady-flower, curves upon her with amorous firm legs, takes his will of her, and holds himself tremulous and tight till he is satisfied."[166] The interspecies dynamic and choreography of Whitman's image is echoed in H. J. Ford's 1892 illustration for the English translation of the comte de Caylus's fairy tale "Rosanie" (fig. 40).[167] The bees' stingers, however, though mentioned in the text as objects of fear for the abducted princesses, are conspicuously absent, presumably because they would have appeared too lewd pressing between the nubile girls' soaring thighs. Looking solicitously over his lady's shoulder to ensure that she is not unduly distressed, the largest bee is further disarmed, or rendered disarming, by his chivalry. Other insects, in their literally shining armor, are the soul of gallantry as they vie for their ladies' favors. One such knight (fig. 41) woos a begonia; wearing one of her blossoms over his heart, he bows before departing on some romantic quest. Even Proust's bird-identified Swann and flowery Odette's act of "doing" or "making a cattleya," narrated step by step, is marked by polite courtesies. As their lovemaking proceeds, Swann is anxious to ensure the comfort of his mistress: "there's a little—I think it must be pollen, spilt over you. Do you mind if I wipe it off with my hand? I'm not doing it too hard? I'm not too rough?"[168]

38

J. J. Grandville, "Un vieux papillon," in P.-J. Stahl, *Scènes de la vie privée et publique des animaux,* 2 vols. (Paris: J. Hetzel et Paulin, 1842), vol. 1, between 136 and 137.

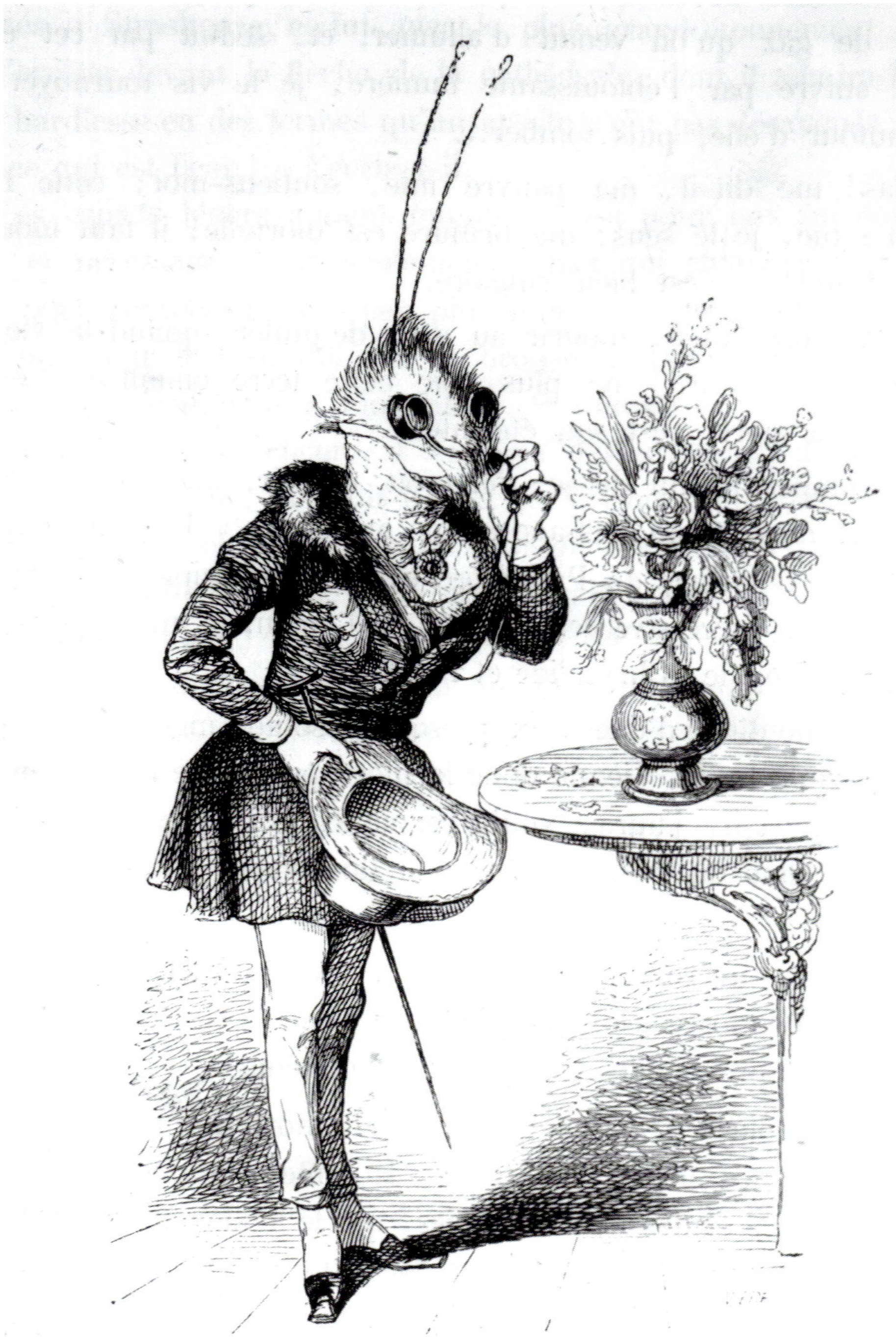

Like interfloral romances, then, the relations of male pollinators and female plants could be innocent or considerably more knowing. James Fenimore Cooper's *The Oak Openings, or The Bee-Hunter,* which follows the relationship of Ben Boden and Marjery Waring, known as "Buzz" and "Blossom," respectively, offers a good example of the virtuous type of vegetable courtship,[169] as does another American novel, Louisa May Alcott's *Rose in Bloom*. "Naturally enough," in this 1876 book, the young men cluster "about the girls like butterflies around two attractive flowers."[170] Even in London, as the *Illustrated Sporting and Dramatic News* comically reports (fig. 42), fashionable butterflies flock to the spring flowers—those just "coming out" onto the marriage market this season—and try to display themselves to best advantage.

The preceding examples all naturalize matrimony, but others represent it as a cage. In *Moths* the heroine Vere is associated with a briar rose, the singer Corrèze with a "nightingale." Pulled apart and worn down by the characters surrounding them (the "moths" of the title, destructive versions of social butterflies), the two are united only when Vere leaves her husband to live, unmarried but in a union sanctioned by nature, with Corrèze.[171] Husbands often inhibit

39

J. J. Grandville, illustration for "Les aventures d'un papillon," in Stahl, *Scènes de la vie privée et publique des animaux,* 1:121.

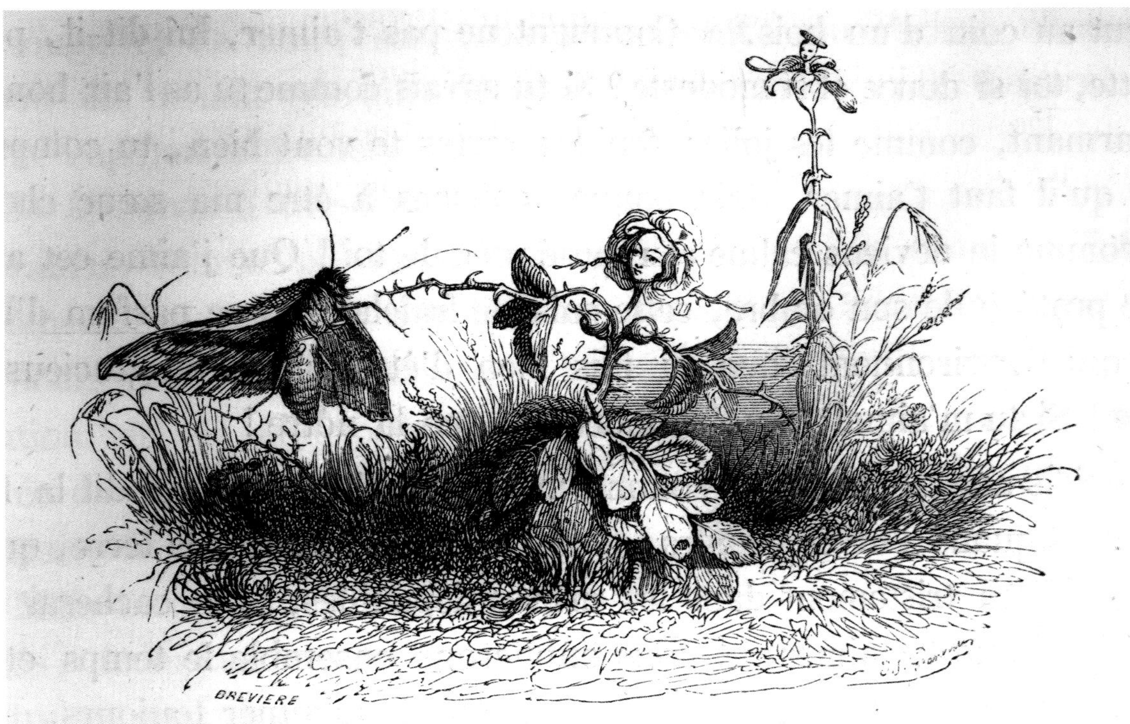

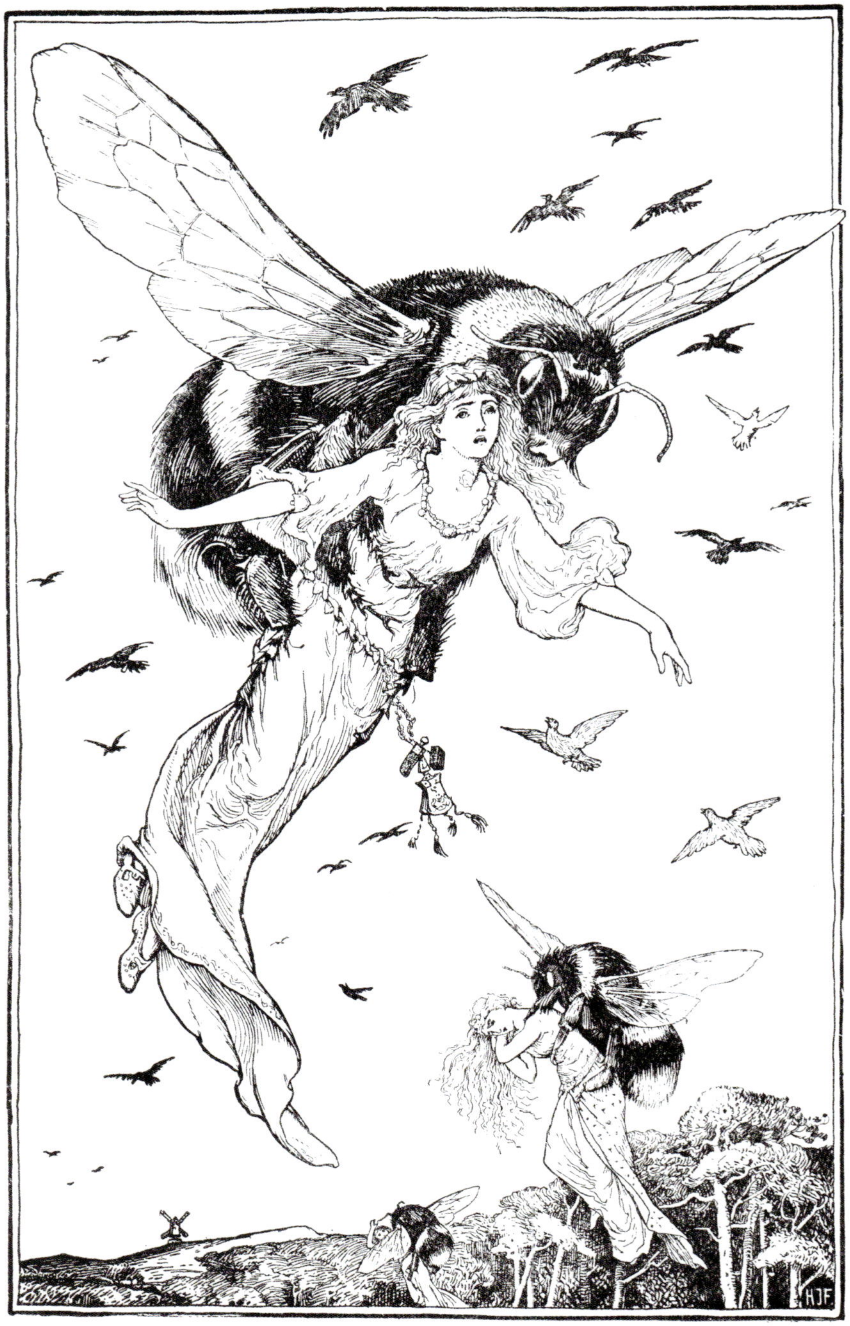

THE PRINCESS CARRIED OFF BY THE BEES

40
H. J. Ford, "The Princess Carried Off by the Bees," in Andrew Lang, *The Green Fairy Book* (London: Longmans, Green, 1892), 53.

41
Chicorée trade card, 1890s, 10.4 x 6.5 cm.

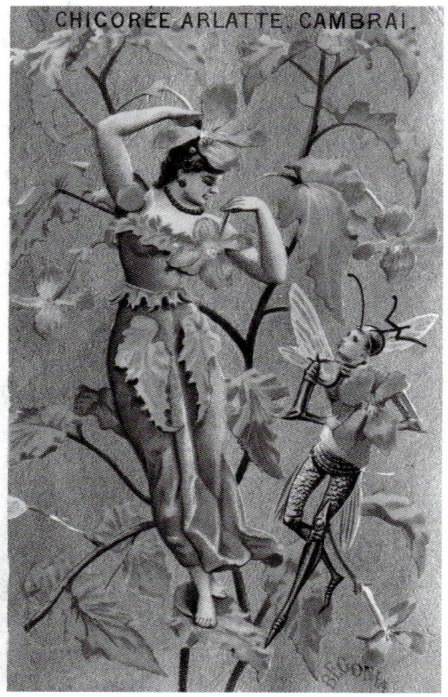

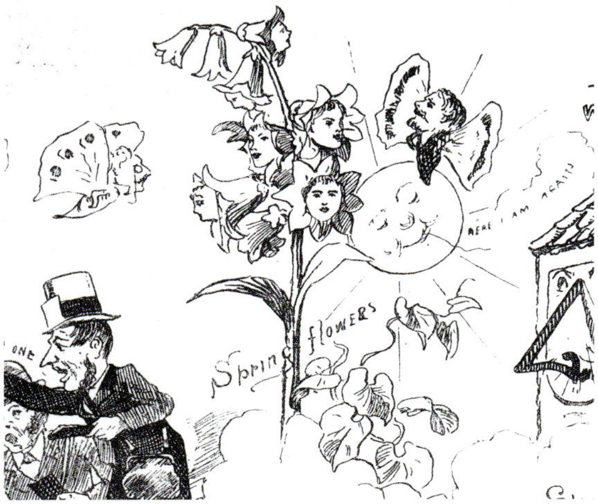

42
"Spring Flowers," ***Illustrated Sporting and Dramatic News,*** **10 April 1879, 108.**

natural desires: Alfred Le Petit's caricature "La rose impériale" (fig. 43) depicts a rosy Empress Eugénie in the aftermath of the Franco-Prussian war (with her husband Napoleon III imprisoned and thus out of the picture), happy to find herself surrounded by a swarm of waspy and would-bee suitors. The extended rosy tongue of Minister Émile Ollivier (a supporter of the empire but also a proponent of civil liberties) is at the ready to please the *femme-fleur* and taste the sweets of her favor. Sade offers an even more cynical vision of marriage; he slyly extols the virtues of anal sex for women in *Justine* with a pollination analogy: unlike vaginal sex, "as soon as the bee has left off sucking the pollen, the rose's calyx closes shut again; one would never imagine it had been opened."[172] Naïve or cunning, however, these activities are all equally natural.

Pollinators, their parts, and their human analogues were often sexed as male—"bird," "antenna," "stinger," and "beak," were slang for the penis,[173] and libertines were understood to be butterflies[174]—but this was not universally the case. In flower and insect spectacles at the theater and ballet, male *and* female performers were subject to pollinator metamorphosis, and dancers dressed as diverse winged creatures—butterflies, dragonflies, birds and fairies, wasps and moths—flitted and fluttered across the stage. The metamorphic potential of standard features of masculine attire was exploited for costumes: Théophile Thomas's designs for *Le roi Carotte* (fig. 44) show how canes became stingers or proboscides, feathered hats antennae. Facial hair could also be shaped to suggest the furriness of pollen brushes or other feelers. Costumes for women changed dramatically over the course of the century, however, and had little relation to everyday attire. In the early nineteenth century the airy lightness of female insects and butterfly-winged fairies was stressed: Alfred Albert imagined Emma Livry's costume with the figure resting on a rose (fig. 45). On point shoes performers could seem to float almost weightlessly, and leaps became flights with the introduction of wirework. In *La sylphide,* first performed in 1832, the winged Marie Taglioni moved with ethereal grace on point, with wires, and in a new style of costume—short and gauzy—that emphasized her aerial existence.[175] The effect was so fantastic that to describe it, according to one contemporary critic, "one would have to dip a humming bird quill into the colours of the rainbow and inscribe it on the gauze wings of a butterfly."[176] Later ballets stressed the sexual appeal of female dancers: over the course of the century the stage costumes for female flowers, insects, and other ephemeral beings became more revealing. In the ballet *La source* (fig. 46), for example, filmier

43

Alfred Le Petit, "La rose impériale," cover of *La Charge* 4 (1875).

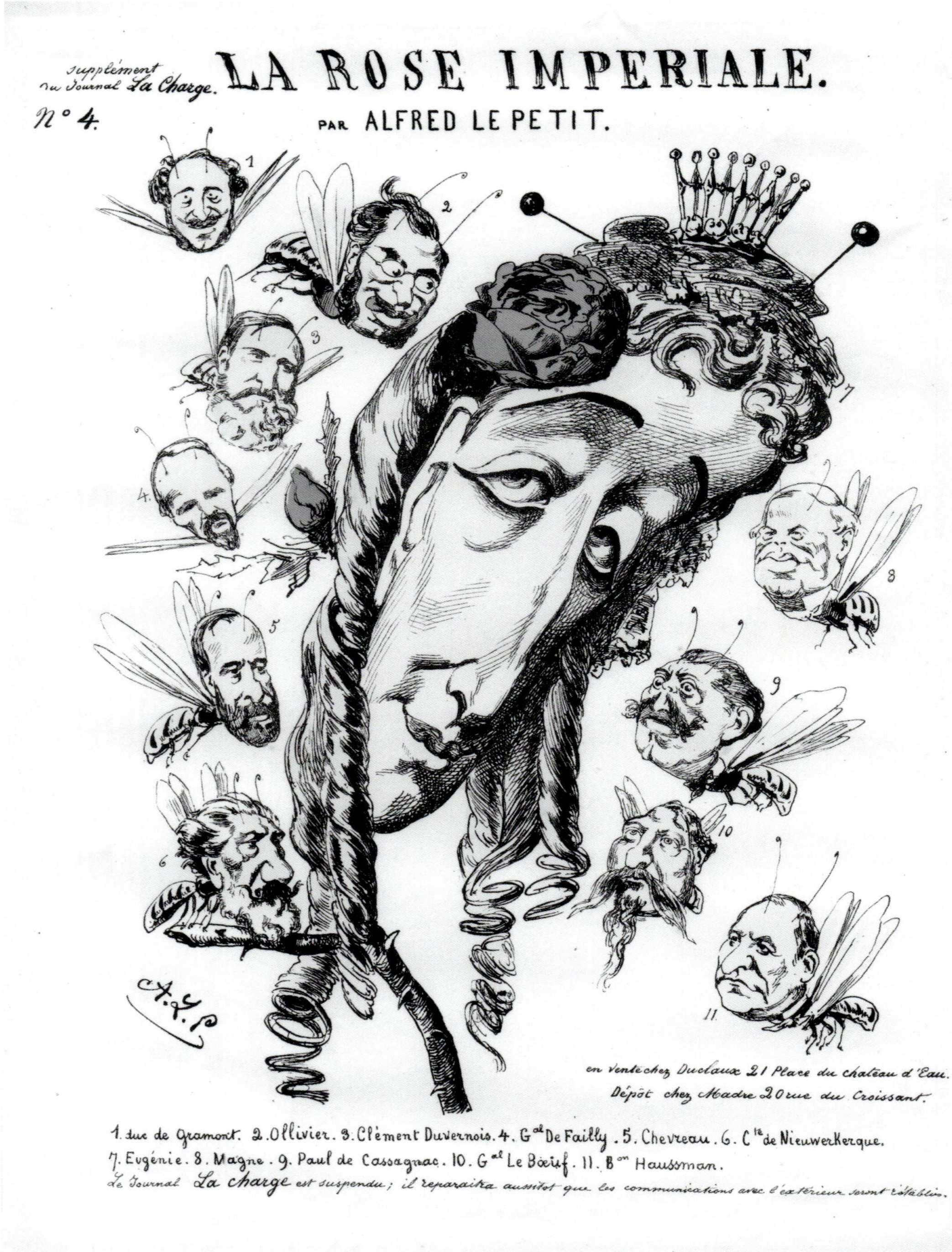

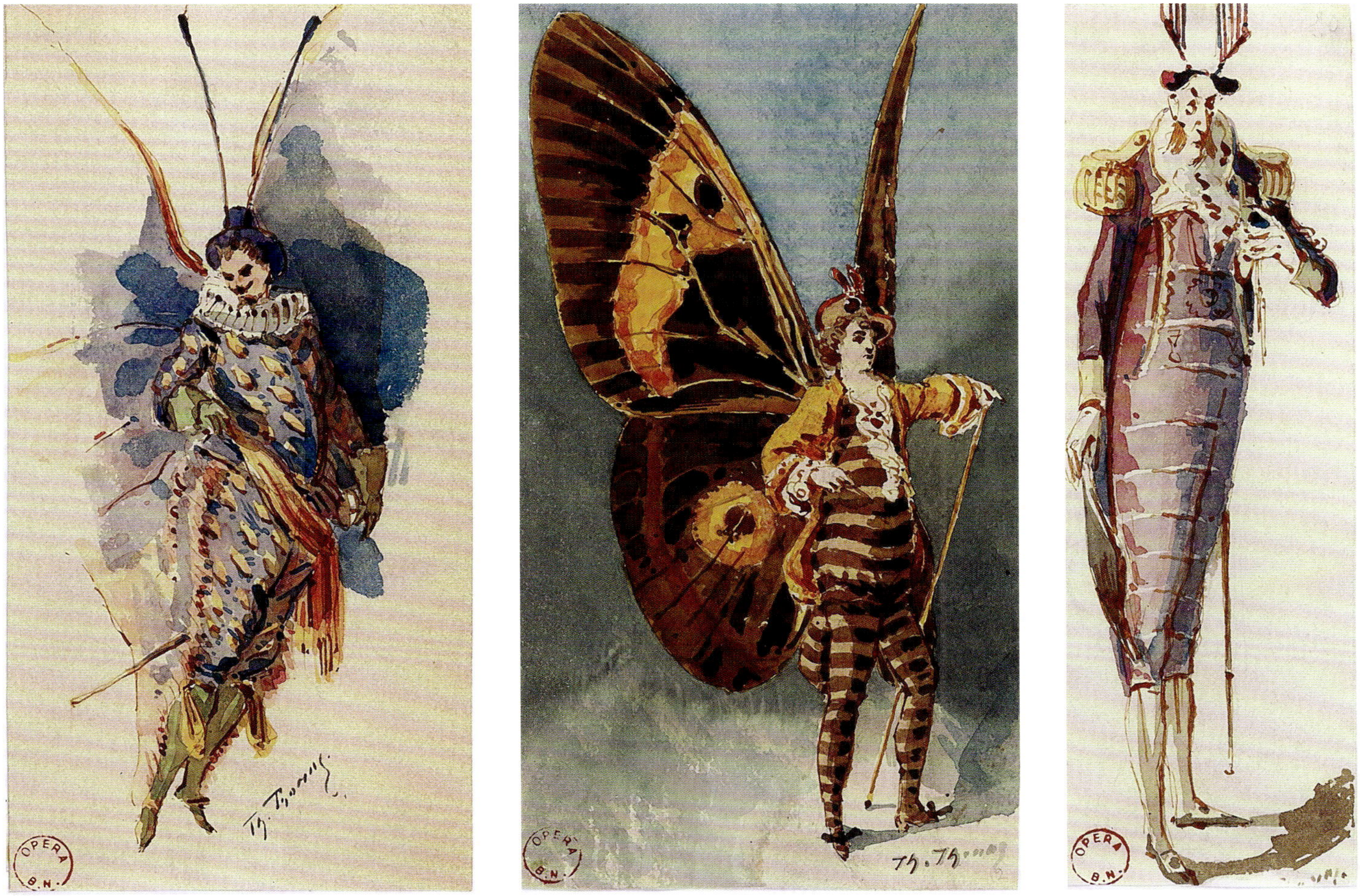

44
Théophile Thomas, maquettes of costumes for *Le roi Carotte*, 1872. Watercolor on paper. Bibliothèque musée de l'Opéra, D-216 (Z1), plate 14.

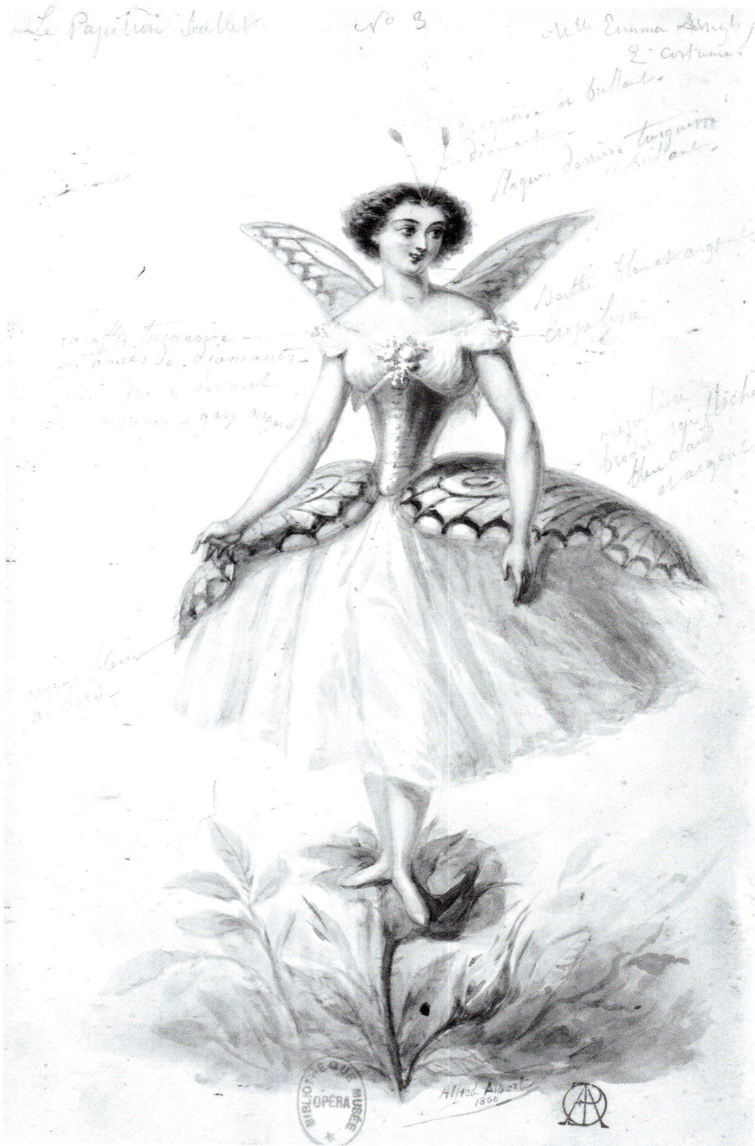

45
Alfred Albert, Emma Livry's second costume for *Le papillon,* 1860. Bibliothèque musée de l'Opéra, D-216 (21), plate 61.

fabrics and low-cut bodices were used to best display the hybrid creatures' sexiest features.[177] Interspecies romance was a popular theme throughout the century; men fell in love with these female pollinators, and like the *femme-fleur,* the seductive lady butterfly could inspire fetishism, even sadism. *Le papillon,* a pantomimic ballet first performed in Paris in 1860, included a scene in which the heroine, metamorphosed into a butterfly, is literally and painfully pinned to a tree.[178]

Female pollinators could be both demure and sexually unsettling. Trade cards frequently depicted female birds, bees, dragonflies (in French, *demoiselles* are girls and dragonflies), beetles, butterflies, and moths (fig. 47) engaged in innocent activities like gathering blossoms, sipping nectar, or even painting. These hybrid creatures occasionally exhibit disturbing features, however, as if the metamorphosed body could not be securely sexed. The lower half of the butterfly adorning herself with flowers in the trade card at the bottom left, for example, displays a large furred, phallic abdomen. Sometimes the sexual ambiguity occasioned by the suggestiveness of insect morphology is literally edited out. The lower half of the body in the advertisement for a stage production of Ouida's *Moths* (at Wallack's in New York) has been eliminated (fig. 48), though Vere herself, the heroine of *Moths,* has a masculine name.

Accounts of female pollinators at work presented them as flighty as well as diligent. The aptly named May Server in James's *The Sacred Fount* "darts from flower to flower, but she clings, for the time, to each."[179] The embraces of Arthur Symons's Bianca are delicately apian: "Tenderly as

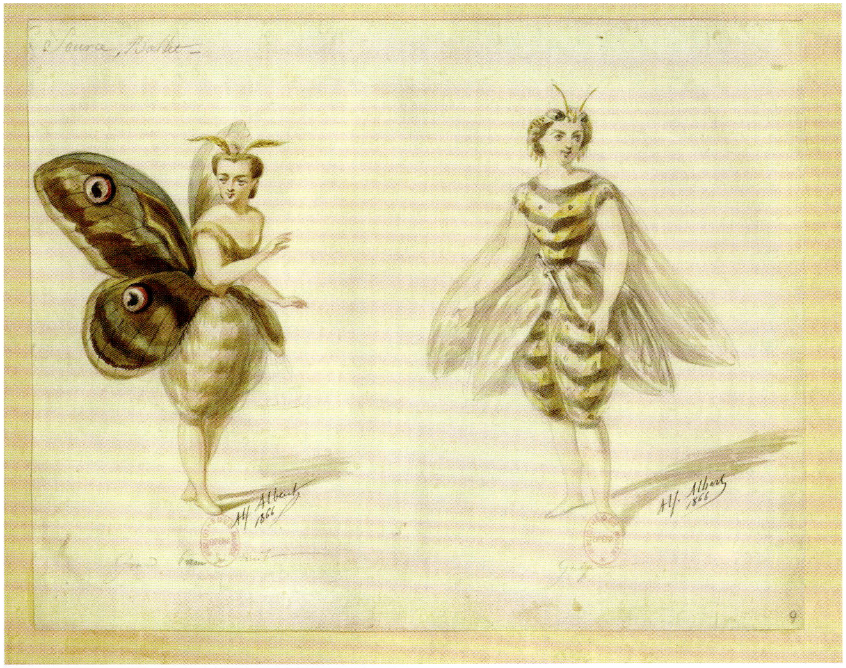

46

Alfred Albert, "Wasp and Large Moth," maquettes of costumes for the ballet *La source*, 1866. Colored pencil on paper. Bibliothèque musée de l'Opéra, D-216 (23bis), plate 21.

a bee that sips, / Your kisses settle on my lips."[180] In *Fanny Hill* Louisa plays "the industrious bee" and gathers "the sweet" of the poor male flower-seller's rare "flower,"[181] and "lady-bird" was common usage for a mistress or a "wanton girl."[182] Actively pursuing their desires, such creatures occasionally sniffed up the wrong tree, as it were. The heroine of Armory's play *Le monsieur aux chrysanthèmes,* Marthe Bourdon (a bumblebee), for example, is drawn to the novelist Gill Norvège, an invert who wears a yellow chrysanthemum in his buttonhole.[183]

If female pollinators could be sexually ambiguous or sexually demanding, they could also be potentially predatory. Grandville's entomological social commentary, for example, included "painted-lady butterflies that receive you with elegance" and "narrow-waisted wasps" that are "dangerous company" (fig. 49).[184] Real woman-butterflies, -wasps, -bees, and -birds could be predatory not just toward man-flowers but also toward other pollinators. Women regularly ornamented themselves with "attractors" derived from creatures: feathers, wings, even whole stuffed birds; beetle-wing embroidery or dried insects; live fireflies; butterfly wings or whole butterflies.[185] Lillie Langtry, known as the Jersey Lily, emphasized her floral charms at a ball by wearing a dress covered in "preserved butterflies of every hue and size."[186] And in Ouida's *Moths* the heroine's mother wears "a gown combined of parrots' breasts and passion-flowers,"[187] a sort of pollination tableau calculated to draw the attention of the insects around her. Mary Haweis, author of *The Art of Beauty,* argued that wanting to cultivate one's attractions is "a healthy instinct not to be sneered down," since Darwin tells us that "conspicuousness" results "in the survival of the fittest."[188] She acknowledged its potentially destructive consequences for other species, however. A *Punch* cartoon from 1892, "A Bird of Prey," makes the point forcefully. A lady of fashion is depicted as a terrifying winged creature swooping down on smaller birds. The text berates the chic for their voracity: "One dealer in London is said to have received, as a single consignment, 32,000 dead humming-birds, 80,000 aquatic birds, and 800,000 pairs of wings."[189]

In Sargent's oeuvre portrait subjects are frequently insects and birds; a lifelong habit of "comparing people with birds and animals"[190] informed his painted "scorpions and butterflies."[191] While sometimes humorous—the portrait *Mrs.*

47
(*Clockwise from top left*) French trade cards, 1880s and 1890s, all between 10.5 and 11.5 x 6.5 and 7.5 cm; (*center*) postcard (cropped), 1900s, 14 x 9 cm.

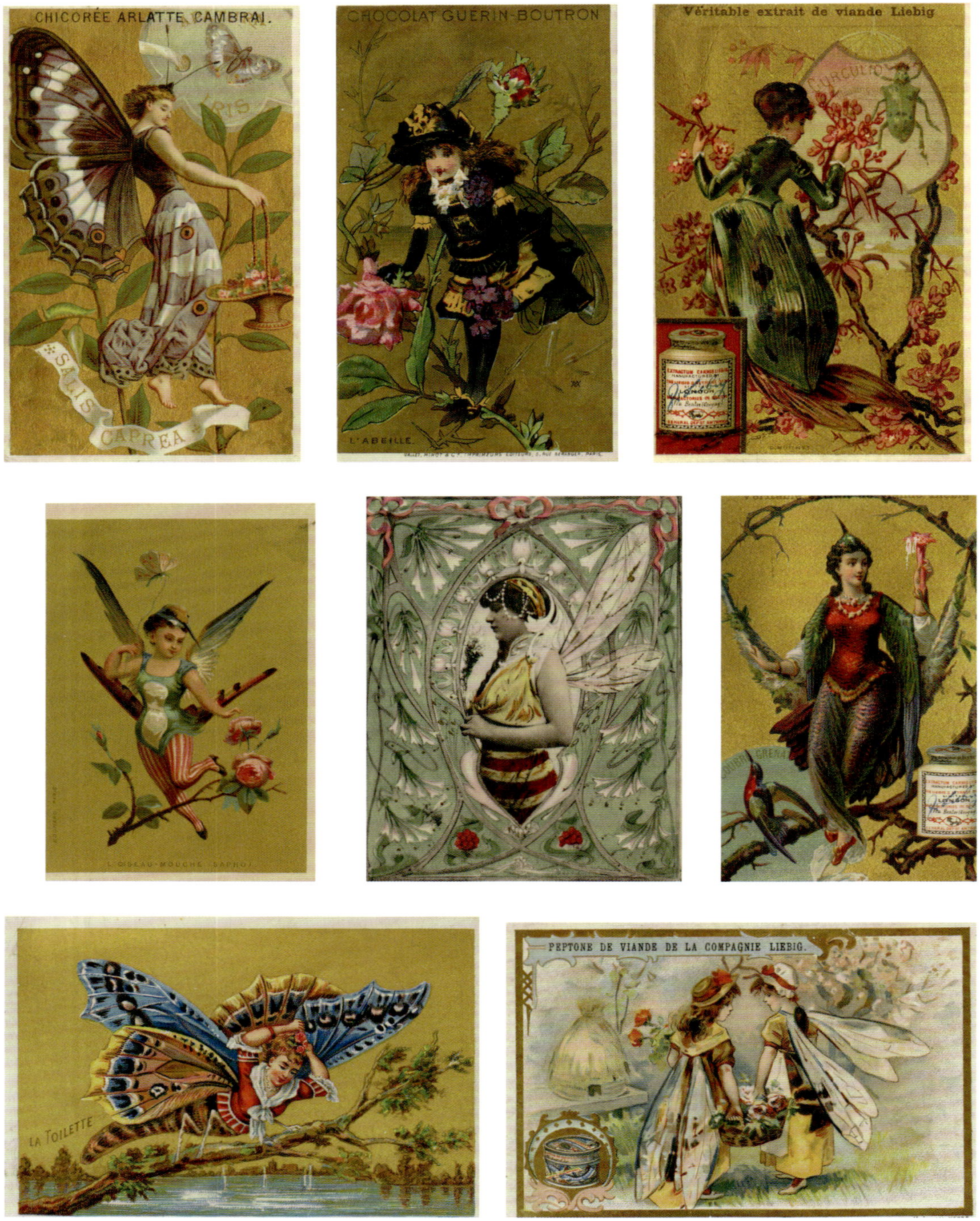

48
Advertisement for stage version of Ouida's *Moths*, 1886, 15.7 x 20.3 cm.

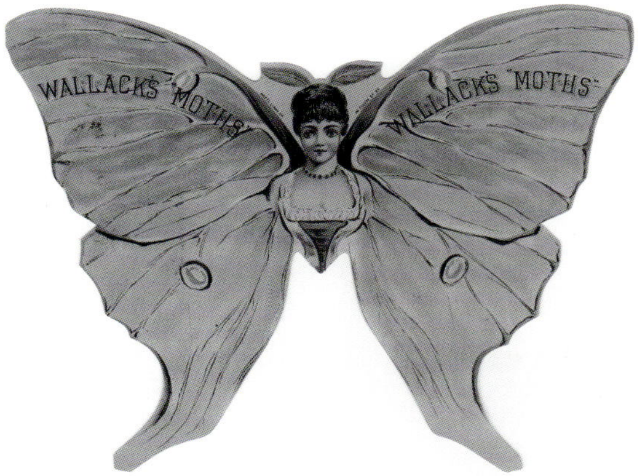

49
J. J. Grandville, "Guêpe," in Stahl, *Scènes de la vie privée et publique des animaux*, vol. 2, between 116 and 117.

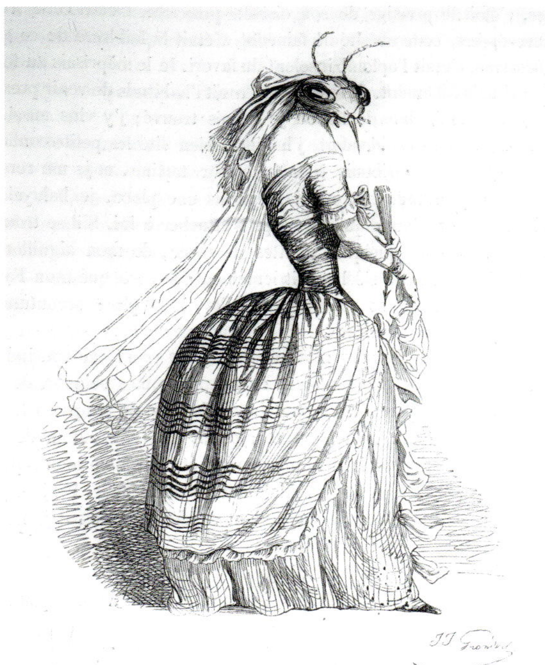

Edward Darley Boit depicts the plump matron as a rooster, in spotted attire with a coxcomb headdress—the artist's female pollinators are more often rapacious. When Sargent saw Ellen Terry's beetle-wing-covered dress in *Macbeth* he insisted on painting her in it; holding a crown, the green, grasping Lady Macbeth is a frightening, almost mantislike creature.[192] Lady Sassoon (fig. 50), with her round body and enormous ostrich-feather crest, is a giant bird. Her plumage, with its feathery strokes, is a gleaming jet, relieved by a cream-colored breast and the orange-pink tips of her winged cloak. Her clawlike hands, sharp beak, and piercing eyes, however, suggest that she is as much harpy as feathered friend. Sargent often depicted male subjects as predatory pollinators, as we shall see in chapter 5. Many of his images, though, eschew such threatening connotations. His men are frequently dandy long-legs; he often emphasized their bent, stick-insect limbs and the angle of the knees.[193] In a sketch of two figures in the Corcoran (fig. 51), the thin, angular legs are light and delicate below their rounded abdomens; the smoke from the right figure's cigarette unfurls his antennae, rendering his entomological metamorphosis complete. In Sargent's 1879 portrait of his teacher Carolus-Duran (fig. 52), the sitter appears as a painter-pollinator, with a round, puffed body sheathed in greeny-yellow and black; a furry, heart-shaped head with its protruding mustache and beard; and a shiny brown stinger between his legs. Carolus's capacity for pollen-brushing is conveyed by his velvety coat, bushy hair, splayed cuffs, handkerchief, necktie, and long fingers. A contemporary caricature of the painting (fig. 53) makes the insect analogy explicit by adding

50
John Singer Sargent, *Lady Sassoon*, 1907. Oil on canvas, 157.5 x 104.1 cm. Private collection.

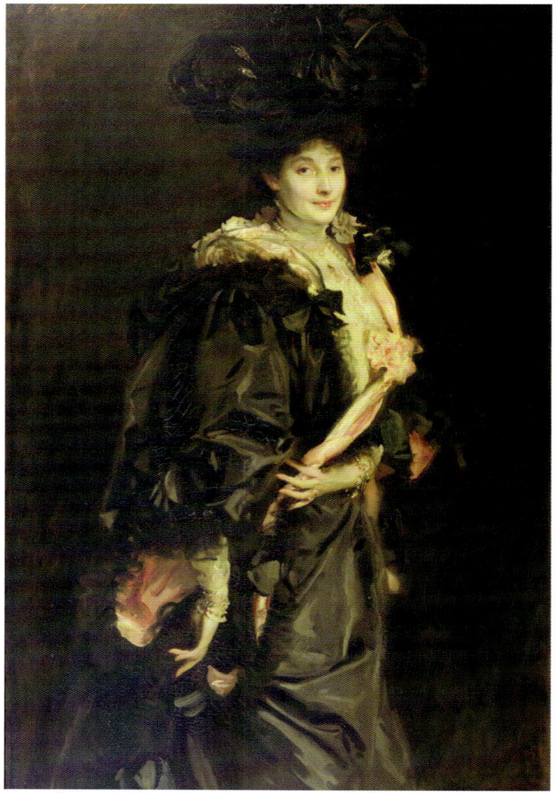

the missing mothy, furred antennae—not demonic horns—to the painter's head.

The flourishing of plant and pollinator identifications in the second half of the nineteenth century attests to the Darwin-inspired "romantic materialism" of the age—what Gillian Beer has shown was a widely shared "desire to substantiate metaphor, to convert analogy into real affinity." The visual manifestations of this "profound imaginative longing"[194] are my primary concern here. Yet beyond the cultivation of formal resemblances, one might ask, how does all of this floral eroticism and passion for pollination relate to the actual practice of painting?

In addition to models, painters found in flowers muses, mates, and ways of life and ways of reproducing worth adopting. Albert Moore "could not paint without the companionship of some bunches of flowers" and considered floral connoisseurship a marker of artistic quality: "when a painter says a rose is his favourite flower I know what sort of painter he is," he remarked witheringly.[195] Flowers were capable of inspiring painterly desire. Jacques-Émile Blanche recalled that Renoir's brush stroked "the geraniums and roses" on his canvases as gently and amorously "as one caresses" a "woman's neck."[196] Blossoms also aroused viewers, sometimes more than conventional, same-species objects of desire. Raffalovich, who declared that he and Moore "cherished" flowers together "with a peculiar and parallel intensity," avowed that he "must have fallen in love with [Moore's] painted flowers to endure with so unruffled an equanimity the girls who shared the same canvas."[197]

If flowers themselves were desirable, Mallarmé argued in 1876 that painting "concerns itself more" with "flesh-pollen than any other human attraction."[198] That the subjects of so many of Sargent's portraits have been perceived as floral is an indication that we should take pollinating metaphorics more seriously into account when considering his oeuvre. The painter's interest in metamorphoses included their visualizations both past and present; he owned a sixteenth-century illustrated edition of Ovid,[199] and he was fond of fairy, flower, and other garden spectacles at the

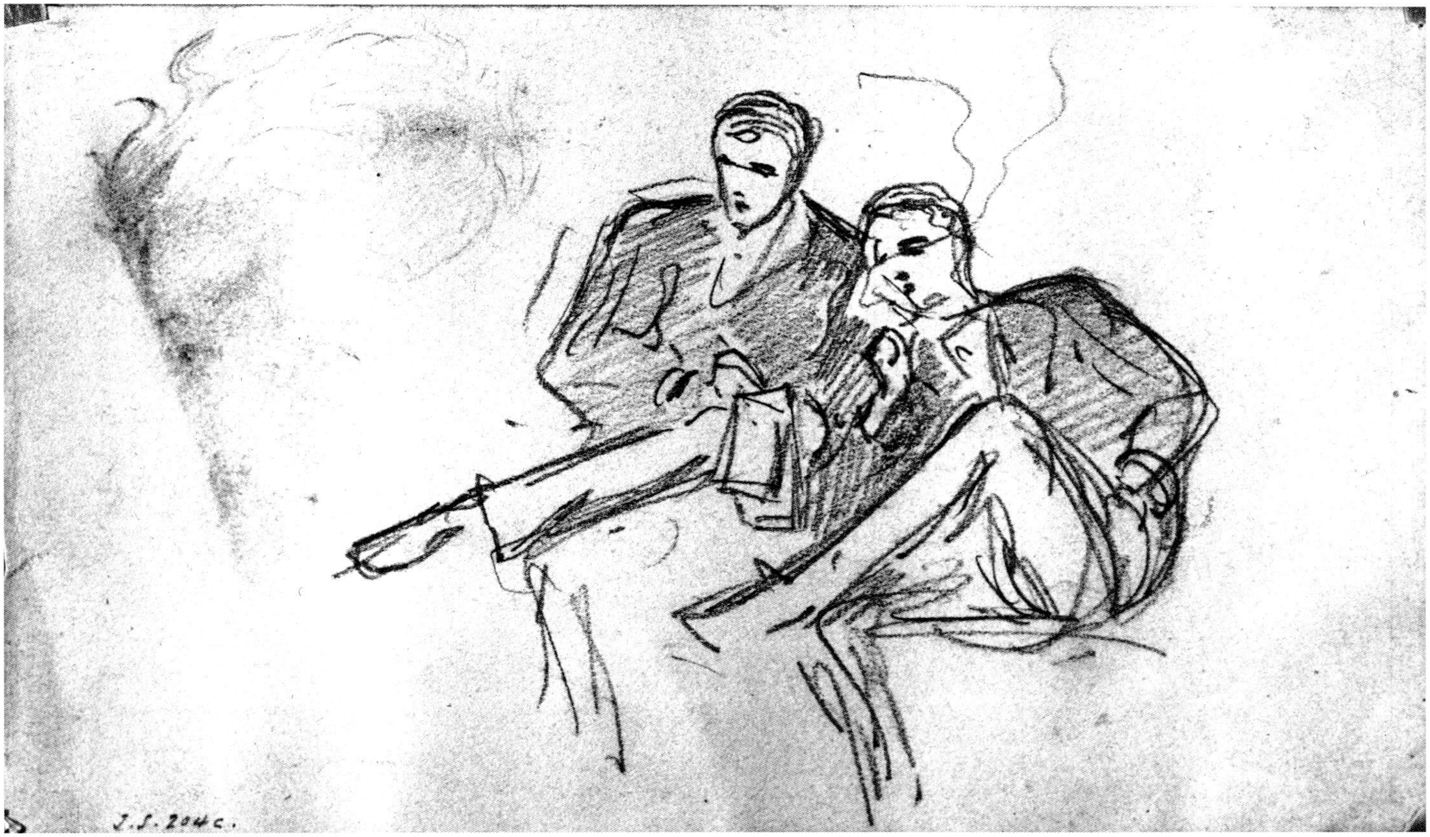

51
John Singer Sargent, *Two Men Smoking and Writing* (?), c. 1874–77. Pencil on beige paper, 10.16 x 16.83 cm. Gift of Miss Emily Sargent and Mrs. Francis Ormond, sisters of the artist, 49.148d. Collection of The Corcoran Gallery of Art, Washington, D.C.

theater.[200] One incident from Sargent's biography particularly graphically illustrates the relevance of Mallarmé's dictum to his work. At Broadway in the summer of 1885, Edmund Gosse's hair, which "no one had ever thought beautiful before," became lilac in the sun. As he approached Sargent, who was sketching outdoors, the painter "gave a convulsive plunge in the air with his brush, and said 'Oh! What lovely lilac hair, no one ever saw such beautiful lilac hair!'" He painted Gosse, who "stood laughing," and "ejaculated at intervals, 'Oh! What lovely hair!'"[201] Sargent here is the prototypical apian or otherwise pollinating artist, whose amorous and creative drives are aroused by the floral quality of the painterly subject. Drawn into delicious promiscuity by colors and bouquets, Sargent's painterly proboscis dipped, brushed, delectated in sitters' nectaries; through these familiar encounters, flowers of art were reproduced.

While painting, Sargent assumed by turns or at once the roles of apian artist and pregnable flower according to the metamorphic inclinations of his subjects, as we shall see. Whistler, too, assumed various roles in his artistic encounters, an ability his hybrid emblem signals. Like Sargent's, Whistler's portrait subjects were perceived by

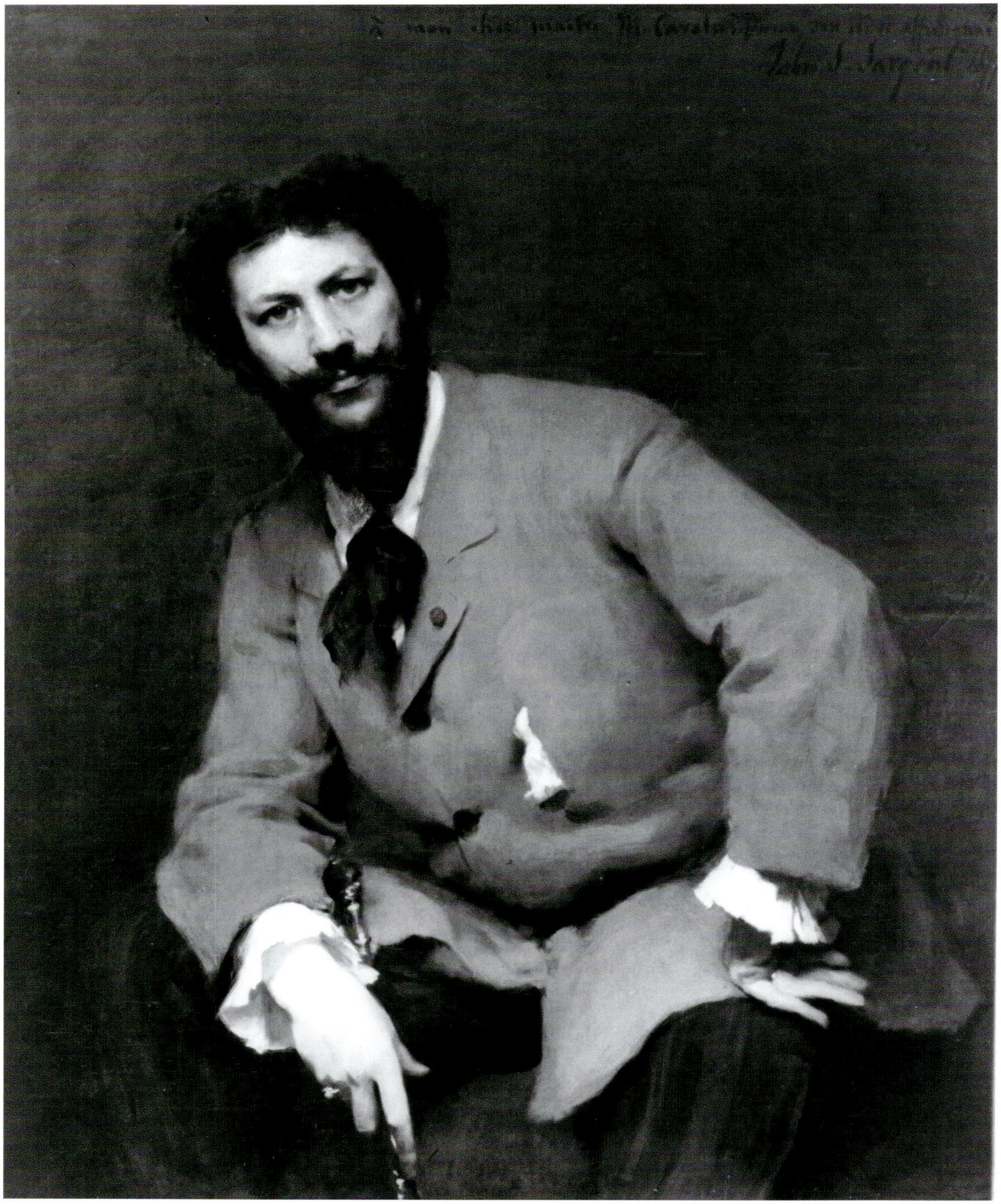

52
John Singer Sargent, *Portrait of Carolus-Duran,* 1879. Oil on canvas, 116.8 x 96 cm. Sterling and Francine Clark Art Institute, Williamstown, Massachusetts, 1955.14.

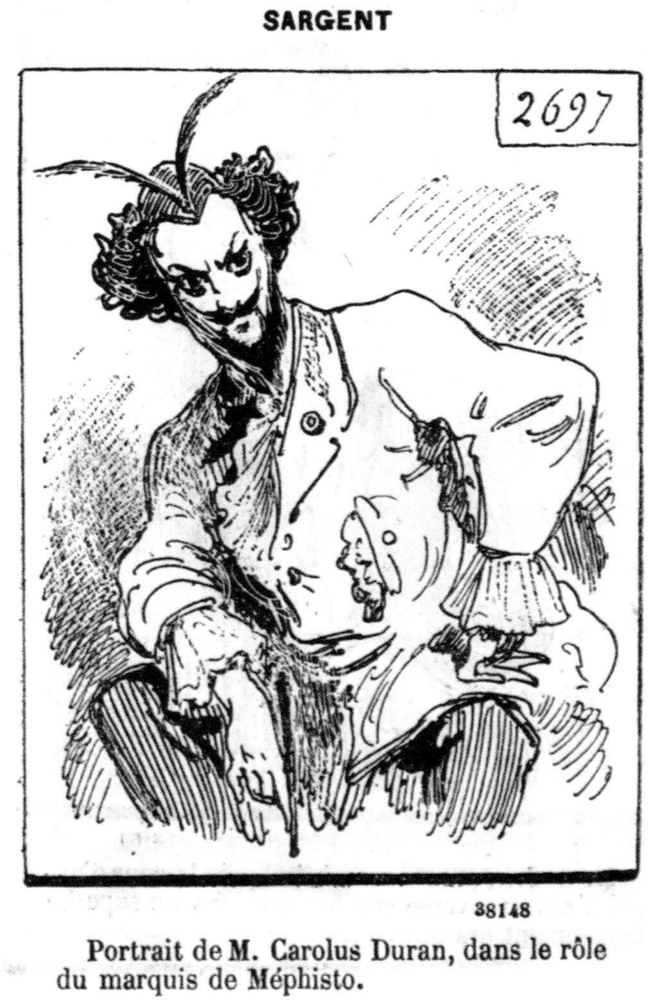

53
Stop [L. P. G. B. Morel-Retz], "Portrait of M. Carolus Duran, in the Role of the Marquis de Méphisto," *Journal Amusant*, 5 July 1879, 5.

contemporaries as both blossoms and pollinators: often his "faces exist softly, like flowers . . . offering their most secret perfumes"; at other times "Whistler has caught him, poised him, posed him, another butterfly, and alive."[202] As previously discussed, in the act of cross-fertilization inhabitable subject positions and identifications multiply. Visits to the pictorial gardens could thus stimulate the metamorphic motility of viewers as well as artists. When "all the picture shows in London" were "in full bloom,"[203] for example, spectators could be inspired to flights of fancy like those pictured in Du Maurier's cartoon for *Punch* (fig. 54), where skirt-tails and fanned wings allow airborne creatures to thrust their organs of perception into the flowery world of nineteenth-century painting.

Whistler offers an example of the way the metaphorics of cross-fertilization could be used to naturalize alternative sexual and artistic practices. Laurie Schneider has argued that the artist's butterfly symbol—often equipped with a scorpion tail and stinger—attests to his bisexual identifications. Although the "recorded facts of Whistler's biography hint only indirectly at homosexual activity," the painter's sexual proclivities have long been a subject of speculation.[204] Edmond de Goncourt remarked on his "air of a fantastic and macabre pederast,"[205] and the painter's butterfly/pansy signatures occasionally provide quite graphic images of his anal orientation: in one drawn to adorn *The Gentle Art of Making Enemies* (fig. 55), the butterfly, endowed with human proportions and a dandyish air, is copulating with a distinctly sphincterous pansy.[206]

54

"Edison's Anti-Gravitation Under-Clothing," *Punch's Almanack for 1879*, 9 December 1878.

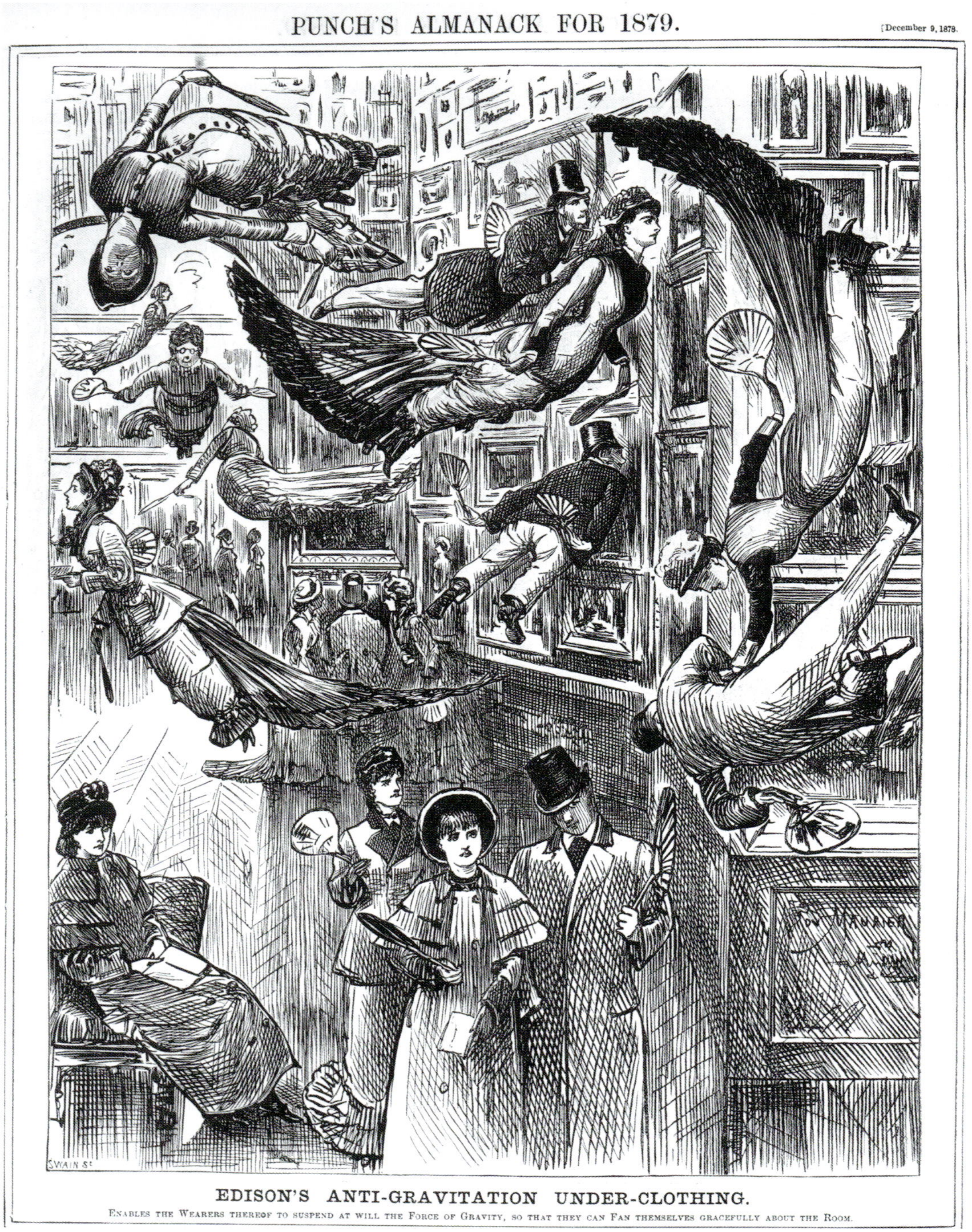
PUNCH'S ALMANACK FOR 1879. [December 9, 1878.

EDISON'S ANTI-GRAVITATION UNDER-CLOTHING.

Enables the Wearers thereof to suspend at will the Force of Gravity, so that they can Fan themselves gracefully about the Room.

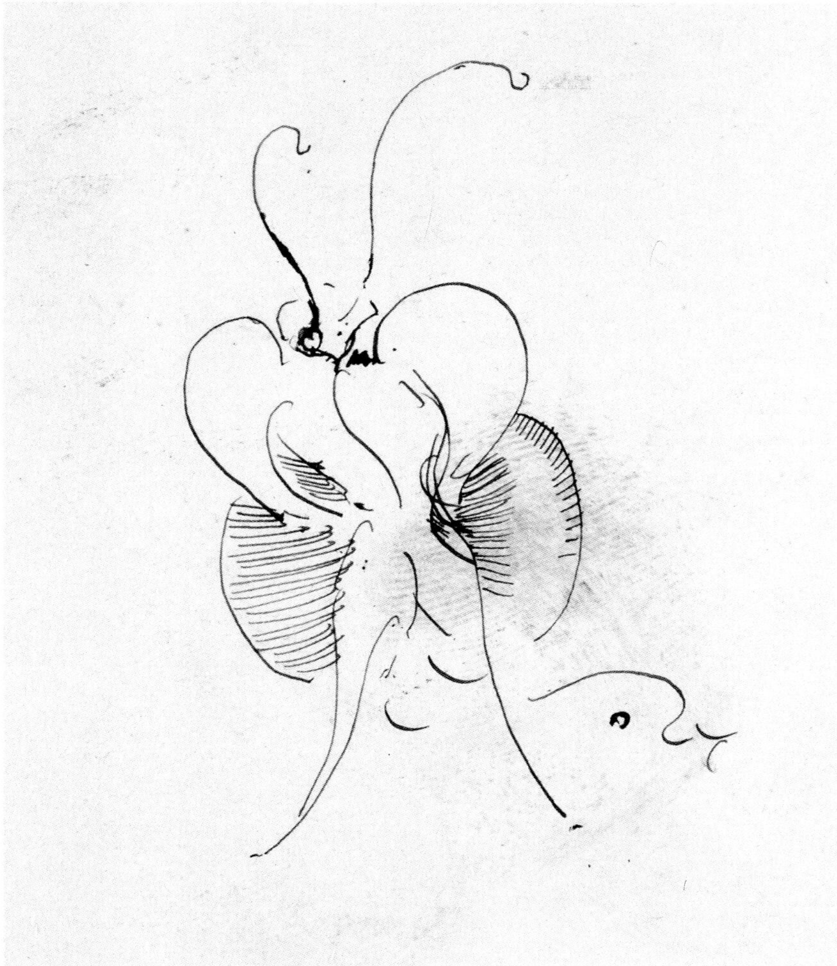

55
James McNeill Whistler, *Butterfly*, c. 1899. Pen and black ink on cream wove card, 12 x 15.8 cm. Gift of Mr. Walter S. Brewster, 1933.291, The Art Institute of Chicago.

By now it should be clear that despite being rooted in a variety of affinities, the human-flower analogy was neither as anatomically grounded nor as stable as Linnaeus would have liked. If the eighteenth century in general tended to trope the whole person as a flower, or the floral calyx as a stage where a number of individual players performed, in the nineteenth century this holistic approach came to be rivaled by a mobile system of cathected part-objects and equivalents. Any anatomical fragment, in addition to the more obvious genital apparatus, was susceptible to floral troping and botanical eroticization. The effect of this vegetable blazoning was to decentralize and de-essentialize the floral body, transforming it into a garden of surrogate and compatible members, orifices, and sensitive surfaces. Pollinators received the same treatment, and so it is to part-objects rather than whole *femmes-fleurs* or moth-men that we shall turn. Sargent's interest in the affinities of human and floral reproduction played itself out particularly in depictions of the counterparts to his creative organ: his subjects' floral hands, whose polyvalent figuration as flowers, pollinators, and sexual organs is the focus of the next chapter.

2

LASCIVIOUS DIGITATION,

OR THE IMPORTANCE OF MANUAL STIMULATION TO THE INVERT ARTIST

His cool, white, flower-like hands . . . had a curious charm. They moved, as he spoke, like music, and seemed to have a language of their own.

— OSCAR WILDE, *THE PICTURE OF DORIAN GRAY* (1891), 44

There is a peculiar relation of hands to flowers, and fingers to petals. The similitude has long been marked—think of plant names like lady's fingers, foxglove, and palm[1]—and is more than superficial. Hands and flowers share parts and physiological processes: a phalanx is both a finger or toe bone and a bundle of stamens, and an unguis is the narrow base of a petal, a nail, or a claw. The carpal bones of the hand etymologically derive from the Greek *karpos,* which means both wrist and fruit, and the French *main* signifies both hand and botanical tendril.[2] The unfurling of petals, or opening of the bloom, is termed digitation. Flowerlike hands crop up throughout literary history, but they grew to unprecedented numbers in the nineteenth century. Such appendages appear in all stages of development, from fists that are "still unfolded buds,"[3] to "finger-tips like honeysuckle buds / Five-parted,"[4] to hands that wave like "the blowing of a blossom-loaded spray of creeper in a breeze."[5]

Sargent shared his contemporaries' fascination with hands, which he painted in all degrees of openness or closure, and in every imaginable position with respect to wrist rotation, finger splay, and interplay with other hands through the manifold possibilities of digital intertwining. His hands are almost infinitely pliable; some seem contorted, others rather disturbingly boneless in their malleability or overattenuation. They often hold, touch, or finger flowers; some metamorphose into them; and almost all of them, in their cupping, curling, splaying, folding and drooping, springing and brushing, resemble flowers. Floral hands, as the epigraph by Wilde suggests, have an alluring language of their own. They also allow something of their subjects' hidden interiority—Maeterlinck described his hands as "the lilies" of his "soul"[6]—to unfurl, voluntarily or involuntarily, into the field of vision. As we shall see, Sargent's exploration of such vegetal feelers reflects his own curiosity about their silent signage and the secrets they unfold.

Through his metaphoric and metamorphic investigations of hands, Sargent attempted to grasp the nature of identity, perception, and artistic creation, to master the hand's capacity both to represent and to betray. In so doing, the painter was sharing in the widely held notion that hands were especially important sites of identity formation and signification, to be carefully watched or observed. The writers of etiquette manuals

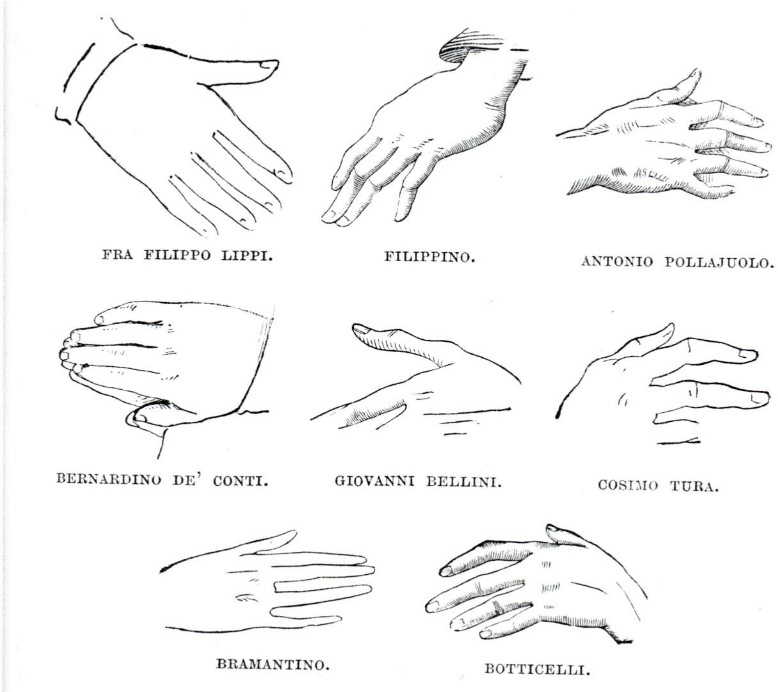

56

Giovanni Morelli [Ivan Lermolieff], "Typical Hands," in *Italian Painters: Critical Studies of their Works*, vol. 1, *The Borghese and Doria Pamfili Galleries in Rome* (London: John Murray, 1892), 77.

cautioned that the hand "is a tell-tale member, and reveals all the character which the face tries to conceal,"[7] and criminologists concurred. Sherlock Holmes chastises Watson for never recognizing "the suggestiveness of thumb-nails,"[8] and tellingly, after Hyde irrupts into being in Stevenson's *Strange Case of Dr. Jekyll and Mr. Hyde,* Jekyll identifies his two distinct incarnations by the sight of both his physical hands and his different styles of handwriting. These latter "hands," although they appear "in many points identical," are "differently sloped," betraying their owners' different inclinations.[9]

Painted hands helped connoisseurs finger their artists. Giovanni Morelli's *Italian Painters* made famous the examination of details for clues to their creator's identity. "Every important painter," Morelli argued, has a "type of hand" that is "peculiar to himself,"[10] the brushwork and shape of which signify much to the eye of the initiated beholder, often revealing technical or personal shortcomings.[11] *Italian Painters* is almost a literal handbook, as a sampling of its observations will attest: Raphael's typical hands have a "broad metacarpus and somewhat stiff fingers, the nails extending to the tips only"; the balls of Titian's thumbs are unfortunately "too strongly developed." Botticelli's hands are "very bony and plebeian, and the nails broad and square," but they are at least vaguely natural. Fra Filippo's fingers, on the other hand, are of a structure "both peculiar and unpleasing. The juncture with the metacarpus is so sharply defined that it has not the appearance of a natural growth; the fingers look as if they had been screwed into their places, and are long, wooden, and nerveless."[12] A chart is provided for easy reference (fig. 56). These hands, like Poe's purloined letter, lie on the surface; the connoisseur detects and deciphers them. Freud himself likened Morelli's technique to that of psychoanalysis, in which even the most minor signifier is potentially a symptom and a clue.[13]

If they could point to the identity of their painters and the latter's foibles and faults, hands and their movements also promised to reveal something of their owners. Proust's description of the slender and delicate hands of Andrée attests to the fascination the enigmatic life of hands holds for the artist. They "had as it were a private life of their own, obedient to the commands of their mistress, but independent, and would often stretch out before her like thoroughbred greyhounds, with lazy pauses, languid reveries, sudden flexings of a finger-joint, seeing which Elstir had made a number of studies of these hands; and in one of them, in which Andrée was to be seen warming them at the fire, they had, with the light behind them, the golden diaphanousness of two autumn leaves."[14] Capricious, constantly metamorphosing, beckoning, and beguiling, these hands are eventually rendered, through the artist's labors, more transparent. There is ample evidence that Sargent felt the pull of such mysterious digital motions. The airy manipulations of the hands of the Javanese dancers at the 1889 *Exposition universelle* inspired Sargent to make a number of oil and pencil sketches. One of these leaves the ghostly pentimento of a fluttering hand visible,[15] a reference to the chronophotographical motion studies of Muybridge and Marey that functions as an index of the painter's desire to capture those unconscious motivations and revelations of the hand so often elided by normal vision.[16]

57
John Singer Sargent, *The Misses Vickers*, 1884. Oil on canvas, 137.8 x 182.9 cm. Graves Gallery, Museums Sheffield.

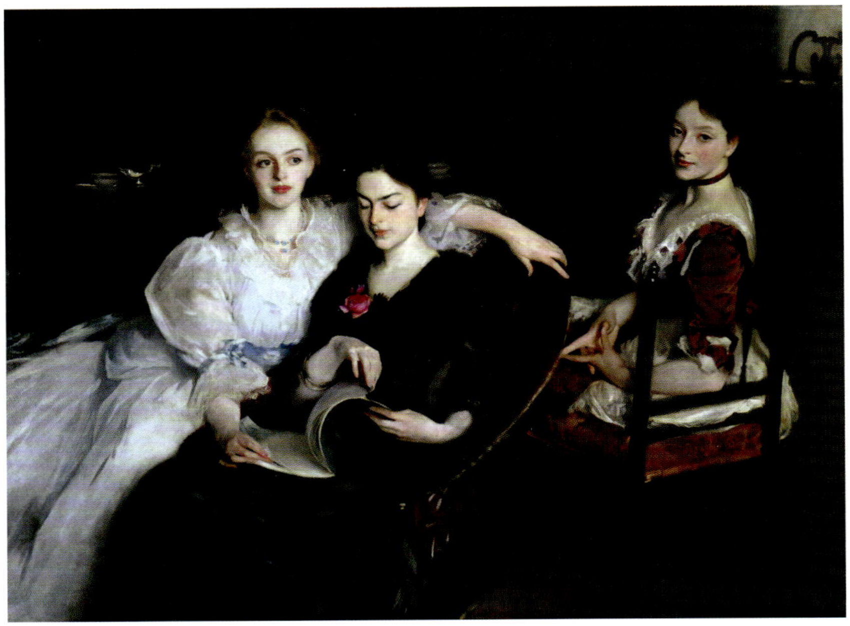

Hands can disclose or let slip much about their humans; we can begin with their social position. There is a long history of *classing* hands according to their vegetable morphology: in "Elegie VIII," for example, Donne contrasts his mistress's arm and hand, "like that slender stalke, at whose end stands / The wood-bine quivering," to those of a woman whose inferiority is marked by her tanned skin and "the short swolne fingers" of her "gouty hand," which "like a bunch of ragged carrets stand."[17] The hands of the well-to-do Andrée are "noble"[18] as well as leaflike. In this sense, the semiautonomous appendicular life forms Sargent paints are often in keeping with the nature of their owners' identities: the hands of his subjects are mostly of a floral cast(e), lily white and naturally graceful, which happen to be those of *la fine fleur de la société*. The Wyndham sisters' (fig. 23) are soft, pale, and succulent, like the magnolias surrounding them. The fingers of the two rightmost *femmes-fleurs* gently unfurl at the edge of the canvas, reaching out to their delicate sister blossoms. In the group portrait of the three Misses Vickers (fig. 57), the pair of sisters on the left, frothily floral in their wicker basket chair like Snow White and Rose Red (one blond and all in white, the other dark with a red rose on her breast), extend gracefully choreographed, arching, falling hands; their limbs even leaf (through the pages of a book). Those of the lower born, when they appear in Sargent's oeuvre, manifest their humble roots.[19] A Venetian onion seller (fig. 58), whose features are on the whole rather bulbous, has a large root vegetable for a hand. Where wrist and dorsum

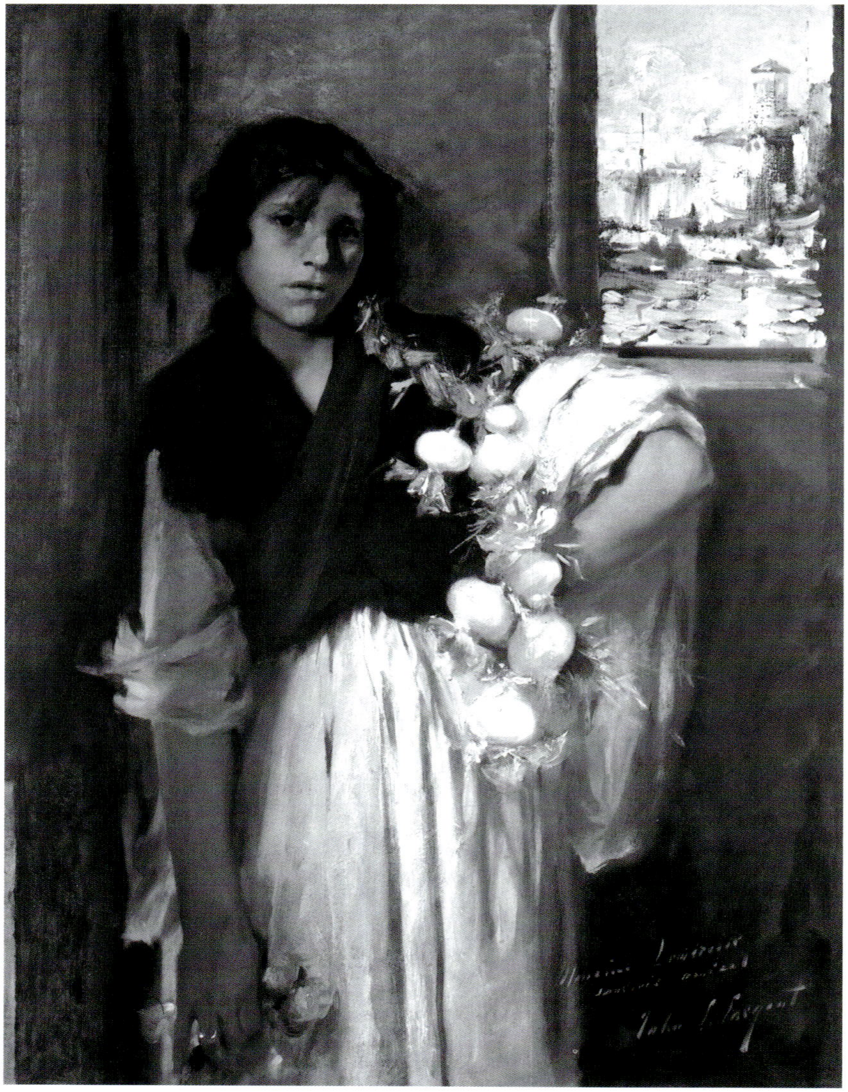

58

John Singer Sargent, *Venetian Onion Seller*, 1880–82. Oil on canvas, 95 x 70 cm. Museo Thyssen-Bornemisza, Madrid.

should be, her left forearm suddenly sprouts into a pair of onions of increasing magnitude, and spreading out in place of her bent-under fingers are frayed and straggling roots. The light renders these visible, but the digits of the onion seller's other hand are obscured in the dim, earthy brown, cellarlike space. Dark, perhaps dirty, and barely articulated, this appendage seems to shoot down toward the earth, though the small bulbs it grasps, like the string of onions, angle up toward the light and air let in by the window.

Ruskin maintained that the "rank scent" of "squills, garlics, and onions" was responsible for separating peasant life "from that of the higher classes,"[20] but other vegetable features—their literal earthiness, suggestive shapes, and lascivious associations,[21] for example—played a part in this exclusion. Edwin Blashfield was relieved that Sargent's Venetian paintings were endowed with sufficient "style, style, style!" to sublimate the "onions and garlic"[22] they contained, but the painter found these vegetables handsome enough (literally, better adapted to the hand, easily grasped or manipulated). Margaretta Lovell describes Sargent's onions as "even handsomer" than his Venetian women, and this may have been the painter's order of preference.[23] Unashamed of his manual labor as well as his vegetable proclivities, he proclaimed himself a "horny-handed son of toil," loath to subject his *griffe* (claw or signature) to the aristocracy's "mawbid wish to civilize."[24]

Artists' hands were often described as fine, sensitive, almost feminine features. "If ever a painter had painter's hands it was Whistler," wrote Arthur Symons: they had "the most sensitive fingers" he had "ever seen."[25] Lillie Langtry noted that "with their delicate, tapering fingers," his hands "were thin almost to transparency."[26] Despite his protest, Sargent's hands were apparently "strong, supple, long-fingered,"[27] "delicate," and "dexterous."[28] Like Whistler's, they were admired by his fellow artists. While he was a student, Carolus-Duran had "especially liked to paint" them. Even after Sargent had left his teacher's studio, Carolus "used to send for him to come over and pose for hands."[29] When once the young painter had to refuse his teacher's request, however, to reserve his tools for his own use, Carolus in anger broke off their friendship and painted out his former student's face from the Luxembourg ceiling murals they had worked on together.[30] The intimate relation between the hands and the face, hands and identity, is rendered graphically in this anecdote. Among other things, then, Sargent's hands were the occasion of a wound.

If they help signal the identities of their owners and, by Morellian analysis, authenticate those of their painters, painted hands which display by their vegetal blooming a certain capricious autonomy nevertheless register the irruption of the unconscious, of desire, into the field of the visible, and thus substantially estrange their subjects. In Sargent's 1884 portrait of Mrs. Albert Vickers (fig. 59), an enormous magnolia flower grows in place of her right hand, the splay of its petals visually echoing the finger spread of her left. This outsized bloom has prompted a number of inadequate explanations. David Fraser Jenkins has argued that the choice of blossom was dictated by the painter's conscious attempt to cater to "an English taste in painting," the magnolia being "a commonplace of the Victorian country house façade";[31] David Park Curry suggests that the flower expresses Mrs. Vickers's "love of nature" and functions most significantly as the means by which Sargent "counter-balanced the lifesize figure" in a "reassuringly" traditional portrait.[32] Reassuringly traditional is certainly not how the painting and its waxy funereal flower were perceived when first exhibited in the 1885 Salon. To Theodore Child the portrait was evidence that "Mr. Sargent is growing more and more eccentric and his painting more and more unpleasing."[33] "He isn't orthodox," wrote Paul Mantz in *Le Temps*.[34] Charles Ponsonailhe specified the cause of audiences' unease: Mrs. Vickers had been imbued by the painter with "all the unreal and mystical character of the heroines of his compatriot Edgar Poe,"[35] particularly Ligeia, in whose beauty also "there was much of 'strangeness' pervading."[36]

Like Ligeia, who "came and departed as a shadow"[37] even before she returned from the dead, Mrs. Vickers is rendered "ghostly"[38] by the seeming absence of her feet, her pseudo-eighteenth-century grey costume, and the dagger of a brooch plunged into her breast. That Sargent sought to achieve something like Poe's lugubrious atmosphere can be inferred from the changes made from the preliminary sketch (fig. 60) to the finished portrait. The preparatory drawing shows the sitter perched cross-legged in a sunny room, looking out a window, while in the finished painting the enormous nocturnal bloom,[39] dark-paneled room with its door slightly ajar, looming wardrobe, and shadowy face with eyes askew all hint at some mystery. This is not entirely absent from the study: the slightly disturbing hand-flower is present there

59

John Singer Sargent, *Mrs. Albert Vickers,* 1884. Oil on canvas, 208.4 x 101.5 cm. Virginia Museum of Fine Arts, Richmond, the Adolph D. and Wilkins C. Williams Fund.

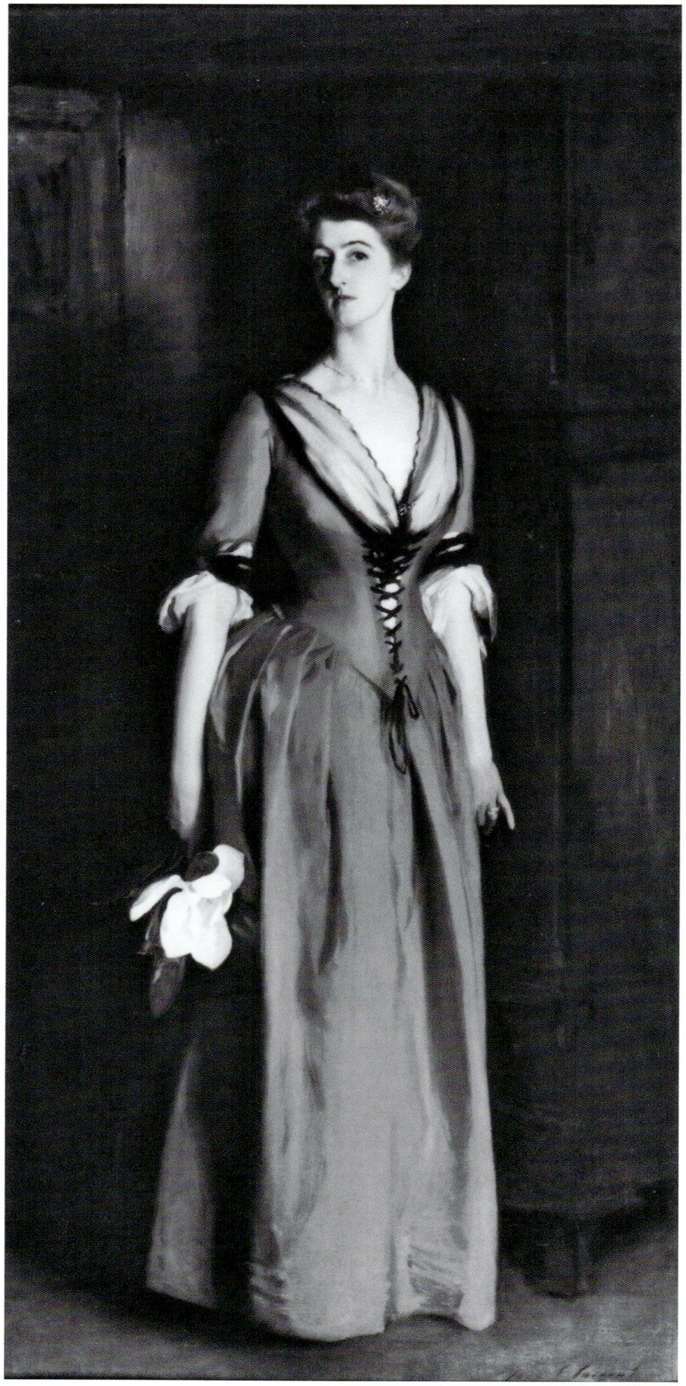

60

John Singer Sargent, *Study for Mrs. Albert Vickers* (Edith Foster), c. 1884. Charcoal on off-white wove paper, 33.3 x 23 cm. Harvard University Art Museums, Fogg Art Museum, gift of Miss Emily Sargent and Mrs. Francis Ormond in memory of their brother, John Singer Sargent, 1931.92.

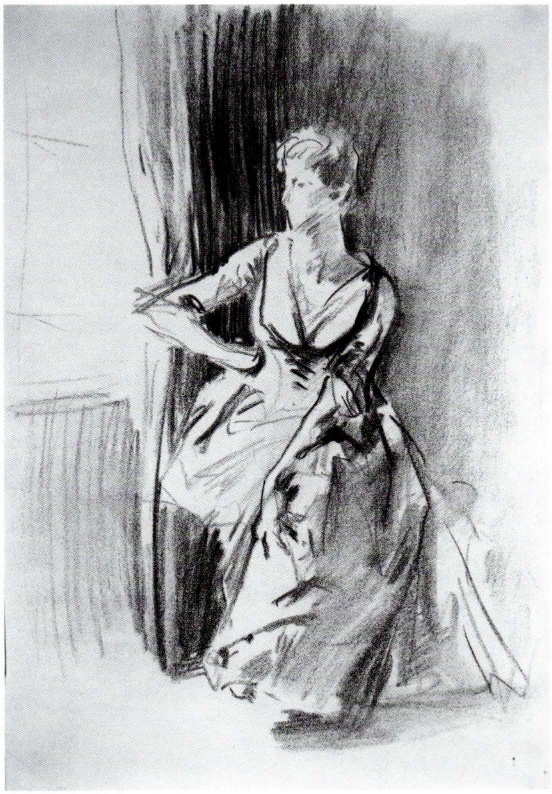

61

Detail of *Study for Mrs. Albert Vickers.*

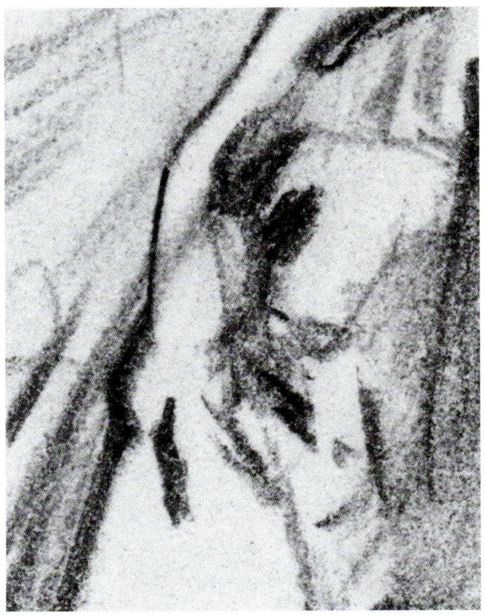

too, appended to the sitter's left arm. Half-spider, half-blossom, the creature creeps as if of its own accord down her calf, seemingly dragging her torso after it (fig. 61).[40] In the finished painting the spider reappears as an ornament in Mrs. Vickers's hair.

Poe's narrator relates that the "outwardly calm, the ever-placid Ligeia, was the most violently a prey to the tumultuous vultures of stern passion," which, during their "long intercourse," failed to give "immediate evidence of its existence" save in the unfathomable expression of her eyes. He can only liken the experience of perceiving her "gigantic volition" to a feeling that also occasionally grips him "in the survey of a rapidly growing vine" or metamorphosing creatures: "a moth, a butterfly, a chrysalis."[41] Sargent's portrait similarly emphasizes the subject's metamorphic potential: her limbs elongate almost vegetably, her stemlike arms and neck twisting toward the light. More than anything else, though, it is the gigantic magnolia blossom here that functions, like Poe's vine, as a vegetable index of the Ligeian, mysterious, passionate, perhaps even morbid, but above all the "gigantic volition" of the unconscious, which insistently *jette sa pousse*.[42]

We can begin to be more specific about what Mrs. Vickers's floral hand reveals if we recall that, at least according to Huysmans, the figure of Ligeia is unsettling because of her sexual ambiguity. In the painting Mrs. Vickers's strangely phallic neck strains upward and, like her literary counterpart, she has a "boyish torso"[43] (the laces at the bottom of her *corsage*—bodice—form a suggestive loop), but I would argue that it is particularly the magnolia's hermaphroditic sexuality—its pistils and stamens invisible but its bisexuality indirectly manifested in the soft and delicate yet vigorous and massive blossom—that gestures to the subject's own sexual polymorphy. In Gautier's *Mademoiselle de Maupin,* famous for its preface's plea for art for art's sake, the protagonist's hands offer the first clue to the enamored d'Albert that there is something strange about "him": "his beautiful hands issuing from their ruffles like the pistils of a flower from the midst of their petals, he looked not the handsomest of men but the most beautiful of women"; his fingernails are like "the inner leaves of a rose." The hero/ine, in fact, belongs to "a third, distinct sex,"[44] a hermaphroditic one that these hands naturally, *comme une fleur,* make visible.

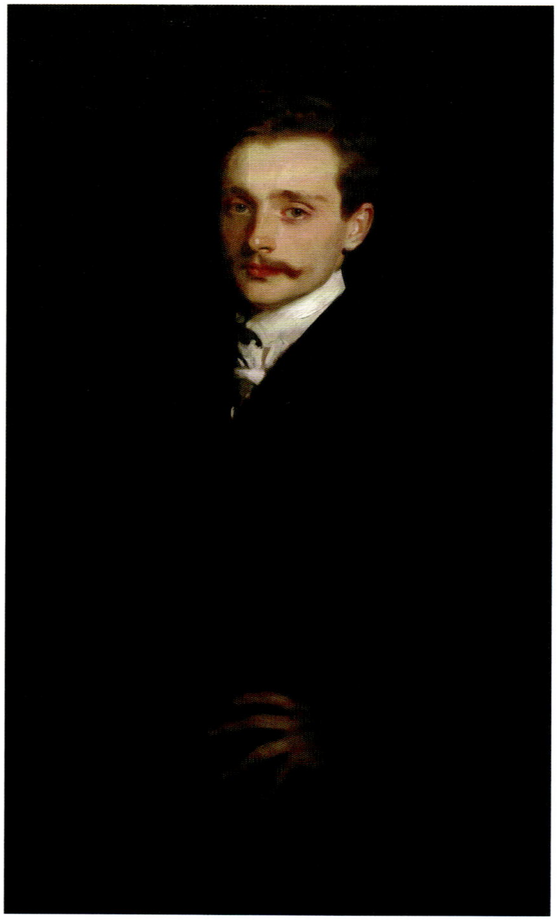

62
John Singer Sargent, *Léon Delafosse,* c. 1895–98. Oil on canvas, 101 x 59.4 cm. Seattle Art Museum, 2001.17. Given in honor of Trevor Fairbrother.

William Cohen has argued that in the Victorian novel, "a genre that forbids direct observation of genitals in action, [a] manual code gives voice to what otherwise cannot be spoken."[45] The same could be said of Sargent's oeuvre, in which the painter's keen interest in sexual ambiguity comes across in the often oversupple digits of his gentlemen and the frequently (and fetishistically) phallic fingers of his ladies. The hands of Sargent's *cher maître* Carolus-Duran (fig. 52), like those of Mademoiselle de Maupin, emerge elegantly from the ruffled calyces of his cuffs, his right pendulous, attenuated, and languid; his left bent but springy. When exhibited they were seen as "rather too effeminate for any man,"[46] and in the caricature from *Le Journal Amusant* (fig. 53) Duran's extended pinky exaggerates his daintiness. The slender, gently spreading fingers of Léon Delafosse (fig. 62), a musician and the model for Morel in *À la recherche du temps perdu,*[47] are of a plantlike delicacy and "fragile tenderness."[48] At the other end of the spectrum, Isabella Stewart Gardner's portrait displays "firm hands, the thumb almost too long, which might guide four horses with the precision of an English coachman," in Paul Bourget's description.[49] Miss Helen Dunham (fig. 63), who belonged to "the most admirable of all the sexes," according to Henry Adams (there being, it was understood, many), seems to be clutching a pet willy in her lap.[50] Demure and white as a lily in the heart of her rose interior—the

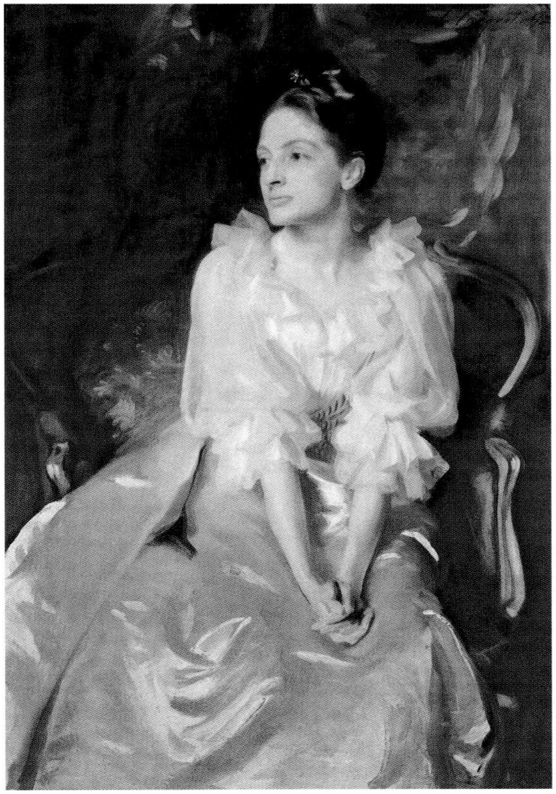

63
John Singer Sargent, *Helen Dunham,* 1892. Oil on canvas, 121.5 x 81.3 cm. Private collection.

wooden trim of the chair almost delineating petals—Helen cups in the folds of her petaline, even labial fingers a stamen that might perk up at any moment. And the thoroughly precocious Beatrice Townsend (fig. 64) single-handedly strokes, clasps, and fingers the hairy antherlike tip of her little dog's tail, which, in the location of its protrusion from her lap, could quite easily be a hermaphroditic appendage of her own (little boys are made of snips and snails and puppy dogs' tails, after all).

The hermaphroditic character of floral hands was not just Sargent's idiosyncratic preoccupation. It is captured in two midcentury sculpted hands belonging to royal progeny, possibly by Abraham Kent, which emerge from the plant life at their wrists (fig. 65). The carved vegetation seems intended to distinguish sex: the appendage on the left, on which a butterfly alights, springs from roses, clover, and thistles; the one on the right from oak leaves and tiny penile acorns.[51] This vegetal differentiation is reinforced by the former's open palm and the latter's closed fist. Stable gendering, however, is exactly what floral hands cannot maintain, and it is likely that these hands were sculpted from a single sitter's pair. Ambiguous floral hands also abound in greeting cards, calling cards, and scrap (fig. 66). In these examples, clothing enables one to label the hands male or female, but the hands themselves are virtually identical in size, color, shape, and softness in every case but the topmost. If the graceful hands emerging from circlets of roses and white, petally cuffs on the sides seem likely to be female, in other instances cues may be confusing. In the calling card second from the top, two feminine hands, one wearing a bracelet, the other a ring, reach out from beneath flowers—a white camellia on the left, a purple pansy on the right; the romantic message, however ("Play not with my heart lest all my joy depart"), suggests that the hands belong to a pair of lovers, confounding any heteronormative or securely gendered reading.

The hermaphrodite condition Sargent's hands often betray reflects the innate bisexuality of humans, which is not generally made so visibly manifest. Occasionally their instinctive doubleness of orientation surfaces so violently as to wrest control from the subjects' arms, suggesting a disturbing, almost schizophrenic, lack of self-

64
John Singer Sargent, *Beatrice Townsend*, c. 1882. Oil on canvas, 81.9 x 58.4 cm. Private collection.

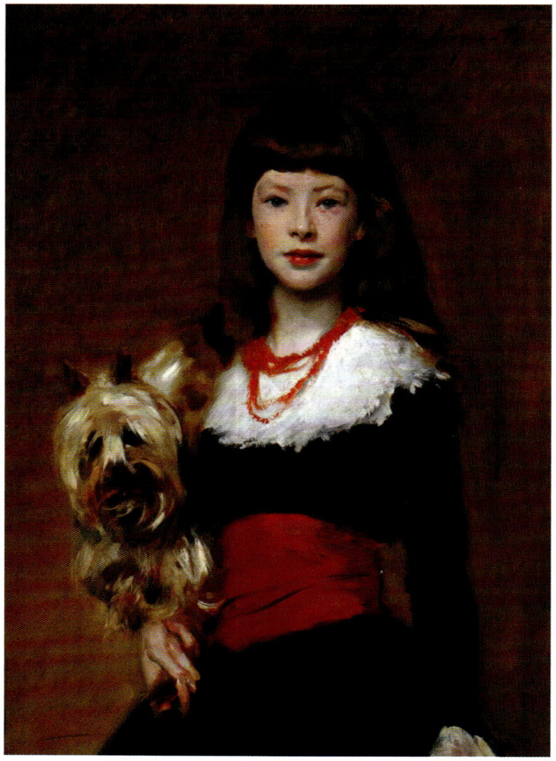

65
Abraham Kent (?), sculpted hands of royal child(ren), possibly Princess Victoria's, c. 1840s. Marble. RCIN 34358, The Royal Collection.

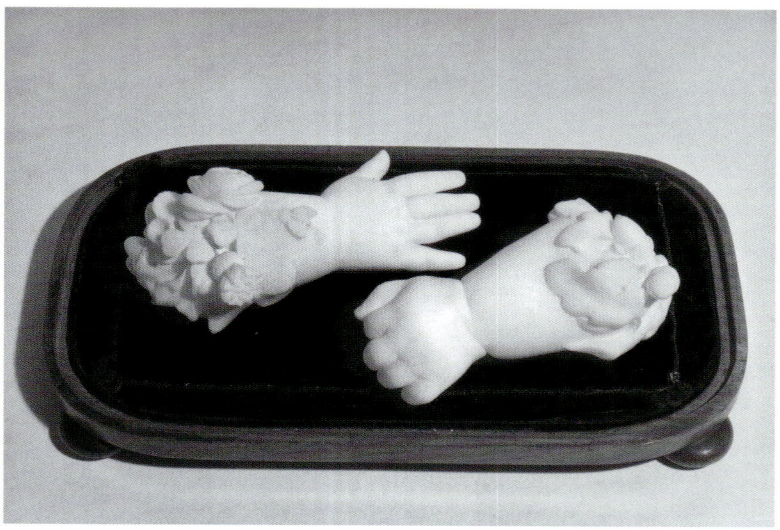

possession (fig. 67). At the ends of her marionettelike upper limbs (her shoulders in particular appear held aloft by invisible strings), the hands of Mrs. Charles Thursby seem, in their antithetical dispositions—one clenched, the other's fingers spread wide like branching twigs—to wring her body in opposite directions. The sitter's instability and androgynous character received contemporary comment: to Clementina Anstruther-Thomson, she looked "like an 18th century coachman . . . flying along."[52] Her indeterminate gender does not deter the black and lilac butterflies, created by bows of ribbon, from alighting on her breast; it constitutes her allure. The hands of Mrs. Kate "Puss" Moore, one clawlike and clenched, the other languid and unfurling, similarly contort her body, her right shoulder appearing almost dislocated. This sitter "enjoyed a European fame for malaprop use of French," of which Sargent had firsthand experience: on one occasion Puss "took exception to a glass of flowers standing between her and the painter. She rang the bell and when the footman appeared said, pointing in the direction of Sargent, 'Otez-moi *ce salaud*' (*sale eau*)"—get rid of this bastard (dirty water).[53] Significations multiply wherever Sargent paints flowers.

The really perverse potential of Sargent's hands stems from the self-fertilization their floral charms make possible. In the portrait of the stiffly erect Brigadier Archibald Campbell Douglas Dick (fig. 68), one naked lily hand strokes the other. These succulent white twiners manipulate his wooden stick, the testicular knob of which is doubled by the pentimento: one arm spirals around the cane while the other hand grips its long, hard

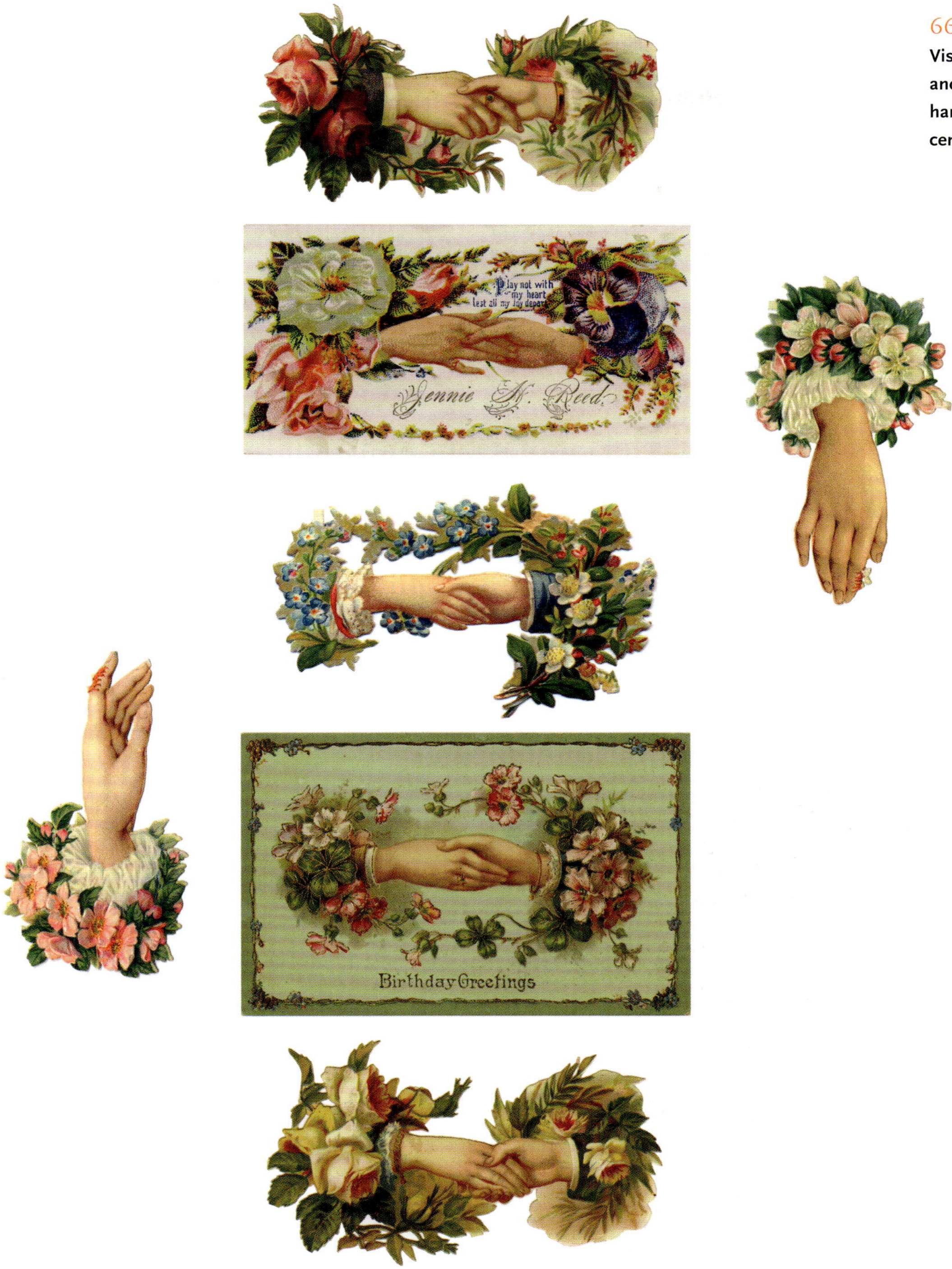

66

Visiting and greeting cards and scrap depicting floral hands, later nineteenth century.

67

John Singer Sargent, *Mrs. Charles Thursby*, c. 1897–98. Oil on canvas, 198 x 101 cm. Purchase by exchange, 1985. Gift of Mr. and Mrs. J. Duncan Pitney, Emilie Coles (from the J. Ackerman Coles Collection), Mrs. Lewis B. Ballantyne, Mrs. Owen Winston, and the Bequest of Louis Bamberger. Collection of The Newark Museum, Newark, New Jersey.

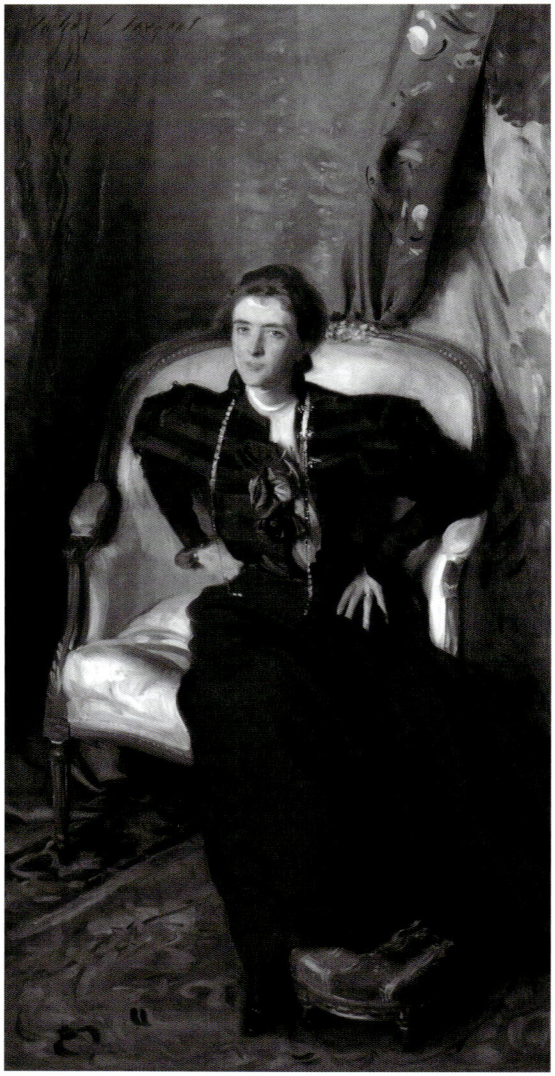

shaft; the left hand steadies the right wrist, and under the touch of fingers held at just the right degree of tension, the scarlet wadding of the brigadier's gloves unfurls like a flower. Through their effleurage, then, these orchidaceous creatures seem to seek to bring his staff to fruition.[54]

If the brigadier's hands evoke Swinburne's description of the lovemaking of hermaphrodites, "like a flower laid upon a flower,"[55] hands are not limited to this same-species love in Sargent's portraiture: they can also function as *both* flowers and pollinators. The pale gold left forearm of the slightly moustached Mrs. Thomas Wodehouse Legh (figs. 69, 70), for example, so nearly merges into the fabric of her dress that one is quite unprepared for the shocking sight of the erect shafts of her penile fingers springing up from between her thighs and of her swollen and very reddened thumb—a Whitmanian "phallic thumb of love"[56] on this mannish woman's hand that *jette sa pouce*. In its golden, polleny meadow the sitter's other hand, metamorphosed through the wing of her fan into a butterfly, approaches the proffered digital bloom between her legs, perhaps to be held there for a moment. This isn't the only example of a pollinating appendage in Sargent's portraiture: the rightmost Miss Vickers in the sisters' portrait (fig. 57) is an alert, stripe-necked brown and white creature. Perched next to the seat in which her floral sisters are planted, she exchanges a knowing look with the viewer, who can see where her beak- or stingerlike fingers point: the heart of the rose. And Miss Townsend's furry feeler (fig. 64) can be read as a pollen brush, since she is, after all, a Bea. Mrs. Legh's portrait metaphorically figures the impending contact of hands as interspecies, but it also constructs an autoerotic subtext that is made even more explicit in another painting.

Lady Agnew of Lochnaw (fig. 71) fingers with her right hand a small white rose, nestled in her lap, whose petals are just beginning to unfurl. The eyes of this polite vegetable onanist are dulled with pleasure, and a faint sheen of perspiration is visible

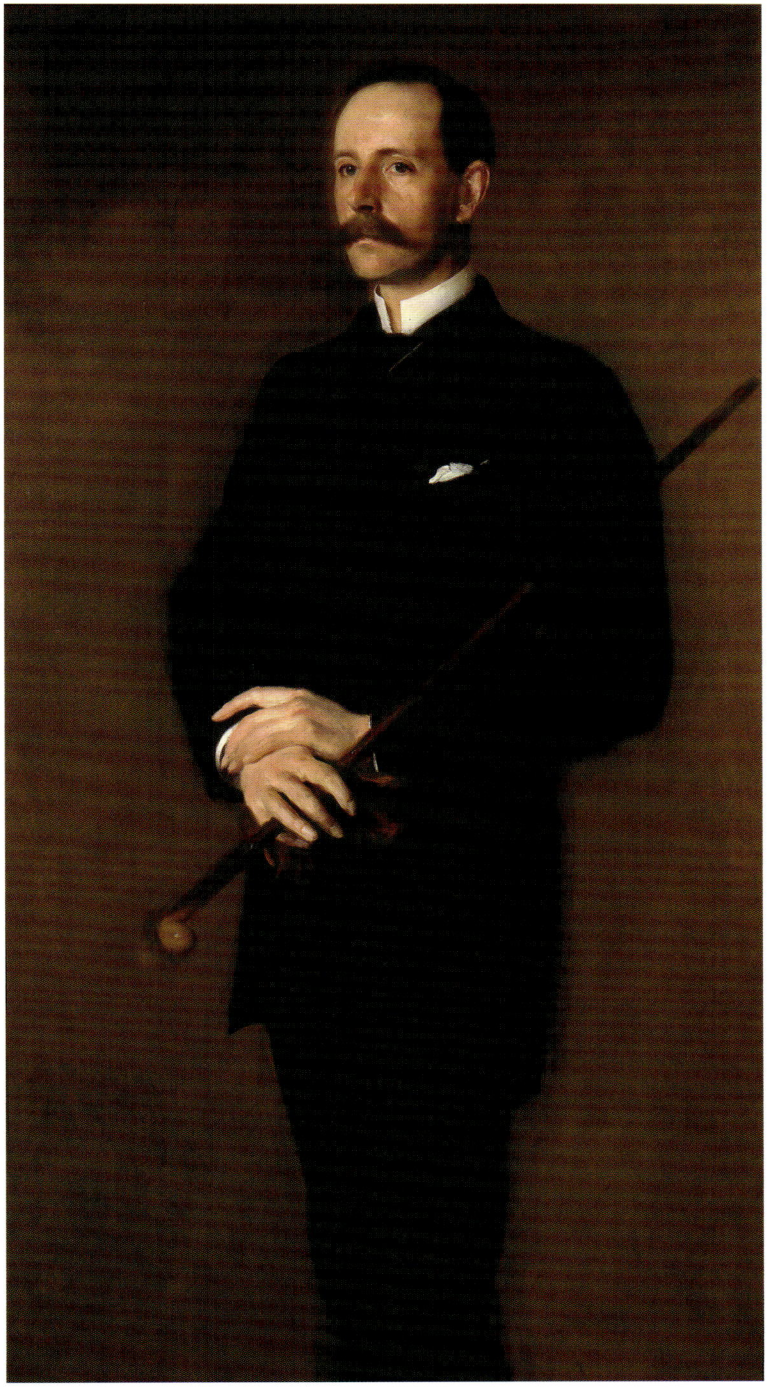

68

John Singer Sargent, *Brigadier Archibald Campbell Douglas Dick,* 1886. Oil on canvas, 160 x 90.8 cm. Private collection.

on her face. She grips with her left hand the wooden frame of her Louis Quinze–style *bergère,*[57] which exemplifies what Huysmans's arch-aesthete Des Esseintes might have had in mind when he opined that "the eighteenth century knew how to envelop woman in an atmosphere of depravity, modeling its furniture after the curves of her charms, mimicking the contractions of her pleasures, the volutes of her spasms, with the undulating and twisting intricacies of wood";[58] the chair arm's little swelling protrusion occurs on a level with the clitoral or penile bud (we can't see if Lady Agnew grasps a stem, but that very ambiguity again foregrounds the hermaphrodite theme). Beneath the sitter's fingered flower, her white dress (painted "like apple blossom, soft and light and about to fall")[59] and lavender sash cascade voluptuously. As though to ensure that there is no misunderstanding of the scene unfurling before us, a lotus blossom on the sky-blue silk hanging at the level of the bud and wooden spasm is lasciviously digitated. Contributing to the aura of hothouse sexuality, golden oriental letters, like "fecundating rains of pollen,"[60] shower down on the sitter in the flowery field of her chair.[61]

Sargent's charged iconography drew on an established paradigm, as the interaction of digits and blossoms was widely figured as amorous in the

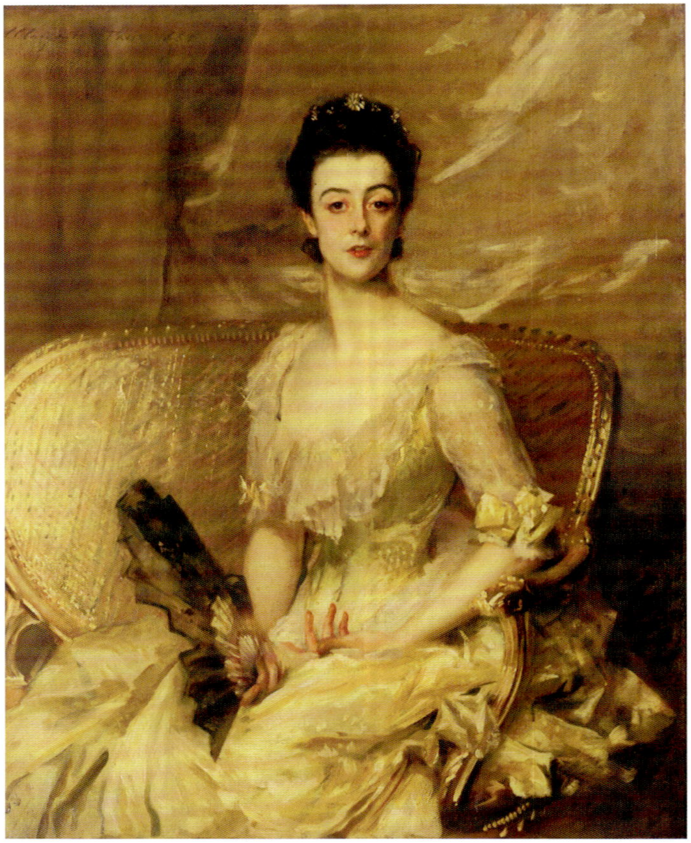

69
John Singer Sargent, *Mrs. Thomas Wodehouse Legh,* 1884. Oil on canvas, 119.4 x 94 cm. Private collection.

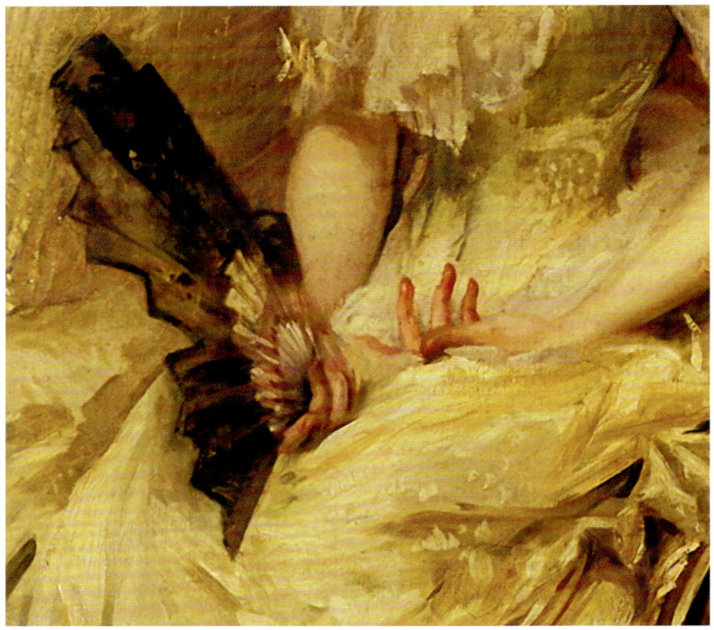

70
Detail of *Mrs. Thomas Wodehouse Legh.*

71

John Singer Sargent, *Lady Agnew of Lochnaw,* 1892. Oil on canvas, 127 x 101 cm. The National Gallery of Scotland, Edinburgh, purchased with the aid of the Cowan Smith Bequest Fund, 1925.

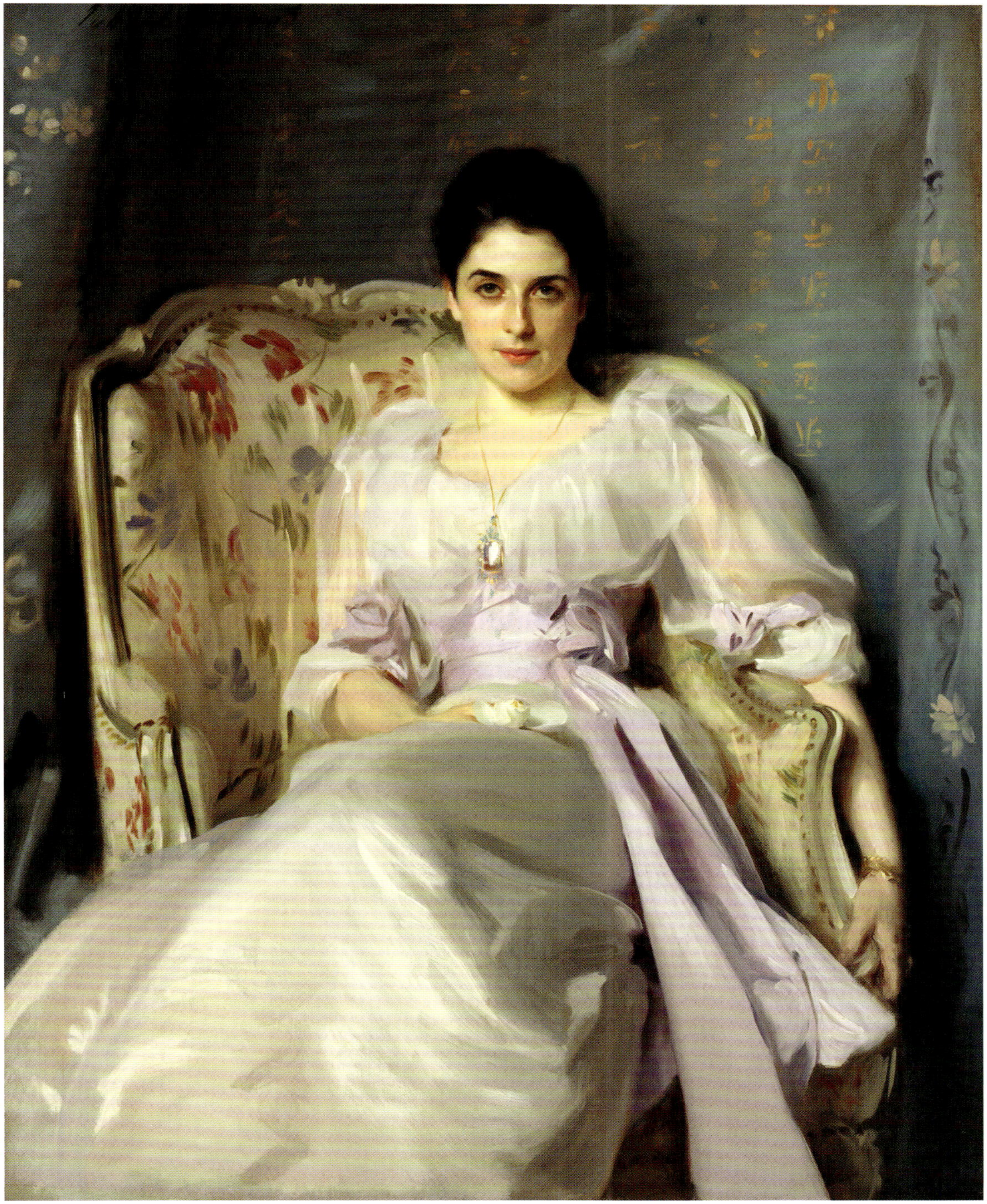

period. Some responded to the stimulus of flowers with gentleness; Edmond de Goncourt, for example, was amazed at Octave Mirbeau's knowledge of and consideration for the flowers he handled: "naming, finding again without turning a hair, enumerating all the names of flowers, and with amorous notes in his voice, when he described them, and caressing gestures, which had the air of fearing to cause them to wither under his fingers."[62] Few could resist the delight of touching[63] or plucking; Maeterlinck wrote that those who see campanula "for the first time, refusing to believe their eyes, want to touch with their finger the bluey marvel."[64] Thomas Wentworth Higginson instructed readers to be gentle and loving when picking water lilies: "coil your thumb and second finger affectionately round [the flower], press the extended forefinger firmly to the stem below."[65] Others were less benign in their approach; some even needed to crush the delicate creatures. Gerard Manley Hopkins's sadism is palpable in his description of bluebells: "if you draw your fingers through them they are lodged and struggle / with a shock of wet heads; the long stalks rub and click and flatten to a fan on one another like your fingers themselves would when you passed the palms hard across one another, making a brittle rub and jostle like the noise of a hurdle strained by leaning against; then there is the faint honey smell and in the mouth the sweet gum when you bite them."[66] Rubbed and bitten, the flowers, "like your fingers themselves," enact the poet's vicarious erotic experience: the long stalks and wet heads struggle and strain until the moment of destruction and release, when sweet gum and a faint honey smell are exuded. Wordsworth's characterization of the typical botanist as "a fingering slave, / One that would peep and botanize / Upon his mother's grave,"[67] aligned such practices with a host of taboos: voyeurism (which leads to touching), incest, and necrophilia as well as interspecies sexual transgression.

As with flowers, contact with floral hands may be sensitive or sadistic, but it is almost always in some way improper. In Maeterlinck's *Pelléas et Mélisande,* Golaud's outburst, "oh! these little hands that I could crush like flowers," is a threat born of sexual jealousy.[68] Floral hands (much like Hopkins's handlike blossoms) may, it seems, inspire the most destructive passions. In Pierre Quillard's 1886 Symbolist drama *La fille aux mains coupées,* a young girl's father fetishizes her hands. In the first act he interrupts her prayers to "scorch her hands" with his "incestuous and brutal" kisses and caresses. Mortified, she asks an old servant to chop off the tainted appendages. He replies,

> O my girl, your hands are delicate corollas;
> Your hands are a bouquet of young hawthorns;
> The breath of spring exhales from your flesh:
> I don't gather flowers with iron.[69]

She insists, "untremblingly" placing her hands, which are prefiguratively "spurting from her sleeves," on a table. The deed done, she "brandish[es] in the half-light the twin torches of her mutilated arms" and has the servant put her severed hands in a crystal ewer, where they float, "terrifying and radiant, a double water lily on scarlet stalks" in a "pink foam," to be carried to her father.[70] The association of floral hands with

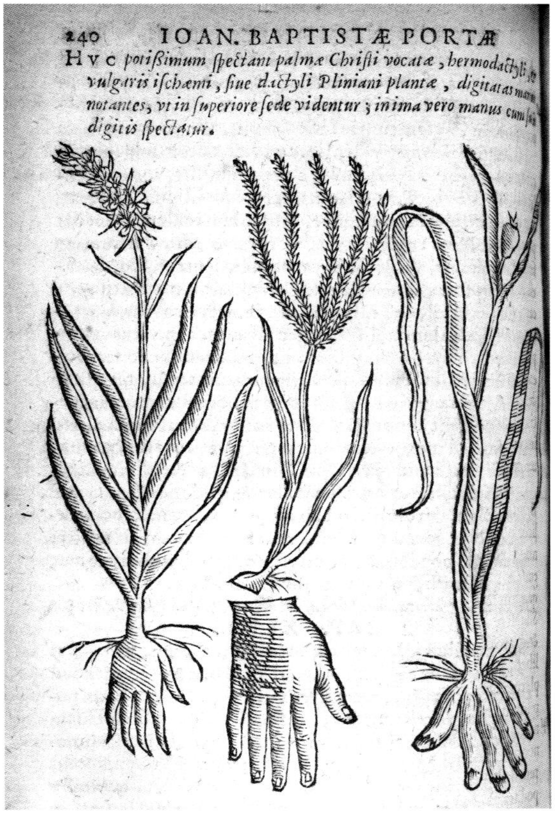

72
Giambattista della Porta, "Orchid," in *Phytognomonica* (Rothomagi: Berthelin, 1650), 240.

sensuality, passion, and crime, as Luigi Berliocchi has shown, is deeply ingrained in the European imagination, which has "fed primarily on those orchids whose plump, fleshy tubers are surprisingly similar to the human hand."[71] Della Porta's woodcut of these disreputable plants (fig. 72) could serve as an illustration to Quillard's play.

Hands, like flowers, then, open out onto multiple perversions. Swinburne, figuring babies' hands as "rosebuds furled / Whence yet no leaf expands,"[72] described their floral fluttering as a "foretaste of fancies and freaks."[73] By grafting onto the hands of his subjects a floral model of exposed "interiority" and hermaphroditic sexuality, Sargent figured them into an iconographic tradition and a network of associations that served to naturalize alternative sexualities. We find this strategy operant not only in the art of Gautier but also in the works of Huysmans, the Decadents, Wilde, Proust, and down through those of the Surrealists and Genet, from the hands of whose conversing queens oft "escape[d] ripples of flowers, in the midst of which they simply stood or sat about as casually as could be."[74]

If flower-finger contact and floral fingers could represent the full spectrum of polymorphous perversity, hands were not innocent in these affairs. Although from at least Aristotle on the hand had been thought to distinguish humans from animals as much as the faculty of reason,[75] in the later nineteenth and early twentieth centuries the privileged position accorded to humanity by the "instrument of instruments" was called into question. Sir Charles Bell's 1833 Bridgewater treatise *The Hand: Its Mechanism and Vital Endowments as Evincing Design* was the last important articulation of a fading conviction. Comparative anatomy—the discipline in which Morelli was originally trained—had shown that the structure of the hand is not a distinctly human feature: the same bone pattern underlies wings and fins, and thus the hand can only superficially, ectodermally, be considered an evolutionary advance.[76] Even nineteenth-century authors who had no intention of displacing "man" from his exalted position vis-à-vis other species were happy to concede that the hand was not a sign of human mastery: Reverend J. G. Wood's popular natural history book *Homes Without Hands,* which describes the wonders of diverse animal architectures, is an apt example.[77] Although Darwin considered the human hand near perfect for its functions, he nevertheless averred that a "man-like animal" with a sufficiently developed hand and arm "could, with sufficient practice, as far as mechanical skill alone is concerned, make almost anything which a civilised man can make."[78] Remy de Gourmont went further, arguing that "the hands of man, marvellous as they appear to us, add almost nothing to his intelligence. . . . The external instrument is secondary: little import whether it be beak, prehensile tail, teeth, trunk, paws, all would have served the function of hands."[79] Gourmont concluded that the hand's adaptability is simply a "wanton" extravagance on the part of nature, wanton being the operative word: when it makes its evolutionary appearance, with the opposable digits of the quadrumanes, it is only to facilitate masturbation.[80] The hand thus appears primarily as a tool for realizing desire, a handmaid of perversion itself.

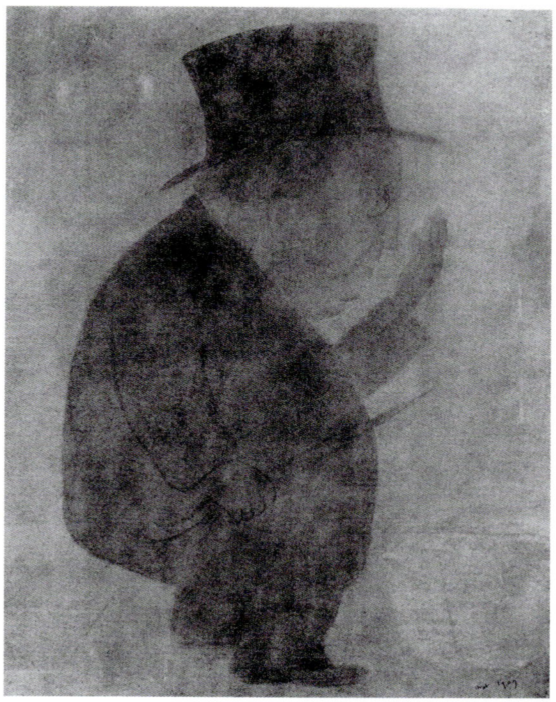

73
Max Beerbohm, "London in November, and Mr. Henry James in London," in *A Book of Caricatures* (London: Methuen, 1907), plate 11.

In this context the appendicular obfuscation of Beerbohm's caricature "London in November, and Mr. Henry James in London" (fig. 73) is particularly amusing, given the studied ambiguity of James's own sexual persona and prose. Beerbohm's caption reads, "It was, therefore, not without something of a shock that he, in this to him so very congenial atmosphere, now perceived that a vision of the hand which he had, at a venture, held up within an inch or so of his eyes was, with an almost awful clarity, being adumbrated."[81] The foggy atmosphere in which James immersed himself is not, in Beerbohm's image, enough to prevent him coming to face to face with his own pressing member.

If a hand on its own, as Whitman assures us, is erotic enough—"I merely stir, press, feel with my fingers, and am happy"[82]—when one hand encounters another the sensations awoken may be overwhelming. The catalogue of hand types proffered by Wilde and Co. in *Teleny,* modeled on those found in etiquette manuals, offers a visceral sampling of the range of creatures with whom one can come into contact:

> Many persons seem to bear a temperature of their own about them. They are hot and feverish in mid-winter, while others are cold and icy in the dog-days. Some hands are dry and parched, others continually moist, clammy, and slimy. There are fleshy, pulpy, muscular, or thin, skeleton and bony hands.

> The grasp of some is like that of an iron vice, others feel as limp as a bit of rag. There is the artificial product of our modern civilization, a deformity like a Chinese lady's foot, always enclosed in a glove during the day, often poulticed at night, tended by a manicure; they are as white as snow, if not as chaste as ice. . . . Some hands are coy, others paddle you indecently; the grip of some is hypocritical, and not what it pretends to be; there is the velvety, the unctuous, the priestly, the humbug's hand; the open palm of the spendthrift, the usurer's tight-fisted claw.[83]

A hand's degree of moisture, fleshiness, grip, color, and texture are all clues to its owner's nature. The question of decency arises here not only in the threat of sexual transgression, through wanted or unwanted paddling, but also of moral duplicity, as a hand may not be "what it pretends to be." Potential regression rears its head in the image of the claw. Following the question, "Who has not been sentient of the manifold feelings produced by the touch of a hand?" this passage suggests that touch, like masturbation, is often a gratuitous act. Certainly the nineteenth century perceived the two as equally risky activities.

Descriptions of the act of shaking hands in late nineteenth- and early twentieth-century literature will attune us to the *sensibilité à fleur de peau* (hypersensitivity) essential to perceiving the reach of its perversity. Proust writes,

> plumper than [Andrée's], Albertine's hands would yield for a moment, then resist the pressure of the hand that clasped them, giving a sensation that was quite peculiar to themselves. The pressure of Albertine's hand had a sensual softness that was as if in harmony with the pink, almost mauve coloration of her skin. This pressure seemed to allow you to penetrate into the girl's being, to plumb the depths of her senses, like the ringing sound of her laughter, somehow indecent in the way that the cooing of doves or certain animal cries can be. She was the sort of woman with whom shaking hands affords so much pleasure that one feels grateful to civilisation for having made of the handclasp a lawful act between boys and girls when they meet.[84]

The handshake is not merely synaesthetic, intensely gratifying, and felicitously legal in this account: the plumbing of the sexual depths of all sorts of different beings afforded by an apparently superficial gesture makes this "formality" alarmingly the reverse of what it seems (or pretends) to be. The extravagant variety of experiences to which the lawful handshake lays one open constitutes an essential part of its transgressivity. We should not be surprised, then, to learn that nineteenth-century etiquette manuals and writers declared the handshake an unduly familiar form of address.

Many European visitors to America, confronted with the widespread practice of handshaking, deemed it a "democratic intrusiveness"[85] carried too far in light of the uncertain nature of the attachments formed thereby. One visitor wrote, "you go on shaking hands here, there, and everywhere, and with everybody; for it is impossible to know who is who, in this land of equality."[86] A disapproving American moralist opined that he "may have much respect for a man, and yet not wish to touch him. What a barbarity, then, are these universal handshakings."[87] The practices of the effeminate French illustrate where this custom may lead: "When two Frenchmen are presented to one another, each presses the other's hand with delicate affection. . . . If we are to follow the French, in shaking hands with one whom we have never before seen, we should certainly imitate them also in kissing our intimate male acquaintances."[88] The distance between the "delicate affection"—or affectation—of the handshake and more intimate embraces was disquietingly small. The French, in fact, provided not only an illustration of where seemingly innocent touching could lead but also a linguistic key to the perversions handshaking could introduce. Hand-to-hand, or in French *main*-to-*main,* contact could function as the expression of a more problematic desire for man-to-man contact.

The potential homoeroticism of the handshake was explored in numerous *fin-de-siècle* texts. In *Teleny,* the narrator describes his first meeting with the titular character:

> The pianist stretched forth his ungloved hand. In my fit of nervousness I had pulled off both my gloves, so that I now put my bare hand into his. He had a perfect hand for a man, rather large than small, strong yet soft, and with long, tapering fingers, so that his grasp was firm and steady. . . . How can I express all that I felt from the contact of Teleny's hand? It set me on fire; and, strange to say, it soothed me at the same time. How sweeter, softer, it was, than any woman's kiss. I felt his grasp steal slowly over all my body, caressing my lips, my throat, my breast.[89]

The naked flesh of one man's hand against another here is "sweeter" than "any woman's kiss." Soft yet manly, firm yet supple, Teleny's grasp "steals," quite criminally, like liquid all over and through the narrator's body. Teleny, in fact, has "a supple, mesmeric, pleasure-giving fluid in his fingers" that soaks the narrator through.[90] This coursing body fluid, the particular character of which is left to the reader's imagination, raises the important question of exactly what could be transmitted through the handshake.

The idea of hands as reservoirs of liquids may strain credulity, but the *Bazar Book of Decorum* informs us that it "is a not uncommon affection of the hands" to have "a moist condition, which seems to resist all the ordinary efforts of absorption. Such hands are so constantly dripping with humidity that every thing they wear or touch becomes readily saturated."[91] In Edmond de Goncourt's 1882 novel *La Faustin,* Carsonac chides Planchemol that he ought to have his hands "drained"; they are so moist they make other people's wet, too.[92] An Aesthetic teapot designed by James Hadley the same year (fig. 74) literalizes the idea of the liquid-dispensing hand and firmly attaches it to the sexually deviant body. The hermaphrodite pot, which features a sunflower-sporting limp-wristed aesthete on one side and a calla lily–entwined Aesthetic lady on the other, pours tea from the hand. As Kristeva, Emily Eells, and others have shown, "tea" was an erotic code

74
James Hadley, *Aesthetic Teapot*, 1882. Glazed parian, 15.7 x 17.5 x 8.5 cm. Worcester Porcelain Museum.

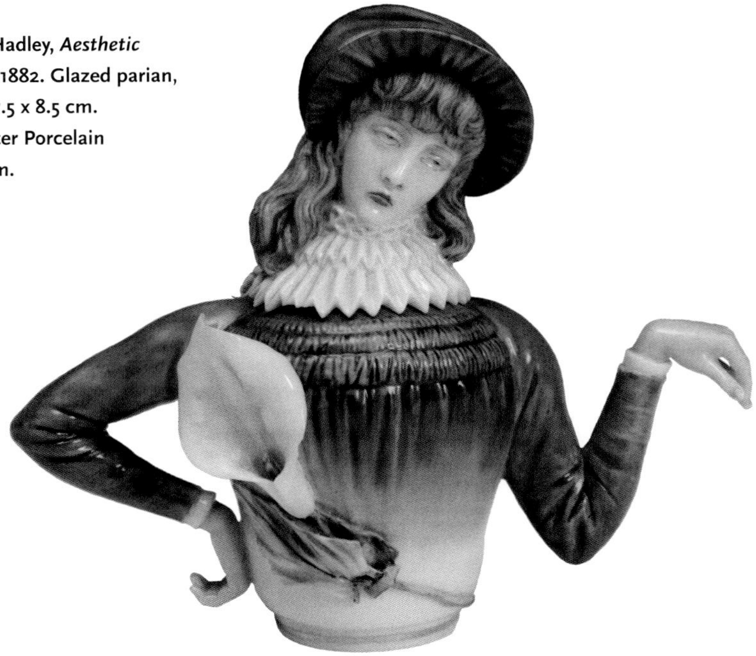

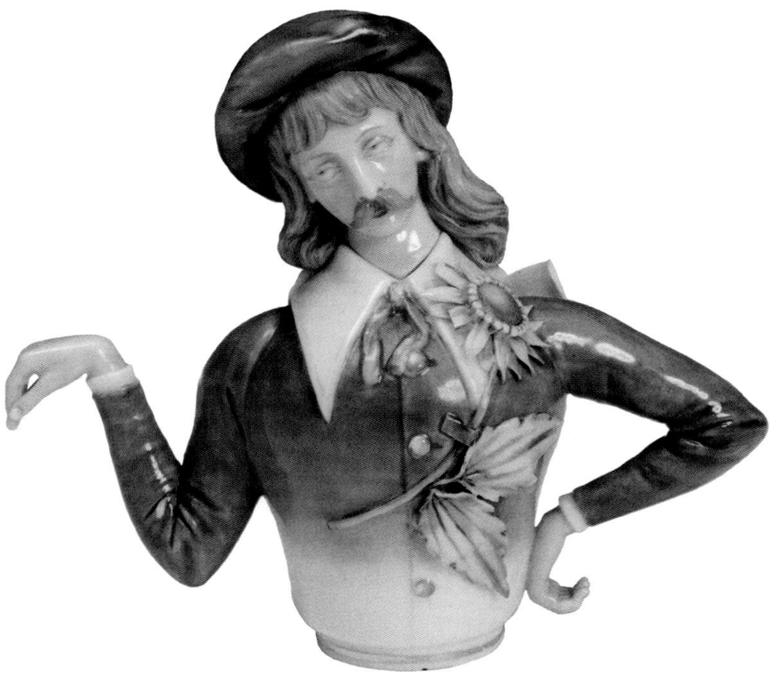

word in French: teapots and teacups were types of urinals, and "having some tea" was slang for a sexual encounter.[93] In English, too, "tea" was slang for urine.[94] In his 1891 study *Perversions of the Sex Instinct,* Albert Moll lists "handle" as one of the typical nicknames of inverts or Uranists, who stood like teapots or teacups with hand or hands on hips.[95] Decanting its decadent drink, then, Hadley's little teapot pours out a body fluid of questionable identity.

Manual secretions were sexually charged, not least because clammy hands were considered a symptom of the (apparently self-lubricating) masturbator.[96] If difficult to identify, the "mesmeric, pleasure-giving fluid" in Teleny's fingers marks him as possessed of one of the most special types of hands—"the magnetic hand, which seems to have a secret affinity for your own; its simple touch thrills your whole nervous system, and fills you with delight."[97] Already a kind of sexual transgression, the illicit charge of an electric touch was a recurring theme in *fin-de-siècle* literature. Such handshakes find their fullest pseudotechnical articulation in Villiers de l'Isle-Adam's 1886 novel *L'Ève future,* in which the American inventor Thomas Edison undertakes to create a sex android for his friend Lord Ewald, one he will endow, notably, with "all the passionate mysticism of Poe's Ligeia."[98] Lord Ewald is introduced to his dream partner through an initial handshake with what appears to be a severed limb. Picking up the proffered appendage, the cold and correct (in other words, closeted) Lord Ewald is shocked and "deeply disturbed" (i.e., thrilled) by its palpable reaction. The mechanical hand responds to his touch with more than warmth and pressure;

it stimulates an "intimate transmission" of electromagnetico-nervous fluid through invisible networks.[99] Edison's android is imbued with a lubricant spread "throughout the magnetic-metallic organism" that is communicated and circulated by a superconducting structure.[100]

The lubricious *main*-to-*main* contact Lord Ewald experiences in the handshake is an explicit avoidance of man-to-woman contact—Ewald comes to Edison disgusted by women to the point of suicide. More to the point, the *main*-to-*main* contact is also a vicarious indulgence of man-to-man contact, one that changes his life. Awakened by the shock of the first handshake intruding upon his hermetic isolation, Ewald finds himself situated in a network of potentially explosive drive-forces.

The osmotic quality of the handshake, wherein membranes are fantastically crossed and the boundaries of the skin, the self, become permeable, facilitates what Merleau-Ponty describes as the sensible union of "the 'inside' with the 'outside.'"[101] The erotic efflorescence that occurs at the moment of the meeting of touches allows the opposition of container and contained to melt dreamily away. There are risks attached to this intoxicating exchange of body fluids: that one might lose or forget oneself, in public, say. Edison's hand, and later his android, functions, then, as a sort of masturbatory safety valve for Ewald, as a reassuringly closed-circuited automatic relay for the drive energy released through *main*-to-*main* (and fantasized man-to-man) contact. Delacroix dreamed of such an instrument of instruments: "I dreamed, this very night, that [George Sand] gave me a present of a little hand made of a delightful substance, which resembled flesh but was in fact no such thing. It was a kind of toy, a trinket to put in one's box of curios; furthermore this hand responded to pressure, was endowed, in short, with sensitivity, and I stopped there with the hand of my dream. When I awoke, I said to myself that a jewel like that would be very handy on journeys."[102] Stopping with remarkable discretion before anything untoward happens, Delacroix's dream hand is decently detached from the painter and from any other person, and can, like Edison's android, be put in a box when not in use.

It is not surprising that in Villiers's text the British subject's emancipation occurs at the electric hand of an American inventor. Since the later eighteenth century, as James Delbourgo has demonstrated, electricity, liberty, and fraternity had been linked in the American imagination. Electricity, like sentiment, was known to be "a communicable property of proximate bodies," and was considered a galvanizing political force.[103] The homosociality it channeled is celebrated in Whitman's body electric and other utopian visions that, significantly, transcend national boundaries.[104] A powerful force, this electricity was not without danger: if political, electricity was also highly erotic. A metaphor for sexual attraction, electricity was often the subject of quasi-scientific demonstrations, in which sparks would be made to fly between lips or hands.[105] Currents could run through couples and crowds, overcoming divisions, building communities or romances, creating new constellations. Associated with the origin of life itself in the scientific world and in texts like Mary Shelley's *Frankenstein,* electricity was linked with the creation of new selves;[106] in *Song of Myself* Whitman describes an electric touch, "quivering me to a new identity, / Flames and ether making a rush for my veins."[107]

Such electric feelings were more than mere metaphor. In the eighteenth century a connection between nerve fluid and "electric fluid" was posited, and theories linking electricity and nervous conduction circulated. Experiments on frogs and electric fish were practiced throughout Europe.[108] In the nineteenth century Italian and German scientists claimed to have found "animal magneto-motors" or "organs of animal magnetism" in the "elliptical, whitish corpuscles" of the "sensible nerves of the fingers." The Italian biologist Pacini compared these "magnetic" corpuscles of the hand to "the electrical organs of the torpedo"[109]—that is, in nineteenth-century terms, the electric eel.[110]

Curiously, we find in the history of the eel, as in that of the human hand, an intertwining of related issues: evolution, perversion, and electricity. This slippery creature's transgressions were manifold. Resembling snakes, fish, worms, and microscopic organisms, eels were popularly renowned for their uncanny ability to cross boundaries, not only, potentially, between the sexes, but between fresh and salt water, and between water and land.[111] Eels are also famous for their seeming ability to cross the boundary between life and death, as they will keep writhing and twitching after death and even, occasionally, after being cooked.

The sex and sexuality of eels had long been held mysterious. An eighteenth-century translation of the Greek poet Oppian's account of eels' reproduction conjures up some of the creatures' enigmatic fascination:

> Strange the Formation of the *Eely* Race,
> That know no Sex, yet love the close Embrace.
> Their folded Lengths they round each
> other twine,
> Twist am'rous Knots, and slimy Bodies joyn;
> Till the close Strife brings off a frothy Juice,
> The Seed that must the wriggling
> Kind produce.
> Regardless They their future Offspring leave,
> But porous Sands the spumy Drops receive.
> That genial Bed impregnates all the Heap.
> And little *Eelets* soon begin to creep.[112]

Aristotle's view that the creatures were spontaneously generated out of mud had not been much advanced by the Enlightenment, and this ancient image of their slimy, twining embraces sloughing off seeds was as compelling as any contemporary one.

In 1876 the young medical student Sigmund Freud spent two months dissecting four hundred eels; he published his findings the following April in the *Minutes of the Imperial Academy of Sciences in Vienna*. In his article, Freud announces that he will attempt to shed light on "the centuries-old, still-unanswered question" of the sexual organs and reproduction of eels. Although ovaries had been found in eels, testicles had not been observed, and Freud's dissections turned up no conclusive evidence.[113] The existence of testiculate eels aside, there was more at stake in the generation of eels than might first meet the eye.

If the eel was a scientific puzzle and a philosophical bone of contention, the polymorphous perversity of the quixotic creature was popularly embraced. In 1779 the pseudonymous "Lucretia

Lovejoy" published an erotic *Elegy on the Lamented Death of the Electrical Eel, or Gymnotus Electricus.* Claiming that the electric eel "Hath been proved by those who dive deep—into the Secrets of Nature, to be That sinful Serpent—that seduced Eve," Lovejoy raises it to mythic status. Significantly, the ambiguous nature of the creature was key to its continuing popularity: the electric eel and its "Extraordinary Parts" were "of such uncertain Quality" that "It was all Things, to all Men, and all Women." It traversed and transcended class boundaries—"It filled the Hands of a Peasant's Wife, and slipped through the Fingers of a Queen"—and undid species hierarchy: "No Creature in Nature was ever so coveted; not Man excepted."[114]

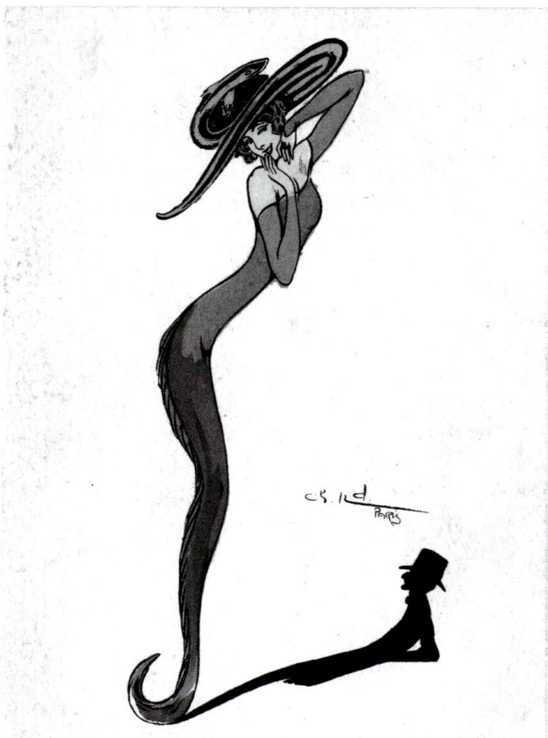

75
"L'anguille," c. 1910, postcard, 14 x 9 cm.

In the *fin de siècle* the eel came to function as a slippery yet handy little figure—far from straightforwardly phallic—of both pleasure and perversion in certain circles. In Jean Lorrain's 1893 story "Les yeux glauques" one character remarks that "honest men whose tastes are admissible and allowed by the law" don't appreciate "the impression of making love to a fish." "Don't insult eels," his companion insists; "they're wonderfully flexible."[115] The eel was still linked to the sexually ambiguous body in a French postcard titled "L'anguille" (The Eel) (fig. 75) from around 1910. It depicts a creature half-eel, half-woman, whose fashionably lithe silhouette snakes upward, interrupted only by her arms bent like fins. The figure balances on its tail and is topped with a hat made of a curled eel. Its seductive femininity is destabilized by its shadow, which takes the shape of a top-hatted man. The caption—"The Eel. Wriggling, slippery fish of a hot temper, recommended to the neurasthenic"—reads like a dictionary entry and declares with authority that a spicy and sexually dubious dish can galvanize the most jaded palette.

Seafood has long been regarded as an aphrodisiac, but such prescriptions were notably *au courant* at the end of the century. "Fishy" was slang for "effete,"[116] visually as well as verbally, as a cartoon in the *Illustrated Sporting and Dramatic News* (fig. 76) suggests. The "queer fish" at the Westminster Aquarium are presumably Gilbert and Sullivan—the fishes' faces resemble theirs and Sullivan conducted concerts at the Aquarium, a

76
"Queer Fish at the Westminster Aquarium," ***Illustrated Sporting and Dramatic News,*** **1 July 1876, 333.**

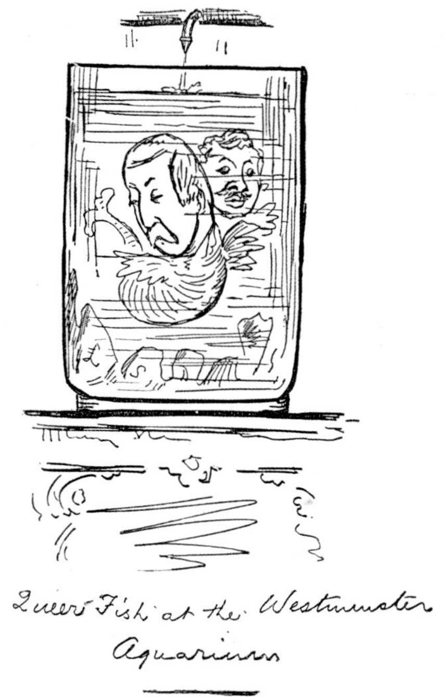

building that housed art galleries, reading rooms, horticultural displays, a concert hall, and a theater. The duo are depicted as hybrid, moustached fish coyly arching their bodies and fluttering their fins beneath a strangely penile fountain. The hothouse world of the theater is imaged as a domain in which pelagic perverts can thrive.[117]

Eel imagery helped Melville render palpable the homoeroticism of the handshake between men in *Moby-Dick*. When Ishmael's fingers, dipped in spermaceti, suddenly feel "like eels" and begin "to serpentine and spiralize," he finds himself overcome by "a strange sort of insanity" and starts "unwittingly squeezing [his] co-laborers' hands." Even after he becomes conscious of his strange activity he persists, "looking up into [his companions'] eyes sentimentally; as much as to say. . . . Come; let us squeeze hands all round; nay, let us all squeeze ourselves into each other."[118] This squeezing "into each other" suggests both bodily and mental merger. Wilde's later descriptions of homoerotic handshakes likewise focus on the intimate exchange of identities that hands' lubricity allows to be rendered fluid.

Descriptions of such dangerous and fishy touching abounded in contemporary literature and memoirs. In Lorrain's *Monsieur de Phocas,* the handshake of the duc de Fréneuse takes on the "lax gentleness" of an "elusive embrace" after he has been introduced to the world of sexual inversion by the artist Claudius Ethal, his "cold and fluid" fingers "gliding between yours like an escaping serpent."[119] Contemporaries recalled that Robert de Montesquiou's hands squeezed and circled "like an infinite number of cabalistic serpents."[120] Sargent liked "little snakes—to watch," and didn't "mind holding them"; his own hands were said to "twist and move all the time."[121] His painting of two octopi lying on a boat deck (fig. 77), their glistening tentacles twining in blind confusion and abandon, captures the polymorphy and merger facilitated by such primal feeling: "Their folded lengths they round each other twine, / Twist am'rous knots, and slimy bodies joyn." Perhaps all sea creatures have this faintly semeny sheen. Michelet described seawater itself as "whitish and somewhat viscous"; according to him, submarine plants and animals are coated in its mucous substance, giving them a "gelatinous" look, "sometimes fixed and sometimes tremulous."[122]

Jacques-Émile Blanche's hands were "as slippery as though he lived in the water," so that

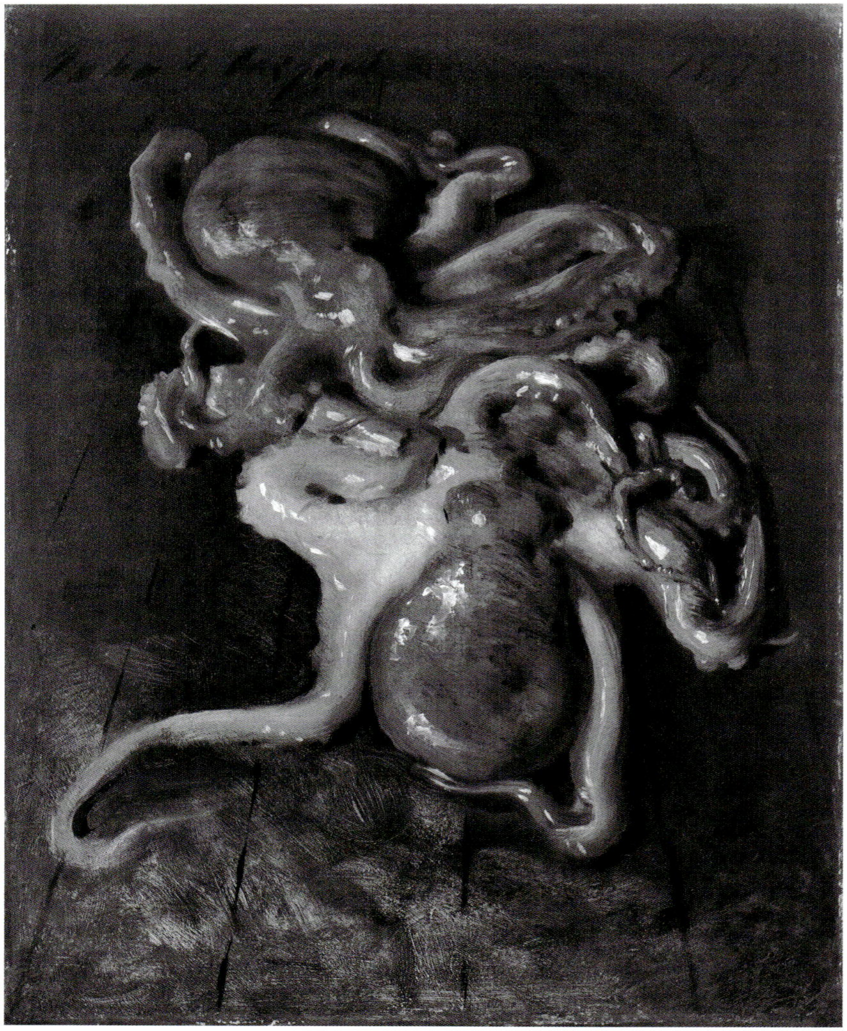

77
John Singer Sargent, *Two Octopi*, c. 1875. Oil on canvas, 40.6 x 32.1 cm. Private collection.

shaking hands with him felt like "shaking the fin of a fish,"[123] and the duc de Guermantes walked through crowds with "a half-open hand floating like a shark's fin by his side, which he allowed to be clasped indiscriminately by old friends and by the strangers who were introduced to him."[124] All of these fins and primal oozing hands point back in time to a world before sexual differentiation, as Oannes describes it in Flaubert's *Tentation de Saint-Antoine,* an *informe* world where "hermaphrodite beasts" slumbered in dark waters and "finger, fin, and wing were confounded."[125]

These slippery hands could lead to sticky situations. Imagine the difficulty of relinquishing a "voluptuous palm that clung like a limpet."[126] Creatural rather than mechanical appendages were not so easily closeted and contained as Delacroix's hand in his box of curios. A dream of Ruskin's, in which he showed his cousin Joanna "a beautiful snake," illustrates this perfectly: "I told her [it] was an innocent one; it had a slender neck and a green ring round it, and I made her feel its scales. Then she made me feel it, and it became a fat thing, like a leech, and adhered to my hand, so that I could hardly pull it off—and so I woke."[127] Ruskin's serpent doesn't respond to his female cousin's touch; his own fingers, on the other hand, have a strange hold over it.

This anatomy of the electric feel and Ruskin yanking off his snake might seem to have taken us far from the delicate brushing of floral hands. The atavistic hermaphroditism of surreptitiously semen-secreting hands and the fact that the lubricant circulating in Edison's androids is "attar of roses,"[128] however, offer clues that we have not left the garden of manual pleasure. To begin with, serpents, eels, flowers, phalluses, and fingers have morphological similarities. That "finger" was slang for penis in French should come as no surprise.[129] More intriguingly, Darwin tells us that in "a multitude of cases" the floral labellum is shaped so as "to render the flower tubular"[130] (fig. 78), like the finger of a glove or a penis. According to Ruskin, this form of labiate flowers betrays "the strange reflex" of their "serpent nature."[131] In *Teleny*, the flower-serpent-genital analogy dominates the protagonist's description of his experience of an orgy at a painter's studio: "it seemed to me as if I were in some rank jungle" where "gorgeous, venomous snakes cluster together and look like bunches of variegated flowers."[132] The *tableau vivant* created by this congerie of fellows figures the mesmeric charms of flowers and serpents as sprung from the same primeval or genital force.

A similar constellation of creatures—hands, flowers, snakes, and art—appears in a late nineteenth-century advertisement for a watering hose (fig. 79). In this bizarre image a hand and serpent combine to make an unusual contrivance for watering flowers. The phallic snake, tumesced with fluid and ready to saturate the garden, is choked off by the "sphincter grip" of a controlling hand; any release will be double and sexual.[133] Evocative of such homoerotic sculptures as the *Laocoön* and Frederick Leighton's 1877 *Athlete Wrestling with a Python*, this illustration showcases another fantasy hand, one made modern and utilitarian: it is "practically indestructible" and capable of wearing out "ten ordinary hoses." Sargent's much later *Hercules and the Hydra* (fig. 80), part of his ceiling cycle in the Boston Museum of Fine Arts, shows Hercules in a similar pose, gripping one of the heads of the hydra, a water beast whose steamy exhalations and coiling limbs create the humid, hothouse atmosphere of the painting.

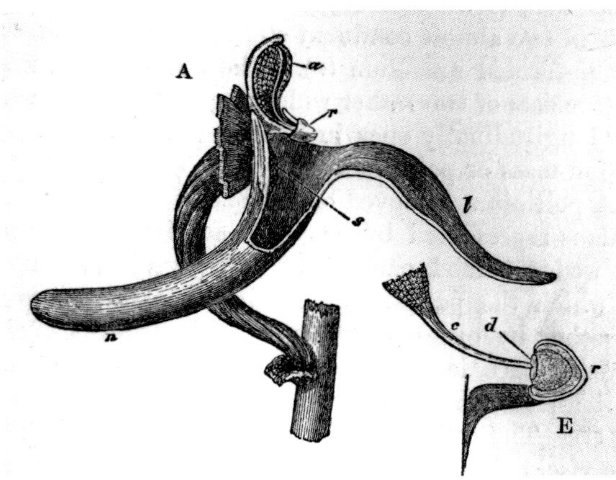

78
Detail of "Orchis mascula," in Charles Darwin, *The Various Contrivances by which Orchids are Fertilised by Insects*, 2nd ed. (London: John Murray, 1877), 8, fig. 1.

Proust made the connection between flowers and marine creatures explicit in *À la recherche du temps perdu*. Just before drawing the famous comparison between the fertilization of the invert and the orchid, the narrator interrupts his chain of thought with this unexpected outburst: "Jellyfish! Orchid! When I followed my instinct only, the jellyfish used to revolt me at Balbec; but if I had the eyes to regard them, like Michelet, from the standpoint of natural history and aesthetics, I saw an exquisite blue girandole. Are they not, with the transparent velvet of their petals, as it were the mauve orchids of the sea?"[134] Evidently for Proust the primordial, formless, rather abject underside of invert sexuality needed to be acknowledged before the sublimatory and classifying rhetoric of floral sexuality could prevail.[135] This felt imperative, while reenacting the evolutionary movement of

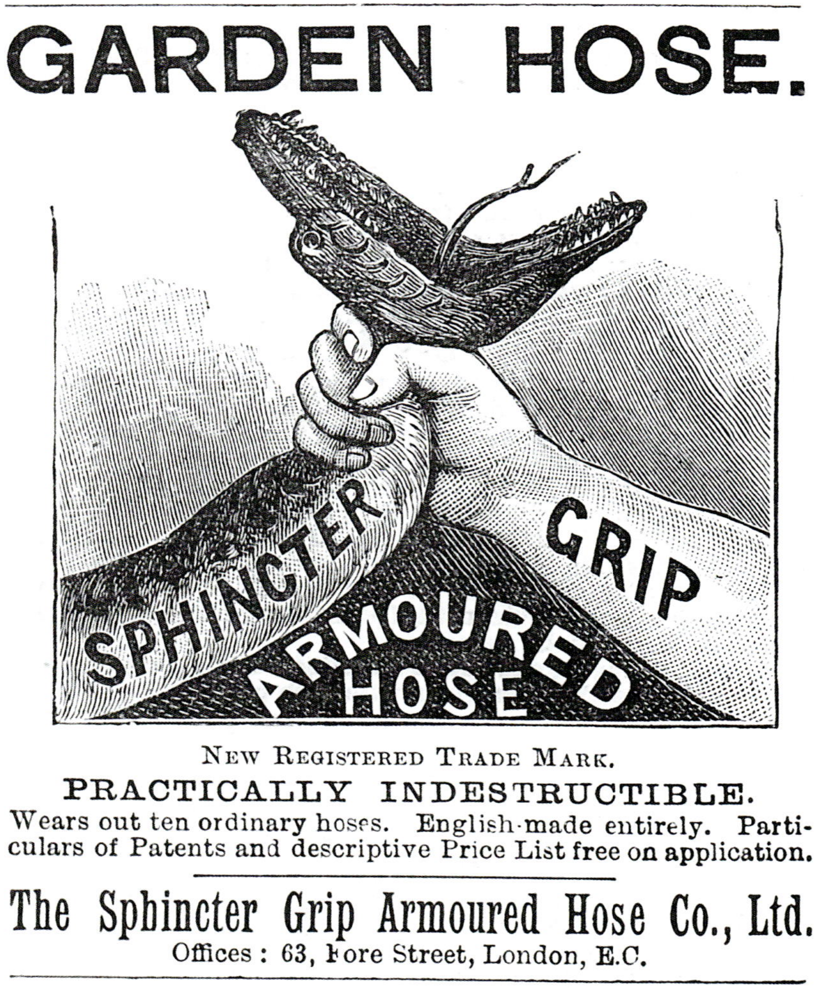

79
Advertisement for sphincter grip armoured hose, in *Garden-Work for Villa, Suburban, Town, and Cottage Gardens*, 29 June 1889, after 307.

plant life from water to land, was perhaps predicated on floral form itself: Bataille explains that while "the visible parts [of the flower] are nobly elevated," its "ignoble and sticky" roots "swarm" and "wallow" "under the surface of the soil," and that it is "impossible to eliminate an opposition as flagrant as the one that differentiates stem from root."[136] The *informe* quality of the underside of flowers can be rivaled by the abjection of liquid—marine rather than floral—hands (Flaubert, as we saw, locates his finger-fins in an *informe* world in the *Tentation de Saint-Antoine*) that dissolve the boundary between inside and outside.

Hands simultaneously floral and serpentine or marine can be found in the art of the period; they tend to be as sexually ambiguous as species-unspecific. In Émile Gallé's glass *Hand with Seaweed and Shells* (fig. 81), for example, sea anemone, hand, and waterweeds are conflated into one hybrid creature. Swirls of liquid blue, yellow, and green glass become branches of seaweed and wavy fingers in an object at once transparent and opaque. Although this has been labeled a female hand, Philippe Thiébaut argues that it is the artist's own; at any rate, a male hand was cast from life to provide the model for the glass piece.[137]

Unsurprisingly, such uncertain appendages are also to be found in Sargent's oeuvre; Robert de Montesquiou professed his horror at the painter's "starfish" hands.[138] Sargent's portrait of Jacques-Émile Blanche (fig. 82)—significantly both a fellow painter and invert[139]—pictures the subject in the dark and slightly swampy atmosphere of a primitive garden. A seemingly boneless arm, sprouting a small flower on the tiniest tentacle, grows liana-like around his torso, as though intent

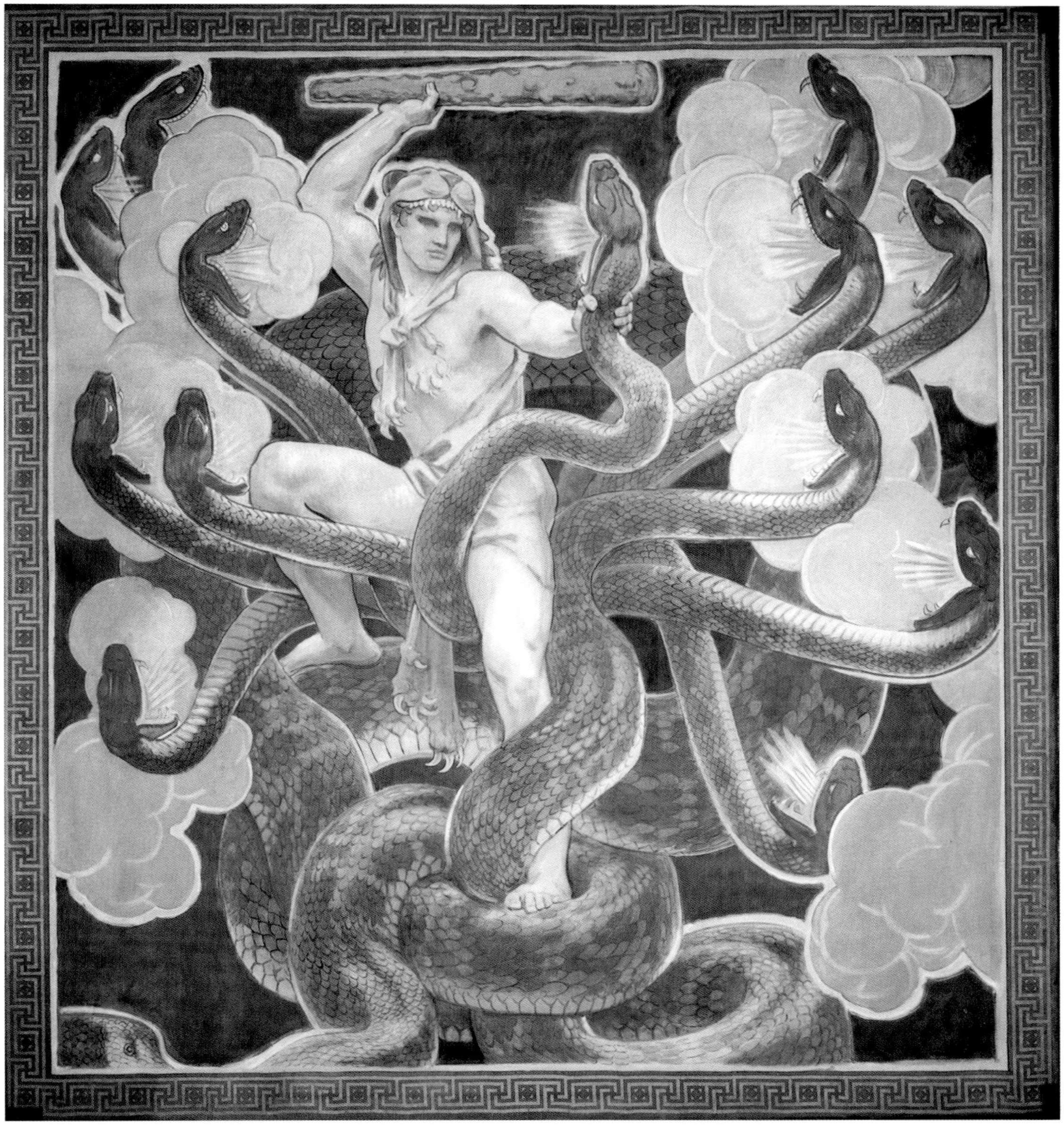

80
John Singer Sargent, *Hercules and the Hydra,* 1922–25. Oil on canvas, 347.98 x 317.5 cm. Museum of Fine Arts, Boston. Francis Bartlett Donation of 1912 and Picture Fund, 25.647.

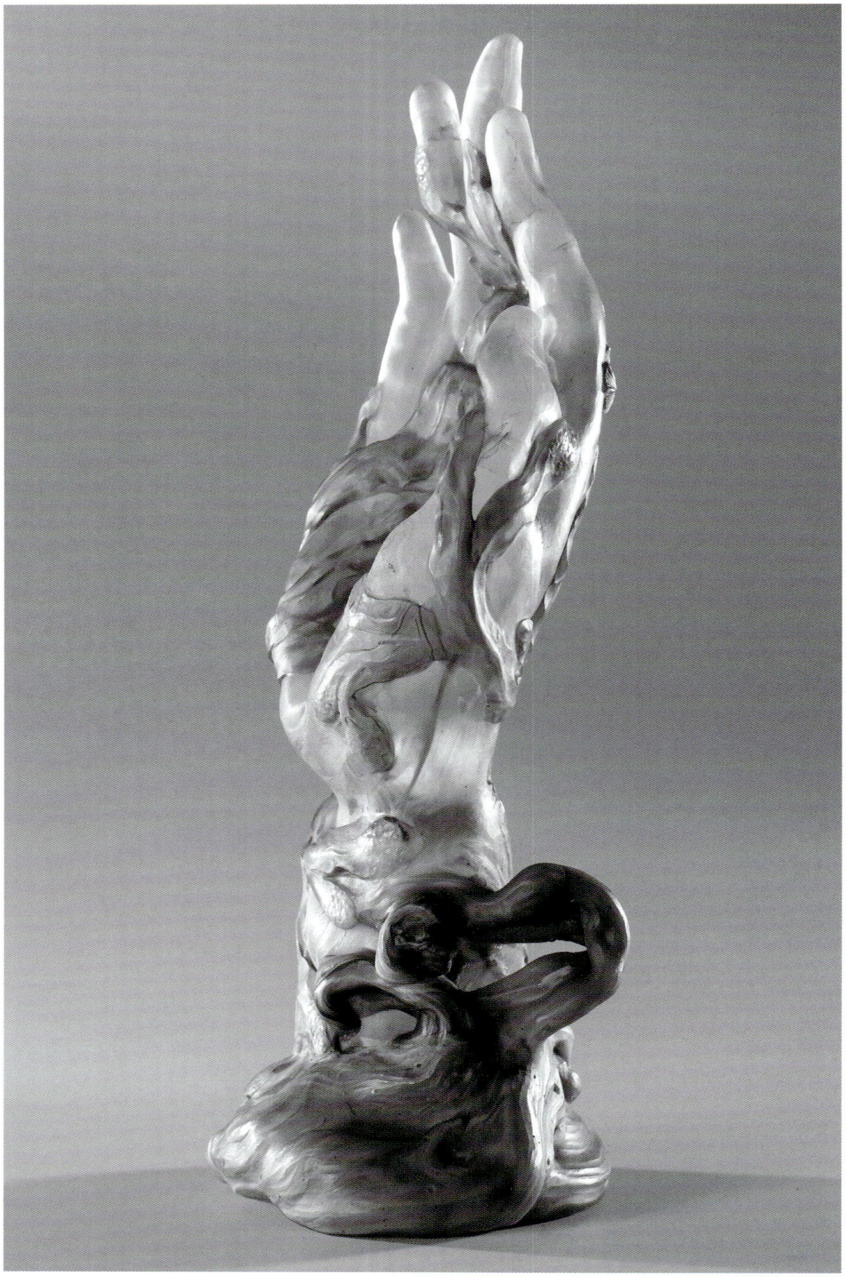

81

Émile Gallé, *Hand with Seaweed and Shells,* 1904. Glass modeled under heat with inclusions of metallic oxides, veins, applications in low and high relief, and wheel engraving, 33.4 x 13.4 cm. Musée d'Orsay, Paris.

on binding him to the tree. At the same time, this regressive, thumbless vegetable limb reads as a snaky or eely tempter pointing to the *main*ly fruits the garden offers, the deep blue-green shadow in the shape of a serpent's head just below it underlining the lure. Folding flower and fish into an invert artist's hand in this primal scene, Sargent links human, vegetable, and painterly creation and vice.

Blanche's own *Self-Portrait with Raphael de Ochoa* (fig. 83), the latter a man who, according to Blanche, had "the same tendencies"[140] as himself, pictures *main*-to-man contact as artistically generative. The long, penile fingers of Ochoa, with their moist red tips, "like the tentacles of an octopus,"[141] slowly snake up the painter's shoulder, the middle one in particular slithering out ahead of its fellow members like a probe. Blanche's cobralike cravat (unlike the flowery one he wears in Sargent's portrait) reinforces the serpentine quality of the encounter, which takes place in a room filled with pictures and in front of a canvas toward which Ochoa's touch propels the painter.

For Sargent, floral and marine metaphors of art production were related, but not identically connoted. Sargent did not care much for Blanche, who was a rather slimy creature all round.[142] A

82

John Singer Sargent, *Jacques-Émile Blanche*, c. 1886. Oil on canvas, 81.9 x 48.9 cm. Musée des Beaux-Arts, Rouen.

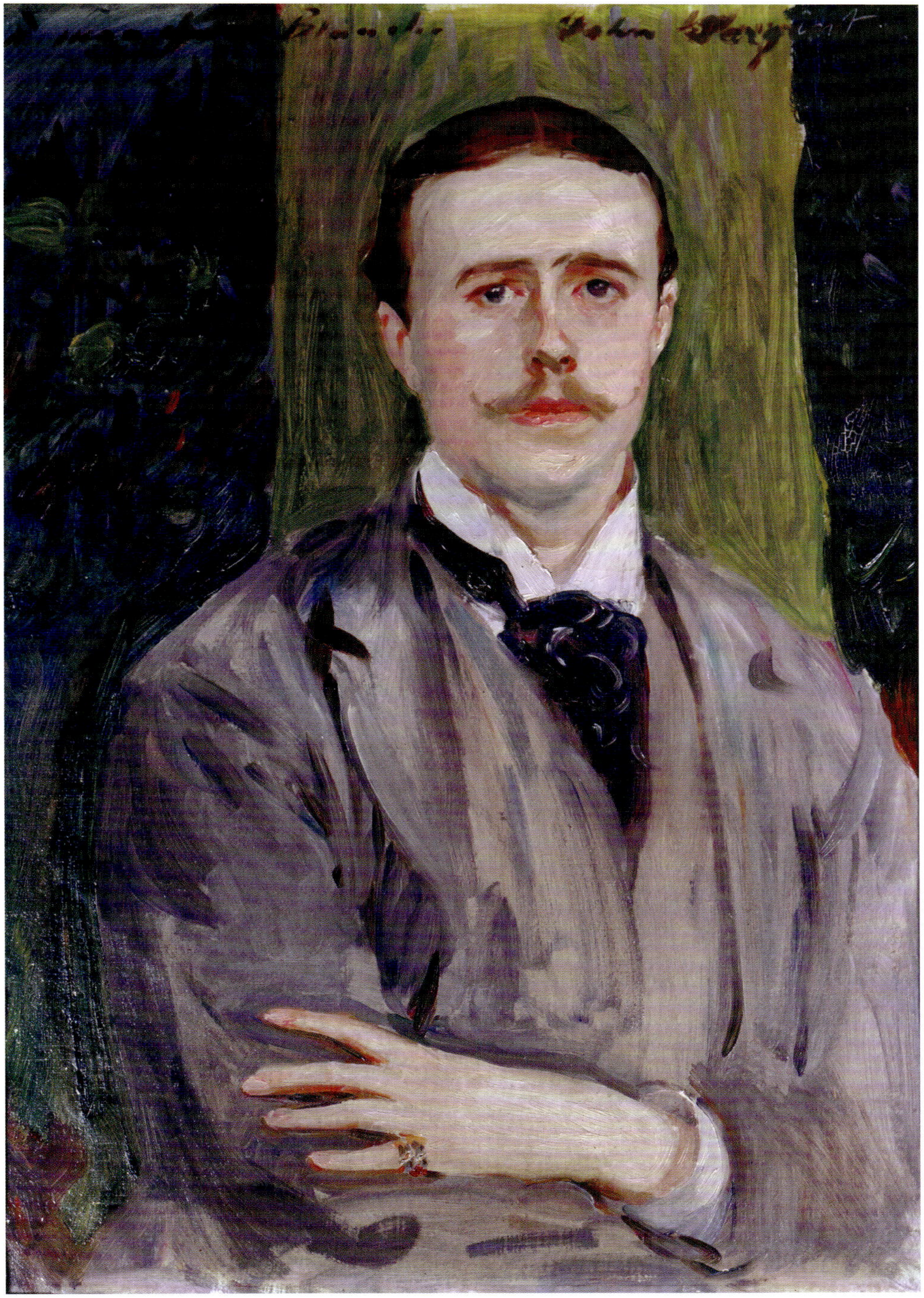

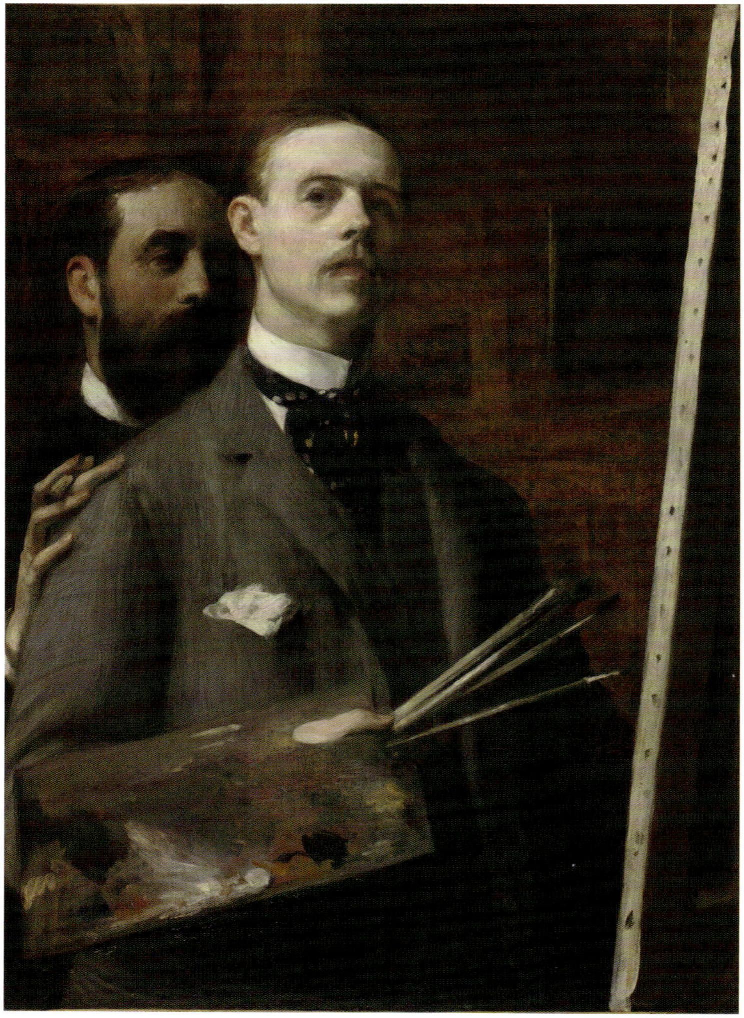

83

Jacques-Émile Blanche, ***Self-Portrait with Raphael de Ochoa,*** 1890. Oil on fabric, 99.2 x 71 cm. The Cleveland Museum of Art, bequest of Noah L. Butkin, 1980.230.

jellyfish with a sting, he cut his former intimate acquaintance Oscar Wilde when Wilde came to Dieppe after his release from prison and was not afraid to out others: he informed Clive Bell that Sargent's sexual exploits "were notorious in Paris, and in Venice positively scandalous. He was a frenzied bugger."[143] The moist, lugubrious atmosphere of Sargent's portrait of Blanche may attest to his rather dim view of the Frenchman.

A pastel of Paul Helleu in the British Museum (fig. 84) hovers between watery and floral realms. The drawing shows the artist lying in a field. With his jacket pulled open to reveal his shapely bottom, his long legs flow amphibiously into their lea-ginous element (the meadow here conforms to Michelet's archetype, "a small undulating sea"),[144] while with a merman's supple dolphin kicks he swells the bottoms of his trousers streaming in the current. The grasses gently billow and swell out from the body of this eely or grass-snaky creature like waves, as if the hybrid nature of this breast-*and* butterfly-stroking artist effected the dissolution of boundaries between ostensibly distinct elements. The musician Charles Loeffler recalled Sargent's "joy" over "the mystery of the propelling power of a little snake in the grass,"[145] and we can sense the painter's delight in this pastoral vision. Following the undulating line of his body from the tail up, we find Helleu drawing a peacock, whose feathers almost blend into the blue-flowering bush above;[146] the artist's own head similarly merges with the dark shrubbery beyond him, which is scored with marks that read as Oriental characters and allude to the Eastern proclivities of the picture's subject. The yellow scribbles in front of Helleu's chest might be a page of his sketchbook,

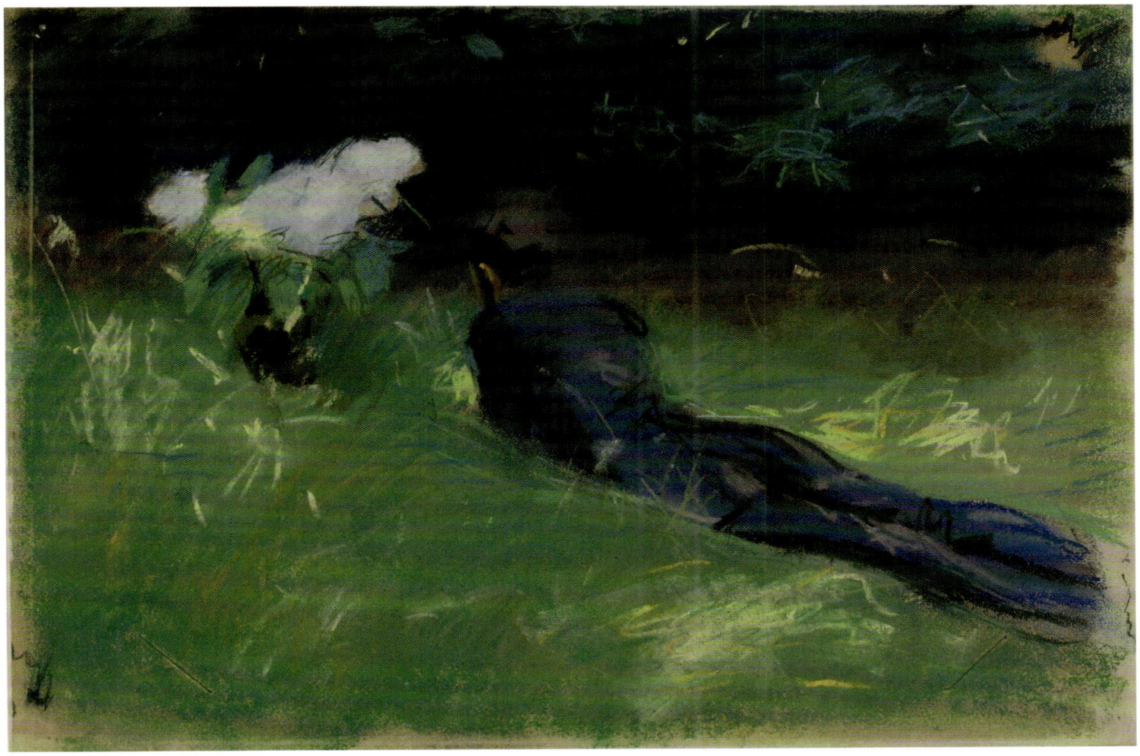

84
John Singer Sargent, *Paul Helleu Lying in a Field*, c. 1889. Pastel on paper, 26 x 37.7 cm. British Museum, London. Presented by Mrs. Ormond and Miss Sargent, 1959, 0102.4.

his hands, or blades of grass, and with this artful move Sargent conflates medium, instrument, and vegetable model. In this drawing, then, Helleu enacts the move from water to land, from potential abjection and dissolution to art.

This tension between abjection and art, between marine and floral worlds, is present to a lesser extent in the extant solo portrait of Robert Louis Stevenson by Sargent (fig. 85).[147] Painted with small, thin, tubular strokes of fluorescent peach and marine blues, Stevenson's undulant hands pulse with an internal, torpedolike electricity in his dim, aquariumlike interior. If submarine, however, they are equally and uninhibitedly vegetal. The writer's left hand softly droops, curls around the trunk of his leg as the epiphytic finger-tendrils of his right quietly creep round to hang from a cigarette that offers light, support, perfume. Soft, languorous, and succulent, the floral hues of these petaline appendages and the implied motion of their "long fingers waving in the air,"[148] as one contemporary reviewer put it (like flora gently stirred by a slight breeze), suggest the writer's plant identification. Stevenson himself wrote, "the life of the plants comes through my fingertips."[149] His trellislike wicker chair is planted on a rug whose tufts of fur also seem, in this queer if natural light, to bend and wave in the wind like

faint, pale lilac smoke, like an alluring perfume, from the unfurling corolla of his hand.

Floral hands inevitably give rise to pollinator erotics, as d'Albert's outburst in *Mademoiselle de Maupin* indicates: when he touches the protagonist's flowerlike hand, his "soul hangs on his lips, like a bee on a flower, to drink in the honey of his words."[157] The colored zones radiating from Helleu's figure against the earthy brown paper ground of the drawing read both as the petals of this *fleur du mâle* and as the faceted, fascinated vision of the insectlike Sargent. Sargent's desire for merger is attested to by the chromatic tension in the work, which is all yellows and blues ready to fuse into an "am'rous" green,[158] and by the stylistic hybridity of the portrait. Pastels were Helleu's medium; the only ones Sargent ever did were of his French friend,[159] and in this work Sargent has adapted the medium for his floral poetics, the oily dust of pastel mimicking the materiality of pollen perfectly, particularly in the golden, Mallarméan flesh-pollen of Paul's face and hand. And in the portrait's literally engulfing and intoxicating, hypnotic atmosphere, we sense that the artist was drawn toward the enticements Helleu's hand held. With his digitation offering so delectable a display, moreover, it would seem that not only Sargent, but Helleu too, *était fleur bleue,* was romantically inclined.

Sargent, as we have seen, exploits the erotic potential of hands to the fullest: they can be chalicelike, cupping and grasping, or languorously spread open; flaccid in their fingers' oddly boneless drooping or erect in the strain of various contortions. The nature of the desire the painter renders with his hands has everything to do with the contemporary associative intertwinings of human and floral sexuality elaborated in chapter 1. Sargent's hands and flowers are strange horizons, indices of a surface tension linked to the perception and constitution of identity that pervades his oeuvre. The connection between palpation, pollination, painting, and perversion suggested here will be more fully unfolded in the next chapter, which explores the floral and marine touch of "Dr. Dieu," Samuel Pozzi: the founder of modern French gynecology, subject of one of Sargent's most celebrated portraits, and a fellow explorer of the intersections of human and floral sexualities.

3

DR. OCTOGYNECOLOGIST

Art, and beauty . . . do you know what they are? Well, my boy, beauty is a woman's belly, opened up, covered in blood and with forceps inside!

—DR. TRÉPAN, IN OCTAVE MIRBEAU, *TORTURE GARDEN* (1898)

In an essay written around 1895, one of his first on painting, Proust argued that just as a woman who has recently given birth would be astonished by a gynecologist's "explaining to her what has taken place in her body, by describing to her the physiological process of the act she has had the mysterious strength to accomplish," so would Chardin, the multiparous, feminine artist, be "astonished" to learn what and how he has created.[1] Thus, well before *À la recherche du temps perdu,* where he would develop the idea, Proust indirectly proposed gynecology as a model for exploring and understanding the origins of perception and art. If Chardin had no need to understand *how* he saw and created, the modern artist, like Proust himself, did. And if a woman had "no need to know medicine to bear a child,"[2] the situation differed for the male invert.

In order to create his chronicle of the descendants of the Sodomites and Gomorreans, Proust, like his narrator M., had to become a gynecologist, for the invert's secret treasure is like the womb's, homosexuality like pregnancy: "Until that moment, in the presence of M. de Charlus I had been in the position of an unobservant man who, standing before a pregnant woman whose distended outline he has failed to remark, persists, while she smilingly reiterates: 'Yes, I'm a little tired just now,' in asking her tactlessly: 'Why, what's the matter with you?' But let someone say to him: 'She is expecting a child,' and suddenly he catches sight of her stomach and ceases to see anything else."[3] At the same time, Proust writes, inversion is like a disease, and the artist-gynecologist must therefore also be, in addition to a reader of surfaces, "a surgeon who beneath the smooth surface of a woman's belly sees the internal disease which is devouring it."[4]

In Proust's case we might find biographical determinants for this phantasmatic identification of artist and gynecologist, for instance in his successful and heterosexual brother Robert, author of a manual of gynecological surgery.[5] This would not be so easily accomplished, however, with other proponents of the same theory. Melville, for example, to draft his "systematization of cetology" and write *Moby-Dick,* had "to grope down into the bottom of the sea after [whales]; to have [his] hands among the unspeakable foundations, ribs, and very pelvis of the world,"[6] in other words, to become a laparotomist.

Felicitously, Merleau-Ponty provides a theoretical link between gynecology, art, and perception by envisioning the space where

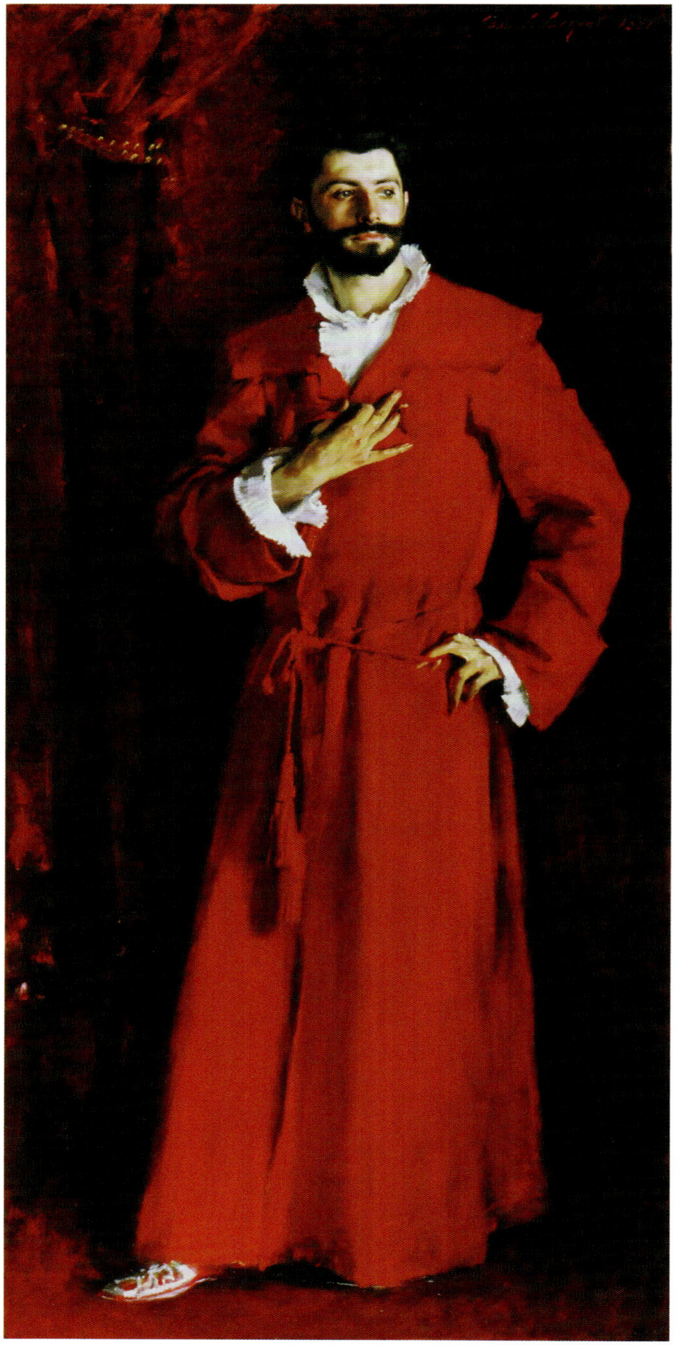

88

John Singer Sargent, *Dr. Pozzi at Home,* 1881. Oil on canvas, 202.9 x 102.2 cm. The Armand Hammer Collection. Gift of the Armand Hammer Foundation, Hammer Museum, Los Angeles, California.

perception and consciousness are "born," which is also the space of the emergence of expression, as a vagina; consciousness, or the eye—essential to the artist—is for the seer an "invagination," a "fold," a "central cavity of the visible."[7] Perceiving the world in one's proper relationship to it and creating art, he argues, require a particular perceptual aptitude: the ability to navigate the horizon of interior and exterior. A "*correct* focusing" allows the seer to discern "pregnancy" *and* opens up the world "as the surgeon opens a body, catching sight, through the window he has contrived, of the organs in full functioning."[8]

As Merleau-Ponty's title "The Intertwining" implies, the visible and the tangible are chiasmatically interleaved.[9] The eye of the modern artist, then, must be able to both palpate and penetrate the visible world, must encompass the two poles of the "conceptual system" Benjamin outlines in *The Work of Art in the Age of Mechanical Reproduction:* it must be the eye of the painter who, conceptually, lays hands on a subject, and that of the camera operator who penetrates it surgically.[10] The implications this dual model of perception has for artistic creation can be elucidated by the investigations of the literal space of creation undertaken by one of Sargent's most alluring subjects: Proust's friend and his brother's colleague Dr. Samuel Jean Pozzi (fig. 88), whose gynecological probings proceeded through touch—his famed "bimanual exploration"—and surgical penetration.[11]

If Pozzi's renown today rests primarily on his seductive portrait in scarlet by Sargent, in his day it was based on his international standing in the medical world, his social success (his first lecture series as chair of gynecology was attended by the comtesse Greffulhe and Robert de Montesquiou, among others),[12] his art collection, his reputation as the lover of a prodigious number of women, including celebrities like Sarah Bernhardt, and his founding of a private sex society called the League of the Rose, whose members "were not only made to answer questions about their sexual preferences but to act them out as well."[13]

Born in Bergerac in 1846, Pozzi came to Paris to study medicine in 1864. Under Paul Broca's influence he branched out into anthropology and comparative anatomy, at Broca's suggestion translating Darwin's *On the Expression of the Emotions in Man and Animals* into French when it came out in 1872. Eventually gynecology became Pozzi's guiding passion, and after completing his training he traveled through Europe and North America to research his nascent specialization. In 1890 he published an influential *Traité de gynécologie clinique et opératoire,* more than a thousand pages long and containing more than five hundred figures; English, German, Spanish, and Russian editions appeared almost immediately, and the French version was revised and reprinted four times during his life.[14] Pozzi opened the first gynecological ward in France at the Hôpital Broca in 1897, and in 1901 he became the nation's first chaired professor of gynecology.

Like Merleau-Ponty, Pozzi stressed the necessity of both sight and touch as investigatory tools for his exploration of the "flesh of the world,"[15] resisting the rule of the speculum. He worked with bare, clean, regularly manicured hands whose sensitivity was legendary; as the auction catalogue for the sale of his art collection after his death put it, "no man knew better than he how to express the sharp, shuddering pleasure of the fingers at the contact of a beautiful, flawless faience."[16] To touch the interior of the living body was, his biographer tells us, a continual "intoxication."[17]

When performing exploratory touch, Pozzi explains in the *Traité,* individual fingers have their special uses: "the index is the most convenient finger" for touching the cervix, for example, and taking account of its direction, "its volume, its form, its consistency, its degree of openness, the state of its commissures."[18] Bimanual exploration, however, which makes use of multiple digits on two hands, is, as Pozzi frequently remarks, among the most precious of the modes of investigation in gynecology.[19] To perform this diagnostic double palpation, the index finger of the right hand is placed in the vagina while "the left hand is placed transversally above the pubis, and the fingers sunk by a soft pressure which pushes the internal genital organs toward the vaginal finger."[20] The investigator proceeds as follows:

> One tries first of all to render an exact account of the position of the uterus in the hypogastric region, then one passes to the lateral parts, and while the abdominal hand depresses the flanks, the vaginal finger goes to meet it by penetrating the vaginal culs-de-sac; thus one thoroughly explores the base of the broad ligaments and the annexes of the uterus abnormally augmented in volume thus. At the same time the sensibility of the parts must be noted; in the healthy state, the pressure on the side of the annexes, the lifting or swaying of the uterus is not painful.[21]

Pozzi's illustration of bimanual palpation (fig. 89)[22] could be a demonstration of the reversibility of hands—their ability to be both perceptual instruments and part of the "flesh of the world," both touching and touched. The doctor's hands are here literally probing the lining of the visible and the invisible, interior and exterior, the fold in the flesh of the world where, for us, perception and being are born. Merleau-Ponty characterizes the relation between the visible, tangible world and our perception as "an intimacy as close as between the sea and the strand,"[23] and Nicolet's drawing seems informed by an analogous understanding: above the doctor's right hand sunk into the flesh of the woman/world, the folds of a sheet billow across the sandy plane of her abdomen, which stretches out laterally farther than the eye can see. The doctor's hands, petite and delicate relative to the engulfing vulva (Pozzi was proud of their fineness), and the amputated, tree-trunk-like thighs, seem to confirm Merleau-Ponty's point that for the seer the visible is always "a cross section upon a massive being."[24]

After touch, Pozzi's other mode of exploration was surgical. His professional success rested in no small part on his innovations in laparotomical procedure. He performed surgery of the abdominal cavity not only in the hospital and before students but also before public audiences and in private homes.[25] Pozzi's biographer describes his entries into the abdominal cavity as the realization of a long-cherished dream: "To open human abdomens, to work there without [causing] pain and to close them again without death ensuing. . . . To see at last, living, the interior flesh promised by their elders, until now inaccessible. . . . What extraordinary opportunity!"[26] Via these two exploratory modes, then, Pozzi investigated the space of biological creation, finding there "pregnancy," both in the conventional sense and in the Merleau-Pontian one of a world of possibilities,

89

Charles Nicolet, "Bimanual palpation," in Samuel Pozzi, *Traité de gynécologie clinique et opératoire* (Paris: G. Masson, 1890), 101, fig. 70.

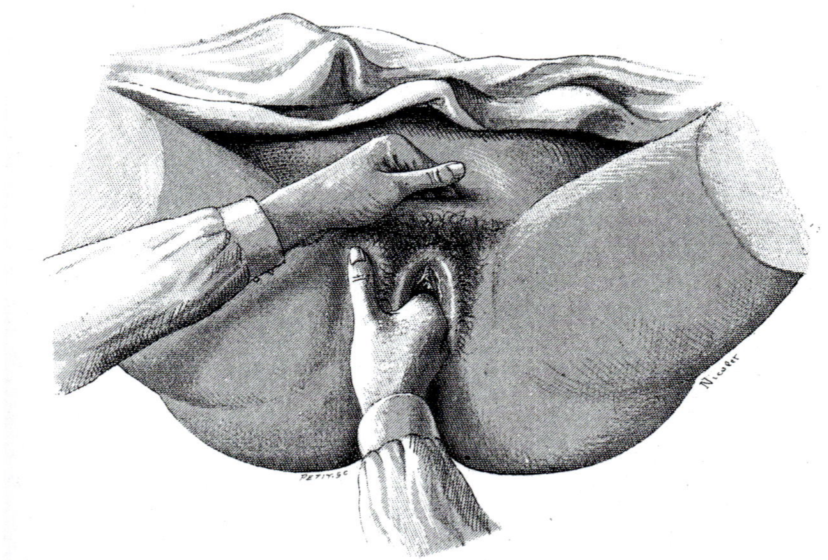

Fig. 70. — Palpation bi-manuelle.

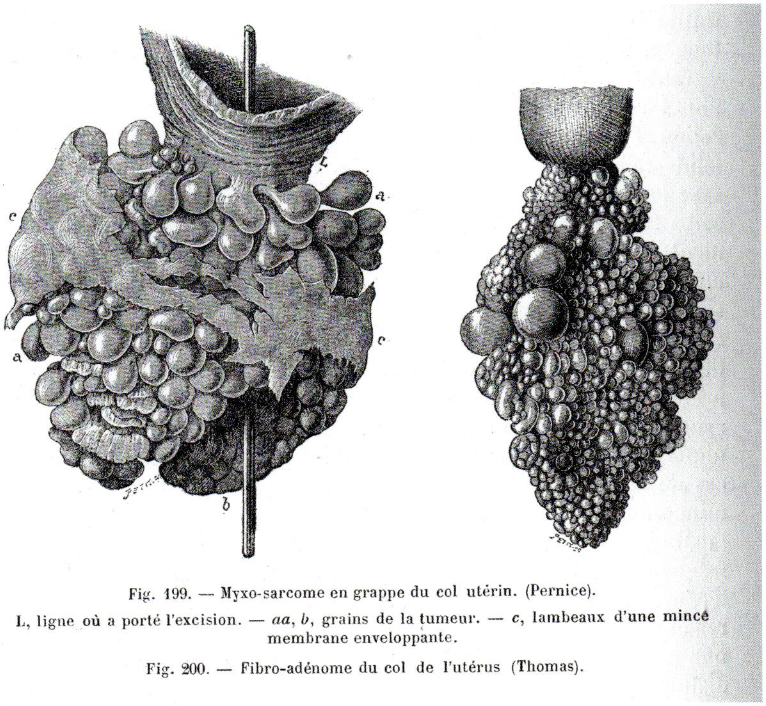

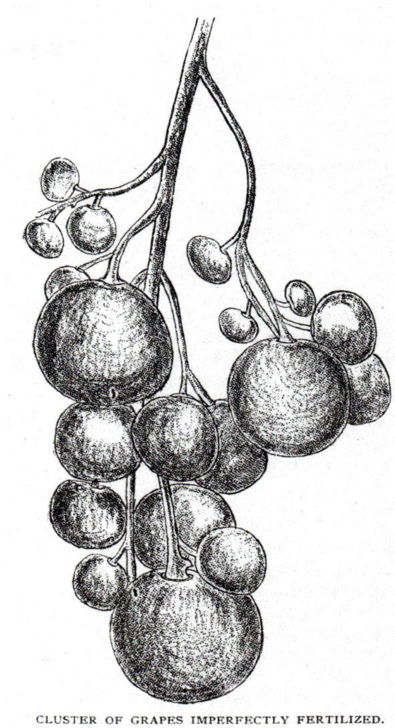

90
"Myosarcoma of the cervix in the form of a cluster of grapes," reprinted from Pernice, and "Fibrous adenoma of the uterus," reprinted from Thomas, in Pozzi, *Traité de gynécologie* (1890), 376, figs. 199 and 200.

91
"Cluster of grapes imperfectly fertilized," in *Vick's Illustrated Monthly Magazine,* December 1883, 381.

and "the organs in full functioning."[27] The sense of a promised land recovered expressed in the quotation above was more than fulfilled by Pozzi's findings.

If menstrual flowers had always bloomed and fallen from the womb, itself conceived of as a flower (fig. 8), Pozzi's explorations revealed a luxuriant garden teeming with life. Nineteenth-century gynecology termed diseases of the genital organs "vegetations," taking its cue from the striking resemblance of diseased genital tissue to various plant forms.[28] Citing a case noteworthy for the way in which a tumor displayed not only the form but also the lifelike color and texture of fruit, Pozzi included in his treatise a case previously observed and illustrated by Ludwig Pernice

92
"Ovarian dermoid containing 3,930 epithelial balls," in J. Bland-Sutton, *Surgical Diseases of the Ovaries and Fallopian Tubes, including Tubal Pregnancy* (London: Cassell and Co., 1896), 65, fig. 38.

93
"Fibro-myxoma of the tubal fimbriae," reprinted from Martin, in Bland-Sutton, *Surgical Diseases of the Ovaries and Fallopian Tubes,* 230, fig. 103.

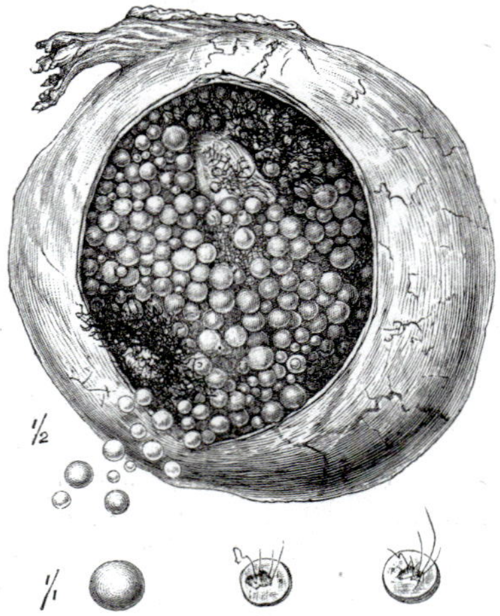

Fig. 38.—Ovarian Dermoid containing 3,930 Epithelial Balls. (Dr. W. Walter's case.)

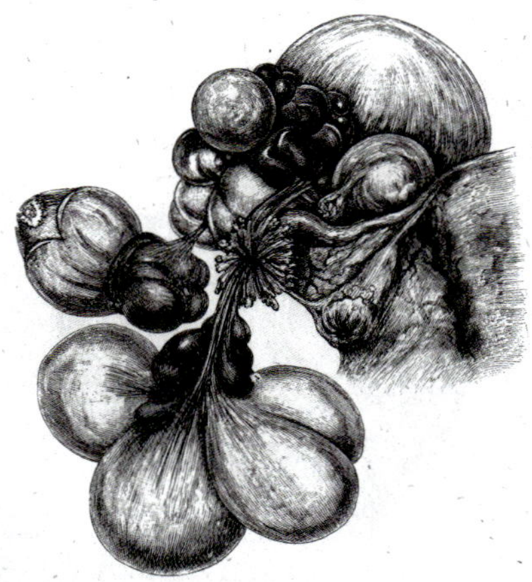

Fig. 103.—Fibro-myxoma of the Tubal Fimbriæ. (After *Martin.*)

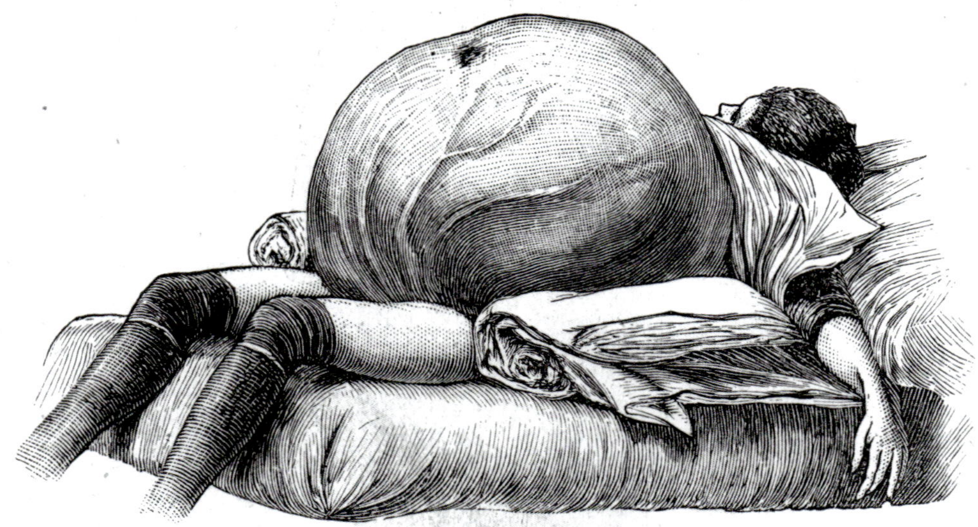

Fig. 50.—Enormous Ovarian Cyst in a Girl Seventeen Years of age ; it contained 78 litres of fluid. (After *Dayot.*)

94
"Enormous ovarian cyst in a girl seventeen years of age," reprinted from Dayot, in Bland-Sutton, *Surgical Diseases of the Ovaries and Fallopian Tubes,* 82, fig. 50.

(fig. 90, left), "of a strio-cellular myosarcoma of the uterus in cluster-form, observed in a nulliparous woman, subject for six months to hemorrhages. The tumor, sprung from the external orifice of the cervix, had the volume of a fist and a half and strongly resembled a bunch of grapes, with violet-colored fruits containing a gelatinous liquid."[29] Another illustration of a fibroid adenoma, reproduced to the right of Pernice's in Pozzi's treatise (fig. 90, right), picks the specimen almost entirely out of its natural habitat, heightening the morphological similitude of carnal and botanical growths and even rendering the bloom on individual grapes. The variation in size of the polyps visible in both examples is common to "imperfectly" fertilized vegetable grapes (fig. 91), according to a contemporary gardening magazine.[30]

Other fruits ripened in women's fertile interior landscapes and were studied by students of uterine botany. The English gynecologist John Bland-Sutton reproduced numerous examples in his 1896 *Surgical Diseases of the Ovaries and Fallopian Tubes*. An image of an ovarian dermoid filled with epithelial balls (fig. 92) resembles an overripe pomegranate split and spilling seeds or another sort of dehiscent pod, while a fibroid myxoma of the tubal fimbriae (fig. 93)—soft tumors filled with jellylike contents springing from the frilly borders of the fallopian tubes—evokes a cluster of plums, grapes, and gourds, their arrangement suggesting the cornucopian excess of baroque still lifes. Ovarian cysts could reach country fair prize–winning proportions. Bland-Sutton reproduces one ripening in a seventeen-year-old's belly (fig. 94): huge and juicy, it looks like a giant peach surrounded by leaflike sheets and springing from twiggy legs; it may also be seen as a large cabbage, with its visible thick, branching veins.

Botanical references in medical texts usually function as simple indicators of size and shape. Pozzi cultivated such vegetable analogies, though, rather more rigorously than most of his peers. More than half of his treatise is devoted to tumors, polyps, cysts, and other physiological luxuriations that, in his words, resemble and grow to the size of hemp-seeds, grains of millet, peas, big peas, hazelnuts, grapes, cherries, walnuts, beans, mushrooms, little pears, apples, oranges, coconuts, cauliflowers, and so on.[31] These "succulent"[32] plants can bloom in isolation, spread like moss, cluster like buds on a stalk, or swell into giant bulbs. The gynecologist's descriptions of these parasitic growths strangely conjure up a pastoral idyll. In the lush garden of the womb, the liquid filling parovarian cysts is as "clear as springwater,"[33] and other cysts overflow with juices ranging from the color of "barley sugar" to "lemon yellow" and "apple green, to the tint of coffee or chocolate."[34] These fluids can also be "milky" or held in "honeycombs."[35]

Honeycombed, mushrooming cysts have a curious affinity with Melville's ambergris, which resembles "fragments of mushrooms growing at the bottom of the sea" or "the honey-comb of bees, fallen from flowery cliffs into the brine," despite being "a morbid secretion" of the "inglorious bowels of a sick whale."[36] Lumps of spermaceti in *Moby-Dick* are fruity, too: squeezed, they "richly broke to my fingers, and discharged all their opulence, like fully ripe grapes their wine."[37] The equation of whales and cystic women was, in fact,

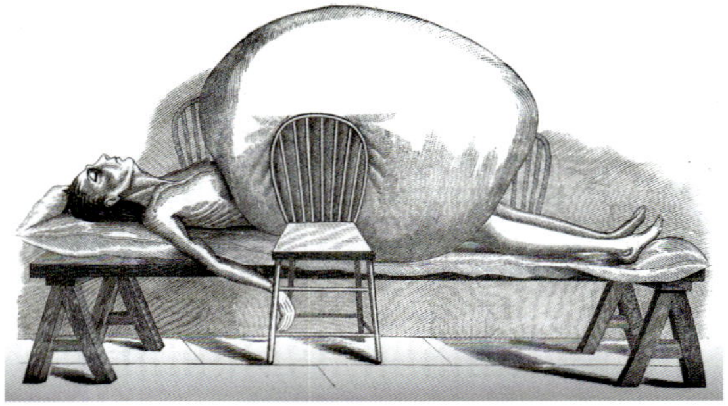

95
"Polycystic ovarian tumor," in Louis A. Rodenstein, "A Case of Enormous Polycystic Ovarian Tumor," *American Journal of Obstetrics and Diseases of Women and Children* 12 (April 1879): 304.

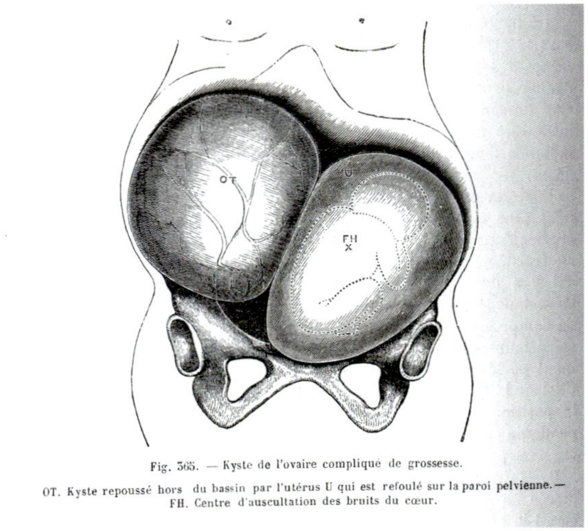

96
"Ovarian cyst complicated by pregnancy," reprinted in Pozzi, *Traité de gynécologie* (1890), 746, fig. 365.

made explicit by Pozzi, who fancied himself a whaler and hunter.[38] In a speech he gave in 1906 he lamented the passing of the age of giant ovarian cysts, which indeed made their hosts cetaceous (fig. 95): they were disappearing like "the whales in the ocean or, in the center of Africa, the gigantic elephants and immense hippopotami, tracked by too many hunters."[39] The largest one Pozzi bagged contained ninety-four liters of fluid.[40]

The fruits of whaling, Melville tells us, "may well be regarded as that Egyptian mother, who bore offspring themselves pregnant from her womb,"[41] and Pozzi, who collected Egyptian antiquities,[42] found a similarly originary hyperfertility in his gynecological investigations. "Not only have all cysts of the ovary a congenital origin, but many of them are of congenital date and may either remain stationary or develop sooner or later."[43] In other words, the genital tissues themselves are possessed of an innate fecundity, not requiring cross-fertilization by an *external* agent in order to create life.[44] The conceptual equivalence of cysts and pregnancy in Pozzi's treatise is encapsulated in an illustration he reproduces (fig. 96), where only the diagrammed outline of a fetus in the swollen sac on the right enables us to distinguish it as the womb. To return to our point of origin, however—the uterine garden—this internal landscape dotted by pavilions,[45] kept fresh by springwater, where the vegetation grows luxuriantly and where the fingers can "promenade," "glide," and "explore,"[46] is a pastoral utopia.[47] Or rather, in the words of Lebert, endorsed by Pozzi, a "plastic heterotopia."[48]

The gynecologist's role in this *hortus inclusus* is that of a gardener: pruning, culling, rearranging, irrigating, taking cuttings, and occasionally hand-pollinating. The procedure Pozzi developed for colpoperineoplasty (reparative surgery involving the vagina and perineum), for example, detaches a piece of the vaginal lining so that it resembles the frilled labellum of a Cattleya (fig. 97). Pozzi's language, always rife with loose vegetable analogy, becomes quite precise when describing his operations. After removing a tumor through the abdomen, Pozzi suspends its pedicle outside of the abdominal walls to ensure its mummification; for some days (the blooming season), he tells us, there "it remains isolated, like a pistil in the center of

97
Charles Nicolet, "Colpo-perineoplasty by flap-splitting," in Pozzi, *Traité de gynécologie* (1890), 541, fig. 305.

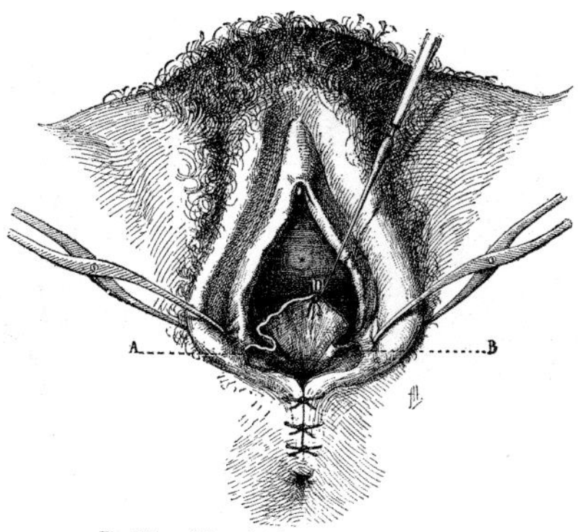

Fig. 305. — Colpo-périnéoplastie par glissement.
Les fils sont serrés ; le lambeau A, B, D est soulevé, prêt à être réséqué suivant la ligne A B.

98
Charles Nicolet, "Suture of the abdominal walls about the pedicle," in Pozzi, *Traité de gynécologie* (1890), 310, fig. 161.

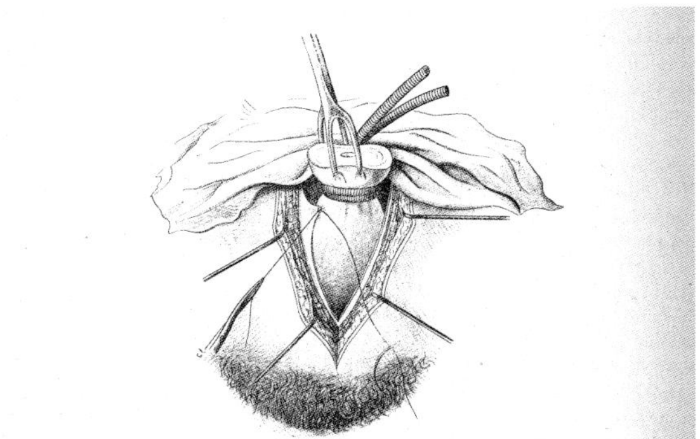

Fig. 161. — Suture des parois abdominales au niveau du pédicule dans l'hystérectomie supra-vaginale. Méthode extra-péritonéale.

99
Detail of "Cypripedium sedeni," in Benjamin Samuel Williams, *The Orchid-Grower's Manual,* 6th ed. (London: Victoria and Paradise Nurseries, 1885), after 256.

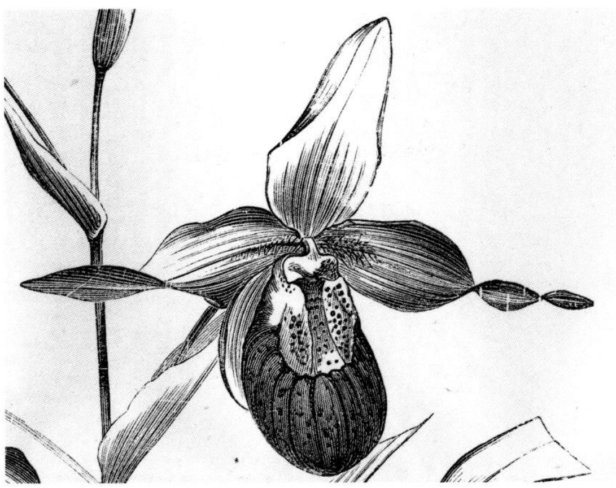

the calyx of a flower."[49] Nicolet's drawing of the procedure (fig. 98), complete with ligatures, clamps, and hooks, might at first suggest that this simile is the hallucinatory product of a decadent and possibly sadistic imagination. When examined with the eye of a botanically savvy, orchid-loving aesthete of the *fin de siècle,* however, the analogy becomes less implausible. The illustration resembles a slipper orchid (fig. 99), the morphology of the flower re-created by the crescent of pubic hair neatly edited to resemble the hairy lip of the flower and the strangely miniaturized sheet that takes the form of twisting lateral sepals.

Central to Pozzi's conception or creation of this pastoral idyll was the notion of the polyp. A primary, polymorphous, and polysemous creature, a polyp is any organism having many feet or footlike processes, including octopi, squid, cuttlefish, starfish, jellyfish; anemones, sea pansies, coral, and other animal-flowers; and, significantly for us, anatomical growths. Polyps transgress species boundaries—they are part animal, part vegetable, sometimes part mineral—and are often able to change form and color.[50] They are also possessed of particular powers of self-renewal: whole organisms, for example, can regenerate from severed limbs. To eighteenth-century thinkers, this faculty of regeneration, a kind of immortality, was uncanny—James Elkins has examined the philosophical quandaries induced by the idea that "Polype 'children' would be 'sons' and 'daughters' of *arms.*"[51] For John Needham and other Enlightenment biologists, polyps offered evidence that forms of life did not have to spring from "parent organisms"; they

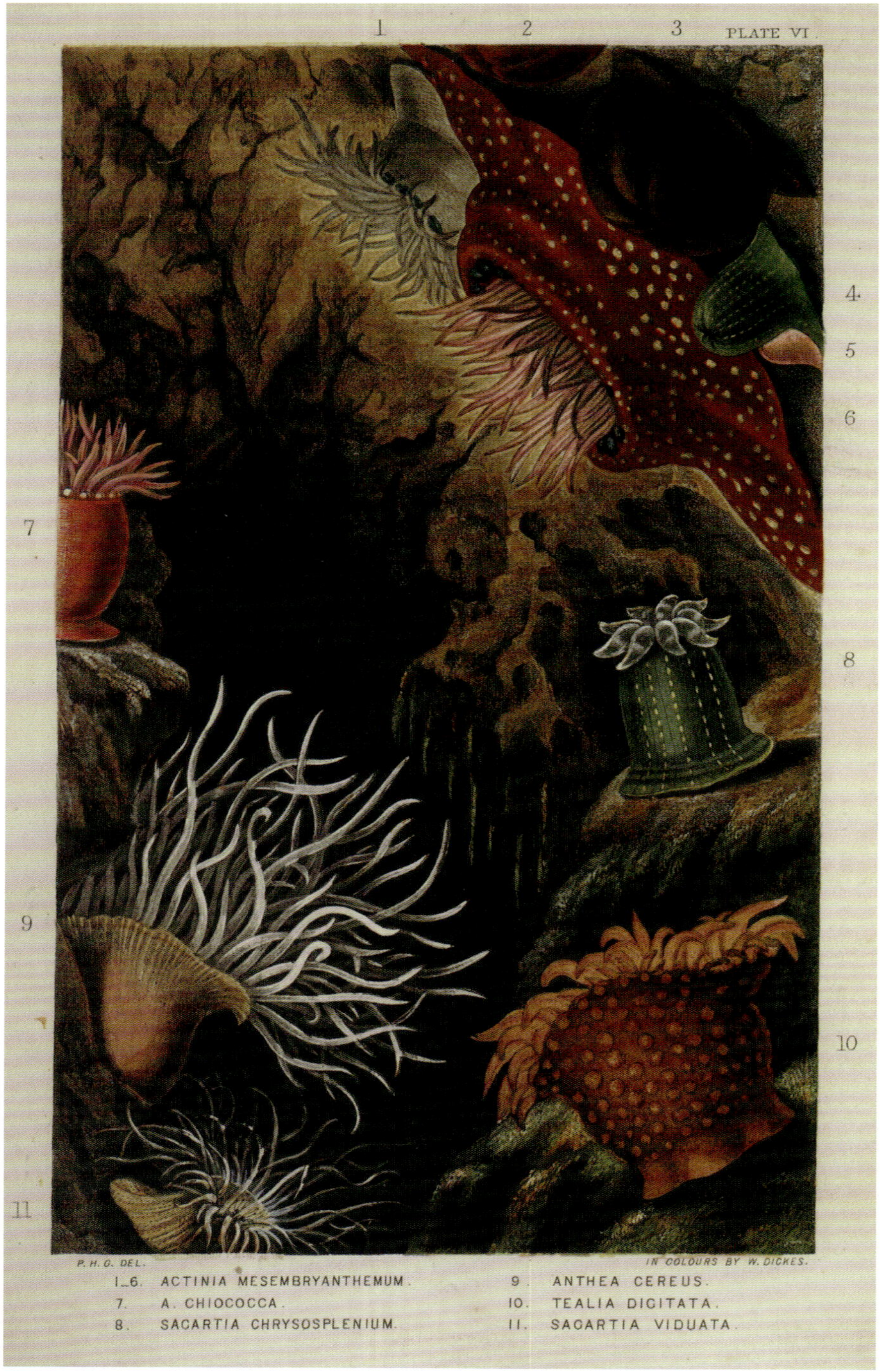

100

"Actinia mesembryanthemum," in Philip Henry Gosse, *Actinologia Britannica: A History of the British Sea-Anemones and Corals* (London: Van Voorst, 1860), plate 6, between 216 and 217. Facsimiles of Gosse's drawings made by W. Dickes.

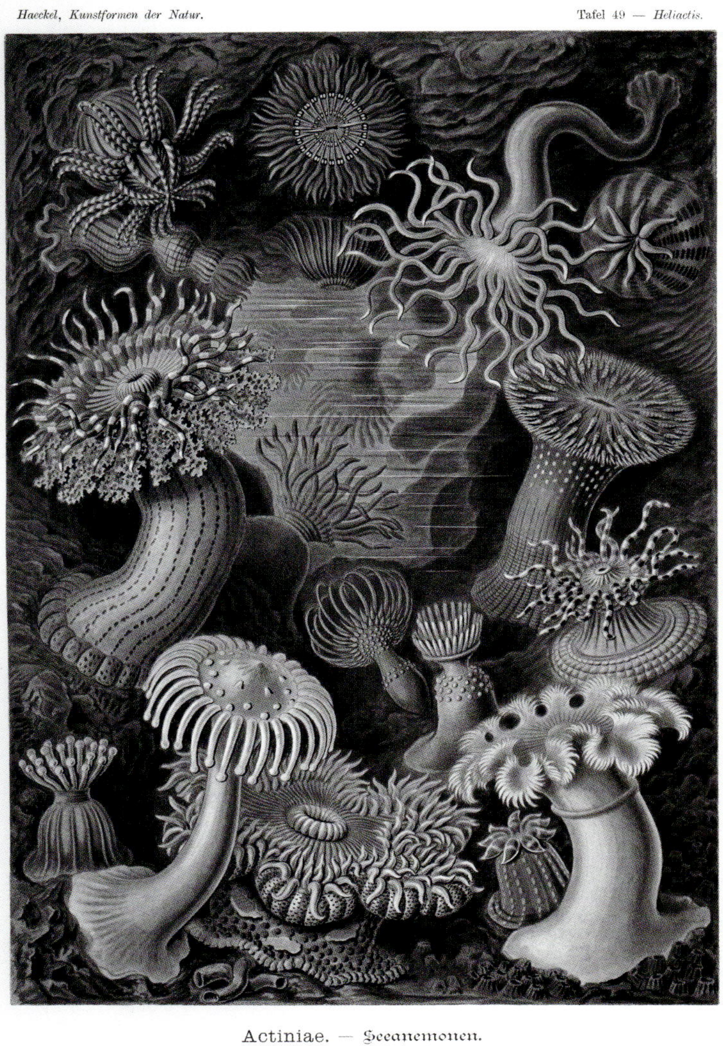

101
"Actiniae," in Ernst Haeckel, *Kunstformen der Natur* (Leipzig: Verlag des Bibliographischen Instituts, 1904), plate 49.

offered a challenge "to traditional views about the nature of generation."[52] For the nineteenth century, though, the peculiar restorative potential of the polyp lay elsewhere.

The conceptual significance of the polyp for the later nineteenth century is difficult to overestimate. Darwin's South Pacific voyages had revealed the origin of the inhabitable world in the work of polyps, and these "unseen architects of the globe," in Michelet's phrase, seemed to offer a direct link to our evolutionary origin.[53] As Philip Gosse put it, "they stand at the very confines, so to speak, of the vital world, at the lowest step of the animate ladder that reaches up to Man; aye, and beyond him."[54] Edmund Gosse spent his childhood exploring the polyps and anemones of the rock pools of the British coast with his father, discovering a "fairy-land" there, "undisturbed since the creation of the world" (fig. 100):

> Half a century ago, in many parts of the coast of Devonshire and Cornwall, where the limestone at the water's edge is wrought into crevices and hollows, the tide-line was, like Keats' Grecian vase, "a still unravished bride of quietness." These cups and basins were always full. . . . They were living flower-beds, so exquisite in their perfection. . . . The antiquity of these rock-pools, and the infinite succession of the soft and radiant forms . . . used to occupy my father's fancy. We burst in, he used to say, where no one had ever thought of intruding before.[55]

In this description, Keats's funerary urn becomes a vase for living flowers, for springing life and creation. The "infinite succession" of forms that occupied Gosse's "fancy" suggests the allure the polyp might have for the artist; Ernst Haeckel's illustration of anemones in *Kunstformen der Natur* (fig. 101)—one of the few plates that present organisms in their natural habitat[56]—affirms their paradigmatic status for the arts.

Although the earthly paradise Gosse described would shortly be destroyed by "an army of 'collectors'" inspired by his father's books, for a little while its still waters, "almost as pellucid as the upper air itself, thronged with beautiful sensitive forms of life,"[57] like Melville's seas and Pozzi's cystic women. Haeckel's softly undulating, tubular rock is fleshlike, conjuring up a bodily interior; conversely, a cross-section of a woman's torso in Robert Barnes's 1873 *A Clinical History of the Medical and Surgical Diseases of Women* (fig. 102),[58] which depicts frilled tubes and wormy sacs floating in a cavernous space, evokes Edmund Gosse's description of "submarine gardens" in a "windless pool."[59]

In Philip Gosse's illustration colorful polyps inhabit a dark chasm that stretches back indefinitely in time as well as space. The polyp not only offered a return path to the origin, however, but the promise of mastery over it, a promise directly related to the polyp's capacity for inversion.

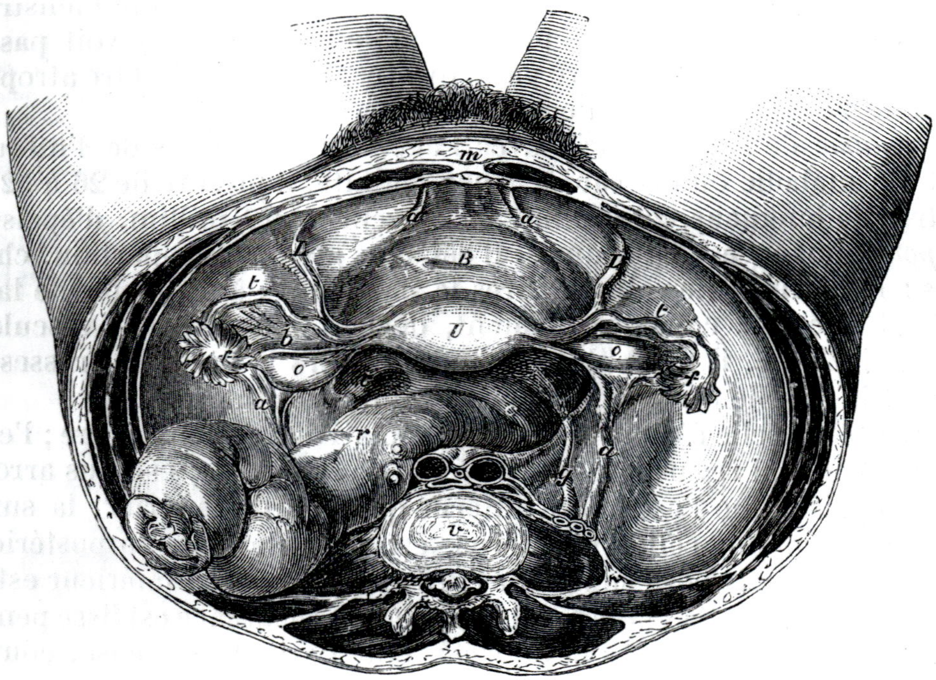

Fig. 1. — Section transversale passant au-dessus du détroit supérieur ; rapports des organes pelviens vus d'en haut (d'après Savage).

102
"Transverse section," in Robert Barnes, *Traité clinique des maladies des femmes*, trans. A. Cordes (Paris: G. Masson, 1876), 3, fig. 1.

Merleau-Ponty's description of what the perceptual reversibility of our hands enables us to fantasize—the presentiment we experience of being able to touch ourselves touching (although this perceptual coincidence always "eclipses at the moment of realization")[60]—is relevant here. Nicolet's drawing of bimanual exploration (fig. 89), while a perfect illustration of the reversibility of hands able to touch and be touched, equally demonstrates the impossibility of touching ourselves touching: we are always thwarted by a layer of skin. Our incapacity to feel our hands touching and touched simultaneously is mirrored by our inability to see ourselves seeing, which too can be experienced as imminent.[61] Merleau-Ponty likens the feeling of imminent seeing ourselves seeing to the sense of being able to turn our invaginated consciousness inside-out like the finger of a glove. This prescience seems to affirm the subject's illusory mastery of its envelope—that which surrounds the hole where it "sees *things*"—and the concomitant fantasy of its being able to "right" its "indirect and *inverted* grasp of all things."[62]

Melville proposes that our "indirect and *inverted* grasp of all things" is perhaps due to a physical inversion undergone "in a remote age," for, "as the lining of our interiors is nothing more than a continuation of the epidermis, or scarf-skin," we "must have been turned wrong side out: an hypothesis, which, indirectly might account for our moral perversities."[63] The author not only posits literal perversion as the condition of the species, he relates it specifically to the problem of bodily mastery, recounting the following fable in *Mardi,* which brings us back to the polyp: "A fresh-water Polyp, despising its marine existence, longed to live upon air. But all it could do, its tentacles or arms still continued to cram its stomach. By a sudden preternatural impulse, however, the Polyp at last turned itself inside out; supposing that after such a proceeding it would have no gastronomic interior." Like the polyp, men are condemned to have an interior morbidly given over to digestion rather than procreation or sublimation, seemingly forever fated to embody proof of their incompleteness. "An inferior species of plant," *Mardi* continues, they "cram" their "insides; and are loaded down with odious sacks and intestines."[64] Creation is the issue here: as Merleau-Ponty puts it, the impossibility of touching oneself touching is an "impotency"; the experience of its imminence is "always abortive."[65]

Although Melville's fabled polyp's attempt to overcome its digestive system fails—"its body proved ventricle outside as well as in. Again its arms went to work; food was tossed in, and digestion continued"[66]—Melville nevertheless posits the possibility of artists overcoming their own inversion and becoming creative. He wrote to a friend while composing *Mardi,* his first imaginatively unfettered novel, that he was just beginning to "feel [his] hand," and later to Hawthorne that, from the day he took up writing, "three weeks have scarcely passed, at any time between then and now, that I have not unfolded within myself."[67]

The male artist's or scientist's imaginary unfolding transforms the intestinal tract into as creative a zone as the womb. Michelet describes the intestines as a "coral-tree"—a polypary—whose feminine "exquisite beauty," maternal "delicate tenderness," and "pour[ing] out" of milky "chyle" are so made by nature to "contradict and confound our ideas."[68] More bluntly, Wilde fuses sodomy and pregnancy in a vision of the anally penetrating penis wriggling "in its sheath like a baby in its mother's womb."[69] If intestines become maternal through inversion, the female body's genital unfolding conversely transforms the apparatus of reproduction into one of digestion. Pozzi's illustrations of uterine prolapse (fig. 103)—the female genital tract turning itself inside out—resemble nothing so much as fecal expulsions.[70]

The fantasy of the invert's procreativity was facilitated by the polyp's metaphorical connection to the soft, pink, seductive polymorphy of female sexuality. *Teleny*'s Odillot describes the female genitals as "two moist lips like those pink shells which opening with awakening lust, reveal a pulpy luxurious world, only of a deep coralline hue."[71] In Zola's *La curée,* the comtesse Vanska, who lends "her dark ardour to a Coral" in a *tableau vivant,* resembles "a monstrous, seductive polype" displaying "a woman's flesh amidst the yawning, pink pearliness of its shell."[72] A hand-colored postcard titled "La perle" (The Pearl) (fig. 104) could have been made to illustrate the book. A nude woman in a shell proffers her strangely cartilaginous flesh to the viewer. She is surrounded by other polyps: an octopus above, starfish at the bottom right, anemones at the bottom left.

As Charles Kingsley, author of *The Water-Babies,* tells us, the "simplest form of polype is that of a fleshy bag open at one end, surmounted by a circle of contractile threads or fingers called tentacles,"[73] this combination of penetrable orifice and myriad supple, stimulating digits creating even more of a male fantasy object than Delacroix's dream hand. The connection between the inverting polyp and the female genitals is suggested by Lacan, who describes the effect of masculine consciousness peeling itself inside-out in terms of Merleau-Ponty's figure of the glove, the glove metamorphosing from a smooth, self-contained phallic symbol into a vagina.[74] While for Lacan this is a dire consequence, for the invert artist or sterile scientist of the late nineteenth century, "invagination" was the desired result.

Sargent painted the male artist's imaginary turning inside-out in his portrait of Pozzi (fig. 88).[75] Although it is usually translated as *Doctor Pozzi at Home,* the original French title of the painting, *Docteur Pozzi dans son intérieur,* or *Doctor Pozzi in His Interior,* articulates more precisely what the painter and bimanual palpator were

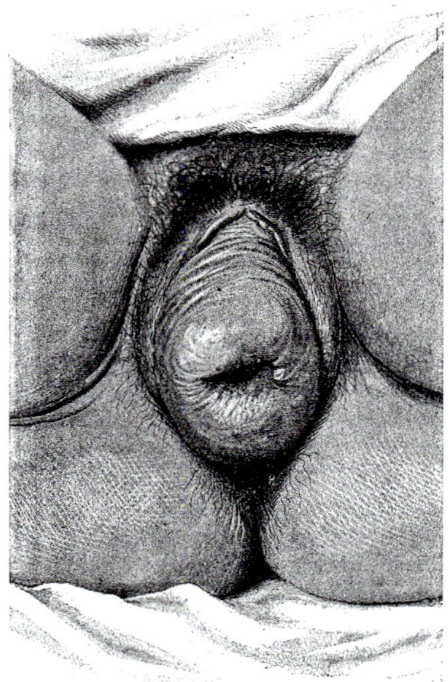

Fig. 282. — Prolapsus de l'utérus ; allongement hypertrophique considérable du col ; cystocèle.

103
Charles Nicolet, "Prolapse of the uterus," in Pozzi, *Traité de gynécologie* (1890), 518, fig. 282.

104
"La perle," c. 1900, postcard, 14 x 9 cm.

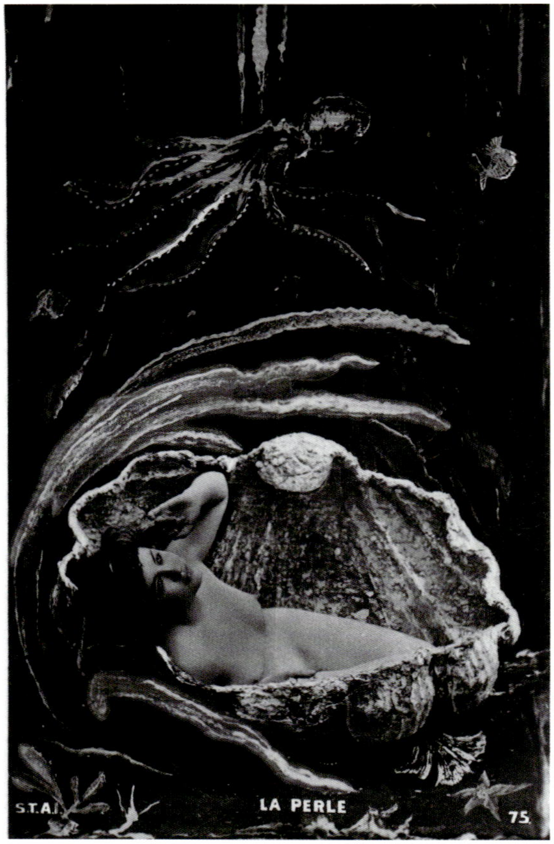

encompasses the entirety of the space of pictorial creation and self-fashioning. There are no sharp contours anywhere in this inner world, which gently expands around the body of Pozzi. An organic cell, its flexible walls are responsive to his movements; a deep fold hollows itself in the curtain on the left in anticipation of an approaching elbow.

The uterine quality of Pozzi's supple, velvety cavity is reinforced by our sense of its liquid saturation. Splashes of red paint in the area surrounding the figure of the doctor, particularly at the left, make the canvas a strange Whistler nocturne in scarlet, giving us an impression of particles suspended in a moist atmosphere, perhaps even one saturated with amniotic fluid. In the lower left corner Pozzi's slippered foot has kicked up sediment, which floats up into view.

If uterine, the atmosphere is also marine. Squidlike tassels hang from Pozzi's belt; spiny hands and fingers extrude from cuffs like the frilled rims of sea anemones; and almost becoming polyp, the one visible foot, with its rootlike embroidery, is a pedicle prepared to attach itself to a frame. Pozzi's pose is vaguely reminiscent of Botticelli's *Birth of Venus,* yet freed from the laws of gravity in this environment of liquid suspense, the doctor's dimensions become strange, unhuman—his head a wormlike thing, his body elongated under his heavy yet gently billowing gown.

In this polypoid world, under his unseeing gaze, Pozzi's delicate hands emerge like branches of coral ("flower of blood," in Michelet's phrase)[76] from his cloak to gently probe, finger, the folds and crevices of his warm, living interior. Through the chiasmic circlings of these touching hands, the

trying to capture. In the portrait, the doctor's red gown reads as the inner lining of his body, as though the gynecologist had turned himself inside out.

Pozzi's robe envelops him in a soft, sanguine tissue. A germinal membrane, it seems to ceaselessly exfoliate: our view of the doctor's right forearm reveals at least three layers, and what we read as his garment blends seamlessly into the surrounding fabrics through transparent layers of red paint—merging into the velvet curtains that open only to unveil more velvet—until it

doctor's veiled sight, the texture of the fabrics represented, and the quality of the paint, Sargent's tableau works on every level to figure palpation (both the artist's and the doctor's), and trigger the viewer's haptic response to the velvety layers of Pozzi's interior. As though something intensely private or even vaguely obscene were represented here in this image of the doctor touching himself inside, with its unspoken invitation to the viewer to do the same, Pozzi, like Dorian Gray or the various owners of Courbet's *Origin of the World,* kept his portrait hidden behind a curtain, a red rather than a green one.[77]

In addition to signaling his professional identity through their bodily investigations, the gynecologist's hands also betray his sexual identification: the fingers of his left resemble the spread legs of a woman, and the thumb and index finger of his right indicate a circle, a hole, notably just at the juncture of exterior and interior. Pozzi's fantasy of being a woman was most directly articulated in his private letters. "I am infinitely more woman than you," he wrote to one of his lovers, in the tone of the professional who would know.[78] He experienced the cravings of a pregnant woman, and in other ways identified with his patients.[79] He desired "to be dissected," but without giving up his position of mastery: "we will anatomize each other more easily privately, scalpel to scalpel."[80] He frequently wrote in red ink, asking his correspondents to excuse this "professional taste" on his part, which reads as vicarious menstruation as well as sadism.[81] Doubtless he believed that he belonged to that select group of men Michelet described as possessed of "the gift of incubation"; only "a few men (who possess both sexes, and are yet the most powerful of males)" have it.[82] Juliet Bellow has argued that the signifiers of domestic interiority in the portrait—the dressing gown and slippers—reinforce those of bodily interiority, both contributing to the painting's undoing of the opposition between masculine and feminine.[83] I would go further: sex itself, rather than gender, is at issue here for both the gynecologist and the painter.

The physical invertibility of the polyp enables the fantasy of literally *turning* the invert's morbid, devouring digestive tract into a fertile, blooming garden. This is what Chardin's *jardin*-esque painting achieves unconsciously: by picturing sights that at first "turn your stomach,"[84] according to Proust, he shows "dead nature" becoming "above all living nature."[85] The painter's gift of life or animation to his objects is more important than beauty or digestive appeal; Chardin's lesson is that "a pear is as alive as a woman" and, presumably, a child.[86] His art, then, reconciles human and vegetable models of procreation. As important, Proust identifies the homoerotic nature of Chardin's work in the way it draws the artist's and the implied male viewer's pleasure together: they become "inseparable." The viewer recognizes this, "and for the first time" in his life "enjoys" it, or, more literally, recognizes his taste for it.[87] Chardin's model of art thereby also validates the male invert's same-sex desire and anal orientation—via the fantasy of rectal or intestinal pregnancy—and naturalizes his offspring.

To return to the lush garden world of Pozzi's gynecology, the doctor's investment in an Edenic vegetal metaphorics led to his decorating the

gynecological ward at the Hôpital Broca with restorative murals on a pastoral theme (fig. 105). He intended these to "nourish [his] patients with beauty,"[88] though Robert de Montesquiou was probably more to the point in saying that Pozzi decorated his hospital "to render illness beautiful and suffering almost happy."[89] The murals were the work of Georges Clairin, a renowned Orientalist and decorative painter. The centerpiece of the ward's decorative scheme, *Health Restored to the Sick* (fig. 106), depicts an allegorical figure of health—modeled by Sarah Bernhardt, one of Pozzi's many celebrity patients—presiding over a symbolic landscape. The dark, tangled, bare-branched woods of illness on the left are populated by three wan patients, Dr. Pozzi, and a mourner, while on the right, fit, floral-crowned *femmes-fleurs* gather in a light-filled bucolic landscape of flowery rolling hills. The patients, one recumbent, one sitting or starting to rise, and one standing, look up hopefully to the divine Sarah. Having mounted an illuminated, overgrown grotto, she reaches out a delicate hand to a sere branch that, beneath her potent yet delicate touch, bursts into blossom and begins to rain down flowers. Like Désirée in *La faute de l'abbé Mouret,* whose fingertips secrete "a procreative dew,"[90] Sarah restores flowers to those who have lost theirs—in other words their menstrual, and thus generative, abilities. Notably, Pozzi too stands under the fecundating rain of her floral touch. On the right, the orange *femme-fleur* whose bouquet is overflowing holds her skirts up

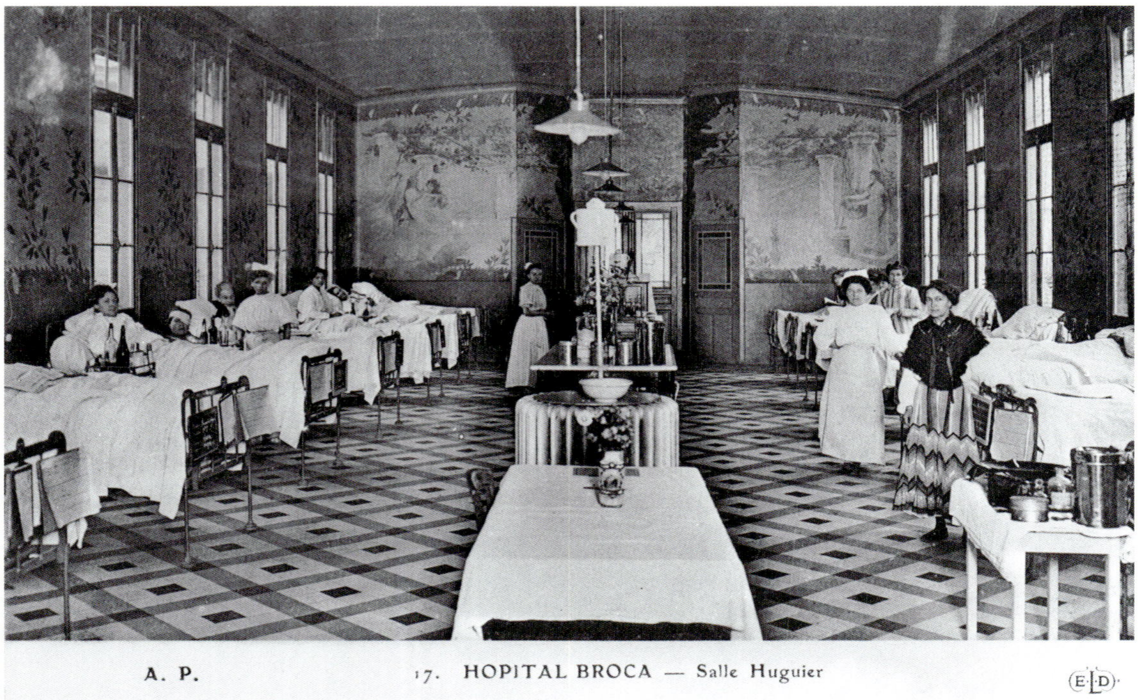

105
Photograph of Pozzi's ward, Hôpital Broca, c. 1900. Archives of the Musée de l'Assistance Publique–Hôpitaux de Paris.

106

Georges Clairin, *Health Restored to the Sick,* 1898. Oil on canvas, 275 x 440 cm. Musée de l'Assistance Publique–Hôpitaux de Paris.

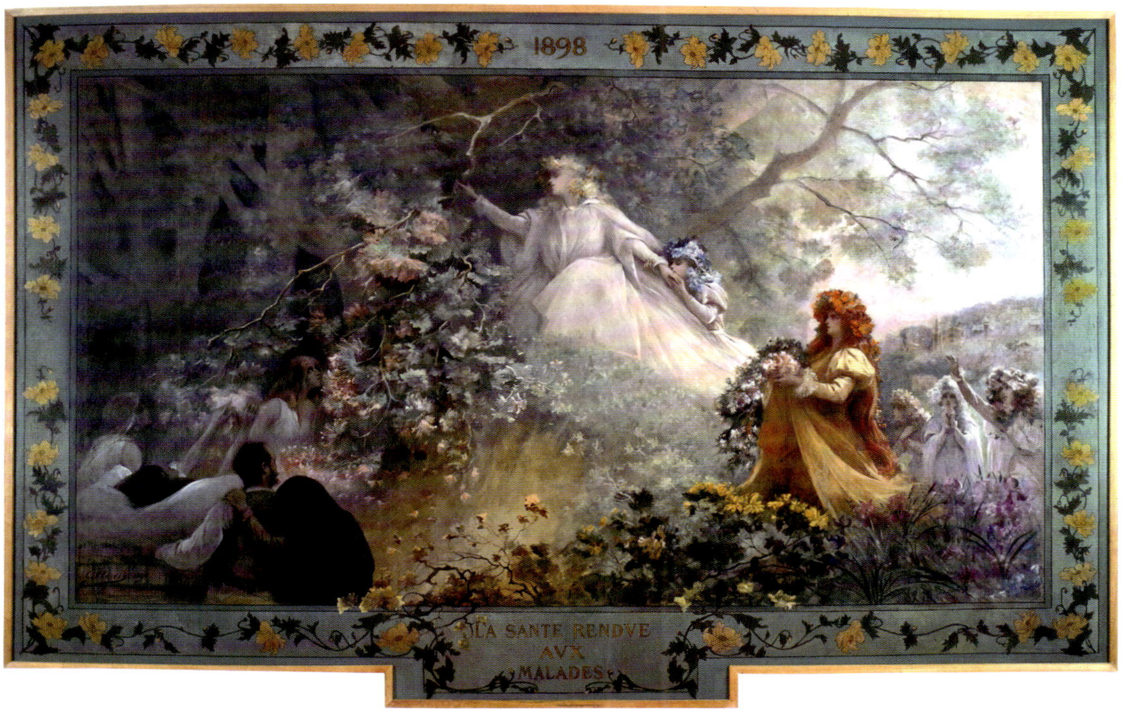

and open, as though to display her attainable genital health.[91]

The fantastic landscape in Clairin's painting evokes the grotto, almost lost beneath the teeming flowers, visited by Albine and the Abbé in Zola's novel. The comparability and compatibility of human and vegetal sexualities are dominant themes in Zola's description of the cavern: "It was an immense mane of greenery, spotted with a rain of flowers, from which stray locks overflowed on all sides, escaping into a mad dishevelment, making one think of a sort of giant maiden, swooning back with pleasure from her loins, her head thrown back in a spasm of passion, in a cascade of superb hair, spread out like a pool of perfumes."[92] The glowing vaginal cavern in Clairin's painting may have been directly inspired by Zola's description, though cave, cavern, crevasse, and grotto were also French and English slang for vagina.[93] The perspective we are given of the bushy grot in Clairin's painting—a close-up of the vagina of a giant creature part animal, part vegetable, part stone—invites comparison both compositionally and in terms of metamorphic motility not only with Zola's tableau but also with several of Pozzi's gynecological illustrations (fig. 107), wherein lichenous or budding plant life crawls over seemingly Paleolithic thighs of a scale to dwarf the viewer. Such deep romantic chasms in their various forms—gigantic mossy caves or intimate, luxuriantly polypoid crevices—thus constituted a key image in the *fin-de-siècle* botanico-gynecological imaginary.

Pozzi acknowledges that the "vegetations" found in such sites "have for a long time been considered a reliable index of venereal infection, either gonorrheal or syphilitic," and through them we enter into the flowery world of venereal disease.[94] The metaphoric and sublimatory excess of Pozzi's descriptions of symptoms can be explained in part by the contemporary cultural resonance of syphilis, which embraced the long-standing association of the disease with luxuriant plant life.[95] Cullerier's 1866 *Précis iconographique des maladies vénériennes,* translated as *Atlas of Venereal Diseases,* illustrates some of its verdant, mossy growths (fig. 108), and earlier in the century the dermatologist Jean-Louis Alibert had catalogued the disease's lentil-, raspberry-, strawberry-, grape-, fig-, and cauliflowerlike pustules and vegetations.[96] Barbara Stafford has shown that Alibert took a peculiarly "botanical" approach to dermatological conditions,[97] but the common name for the infection was "garden gout,"[98] and its symptoms, "the whole gardenful of cutaneous manifestations" and "a pretty bouquet of [other] infirmities,"[99] as one turn-of-the-century author put it, were frequently troped as vegetal metamorphoses in literary as well as medical discourse. The flower-girls of Wagner's *Parsifal,* for example, were associated with syphilis.[100] The disease is personified as "the Flower" in Des Esseintes's nightmare,[101] those who spread it as "mancenilles" (or manchineels, poisonous trees) in a "clinical" novel by André Couvreur.[102] On a trip to Italy in 1850 Théophile Gautier came across syphilitic French troops with tibias "exfoliating in exostoses like ancient columns of greenery in a Roman ruin,"[103] and Alphonse Daudet described the progress of his own syphilitic infection in Ovidian terms: "I feel like some creature from mythology, whose torso is locked in a box of wood or stone, gradually turning numb and then solid. As the paralysis spreads upwards, the sick man changes into a tree or a rock, like some nymph from Ovid's *Metamorphoses*."[104] And over time Jean Lorrain came to resemble, like one of his fictional characters and one of his former lovers, the figure of Flora in Botticelli's *Primavera* (fig. 109).[105]

That the disease was widely understood to act as a metamorphic catalyst is attested to by accounts of the effects of syphilis on fictional flora. If

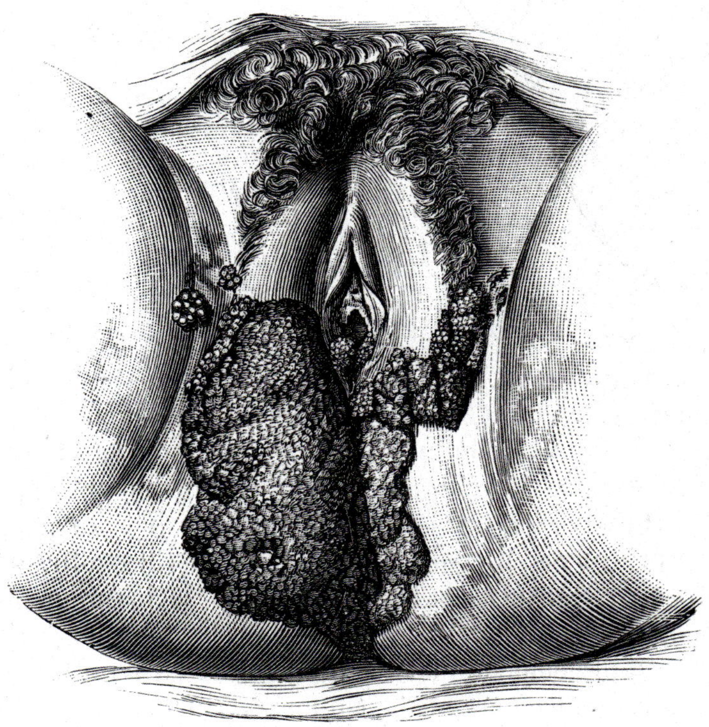

Fig. 456. — Végétations simples de la vulve (Tarnier.)

107
"Simple vegetations of the vulva," reprinted from Tarnier, in Pozzi, *Traité de gynécologie* (1890), 1014, fig. 456.

108

"Bullous syphilide rupia of the face and chest; bullous syphilide rupia, with cicatrices of tubercles," in Auguste Cullerier, *Atlas of Venereal Diseases,* trans. Freeman J. Bumstead (Philadelphia: Henry C. Lea, 1868), plate 23.

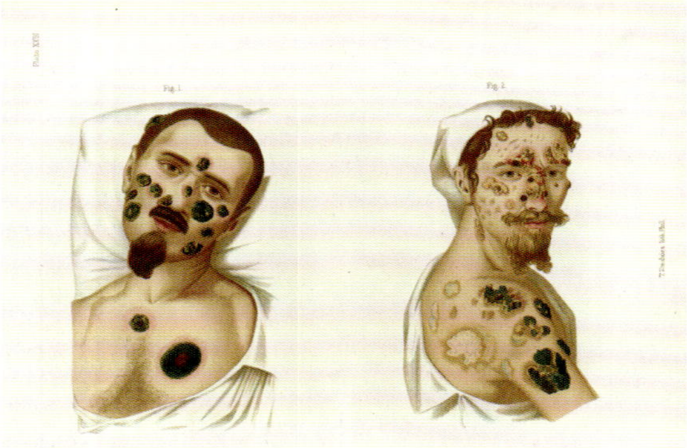

109

Sandro Botticelli, detail of Flora in *The Primavera,* 1477. Tempera on panel, whole 203 x 314 cm. Uffizi, Florence.

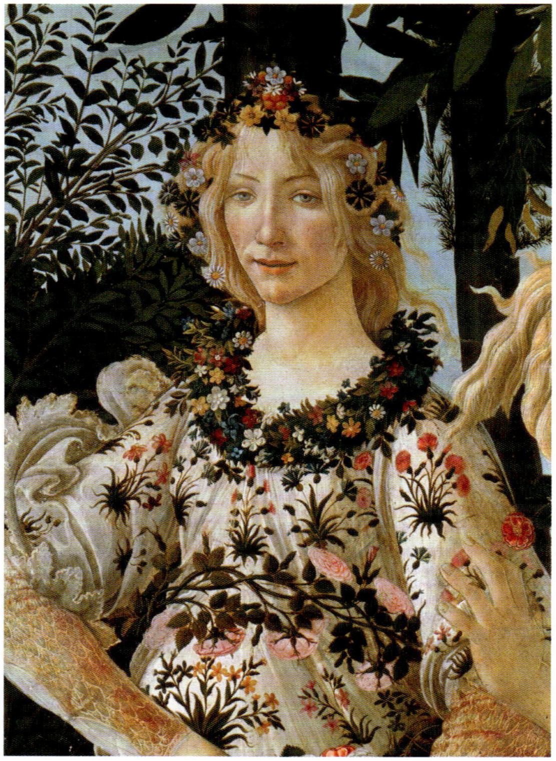

afflicted animals sprout mutant, parasitic foliage, plants, inversely, bear wounds of the flesh. Huysmans describes plants, "as though eaten away by syphilis," exhibiting "livid flesh marbled with roseola and damasked with dartres" and "hairy skins pitted by ulcers and embossed with chancres."[106] For Zola the "shameful vegetations" of Cereus form "huge polyparies," and nipple-cacti are a mass of "live pustules."[107] The lower petal of *Loroglossum hircinum,* Maeterlinck tells us, is "embellished at its source with bronzed carbuncles, huge mustachios and sinister-looking lilac buboes"; it "stretches out interminably, madly, unreally, in the shape of a corkscrew riband of the color assumed by a drowned corpse after a month's immersion in the river."[108] And in *Le jardin des supplices* the "funnel-shaped cornets" of spars are "the dark violet of putrefaction inside, and the greenish yellow of decomposition outside, like the open thoraxes of dead animals"; tigridia open "up from mutilated throats"; and anthurium "like bleeding pleura" greet promenaders.[109]

Many of these late nineteenth-century literary descriptions of syphilitic plant life speak more directly of hospitals, medical procedures, and diseased flesh than Pozzi's *Traité* does. In *À rebours* the flowers of echinopsis "emerging from compresses of cotton wool" are "the vile pink of amputated stumps"; others in Des Esseintes's collection are "the bright pink of scars that are healing, or the brownish tint of scabs in the process of forming; others were blistering from cautery or puffing up from burns."[110] The occasionally astonishing comparability of contemporary botanical and medical illustrations likewise attests to the elective affinities of diseased

IIO

Detail of "Cypripedium," in Darwin, *The Various Contrivances by which Orchids are Fertilised by Insects*, 2nd ed., 227, fig. 35.

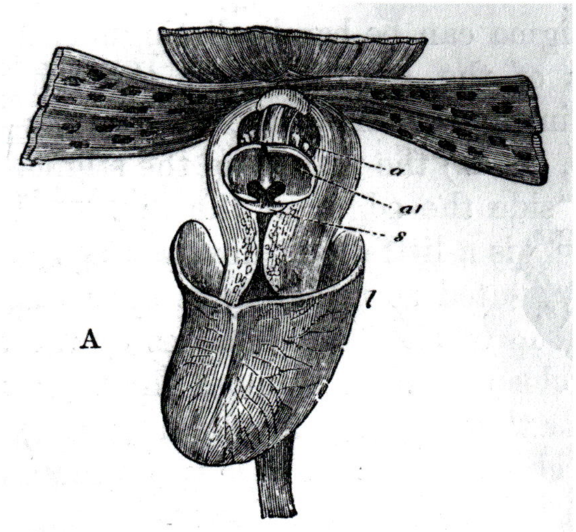

tissues. Huysmans describes the labella of the Cyprideium as "resembl[ing] a clog, or a small oval bowl, with a human tongue curled back above it, its tendon stretched tight just as one sees tongues drawn in the illustrations to works dealing with diseases of the throat and mouth,"[111] perhaps with Darwin's illustration in mind (fig. 110).

Given the broad cultural awareness of interspecies desire in the nineteenth century, it is unsurprising that syphilis should have been imagined as affecting plants. For Huysmans it is "the everlasting legacy, the eternal disease," which, "since the beginning of the world, from father to son, all creatures [have] passed on to one another." "It all comes down to syphilis," he concludes,[112] and the celebrated syphilis specialist Philippe Ricord would have agreed—he frequently joked that "in the beginning God created the heavens, the earth, man, and venereal diseases."[113] Of "SYPHILIS," Flaubert's *Dictionnaire des idées reçues* informs us, "everybody is more or less infected with it,"[114] a notion that did not seem unlikely to nineteenth-century scientists, who deemed it a "contagious disease which can be caught in so many ways, and which manifests itself in forms that are so varied and numerous, that it defies philosophical definition."[115] Diseased biomorphing plants and vegetating humans, straining toward reciprocal anamorphosis, were traceable, then, to evolutionary origins. Des Esseintes's plants displaying a "primal splendour"[116] are all monstrous atavistic blooms manifesting a phylogenetic inheritance reaching back to inorganic matter, the original infection instinct to and coeval with all species.

Like retrogression, sexual deviance was strongly linked to syphilis. Freud thought the abnormal sexual lives and overt and covert behavior of his neurotics and perverts, in all their ineffable, teeming variety, were cases of pronounced expression of our common "syphilitic heritage."[117] Darwin's conclusion that "monstrous" syphilitic flowers are the result of too many generations of self-fertilization,[118] then, should come as no surprise, "the insemination of a flower by itself," leading, according to Proust, "like a succession of intermarriages in the same family, to degeneracy and sterility."[119] Of greatly decreased constitutional vigor, plants that bear mutant progeny, such as double-flowers (the vegetal equivalent of Siamese twins), "irregularly split" corollas that do not open "properly," and "foliaceous" stamens that cohere to the petals, are significantly shorter and productive of fewer and weaker seeds than their cross-fertilized counterparts.[120] These invert vegetables, lacking drive toward development, expansion, change, and reproduction, oriented toward a mode of reduced existence, would seem to manifest the undeflected operation of the "urge inherent in organic life to restore an earlier state of things,"[121] that state before sexual division—in other words, to betray the inertial influence of the death drive.

Atavistically vegetating, syphilitic humans partake of the same phenomenon of reversion, as we have seen. Retrograde genetic expression in humans, however, is limited neither to any particular anatomical locus nor to any one stage of evolutionary development. When Flaubert's syphilis once manifested itself in his tongue, the organ "began to swell until I thought it was transmuting itself into that of an ox. It protruded from my mouth: I had to hold my jaws open."[122] In *The Marble Faun* a "supernumerary finger," or faun's ears "covered with a delicate fur, and shaped like a pointed leaf," are emblematic instances of hereditary peculiarities that surface periodically, in a "wayward fashion."[123] Regressing far beyond domesticated creatures or even mythic ones, Stevenson's Mr. Hyde reverts all the way back to "the slime of the pit."[124] Nordau affirmed that "the relapse of the degenerate may reach the most stupendous depth."[125]

Haeckel's theory that ontogeny recapitulates phylogeny—that the embryo "passes, in the course of its development, through the form of inferior animals"—was widely accepted in the scientific community and other disciplines.[126] One very typical troping of inverts in the later nineteenth and early twentieth centuries fixed their slip-sliding or schizoing on the evolutionary scale at the level of the lower primates. When "the woman is not only inwardly united to the man but hideously visible, convulsed as they are by a hysterical spasm, by a shrill laugh which sets their knees and hands trembling," according to Proust, inverts look "no more like the common run of men than those apes with melancholy ringed eyes and prehensile feet who dress up in dinner-jackets and black ties."[127] Similarly, with the laugh of "an obtuse boy," the active, modern (and therefore masculine or unsexed) woman betrays her perverse self-sufficiency and appears to Rachilde's Léon Reille as a "big ape" only "moderately female."[128] This is a relatively tiny step backward, as the famous cartoon that pictures Oscar Wilde and a monkey and asks "How far is it from this to this?" (fig. 111) implies, but one mightily resonant in the age of Darwin. Without identifying any sitters in particular, Montesquiou reviled Sargent's painted "chimpanzees in frock-coats."[129]

If inverts were linked with apes and fauns, they were also imagined to be strange portals to earlier evolutionary stages, even to the very origin, like the polyp. Proust theorizes that "the race of [human] inverts, who readily link themselves with the ancient East or the golden age of Greece, might be traced back further still, to those experimental epochs in which there existed neither dioecious plants nor monosexual animals, to that initial hermaphroditism of which certain rudiments of male organs in the anatomy of women and of female organs in that of men seem still to preserve the trace."[130] Hermaphroditism was Pozzi's subspecialization,[131] and although he never encountered a case of "true hermaphroditism, of the coexistence of ovaries and testicles," he would not rule out the possibility categorically, as "this disposition is common among fishes and the batrachians," and even "constitutes the rule with the toad."[132] He posited the plausibility of attributing sexual malformations to "a superior cause, atavism, reproducing sporadically in one species the forms of another species, by the effect of what Darwin has called a phenomenon of reversion,"[133] and other evolutionist sex investigators concurred.[134]

Pozzi's illustrations of the genitals of pseudohermaphrodites highlight this nostalgic pathology. The external genital organs of one hermaphroditic patient (fig. 112), for example, seem a form of plant life: Nicolet gives us a perspective, as though through a glass pot, of a young bean sprout pushing up through the grass-tufted earth. The illustration of an anomaly of the hymen (fig. 113) resembles a ciliate amoeba, perhaps alluding to the unicellular origin of life. In Flaubert's *Tentation de Saint-Antoine,* when the saint, subject to a Darwinian temptation (offered evidence of evolution), sees "little globular masses, no bigger than pin-heads, and garnished with hairs all round," across which "a vibration quivers," he cries out, "O happiness! happiness! I have seen the birth of life, I have seen the beginning of movement."[135] The unicellular genital here, a globular mass garnished with hairs, suggests the allure of a similar possibility—the true hermaphrodite's capacity for reproduction without others and thus the birth of life.

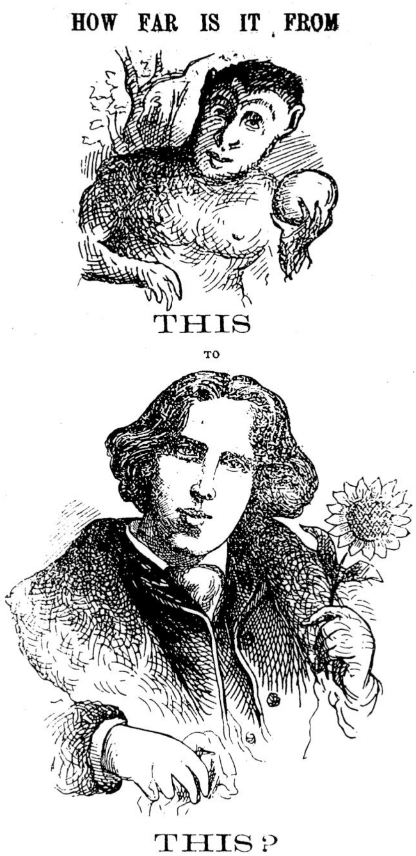

111
"How far is it from this to this?" ***Washington Post,*** **22 January 1882, 4.**

Sexual anomalies, like perversions, then, were associated with syphilis. While the disease was known to lead to malformations, madness, and metamorphoses, however, it was nevertheless also believed to have a beneficent effect on its host species occasionally. Syphilis, like inversion, was commonly associated with genius and creativity.[136] Léon Daudet, a doctor and the son of the metamorphosing Alphonse, who therefore had some investment in the disease's hereditary transmission, declared syphilis "as much the power behind genius and talent, heroism and wit as that behind general paralysis, tabes and almost all forms of degenerescence."[137] He also deemed it the agent of transmission of artistic temperament and a social refiner.[138] Most important, the syphilis spirochete was "as thrusting and as acrobatic as the propagator of life which is associated with it in many a conception by hereditary transmission"; if

112

Charles Nicolet, "Partial pseudo-hermaphroditism in a male (androgyny)," in Pozzi, *Traité de gynécologie* (1890), 1070, fig. 469.

113

Charles Nicolet, "Anomaly of the hymen," in Pozzi, *Traité de gynécologie* (1890), 1064, fig. 466.

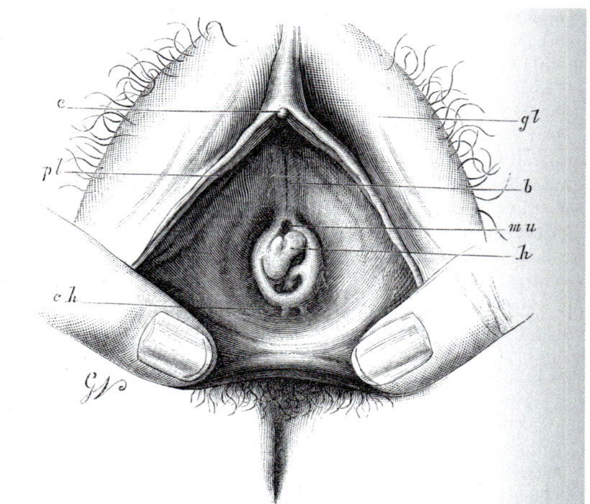

Fig. 466. — Anomalie de l'hymen. Hymen charnu et godronné chez une jeune fille vierge.

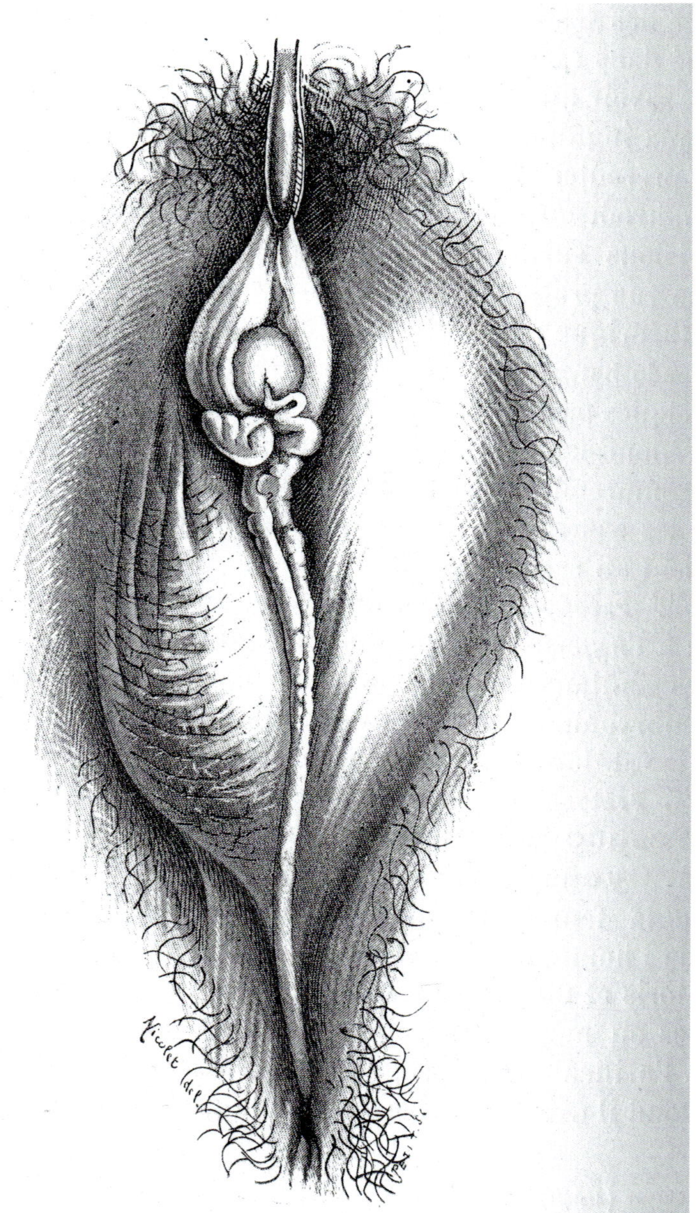

Fig. 469. — Pseudo-hermaphrodisme partiel chez l'homme (Androgynie).

excessive, and the species develop out of all proportion; then, as an antitoxin protects us against disease, as the thyroid gland regulates our adiposity, as defeat comes to punish pride, as fatigue follows indulgence, and as sleep in turn brings rest from fatigue, so an exceptional act of self-fertilisation comes at the crucial moment to apply its turn of the screw, its pull on the curb, brings back within the norm the flower that has exaggeratedly overstepped it."[141] Our common syphilitic inheritance, then, as a cause of metamorphosis and/or inversion, can occasionally come to the rescue of species as well as stimulate creativity. Disease, polymorphic perversity, inversion, and interspecies desire have an essential function and are even implied by nature.

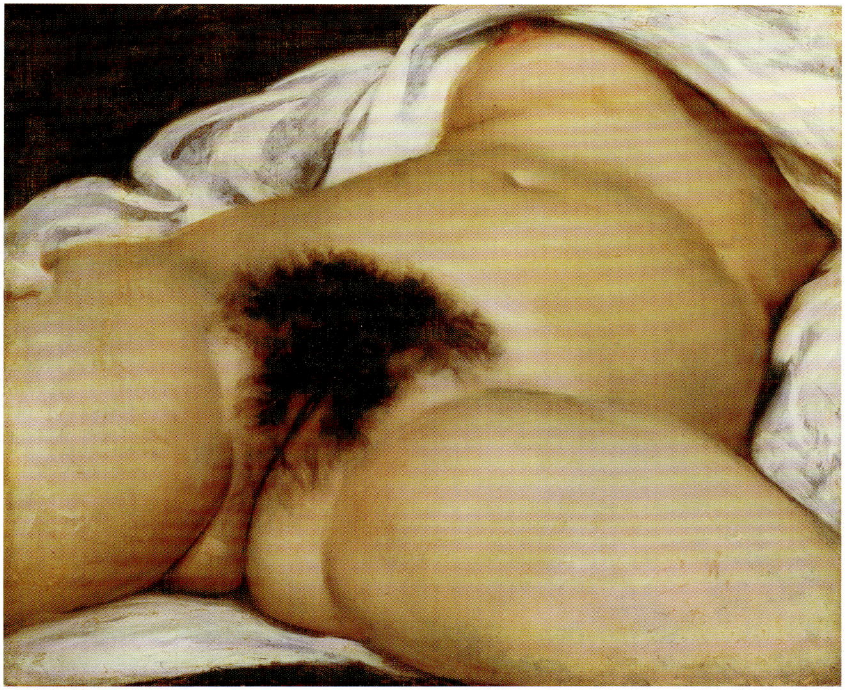

114
Gustave Courbet, *The Origin of the World*, 1866. Oil on canvas, 46 x 55 cm. Musée d'Orsay, Paris.

it could induce sterility, it could also induce "its opposite"—"dramatic intensity of life."[139] Thus, like the fantasy of the male body turning itself inside-out, syphilis was another way of overcoming the sterility associated with inversion, of attaining a vegetable fecundity.

In general, cross-fertilization was equated with the progressive thrust of the libido, evolution, and constitutional vigor, while self-fertilization was associated with the undeflected exercise of the death drive's atavistic pull, with weakness and sterility. Darwin nevertheless conceded that there is "reason to suspect that self-fertilisation is in some peculiar manner beneficial to certain plants."[140] He did not specify in what "peculiar manner," but Proust supplies us with a working hypothesis: the "invigoration" of cross-fertilization may "prove

Pozzi's illustrations of genital vegetations (fig. 107), like Clairin's mural for the ward at Broca, call to mind another nineteenth-century painting that makes visible the connection between atavism, syphilis, and art: Courbet's *Origin of the World* (fig. 114). The truncated thighs of the figure, the white sheet concealing her identity, and the picture's unusual composition all seem inspired by medical illustration,[142] and Michèle Haddad has argued that the painting represents the threat of originary syphilis.[143] We know that Khalil Bey, the painting's first owner and possibly its commissioner, came to Paris for treatment of the syphilis he had contracted in Russia, and received it at the hands of Dr. Ricord. A poem by Castagnary, "On a Painting in the Collection of Khalil Bey," may also allude to the work's syphilitic signification. The writer warns the viewer/reader not to lift the curtain, for the "formidable object" represented behind it "bows you before your time, and turns your black hair

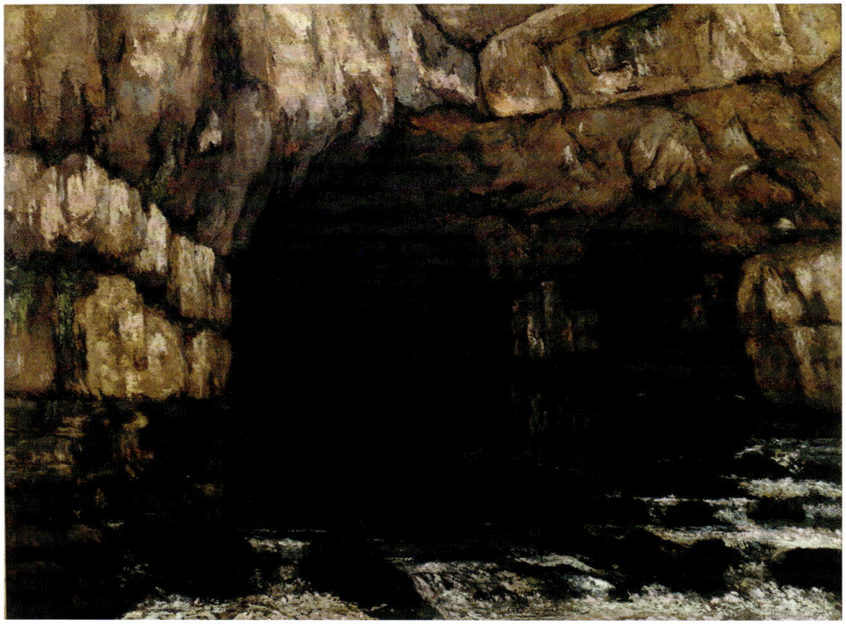

115
Gustave Courbet, *The Source of the Loue,* c. 1864. Oil on canvas, 107.3 x 137.5 cm. Albright-Knox Art Gallery, Buffalo, New York. George B. and Jenny R. Mathews Fund, 1959.

white, it rots your teeth, makes your face livid"; "it has a vengeance ready for any who affront it."[144]

In light of this reading, the long-remarked resemblance of Courbet's *Origin* to his paintings of the source of the Loue may be linked with his awareness of the metamorphic potential of syphilis that Pozzi's rocky crotches embody. On the canvas in the Albright-Knox Gallery (fig. 115), bright peach patches and bluish shadows mottle the flesh-colored rock. Green lichens and weedy vegetation sprout from damp folds and crevices, while along the bottom dirty white frothing waves resemble crusted oozings or flaky, infected skin. Courbet's sketch *La dame verte* (fig. 116), with its flesh-colored vaginal cavern labeled in pencil by the artist, invokes a more explicit cluster of associations. The title refers to the caves dotting the Franche-Compté where *la dame verte,* a beautiful local fairy, was supposed to live in winter.[145] Like the creature of Castagnary's poem and Courbet's painting, however, the legendary wood siren sometimes lured men into her lair only to metamorphose into her real shape, that of an old hag. Furthermore, *la dame verte,* like *la fée verte,* was another name for absinthe, another agent held responsible for both mental degeneration and artistic inspiration.

If syphilis was part of Courbet's conception of the origin, it was for him, as for his contemporaries, linked with atavism. Françoise Gaillard argues that in the age of Darwin the pubic hair so shockingly displayed in the painting was a comment on the bestiality of our origins: the "hairy hole" or "cave" at the center of Courbet's crotch "opens onto abysses of animality"; like syphilis, it is "the agent" of animality's "transmission."[146] The pubic hair, though, is shaped like a Cattleya blossom on the stalk created by the continuous line of her vaginal and anal clefts, and thus seems to offer a link to vegetality as well. This "bottomless gulf" that communicates "with the origin" is, furthermore, the source of sexual inversion: Gaillard argues that it invokes "the history of our terrifying ancestor, the hominoid of the ice age," an ancestor who was, according to Freud, "responsible for our neuroses and our sexual perversions."[147]

In the face of diseased and dying women—Jean Lorrain called Pozzi's clinic a "butcher shop" and a "house of death, where four women died during the time he [Lorrain] was there" for treatment of his anal fistulae[148]—tensions occasionally ruptured the surface of the plastic heterotopia conjured up by the visual and verbal vegetal rhetoric of Pozzi's text and his ward's decorations. A gift from René

Lalique to Pozzi represents a conception of *gynécologie* rather more sinister than that of Clairin's murals. Pozzi had treated Lalique's wife, who later died, and in thanks the artist carved an ivory plaque to serve as the front cover to a book, bound in red morocco, containing transcripts of the talks Pozzi had given at the Hôpital Broca (fig. 117). The inscription explains the artist's intention: "I wanted to figure the illness . . . in all its tragic horror. R. Lalique, 16 March 1901."[149]

Partially covered with a sheet, like the subjects of gynecological illustrations or depictions of ancient furies or mourners, "la maladie" is an emaciated nude hag grabbing at her nonexistent vulva in front of an ivy-covered wall. Although the malady is unnamed, the knotted body of the sere, bony creature, her interlocking limbs echoing the interlocking chain of letters above, suggests the imprisonment of gradual paralysis and, along with the flatness of the splayed body and compressed extramural space, embodies a vision of syphilis opposed to Pozzi's deep, moist grottoes teeming with life. The figure's abdomen is not lushly vegetal and plentiful but emaciated and threatening, forming a ferocious, fishy maw devouring her from the inside: her shrunken breasts are eyes above the piscine jawbones created by the sharply protruding rib cage and pelvis.

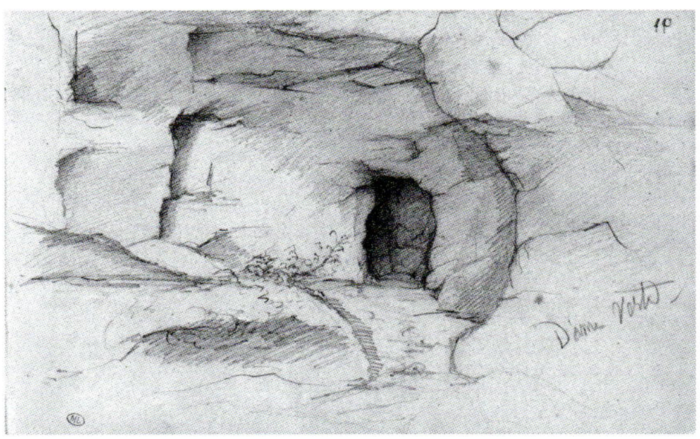

116
Gustave Courbet, *La dame verte*, early 1840s. Pencil on paper, 14 x 21.7 cm. In album "paysage rocheux," RF29234, folio 10, recto. Musée du Louvre, Paris.

The medium itself, the *plaque,* is suggestive, since the word "plaque" in both French and English means patch, blotch, and skin lesion in addition to tablet, and can thus evoke "the flaky epidermis of a syphilis victim."[150] Indeed, the pattern of ivy leaves covering the organic surface of the ivory can easily be read as scaly skin or exostotic bone. Given the dominance of the carved text, however, "la maladie" can also be read as the figure of "gynécologie." While Pozzi himself was extremely careful about sterilizing his hands and instruments, other gynecologists and doctors, like Ricord, were not, and iatrogenic infection contributed in no small part to the spread of syphilis and gonorrhea.

If clean, Pozzi's celebrated touch was not always as beneficent as intended. For one thing, a careful distance between palpation and penetration was sometimes difficult to maintain. Benjamin describes touch and surgical penetration as two poles of a haptic continuum, and Pozzi's treatise confirms that the distance between touch and penetration is smaller than one might think. Nicolet's illustrations attest to the somewhat menacing length of the gynecologist's fingernails (fig. 118), as do stories of the doctor's inadvertent cyst popping, when "an enormous and fragile pocket of pus [would] burst under the nail searching for the cleavage line."[151] While it may not be surprising that someone who makes a living fingering vaginas, anuses, and other body cavities all day is not a nail biter,[152] Pozzi in fact kept his nails long for use in surgery. After being "cleaned

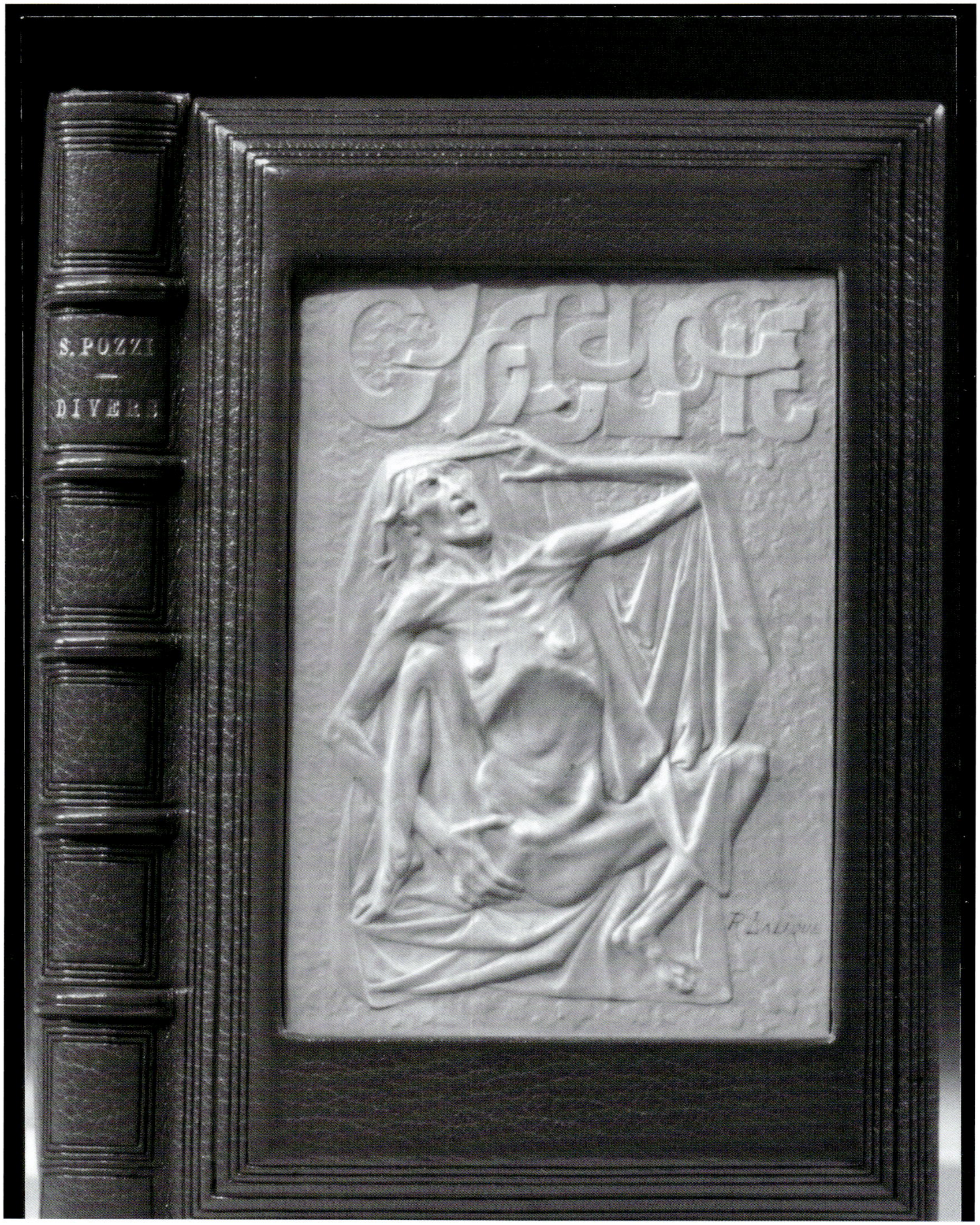

117
René Lalique, *Gynécologie*, c. 1898. Ivory, 17 x 11.5 cm. Robert Zehil Collection. © 2008 Artists Rights Society (ARS), New York/ADAGP, Paris.

118
Charles Nicolet, "Absence of vagina and uterus," in Pozzi, *Traité de gynécologie* (1890), 1090, fig. 475.

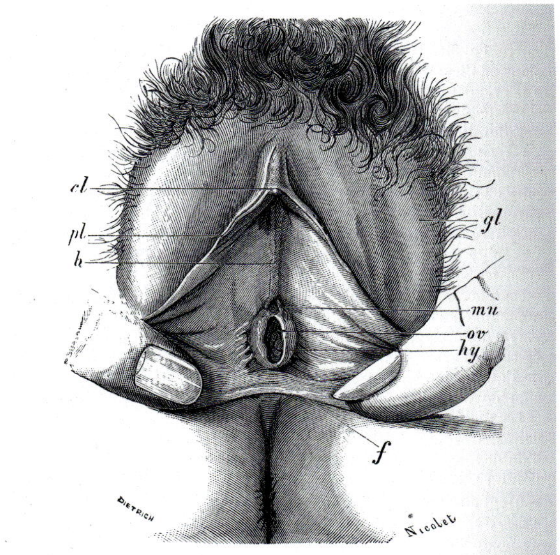

Fig. 475. — Absence de vagin et d'utérus (ou utérus rudimentaire?) avec un hymen bien développé.

119
"Mikschik's instrument for excision of uterine polyps," in F. W. von Scanzoni, *Lehrbuch der Krankheiten der weiblichen Sexualorgane* (Vienna: Wilhelm Braumüller, 1857), 220.

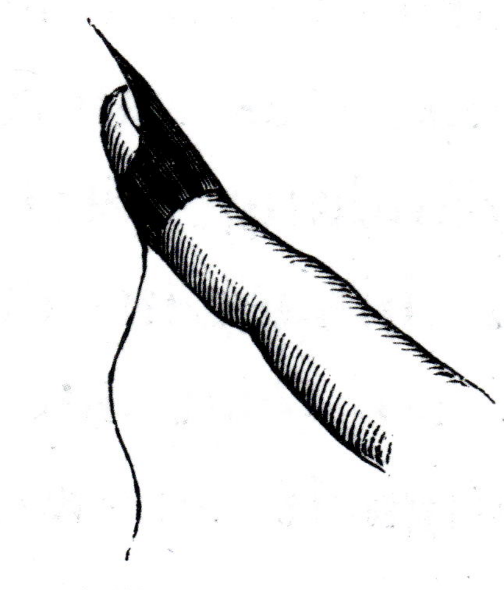

with *great care*, with the aid of a pointed nail file,"[153] they were useful instruments for "scraping"[154] apart resistant tissues. If one's own weren't long or sharp enough, fingertip prostheses were available (fig. 119). Pozzi, fortunately, had "fingers of steel."[155]

The disconcerting slippage between gentle exploratory touch and sharp penetration that cysts often induced, calling into question the mastery of the doctor's hand, made them significant loci in the battle waged for control over the body's germinal membranes. As pseudopregnancies, cysts were for Pozzi fantastic bubbles, the Freudian vesicles of his utopian vision, despite the fact that they frequently erupted under his hands, launching fountains of yellow or green pus into the air.[156] They are jewel-like in his treatise—"pearly-white," "marbled," "crystal clear," and so on—he plucked them from their settings with the loving eye of a collector.[157] In a postoperative note to Robert de Montesquiou, Pozzi wrote of Sarah Bernhardt's cyst that it "was not vulgar: elegant, tenacious in its multiple roots, hidden, SET in the large ligament where I had been to *pluck* it (excuse the word)."[158] Set like a jewel, plucked like a flower—this description is a triumph of sublimation. The imaginary crystallization of the cyst suggests that Pozzi was seeking the Rajah's Diamond for his collection, his cabinet—with it the gynecologist would be master of the secret of creation, of life.[159] How Pozzi must have loved Sarah Bernhardt's calling him "Docteur Dieu"![160]

Proust compared the "vice" of the invert to a "jewel" that may be lost and found again;[161] Freud diagnosed it as the continued belief in the existence of the maternal phallus. Fetishized equivalents for the missing maternal member, Pozzi's crystalline cysts encapsulate the narcissistic fantasy of plenitude and innate fecundity, the potential for self-reproduction sustained by the fiction of the phallic mother (the hermaphrodite par excellence). The cyst-collecting gynecologist, then, is like Freud's *coupeur de natte* (plait cutter), in whom the two incompatible assertions, that the woman has a penis and that the woman is castrated, instill the need to carry out the castration he disavows.[162] In the same vein, when the female genital apparatus created by the skin's polyplike folding in upon itself proves more protean and mobile—inclined as it is to retroversion, retroflexion, inversion, and prolapse—than Pozzi might like, tools of the most surprisingly mechanical nature may be employed to force its alignment or conformity to his vision (fig. 120). Fetish (or climbing?) gear may also be brought into play when the fantasy of reversibility is threatened by fistulae (fig. 121).

Pozzi's love, his worship, of all his women patients, including Bernhardt, partook of this fetishistic sexual economy.[163] Whether curing, collecting cysts, or softly killing,[164] he exercised a mesmeric allure over his patients, as Bernhardt's effusive notes "to the Being I adore and admire and to whom I would with joy give my life" attest.[165] It is certain, however, that not everyone he treated considered the touch of his scalpel or his hands a blessing. On 13 June 1918 Pozzi was shot in his bedroom (his examining room was, apparently, "under renovation")[166] by a former male patient who attributed his sexual malfunctions to an operation performed by the doctor.[167] Pozzi's intestines were perforated in multiple places by the bullets, yet although entrusted to the care of his friend Dr. Martel at the hospital, it was Pozzi "himself who directed the operation, with great sang-froid." After twelve holes had been found and sutured, "they believed the operation finished," yet just when "the doctor seemed to repose with calm, they saw him start. A jerk . . . a clot of blood left his mouth. . . . He was dead. They supposed that a thirteenth perforation had passed unperceived." "Dr. Pozzi died of the wound he had healed so many times," one newspaper concluded.[168]

Whether by palpation or penetration, the gynecologist sought to know again,[169] to recognize and regain the place of origin, no more on pain of death; moreover, he desired to come and go (inside and out) at will. The irony of Pozzi's end—exploring at last the flesh of his own interior and almost, but not quite, escaping the death consequent upon grasping its forbidden fruits—is palpable.

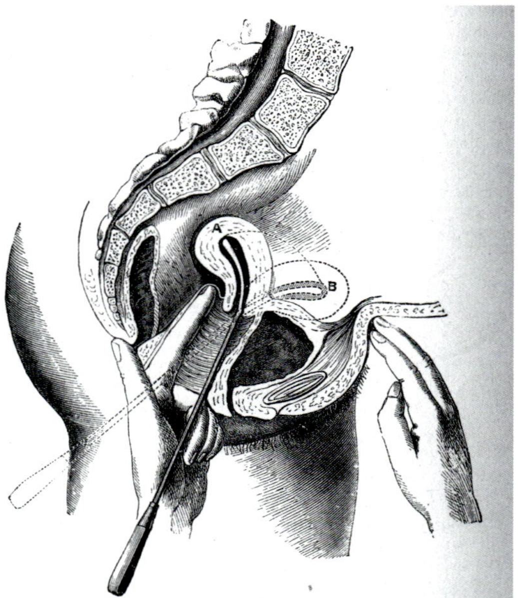

Fig. 253. — Réduction avec la sonde d'une rétroflexion de l'utérus.

120
"Reduction of a retroflexion of the uterus by the sound," reprinted from Barnes, in Pozzi, *Traité de gynécologie* (1890), 474, fig. 253.

The nonhuman fruits of the womb Pozzi studied provided an illustration of the reconciliation of the

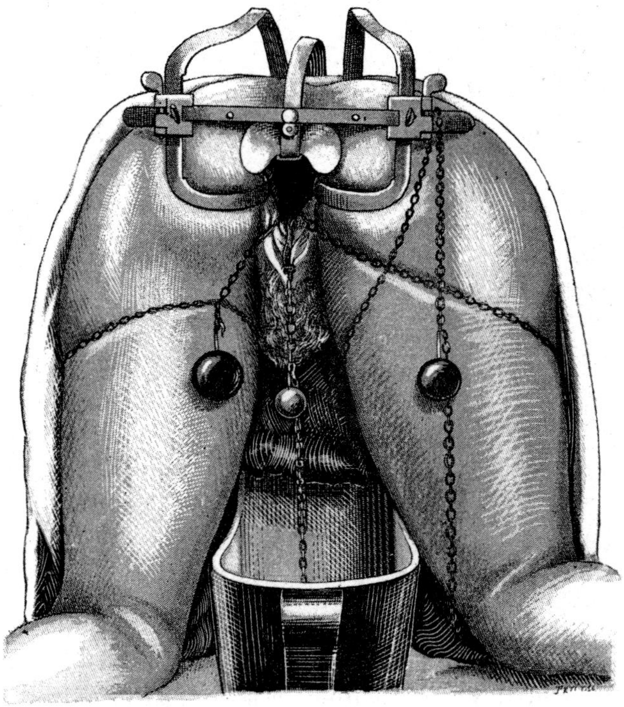

Fig. 395. — Speculum de Neugebauer et crochets rétracteurs, en place.

121

"Neugebauer's speculum and retracting hooks in place," reprinted in Pozzi, *Traité de gynécologie* (1890), 912, fig. 395.

human and vegetable models of reproduction that Sargent was trying to effect. Furthermore, the manner of the doctor's probings of the space of biological creation illuminates Sargent's investigation of the space of artistic creation. Like the gardening gynecologist, the painter used both touch and sight as exploratory tools. Lucia Fairchild described in her diary a visit to the Louvre with Sargent: "We went in to see the Botticellis—the frescoes by him. 'The Madonna with John and Child' he admired enormously, and he twirled his moustaches & bent forward to look at it, frowning a little—'Wonderfully lovely isn't it?' said he—& then touching the red rose, 'Beautiful, that.' . . . The Admiral's portrait he said was a beautiful arrangement too. 'Look at that,' said he, & touched the folds of the stock, 'and that,' touching the coat."[170] Touch was a necessary supplement to sight for the artist, as it was for the gynecologist. And like the gynecologist's, Sargent's own touch oscillated between palpation and penetration; contemporary reviews contended that the painter's "often shrewdly diagnostic"[171] vision could be both "pure tact"[172] and "surgeon-like."[173] His technique was both "akin to sleight of hand"[174] and apt to "violat[e] the secret recesses," his "candor with the brush," like the gynecologist's penetrations, "sometimes even perilous."[175] Sargent referred to his sitters as "patients,"[176] and operated as magician or surgeon according to the degree of affinity he felt with his sitter.

The painter's goal in exploring the horizon of interior and exterior that the floral subjects of his pictorial experiments offered was the same as the gynecologist's: to attain or exert a masterly consciousness through his "correct focusing,"[177] which would endow him with the key to self-inversion and procreativity. "Are human hands powerless to make things come to life?" asks the anguished painter Claude Lantier, model of the failed artist, in Zola's *L'oeuvre*.[178] Many critics insisted on the fecundity of Sargent's painting: Camille Mauclair proclaimed that "each touch of a John Sargent is a contribution to life,"[179] Robert Ross, that his art was "health-giving."[180] Sargent's portrait of *Madame X*, a patient of Pozzi's, however, proved a difficult test case.

The American expatriate Virginie Avegno Gautreau was a celebrated "professional beauty" in the Paris of the early 1880s.[181] A self-made floral work of art with her hennaed hair, rose-tinted ears,

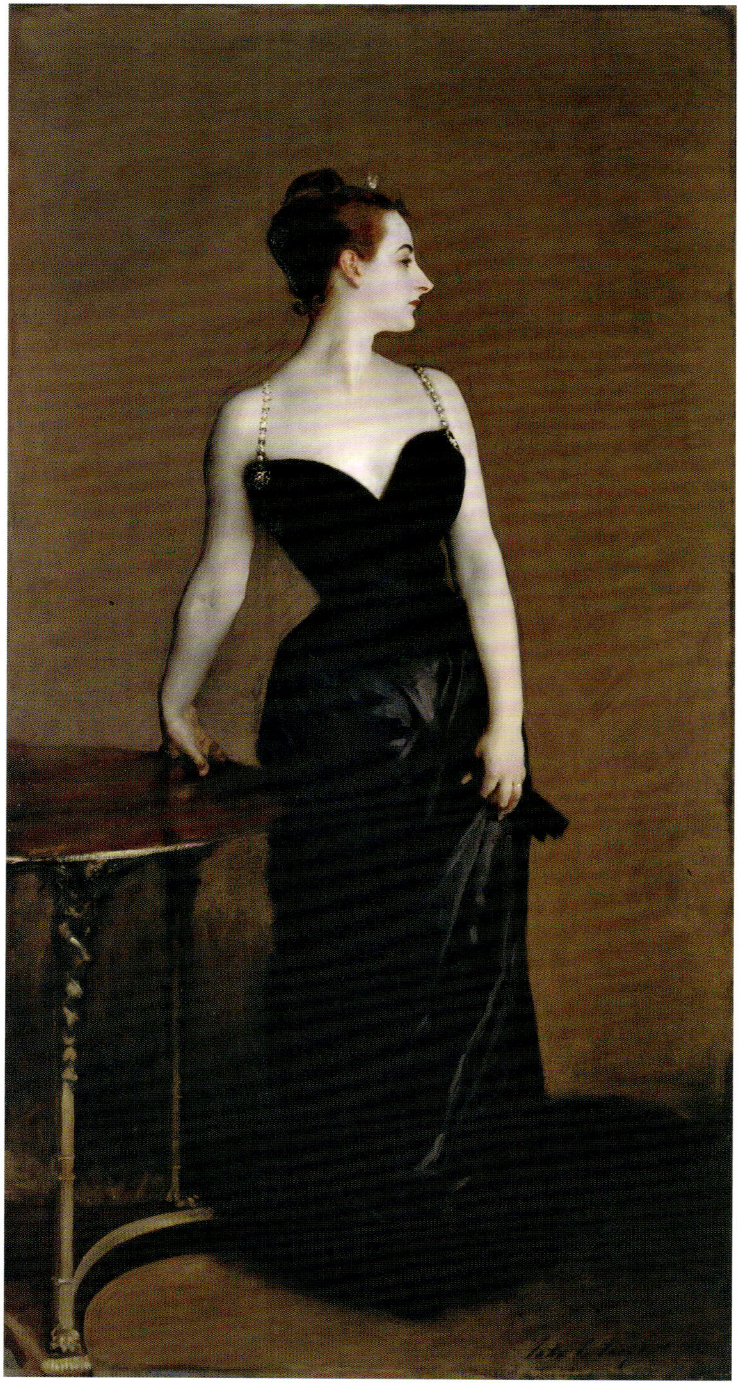

122
John Singer Sargent, *Madame X*, 1883–84. Oil on canvas, 208.6 x 109.9 cm. The Metropolitan Museum of Art, Arthur Hoppock Hearn Fund, 1916 (16.53).

and lavender skin,[182] the "orchidaceous"[183] Mme. Gautreau was rumored to have become pregnant without sexual contact from either husband or lovers, and thus seemed to embody the possibility of alternative artistic and vegetal reproduction. Sargent's portrait of this almost "extrahuman"[184] creature (fig. 122) faithfully records the floral hues of her carefully contrived flesh-pollen, a complexion "where heliotrope is steeped in pink."[185] The painting further insinuates the subject's luxuriant floral nature through the fantastically twisted stem of her right arm, the slipperily breast-bepetalling bodice of her satin nightshade gown, and the hermaphroditic sexuality suggested by the highlighted point swelling at the bottom of her *corsage* (bodice). Just as a female flower "would coquettishly arch her 'styles' if the insect came, and, to be more effectively penetrated by him, would imperceptibly advance, like a hypocritical but ardent damsel, to meet him half-way,"[186] this seductive *femme-fleur* twists her figure to display herself both frontally and in profile and thereby present a broader scope of enticements to the piercing eyes of the crowds swarming the Salon, maneuvers herself, in other words, into a better position to win their favor(s). The composition recalls that of Grandville's "Pensée" (Pansy) (fig. 123): if the *fleur animée* leaning against

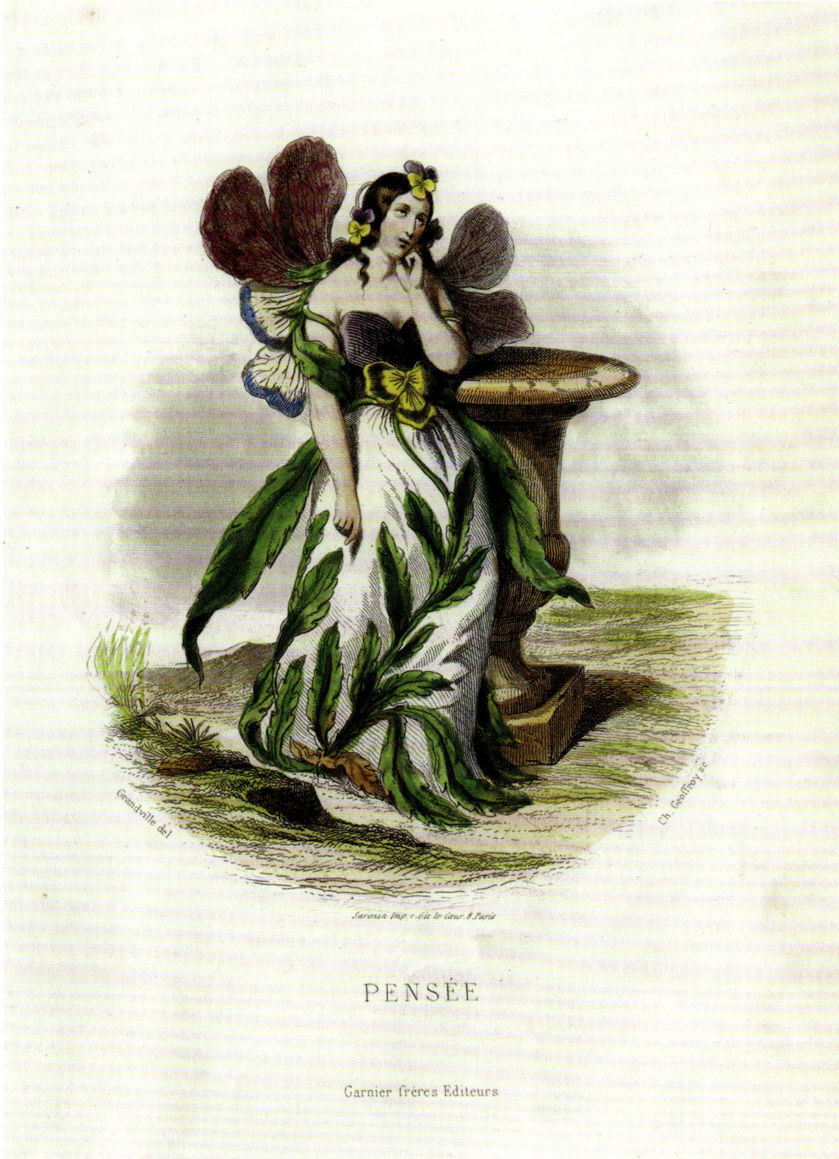

123
J. J. Grandville, "Pensée,"
***Les fleurs animées* (Paris: Garnier Frères, 1867), vol. 1.**

an urn is softer, lost in thought, the two figures share a similar petal bodice with slipping shoulder straps (in the version of *Madame X* exhibited at the Salon, Gautreau's right strap dangled; Sargent later repainted it).[187]

Despite her manifold floral charms, this vegetal subject treated by both Pozzi and Sargent problematized their access to the garden of earthly delights that, on the surface, she appeared to offer. When Pozzi operated he apparently discovered "Madame Gautreau's own unborn twin sister inside her"[188]—a travesty of his dreamed-of double flower, his fantasy of congenital fertility, this unborn progeny ingested rather than conceived through the digestive tract. And at the Salon, despite the hybrid effort of Sargent and Gautreau, who probably planned the pose and costume together,[189] the portrait was received with outrage, perceived not as a delicate, blossoming, lifelike creation but as an "indecent" and "cadaverous,"[190] "coarse" and "vicious,"[191] "cynical" mockery.[192]

Susan Sidlauskas has argued that the scandal surrounding the painting had, in fact, to do with its suggestion of syphilis: "The nineteenth-century viewer had a great need to see *under* the layer of makeup, for in a time of rampant tuberculosis and syphilis, only a vivid, unadorned skin was proof of good health. Not surprisingly, some of her critics saw the diffuse violet of Gautreau's skin as evidence of damage or disfigurement." Her tinted skin read as evidence of infection; Gautreau was seen as "death and decay" incarnate.[193] The painting's reception recalls that of Manet's *Olympia,* whose flesh was also seen as decomposing. Gautreau's skin more resembled that of "a dead than a living body," wrote one critic, for example.[194]

Other contemporary commentaries affirm Sidlauskas's reading while linking the painting more specifically to the decadent garden world of syphilis: a review of the 1884 Salon described the painting's subject as the ideal woman of "an overheated and contrived civilization, with a taste less for fresh, healthy flowerings than for the unnatural plants of the boudoir."[195] The "almost wilful perversion of the artist's knowledge of flesh-painting" that another critic saw might signal the artist's perverse admiration for the fruits of painted, and perhaps even tainted, flesh.[196] But let us consider the writer's "almost."

The importance of bare skin to the artist is elaborated in the Goncourts' novel *Manette Salomon,* where the painter Coriolis views skin as an interior: "all this delicious *je ne sais quoi* of the epidermis of woman, made, one could say, with the underside of a dove's wing, the interior of white roses."[197] These are, of course, not interiors proper, only interiors as they can be "perceived and grasped from the outside."[198] Sargent's exploration of membranes—his interest in petals, glass, fabrics, umbrellas, lanterns, and fans in addition to skins—was rooted in the chiasmic intertwining of interior and exterior worlds they can embody;[199] his investigation into the uncanny horizon of artistic creation and the origins of perception focused on these surfaces where plastic heterotopias may be founded. "By inserting a layer of opaque color between herself and the painter," Sidlauskas contends, "Gautreau violated the conventions of life-affirming transparency and compromised Sargent's primacy as metaphorical life-giver."[200]

Gautreau's resistance to the painter's appropriation of her natural epidermis or violation of her painted one was played out on multiple fronts. She refused to hold her torturous pose for any length of time. In a letter to Vernon Lee from the summer of 1883, when the painter was staying at the Gautreaus' summer home in Brittany, Sargent complained of "struggling with the unpaintable beauty and hopeless laziness of Madame Gautreau."[201] Tellingly, the hands were an important site in the battle of conceptions of art waged in the painting. To quote Sidlauskas once more, "the fleshy inner palm of Gautreau's right hand is articulated with flattened, almost crude, bands of coral and crimson, as though here, at last, Sargent can assert *his* conception of what counts as a vital skin tone. It is his painted resistance to the paint of his subject."[202] Yet this dash of vitality was not achieved with the painter's usual delicate touch. If the hand allowed access to an interior, it was only through forcible penetration: Henry Houssaye saw it as "deboned," evidence of Sargent's surgical sadism.[203] Unlike the softly palpating portrait of Dr. Pozzi, Gautreau's painting as a whole is coldly analytical, her razor-sharp, surgically precise silhouette unusual, and, at the time, critically unsuccessful, for Sargent.[204]

The sirens at the top of the table legs hint that Gautreau's floral charms represent a potential threat to the viewer, as does the crescent moon of Diana crowning this Virginie, a reminder that those who look upon the naked flesh of the virgin hunter-goddess are punished with castration and more; through the act of looking alone, Actaeon, or the "shore-dweller," unwittingly crosses the threshold into a watery zone of nonidentity, of metamorphosis and death, from which he is never to return. Gautreau was perceived by many admirers as both hunter-goddess and prey, Diana and Actaeon. Edward Simmons recalled that he saw "the noted beauty of the day, and could not help stalking her as one does a deer. . . . She walked as Vergil speaks of goddesses—sliding—and seemed to take no steps. Her head and neck undulated like that of a young doe, and something about her gave you the impression of infinite proportion, infinite grace, and infinite balance."[205] Metamorphosis, Venus tells us in the *Metamorphoses,* is "half-way between" death and exile,[206] and Lacan's characterization of the threat of castration that the illusion of conscious pellicular mastery invites could describe Sargent's portrait: "a wild odour emanates from it, providing a glimpse on the horizon of the hunt of Artemis."[207]

Like Courbet's *Origin of the World* and Pozzi's illustration of bimanual palpation, Sargent's portrait demonstrates that the relationship between the visible and invisible is as intimate—as thin as petaline layers of skin—and as potentially treacherous as the relation between the sea and the strand, where depths are uncertain and identities can be washed away. This zone was threatening to Gautreau herself: when she bathed in the sea, crowds "would gather to watch the famed beauty." Emerging from the water, "where she lost much of her make-up," she would not come ashore to be ogled, but waited for "a 'mulatto giant'" to throw "a huge towel over her" and carry her, "completely covered, to a dressing room."[208] This anecdote literalizes the threat of loss of identity that water entails, but the response to the portrait at the Salon was just as stripping.

If the painting of *Madame X* proved that the poetics of vegetable reproduction were not as easy to master as he may initially have thought, Sargent did not give up his horticultural quest, though he did leave "the unnatural plants of the boudoir" behind for a time. In his next major vegetable work, *Carnation, Lily, Lily, Rose,* the painter continued his investigation of the floral, watery, and uterine space of artistic creation, managing in the end to locate his painterly creation more firmly in a terrestrial garden world. The connection between procreation by palpation and procreation by pollination, both consisting of a brushing of surfaces, will be elaborated in the next chapter, as will the painter's difficult negotiation of the boundary between touch and penetration.

4

PORTRAIT OF THE ARTIST AS A YOUNG LADYBIRD

It is natural that a work of art should be a *Hortus inclusus.*

—VERNON LEE AND CLEMENTINA ANSTRUTHER-THOMSON, *BEAUTY AND UGLINESS, AND OTHER STUDIES IN PSYCHOLOGICAL AESTHETICS*, 347

Edmund Gosse, son of the naturalist Philip Gosse, identified one of the pollinator personae Sargent adopted. He recalled that in order to create *Carnation, Lily, Lily, Rose* (fig. 124), the painter each evening "took up his place at a distance from the canvas, and at a certain notation of the light ran forward over the lawn with the action of a wag-tail, planting at the same time rapid dabs of paint on the picture, and then retiring again, only with equal suddenness to repeat the wag-tail action."[1] A wagtail is a bird named for the curious way it twitches its tail feathers. Gosse is offering here what he professed he tried to provide in his literary biographies: close-up portraits of bees "in the mystery of mellification."[2] Honey making begins, as he well knew, with the contact of bees and flowers, with pollination; this artistic minibiography thus begins with an image of a pollinator in action.

Numerous other memoirs describe the way Sargent's artistic conceptions were realized by the painter's darting or flying about, often humming while he worked.[3] In his studio, "the track of his paces" to and fro had worn out the carpet to the point that "it suggested a sheep-run through the heather."[4] Sargent's "rushing back and forth" was to match "tones on his palette from a distance."[5] Like the matchmaking bee flitting between blossoms, the painter's hand's "curious circular motions"[6] even mimicking the spiraling dance of the bee, Sargent reproduced colors on his canvas, and through the correspondence of blooms and shades, pinks, violets, fuchsias, lilacs, lavenders, roses, amaranths, carnations, and greens—colors and flowers—sprang up.

As Elaine Scarry has shown, "the practice of writing" is, in its most material terms, "a laying down of flowers upon flowers," since paper and ink are both vegetable substances.[7] Painting, whether on wood, canvas, or paper, also places vegetable pigments and oils on a vegetable ground.[8] The connection between flowers and pigment has a venerable genealogy in art history, both materially and allegorically, from the story of the child Titian gathering blossoms to make his first attempt at coloring (fig. 125)[9] to Charles Ephrussi's description of Berthe Morisot grinding "flower petals onto her palette, in order to spread them later on her canvas with witty, airy touches, thrown down a little haphazardly."[10] In the late nineteenth century, as James Rubin has argued in *Manet's Silence and the Poetics of Bouquets,* bouquets and the "flowers of the palette" came to function as "paradigms for artistic representation. In

124

John Singer Sargent, *Carnation, Lily, Lily, Rose,* 1885–86. Oil on canvas, 174 x 153.7 cm. Tate Britain.

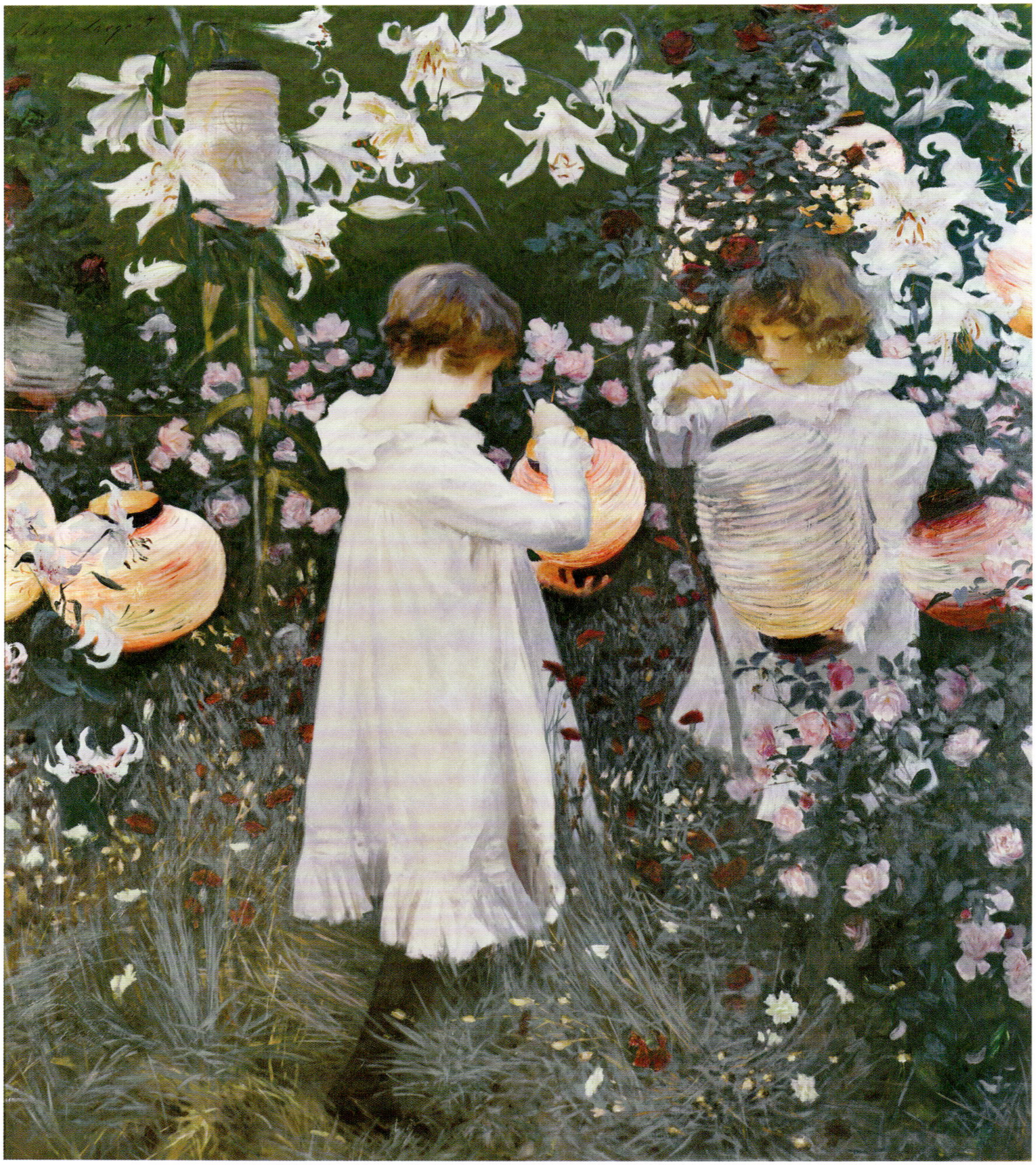

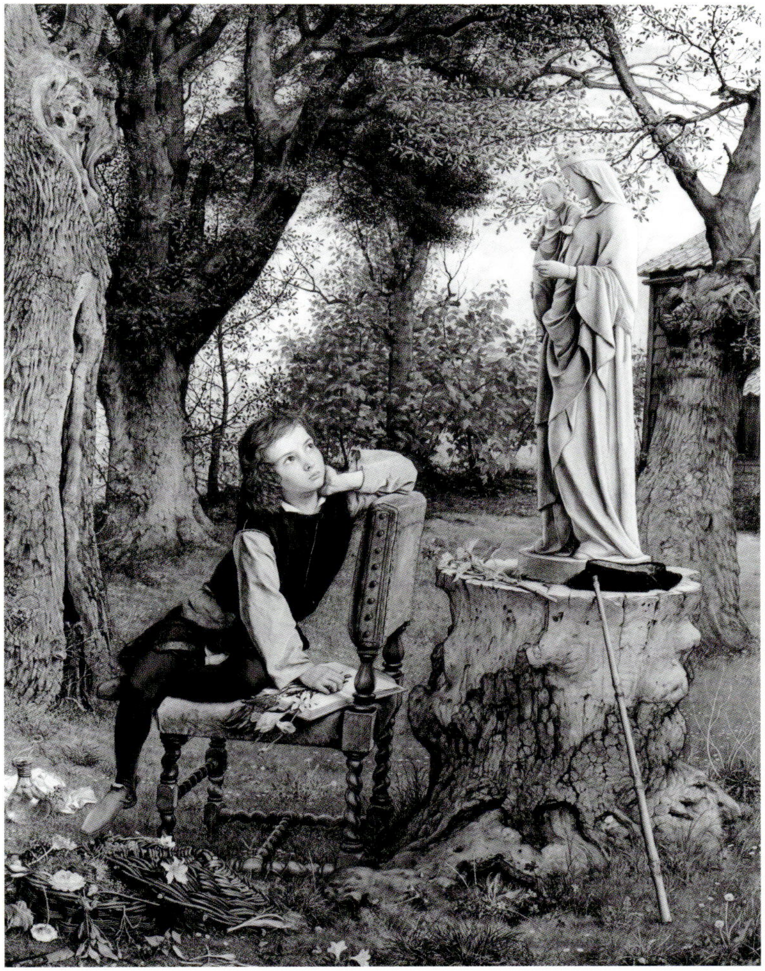

125
William Dyce, *Titian Preparing to Make His First Essay in Colouring*, 1856–57. Oil on canvas, 100.5 x 79 cm. Aberdeen Art Gallery and Museums Collections.

Mallarméan terms, they allude to the merger of self and world that produces art," or in mine, to their cross-fertilization.[11]

While Rubin speaks figuratively of "the flowers of the palette," there are many images from art and visual culture of the period that render these blooms more literally. In Martin Johnson Heade's *Roses on a Palette* (fig. 126), a full-blown, deep pink blossom and its leaves are spread out on an earth-colored palette; pink and green spots of pigment to the right can be read as leaves and a petal (while a yellow one assures us we are looking at paint), and the stem lying across the palette mimics the thin brushes that shoot out ahead of it. Narcisse Diaz's painted palette in the Auberge Ganne—a bouquet in florid disarray—is more difficult to decipher. Emerging from a blue vase, sprays of pinky-white, pale blue, yellow, purple, bright orange, and dark red blossoms fade in and out of volume and legibility, threatening to dissolve into a mess of pigment and literally confusing petals and paint. In the last few decades of the nineteenth century greeting and trade cards were often illustrated with palettes covered in flowers rather than paint (fig. 127). Sometimes, as in the center palette, cross-fertilization seems to give rise to an image—the pastoral scene in the middle. The flower/pigment

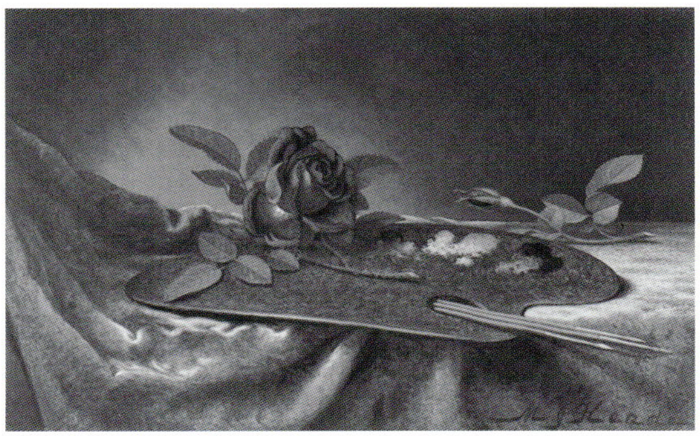

126
Martin Johnson Heade, *Roses on a Palette*, c. 1883–1900. Oil on canvas, 35.7 x 56.2 cm. Questroyal Fine Art, LLC.

127

Trade cards and assorted scrap with palettes, 1870–1910.

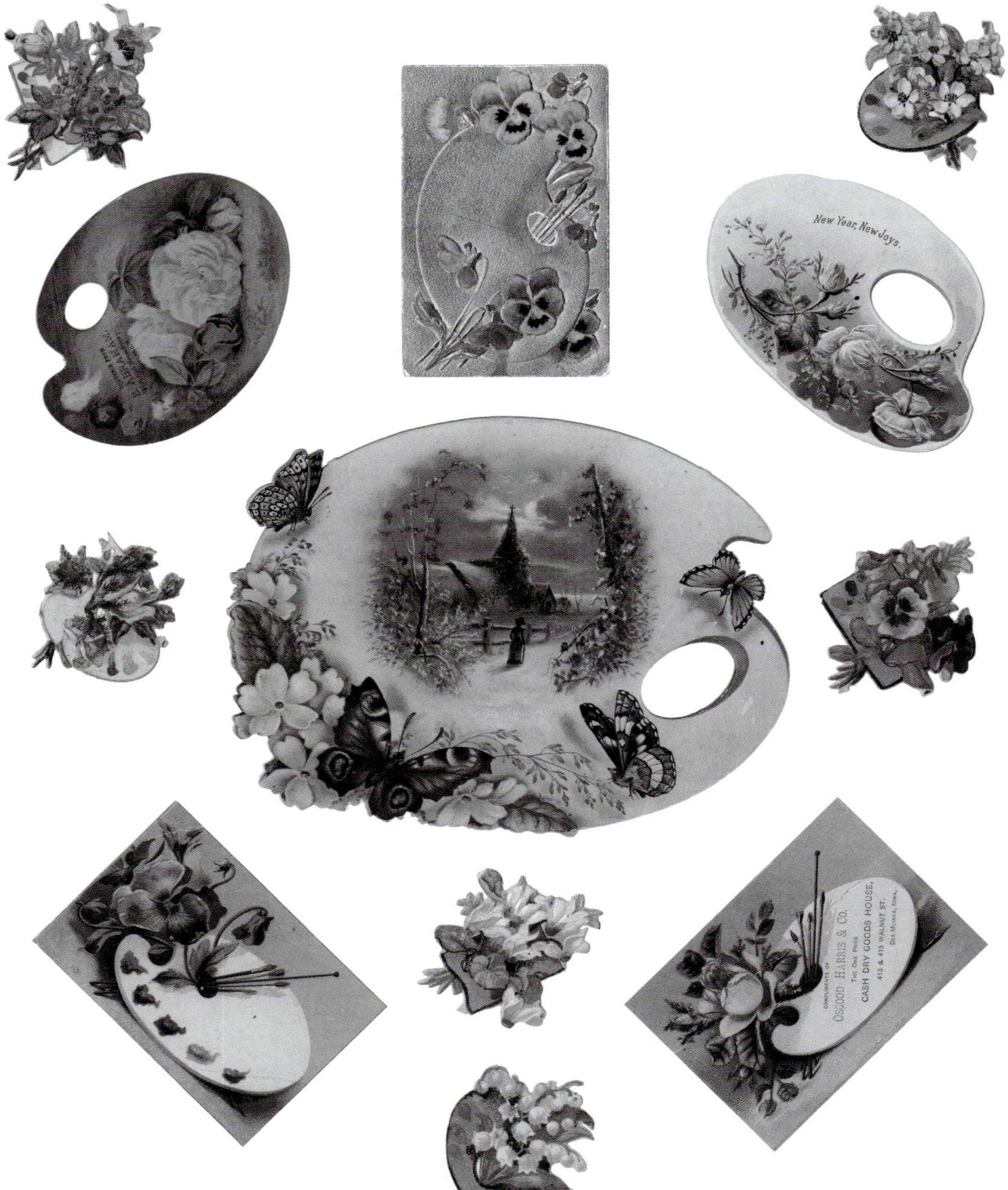

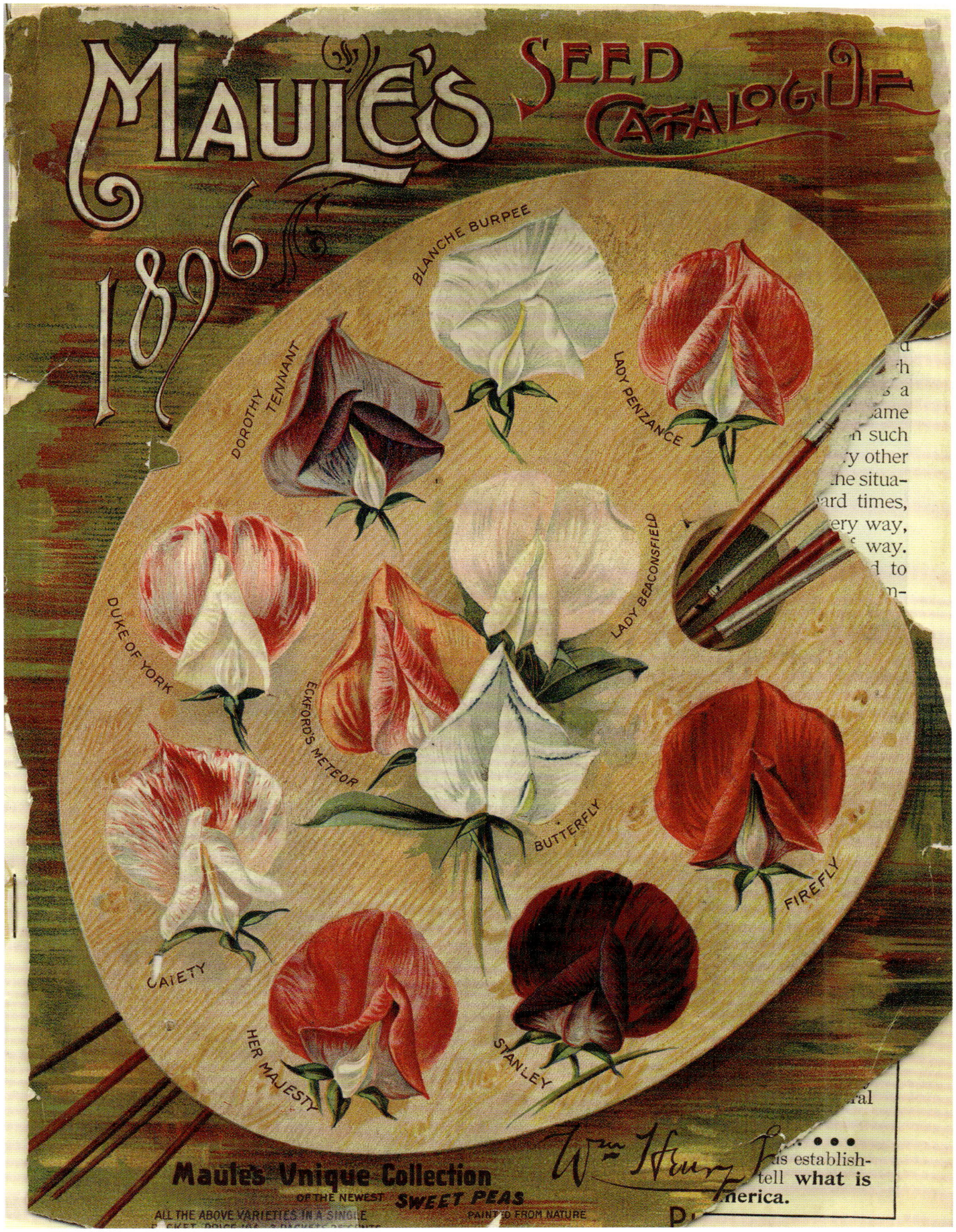

128
Cover of *Maule's Seed Catalog*, 1896, 27 x 19.7 cm.

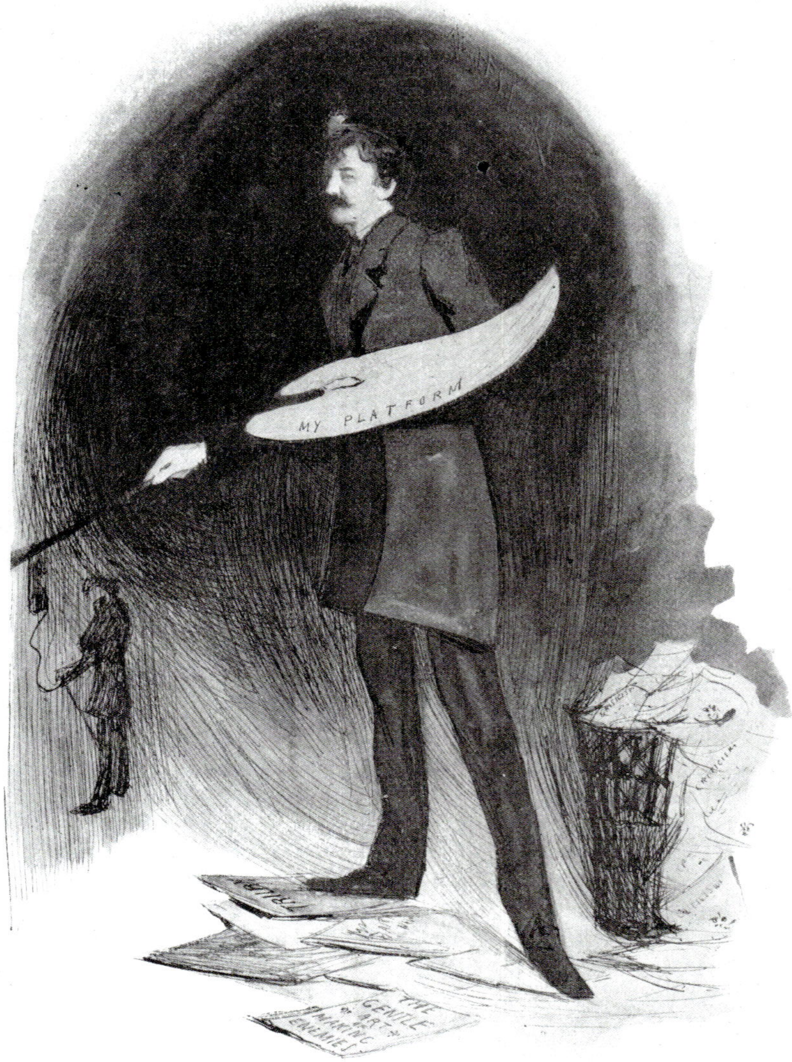

129
"The Man Whistler at the Telephone," *Life* magazine, 8 November 1894, 293.

motif carried over into other media. The brightly colored cover of a seed catalogue from 1896 (fig. 128) likens the gardener's and painter's tasks: it lays out different shades of sweet pea on a palette (white, red, purple, and various shades of pink), the petals' streaks and folds even resembling brush marks. Behind the palette, a green and brown background suggests the space of a literal garden or a landscape painting, for it too is streaked with what can be read as brush marks.[12] In a last example, a caricature of Whistler from *Life* magazine (fig. 129), the artist's palette and thumb form a giant anthurium, a flower whose common name *is* "painter's palette."

Sargent opted for such a literal conflation of palette and bouquet in his portrait of the artist William Merritt Chase. Pictured in the act of painting, Chase is standing in a blue work jacket (fig. 130). Clutched in his left hand, the artist's palette and brushes resemble a bouquet (fig. 131): a pale yellow-bristled brush is a tulip; smears of pigment form orange and white flowers, iris cultivars of some sort; and green leaves shoot off to the right. At the top left, just beyond the painter's maulstick, a pendulous, pale orange, snowdrop-shaped blossom actually hangs off the palette, the edge of which is also overshot by the leaves at the right. The neatness of this arrangement calls to mind Baudelaire's description of Delacroix's palette: "meticulously and delicately prepared," it was "like an expertly-matched bouquet of flowers."[13] Chase's slightly extended right arm holds a probelike paintbrush, as though he is about to reach across the canvas and brush against the flowers for their pollen-pigment before laying down a stroke—about to deliver up a

130

John Singer Sargent, *William M. Chase*, 1902. Oil on canvas, 158.8 x 105.1 cm. The Metropolitan Museum of Art, gift of the Pupils of Mr. Chase, 1905 (05.33).

131

Detail of *William M. Chase*.

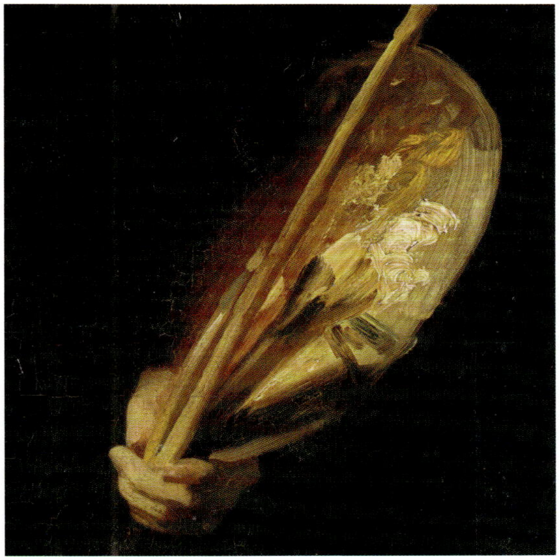

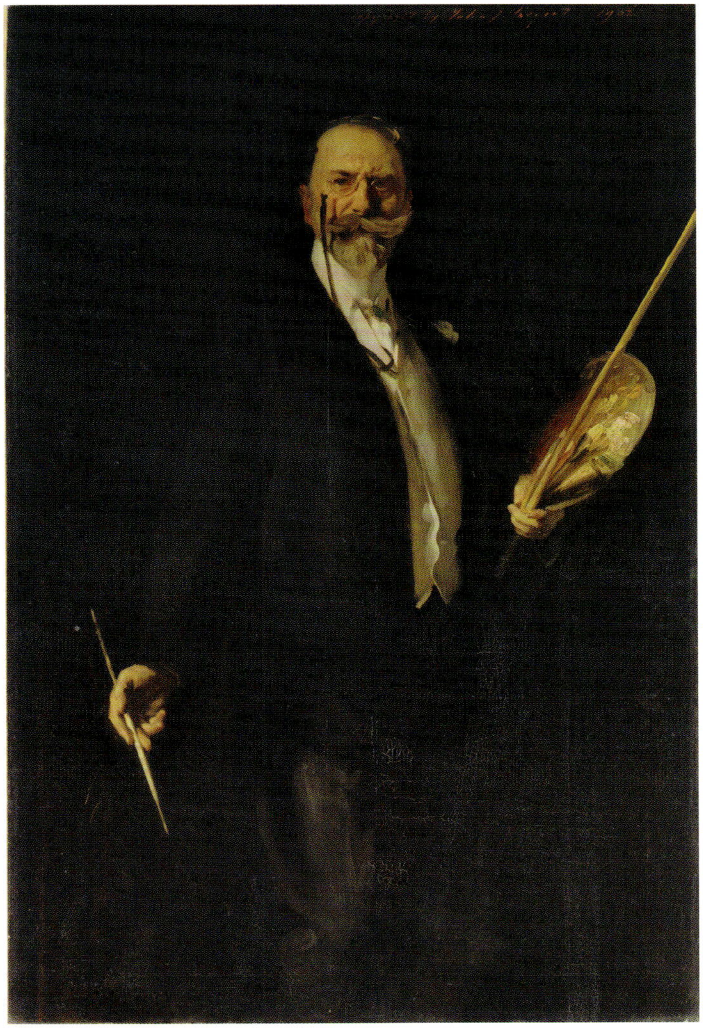

correspondence, to create a work of art through hand-pollination. Sargent's flowers here and elsewhere, while more or less identifiable, eschew traditional symbolism or a Ruskinian fetish for botanical accuracy; his paint-blossoms, in keeping with his flower-paint poetics, function as sites of literal transference: of color and pollen, and affect.[14]

In the country Sargent kept his paints in a fruit basket,[15] though his pigments remained "spots of coloured mud," as Cecil Chard put it, until the artist's touch "infused" them "with the vigour and the actual breath of life."[16] Sargent's colored strokes, like Chardin's in the Goncourts' description, seem mere "smears" or "fragmentations" up close; when viewed from farther away, however, "flowers rise up from the canvas; the leafy design of the carnation, the heart of the flower and its tender shadow, its torn and

132
John Singer Sargent, *Claude Monet Painting by the Edge of a Wood*, 1885. Oil on canvas, 54 x 64.8 cm. Tate Britain.

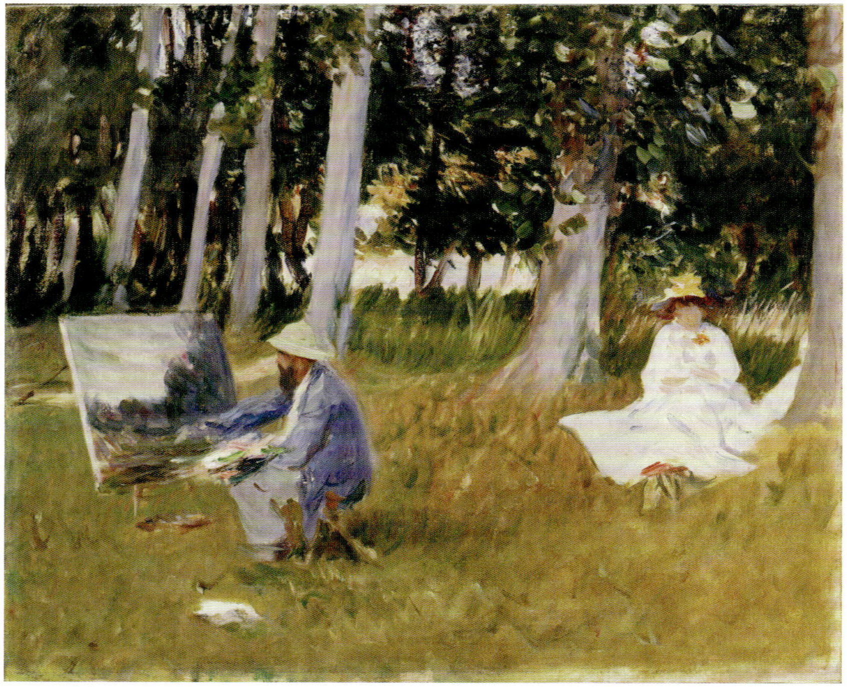

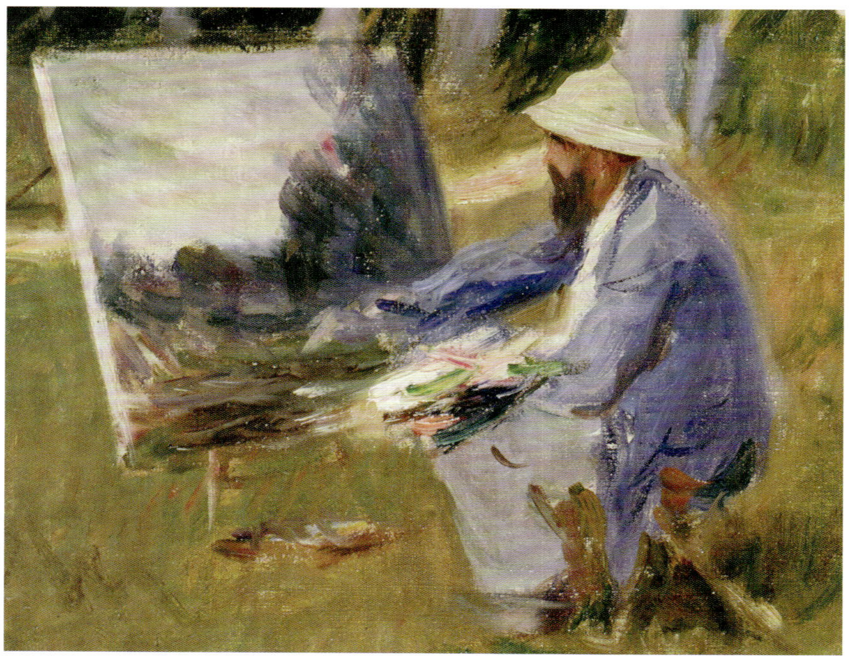

133
Detail of *Claude Monet Painting by the Edge of a Wood.*

ragged forms, everything comes together and unfolds."[17] Mary Hale, the painter's cousin, described his canvases as having "the look of having burst into bloom all at one moment,"[18] though one of his students testified that the blossoming occurred in stages: "as he painted it, the mouth bloomed out of the face."[19] The canvas that reads as "bursting into bloom" was Sargent's ideal. He wrote of Hercules Brabazon Brabazon's watercolors, which he much admired, "immediate sensations flower again in Mr. Brabazon's drawings, with a swiftness that makes one for the time forget that there has been a medium."[20] The viewer's phantasmatically unmediated experience of the work of art and the sensations or impressions that gave rise to it has its counterpart in an ideal of artistic creation grounded in bodily practice, which shuns prostheses of all kinds. Sargent associated this with Impressionism—not surprisingly, given the movement's emphasis on perceptual immediacy and immersion in the landscape, its development out of "naturalism," and of course its predilection for vegetal subject matter.

Sargent had first met Monet at the second Impressionist exhibition in 1876, and having become friends, they spent time painting together in the summers.[21] In Sargent's portrait *Claude Monet Painting by the Edge of a Wood* (fig. 132), Monet seems not even to use brushes. Streaks of pigment on his palette (fig. 133), which is more brightly colored and facturally alive than anything else in the picture, fly off his generative lap and onto the canvas in front of him and elsewhere, of their own accord or as if windblown.[22] The canvas itself, literally sprung from the paint-pollen of the artist's loins, reads as a giant calyx being brushed into existence.[23]

By imagining the work of art as botanically reproduced, Sargent was sharing in the enduring fantasy of artistic parturition, but reenvisioning it for the modern period. A pair of portraits of Sargent's artist friends serves to illuminate his evaluation of the relative merits of the vegetal and human models of creation for the bachelor artist. The canvases are of the same size, were painted at approximately the same time in the Italian Alps (c. 1904–6), and portray the two subjects in almost identical rooms. *An Artist in His Studio* (fig. 134) depicts the painter Ambrogio Raffele amid several of his landscapes; *Padre Sebastiano* (fig. 135), a portrait of a priest reputedly sent to a remote parish because of his too liberal views,[24] represents the subject in the midst of botanizing: a heap of floral specimens lies before him, and his hand is poised to draw. In these images, the pursuits of painting and botany are localized in the bedroom. Tousled sheets seem a part of the process of artistic creation or horticultural study, as though for the bachelor painter and celibate priest these activities allow an alternate form of sexual encounter and mode of reproduction. Both methods *work:* the painter's bedroom is littered with canvases and the priest is in the middle of a book of studies. However, what the painter reproduces, from photographs and not *en plein air* (note the black-and-white print in his hand), are old-fashioned pictures. He has literally backed himself into a corner. In contrast, the priest is forward-looking (and thinking: he will soon give up the priesthood and abandon the Old World).[25] His subjects and his studies are literally fresh.

134
John Singer Sargent, *An Artist in His Studio,* 1904. Oil on canvas, 56.2 x 72.07 cm. Museum of Fine Arts, Boston. The Hayden Collection—Charles Henry Hayden Fund, 05.56.

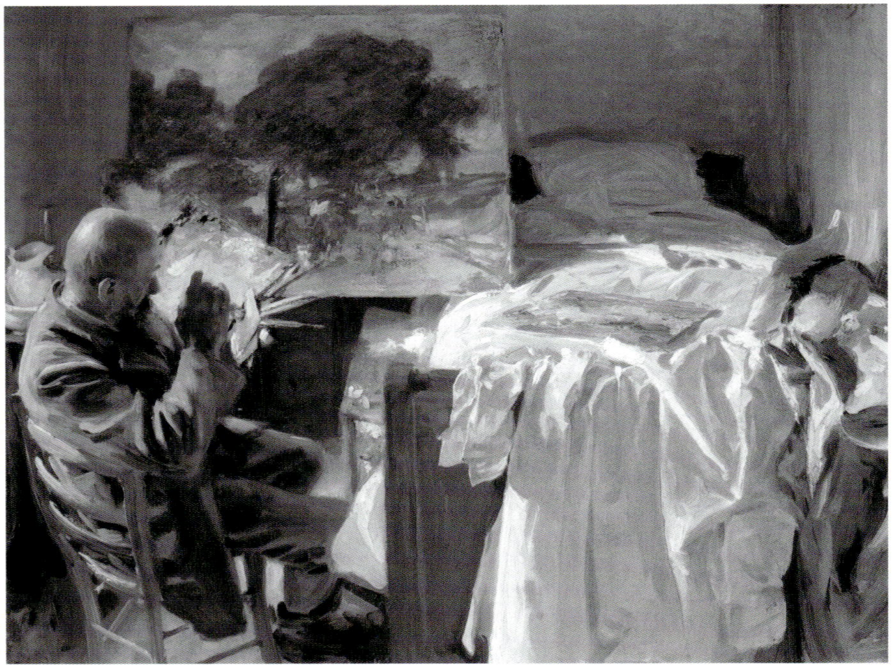

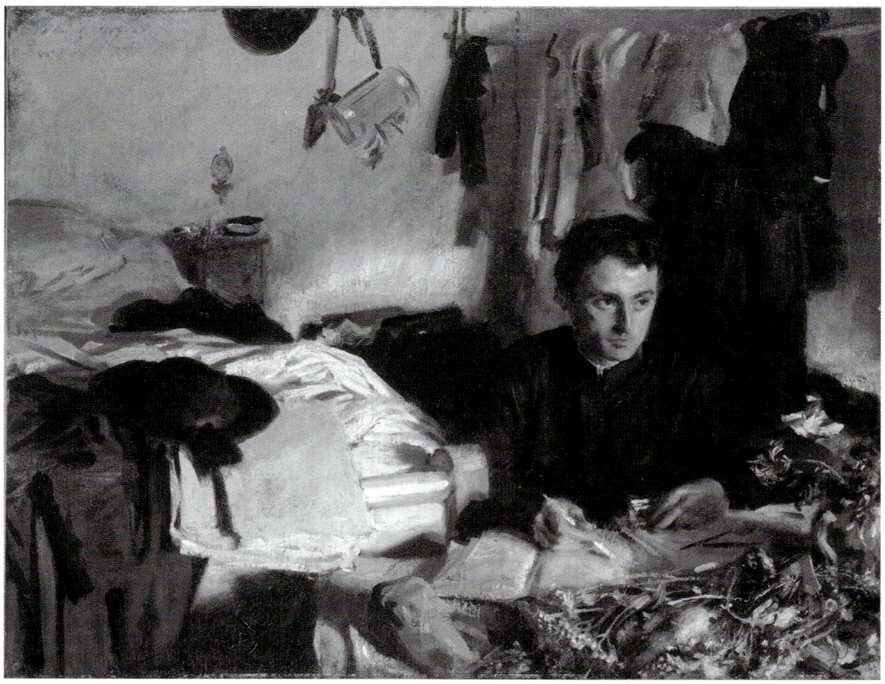

135
John Singer Sargent, *Padre Sebastiano,* c. 1904–6. Oil on canvas, 56.5 x 71.1 cm. The Metropolitan Museum of Art. Rogers Fund, 1910 (11.30).

136

Photograph of Sargent painting *Carnation, Lily, Lily, Rose* at Broadway in Worcestershire, 1885–86. Gelatin silver print on paper, 11.2 x 9.4 cm. Harvard University Art Museums, Fogg Art Museum, gift of Mrs. Francis Ormond, 1937.7.27.1.A.

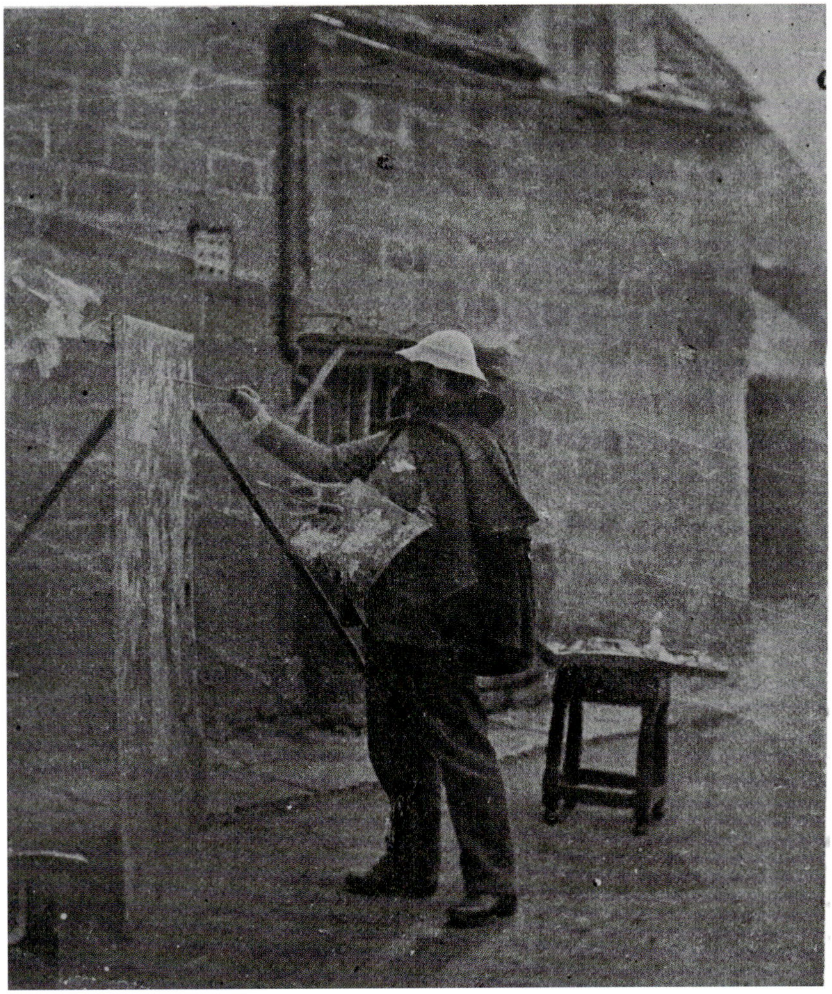

Sargent's adoption of the cross-fertilization model of painterly creation sprang out of an artistic identity crisis, induced by the scandal his entry for the 1884 Salon, *Madame X,* provoked. His painting, as described in the previous chapter, was perceived as too decadent, brutal, contrived, and American, and the "unnaturalness" of the model was equated with Sargent's own. In reaction Sargent launched into his most Impressionist moment, seeking a simplicity and naturalness of painterly practice and subjects he felt certain he lacked. He went to the English countryside to recuperate, and there he began *Carnation, Lily, Lily, Rose.*

Sargent worked on the canvas over fourteen to seventeen weeks of the summers of 1885 and 1886 at the village of Broadway, which was something of an unofficial American artists' colony as well as a haven for amateur naturalists who came to the area to collect specimens of wildflowers.[26] As keenly interested in natural history as his fellows there,[27] he painted as a "naturalist" in the scientific sense *and* in the artistic one. A self-titled "would-be Impressionist"[28] who had been working alongside Monet earlier in the summer of 1885, Sargent took to heart the doctrine of *plein-air* painting; a photograph shows him at work on *Carnation, Lily, Lily, Rose* outside (fig. 136).[29] At dusk each evening the little girls took their places among the flowers and lanterns, and for about ten minutes while the light was right Sargent dashed back and forth painting, often scraping down the canvas afterward and beginning anew the next day, to ensure that the picture was a record of his live impression.[30] He wrote to his sister Emily of the difficulties: "Impossible brilliant colours of flowers and lamps and brightest green lawn background. Paints are not bright enough, and then the effects only last ten minutes."[31]

Carnation, Lily, Lily, Rose, however, tells its own story of creation. Some writers have noted that it abstractly records a kind of sexual encounter.[32] In her 1969 dissertation Martha Kingsbury asked, "What is the significance of young girls in white, in a garden of flowers, absorbed in the solemn action of placing long, stiff, hot-on-the-end things in

137
Saxoléine trade card, 1890s. Chromolithograph, 14.1 x 9.3 cm.

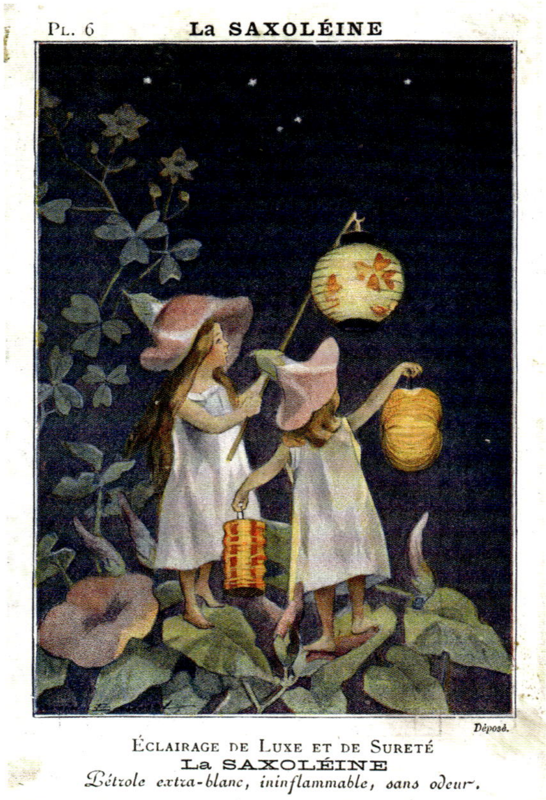

138
John Singer Sargent, *Study for Carnation, Lily, Lily, Rose,* c. 1885. Graphite on darkened off-white wove paper, whole page 24.7 x 34.6 cm. Harvard University Art Museums, Fogg Art Museum, gift of Mrs. Francis Ormond, 1937.7.21.11, verso.

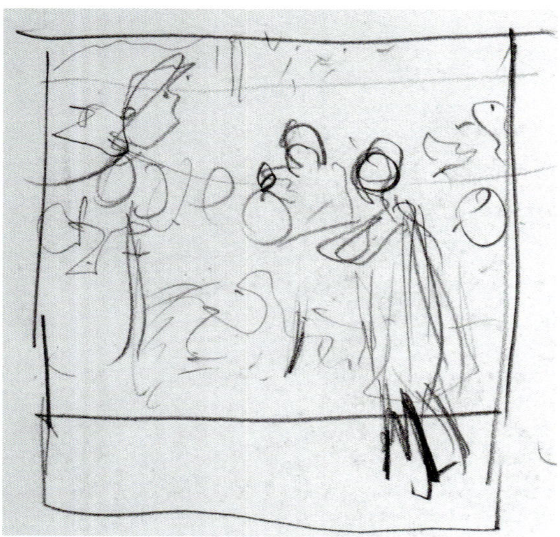

round, hollow things that light up with a warm glow when the proper contact is made?"[33] "Lantern" was slang for vagina in French,[34] and Richard Ormond has described the match as an "erotic metaphor" elsewhere in Sargent's oeuvre.[35] While I agree that the act of lighting lanterns here is sexually provocative, I would argue that the copulation represented is botanical, that in *Carnation, Lily, Lily, Rose* artistic creation is allegorized as cross-fertilization, and painting specifically as hand-pollination with a paintbrush.

In the picture two young children hover in the midst of a floral realm at twilight. The absence of a horizon line and of any indicators of depth or recession, coupled with the green ambient light and the gigantic scale of the lilies, gives us the impression of viewing a world in microcosm. The appropriately diminutive models, Dolly and Polly Barnard, are wearing white dresses and appear to float above the grass. These "flower-like little girls"[36] partake of the nineteenth-century fascination with the *femme-fleur,* and *Parsifal*'s flower maidens in particular;[37] a Saxoléine advertisement (fig. 137) from the later nineteenth century, strikingly similar to Sargent's composition in its depiction of miniature flower girls and lanterns amid foliage in the evening, attests to the popular charm of such visions.

If the girls in the painting are *filles-fleurs,* though, the iridescent dress of the child on the left (Polly) also folds behind her like a pair of wings, and her leg, the only one visible in the picture, is notably not weight-bearing. In early pencil sketches she has spiny mantislike forelegs or wings for arms, and sometimes even four legs (fig. 138). A contemporary critic described the girls as "unconsciously becoming a part of the scene, like so many moths or fire-flies."[38] Both are inserting paintbrushlike matches or candles—some sort of proboscides—into lily-white floral chambers. Lilies were sometimes troped as lamps,[39] a metaphor visually echoed in the painting, where the red-orange stamens and yellow dust of the flowers above repeat the color of the lanterns. In an oil study focusing on Polly Barnard (fig. 139) the match/paintbrush/proboscis analogy is brought to the fore: the figure's right hand dips an unseen implement into an orangey-brown lantern, the shape, size, and color of which make it read as a palette; the flat green surface before the child-artist stands ready to ready to receive a touch of blossom.

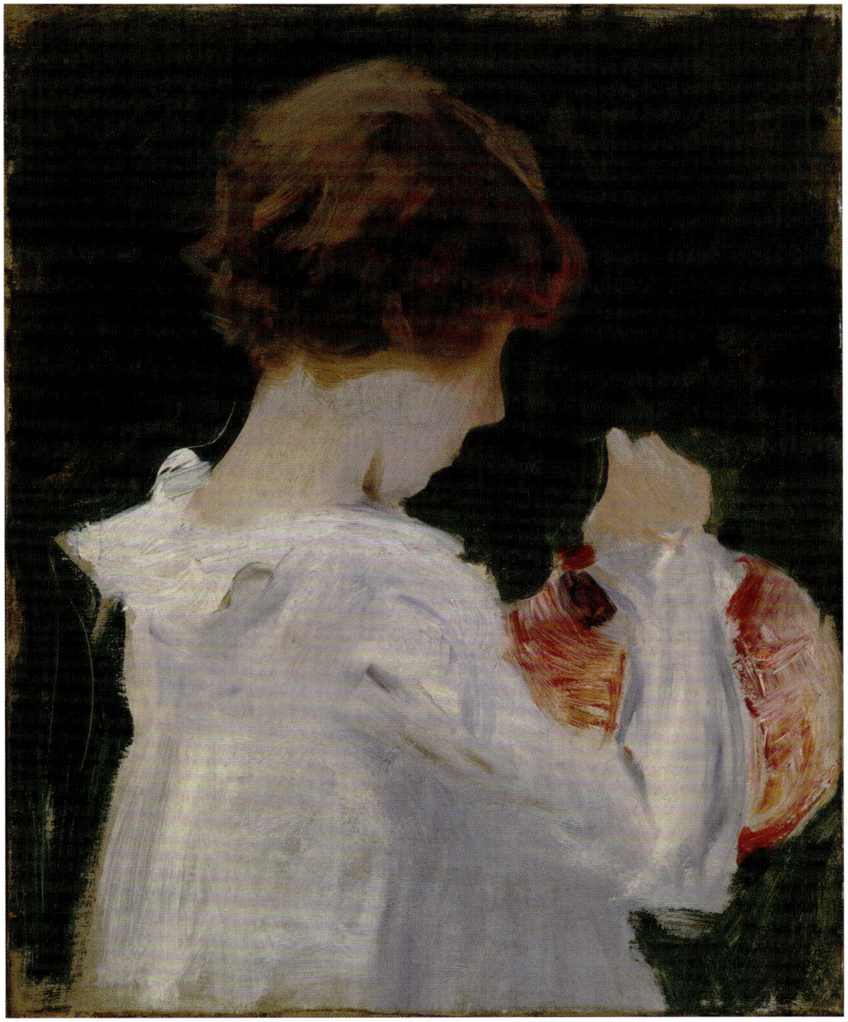

139
John Singer Sargent, *Study of Polly Barnard for Carnation, Lily, Lily, Rose*, c. 1885. Oil on canvas, 49.5 x 39.4 cm. Private collection.

In the finished painting the allegory of painting as cross-fertilization is most explicitly formulated in the lantern held by the figure on the right. When the match is inserted, its light is registered on the white surface of the lantern, a metaphorical substitute for the canvas, in streaks of bright paint. An Impressionist landscape of transient light and color is born: Monet's *Impression, Sunrise* can almost be glimpsed, miniaturized, within (figs. 140, 141). The match's flame, a tiny sun, casts luminous reflections in a sea below, rendered in parallel yellow-orange brushstrokes, while the folds of the lantern punningly suggest that the parallel brushstroke, so important to formalist narratives of modernist painting's evolution, is a structural necessity. The Impressionist project of capturing light itself on a white screen is literalized here, within the lantern and as a whole: "the painting simply *glowed* on the walls of Burlington House, when it was exhibited at the Royal Academy."[40]

When Sargent began painting *Carnation, Lily, Lily, Rose* in the summer of 1885, the two models of artistic procreation outlined in the Introduction—human (giving birth) and botanical (cross-pollinating)—were both available for his imaginary identification. At one point Sargent conceived of the garden he created in the painting as his own offspring. He wrote to Robert Louis Stevenson in

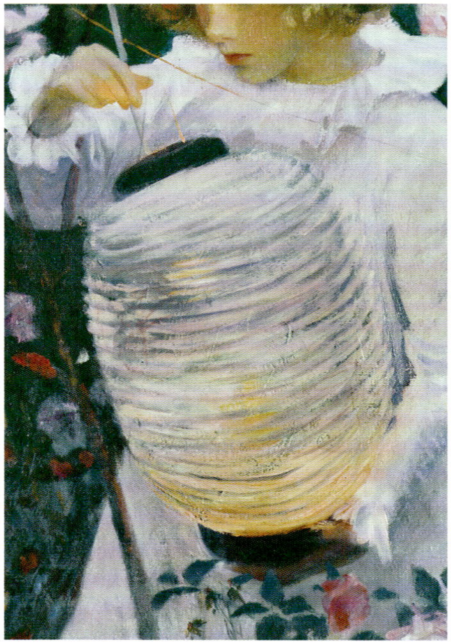

140
Detail of *Carnation, Lily, Lily, Rose.*

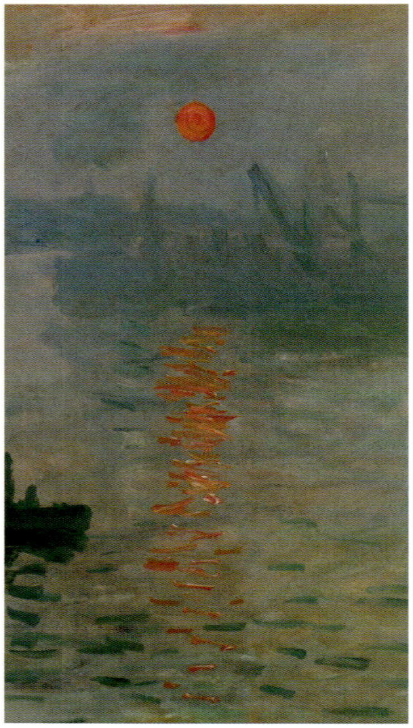

141
Claude Monet, detail of *Impression, Sunrise*, 1872. Oil on canvas, whole 48 x 63 cm. Musée Marmottan, Paris.

November 1885 that he had been "brought . . . to bed of"[41]—he was delivering or giving birth to—the picture, and his fellow artist Edwin Blashfield recalled that after many months of labor this "picture, daughter" at last "came to stay."[42] Over the course of the painting's evolution, however, which can be traced through a series of oil studies leading up to the final work, the vegetal variation of the creation-as-procreation metaphor came to supplant the human.

It was on a Thames boating tour with Edwin Austin Abbey in August 1885 that Sargent first conceived of *Carnation, Lily, Lily, Rose.* Abbey wrote that Sargent "nearly killed himself at Pangbourne Weir [fig. 142]. He dived off the same and struck a spike with his head, cutting a big gash in the top. . . . It was here that he saw the effect of the Chinese lanterns hung among the trees and the bed of lilies. . . . After his head was bound up he knocked it a second time and reopened the wound."[43] As Zeus gave birth to Athena, the painting, like a virgin daughter, sprang eventually from the painter's head.[44] This head birth was not without pain in either case; in Sargent's, the prophecy in the name Pangbourne Weir fulfilled itself in the sharp pang with which the idea for his creation was born. Traces of the artist's head wound may be found in the early oil studies. In one of them (fig. 143), a figure gazes into the depths of a blue-green lantern, while amorphous red congelations, blood and roses, float on and above her head. A mark of fertility (recall that for centuries, in both English and French, menstruation had been termed "flowers"), these blood-flowers protest against Baudelaire's certainty that only "death, soaring like a new sun, will make the flowers" of artists' brains "blossom."[45] An 1871 cartoon by Alfred Le Petit also shows brain-flowers (here *pensées* that bloom and are eventually sublimated into the higher atmosphere as clouds) as the product of a cracked head (fig. 144), as does Gauguin's *Still Life with Japanese Print* (fig. 145), where deep red and blue scabiosa blossom from the bloody crown of the artist's self-portrait jug. As evidence of wounds, however, these blooms affirm that the vegetal artist's flowering can be quite draining.

In the finished painting the roses are identifiable as such, with leaves and stems,[46] but in the study, it is as if we are looking at a reflection in a pool on which flowers drift, whose ripples destabilize the folds of the lantern at the lower left, threatening the vision with dissolution, a curious prefiguration of Monet's water lilies. By picturing a pool that gives rise to flowers, Sargent's study evokes the myth of Narcissus as well as that of Athena. In the *Roman de la rose*, the reflection of roses in the fountain of Narcissus drives the lover mad; he thinks to himself, "if I could have a chaplet of them, I would love no possession as much."[47] The roses hovering about the figure's head in the study evoke the lover's wreath, and may be seen in the context of contemporary illustrations of the poem. In 1884 Sargent visited the studio of Burne-Jones, who was painting scenes from Chaucer's version, *Romaunt of the Rose* (fig. 146).[48] After his encounter with Burne-Jones's illustrations, the *Roman de la rose* would certainly have been on Sargent's mind. The roses in the study, however, evoke not only the chaplet in the poem but also the wreaths worn by the Broadway colony at, significantly, *birthday* parties,[49] and of course the title of the finished painting, *Carnation, Lily, Lily, Rose,* which is a quotation from Joseph Mazzinghi's pastoral art song "The Wreath." The lyrics describe the composition of the circlet of flowers worn by a shepherd's beloved, and the piece was a regular feature in the musical evenings of Sargent and company at Broadway.[50]

In another oil study of the same figure (painted from slightly farther away, the pictorial

142
Francis Frith, photograph of Pangbourne Weir, 1890.

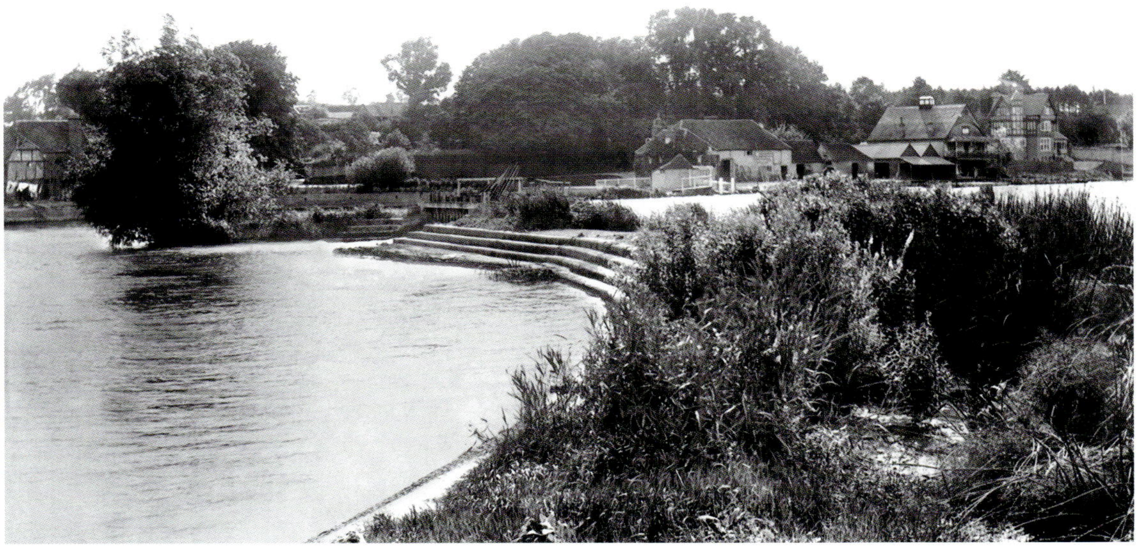

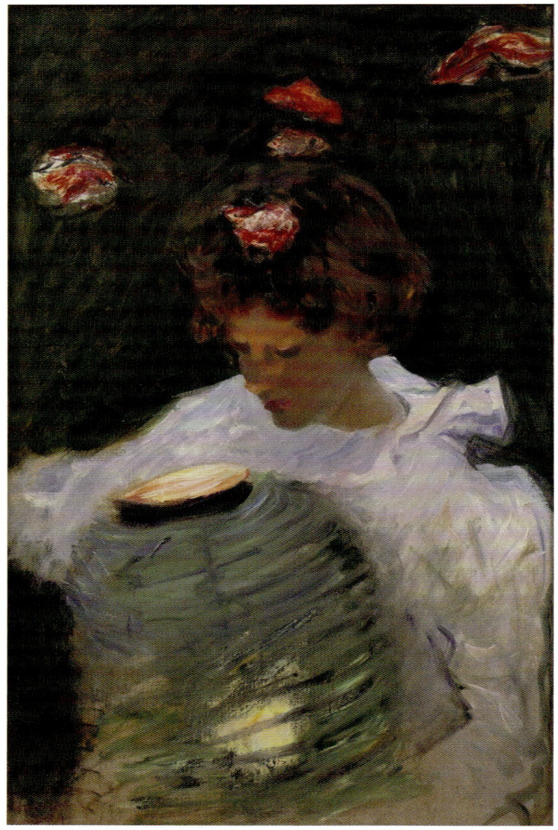

143
John Singer Sargent, *Study for Carnation, Lily, Lily, Rose*, 1885. Oil on canvas, 72.4 x 49.5 cm. Private collection.

field encompassing a cylindrical yellow and orange lantern and a round white one at center left, two or three large white lilies at the bottom left, and a smattering of carnations at the bottom right), the roses, more clearly attached to a stem, have been painted first in red, and then some of them painted over in white.[51] As though trying to erase or contain the wet and visceral origins of the work, the watery rose that figured as a head wound in the first study has undergone the most rigorous process of erasure; only the tiniest spot of red remains at the lower left. If the previous oil study presents the watery surface of the weir, in this one its depths have been contained in the shape of the heavy, almost succulent, dark blue lantern, the contours of which are much firmer than those of the previous study.[52] Where we will eventually see a Monet sunrise there appears a nocturne—on the Thames, of course, the site of Sargent's gash. The facial features of the model have, moreover, been rendered amorphous: the nose and eyes have been brushed out, and the large, dark circular zones where the latter should have been recall insects' compound eyes. Yet if the head wound has been painted out, the carnations at the lower right read as blood splatters on the figure's dress, and the plump wet lantern superposed over her abdomen suggests pregnancy as well as a painting. The lantern-belly conflation is made explicit in a self-portrait caricature by Sargent, which depicts him painting *Carnation, Lily, Lily, Rose* (fig. 147).[53] The folds of his shirt are calligraphically exaggerated, and its long tubular shape echoes the form of the oblong lanterns in the final picture, though this one is not yet convex—swollen or impregnated—with color and light. Spermatic globules of paint, however, fly off the artist's prosthesis and toward his belly. If one lands, it will produce a sun/son. A self-ironizing image, it illustrates the result of the painter's self- rather than cross-fertilization: a stain on the bottom of his shirt.[54]

The painting's originary wound reemerges in a larger two-figure study, where clusters of roses appear displaced but even more prominently, red gashy scribbles and smears of paint that function almost as a scotoma in the field of vision.[55] Aqueous blues and greens still dominate the palette, but the setting is more firmly terrestrial. In the finished picture (fig. 124), lanterns and flowers

FLEURS, FRUITS & LÉGUMES DU JOUR.

PENSÉES.

13

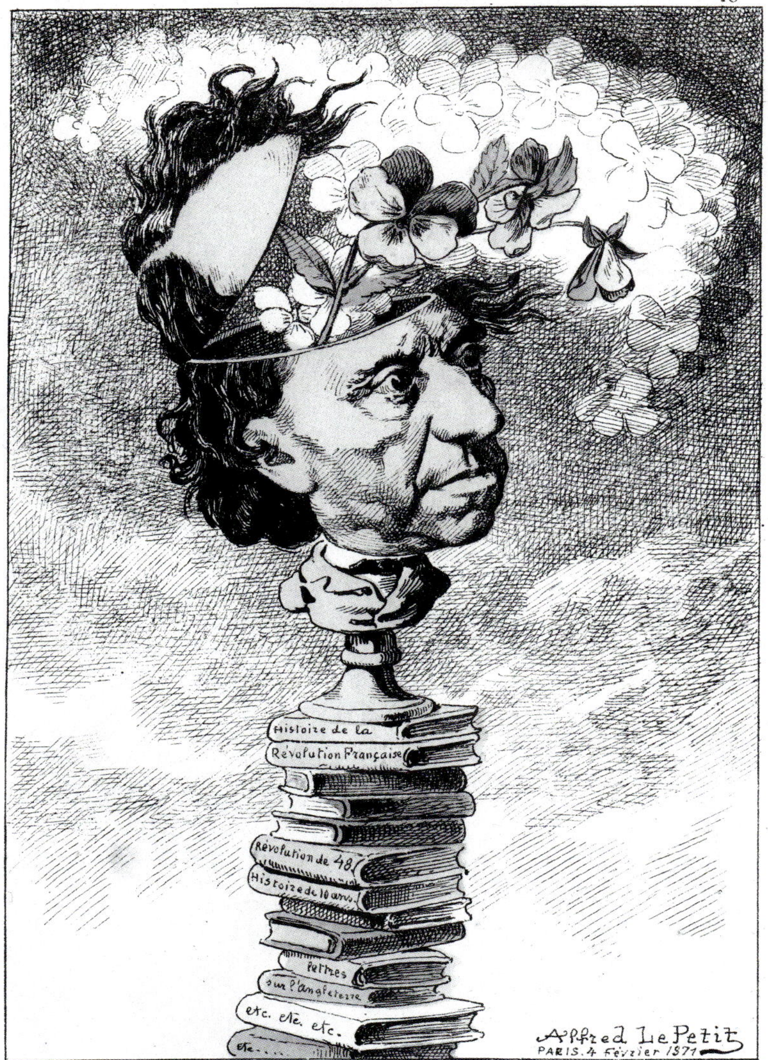

Au Bureau de L'ECLIPSE.16. Rue du Croissant PARIS

Imp. Coulbœuf, Paris.

LOUIS BLANC.

Il montre en tout ce qu'il écrit
Des opinions avancées,
Car, pour le peuple qu'il chérit,
Son front est rempli de pensées.

144
Alfred Le Petit, "Louis Blanc," in *Fleurs, fruits et légumes du jour* (Paris: Au Bureau de l'Éclipse, 1871), 13.

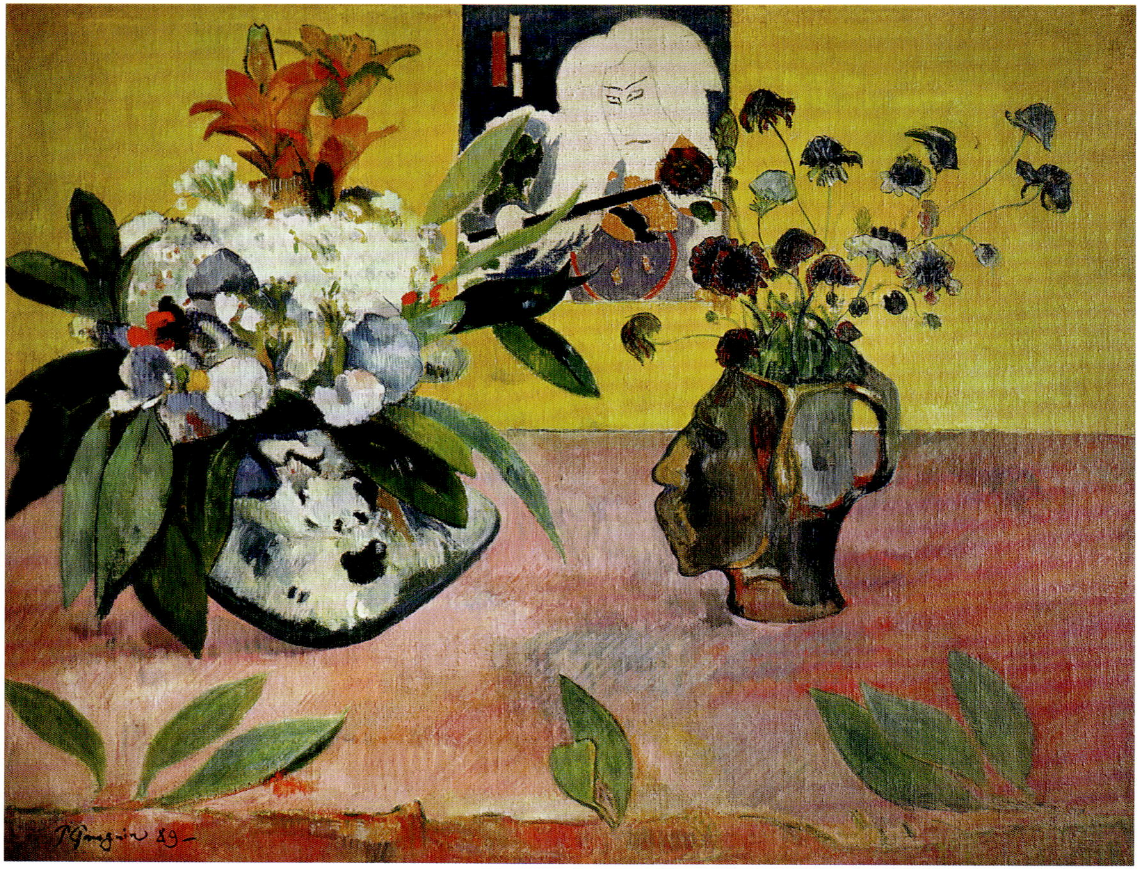

145
Paul Gauguin, *Still Life with Japanese Print*, 1889. Oil on canvas, 85 x 115 cm. Private collection.

throng more densely, and though the roses above the right figure's head have had their redness restored,[56] they have been moved slightly so that there is no mistaking their nature; a drawing in the Metropolitan Museum of Art mapping the exact placement of the flowers attests to the importance of this question to Sargent (fig. 148). The lanterns take on their definitive warm colors, like alchemist's retorts sublimating their original watery contents literally and figuratively. And the flowerlike little girls, if metonymically swollen-bellied, are pregnant only with an innocent art,[57] their impenetrability reinforced through a turning in of the figures toward each other, enclosing the world upon itself. As Dr. Pozzi's exploration of the literal space of creation entailed an erasure of its wet, red, bloody material reality to admit the discovery of a miraculous vegetable world of alternative reproductions, so Sargent's series of revisions allowed the more delicate, if still material, fiction of floral births to surface while the visceral metaphor of human birth and the artist's penetration of depths gave way to a brushing of surfaces.

146
Edward Burne-Jones, *Romaunt of the Rose: The Heart of the Rose,* 1901. Wool and silk tapestry on cotton warp, 150 x 201 cm. Executed by Morris and Company. Badisches Landesmuseum, Karlsruhe.

147
Caricature of John Singer Sargent painting *Carnation, Lily, Lily, Rose* (possibly by Sargent), c. 1885–86. Ink on paper, 8 x 8 cm. Francis David Millet and Millet Family Papers, 1858–94. Archives of American Art, Smithsonian Institution, Washington, D.C.

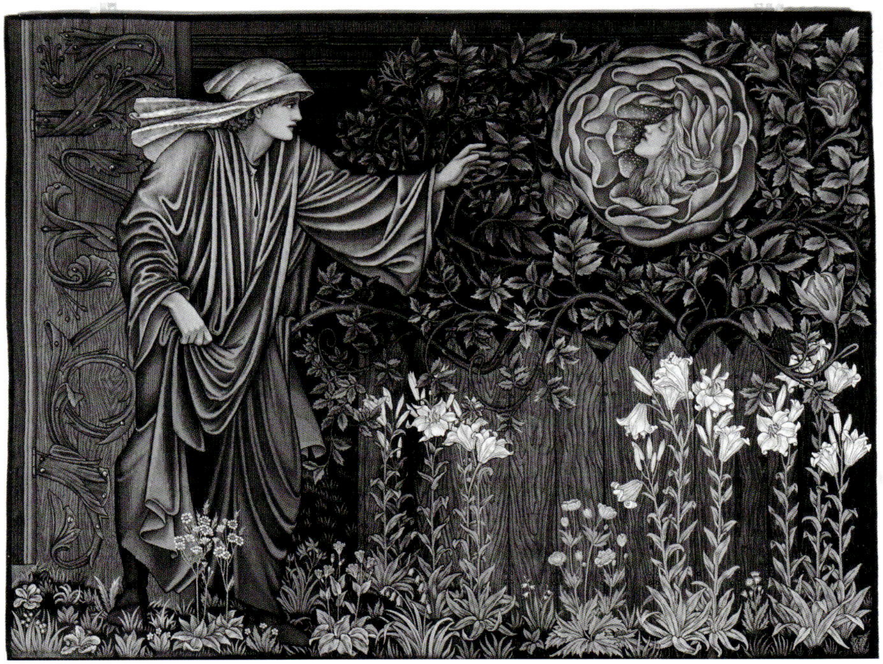

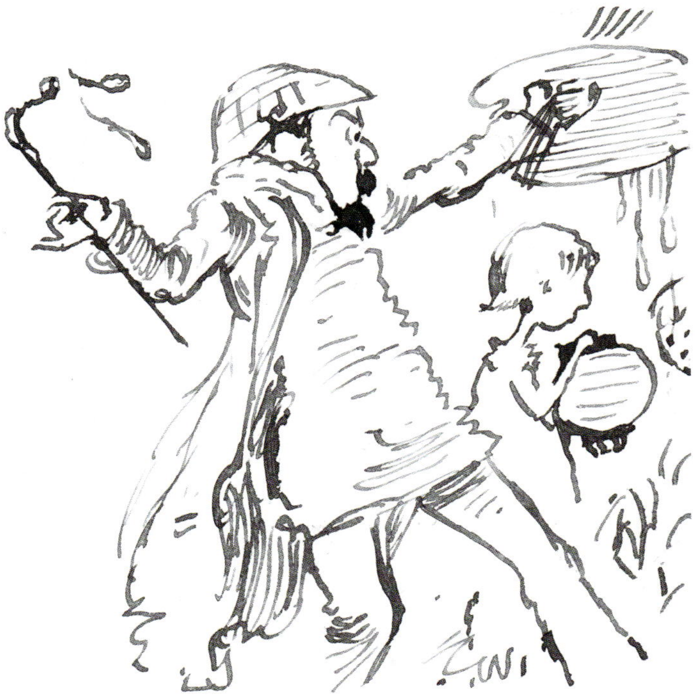

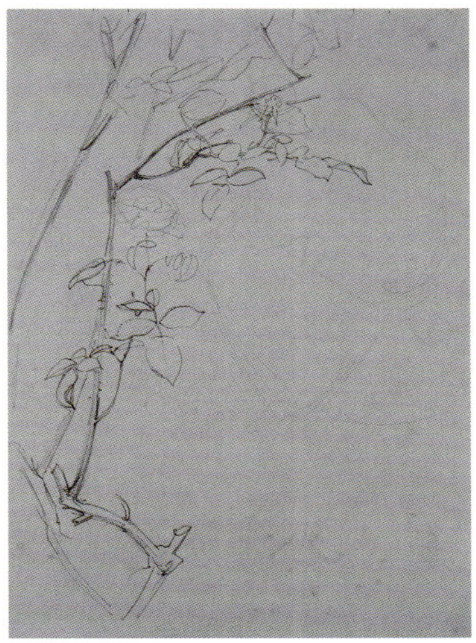

148
John Singer Sargent, *Rose Branch, Study for Carnation, Lily, Lily, Rose,* 1885–86. Graphite on grey-green wove paper, 34.8 x 24.6 cm. The Metropolitan Museum of Art, gift of Mrs. Francis Ormond, 1950 (50.130.142g).

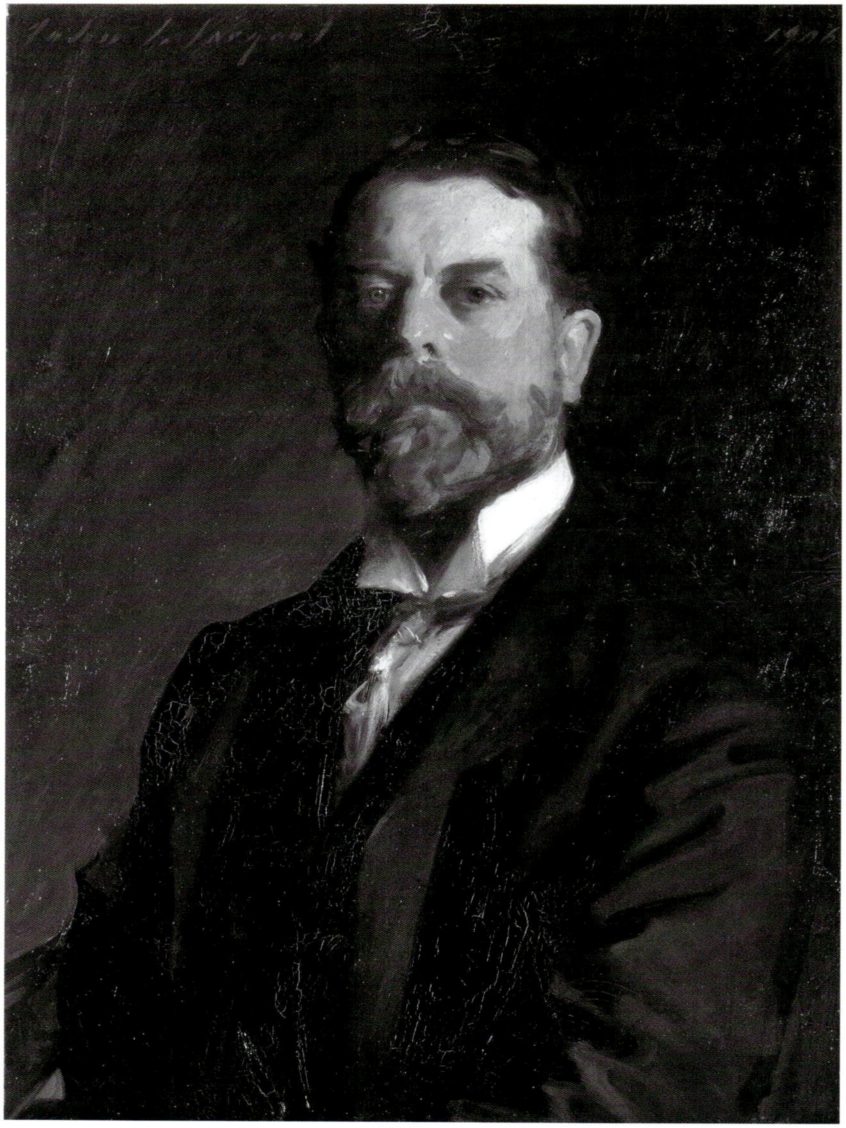

149
John Singer Sargent, *Self-Portrait*, 1906. Oil on canvas, 70 x 53 cm. Uffizi, Florence.

As a real allegory of several years of Sargent's artistic life, *Carnation, Lily, Lily, Rose* can be seen as a self-portrait. Charles Caffin considered it one, declaring that in the painting the artist gives "utterance to himself,"[58] and the first of the oil studies, the Narcissus-like reflection in the weir, lends credence to this reading. The question must be raised, then, of why Sargent chose female children as stand-ins for himself. That he loathed his own image has been amply documented;[59] although he made more than 150 portraits of other artists, mostly his friends, Sargent almost never allowed his visage to be painted by others, though he did permit photographs.[60] He had once willingly posed for a "muscular flower-bearer" (naturally) in Carolus-Duran's Luxembourg ceiling mural, though that experience ended badly.[61] The three self-portraits Sargent did do were painted out of necessity, as entry requirements to various institutions, and the last of these persuaded him to abandon portraiture.[62]

All three self-portraits are linked by a similar pose, lighting, stonewall reserve, and, notably, a lack of hands. Their oddness in comparison with his other images of painters has oft been remarked upon and attributed to his shyness and the privacy with which he maintained his personal life.[63] Leigh Culver has argued that in the 1906 self-portrait for the Uffizi (fig. 149), Sargent "creates the fiction that he was not actually painting" in order to emphasize the picture's staged or disingenuous quality (and thus his distance from it), and furthermore to suggest "that performance, itself, is a necessity of representation."[64] That Sargent considered a certain performance necessary to the creation of art is beyond doubt, but I think a more

150
John Singer Sargent, *Home Fields*, c. 1885. Oil on canvas, 73 x 96.5 cm. Detroit Institute of Arts. City of Detroit Purchase.

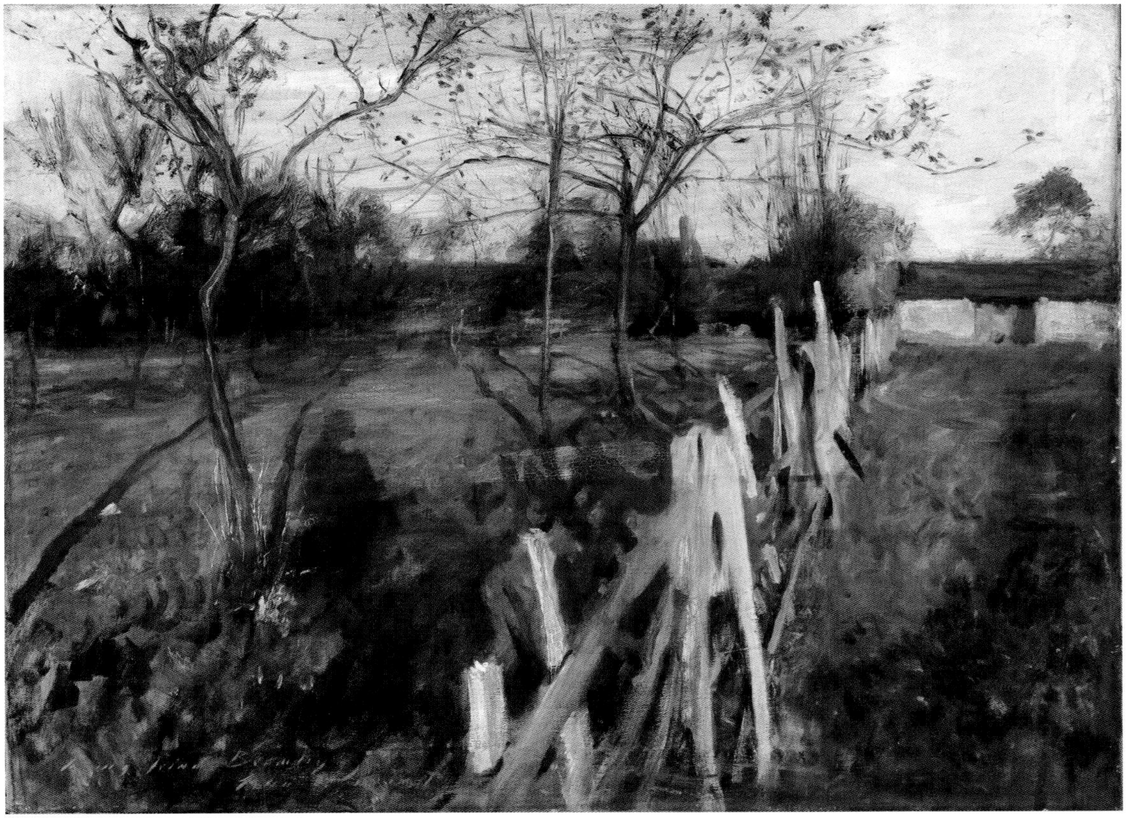

germane explanation is that the representation of himself in the act, as it were, was vaguely obscene to him. The voyeurism of Sargent's images of artists at work, and not just those in bedrooms, has been noted here and elsewhere.[65] Even when, in one or two outdoor paintings, like the 1885 *Home Fields* (fig. 150), Sargent casts himself as a shadow—never a presence—there is a sense of covert and culpable looking.

Given his aversion to direct self-representation, it is not surprising that Sargent would depict himself allegorically. Caffin reasoned that the "deeper level of intimacy he oftentimes [reached] in children's portraits" stemmed from the fact that the painter "seized upon the occasion" to present to us his own personality, as though the sitters were merely ciphers.[66] Similarly, both Sally Promey and Susan Sidlauskas argue that the image of the child operates as "a symbol for interiority," and specifically for "adult" interiority, in other of his works.[67] Many other critics, both Sargent's contemporaries and recent scholars, have posited that the artist was "singularly in sympathy with children,"[68] especially sensitive female ones.[69]

Kingsbury suggests that the very Jamesian mixture of innocence and precocity in Sargent's portraits of children attests to his identification with them.[70] Whether cipher or *semblable,* however, at a moment when Sargent was trying to paint with less art, with a more natural eye and hand, the artist-child was certainly a serviceable fantasy ego. Not a guileless one, of course: rather than simply standing for innocence and purity,[71] Sargent's children exhibit a precocious, polymorphously perverse sexuality and interspecies desire that facilitate the figuration of the artist as hermaphroditic plant and pollinator through them, and consequently, the fiction of fertilization and floral births.[72]

Multiple factors overdetermined Sargent's child-plant-pollinator conflation. From the 1860s onward, "evolutionary psychology" described child development as a recapitulation of human evolution.[73] Thus the trajectory of growth embodied in the child, the idea of a continual process of becoming,[74] sustained diverse possibilities of gender and species identification. This growth and children's receptivity, argued Baudelaire (a master gardener), essentially make them plants.[75] As a child, little "Johnny" Sargent had identified himself with flowers—he drew "a bunch of 'Johnny Jump Ups' [pansies] which he took from nature" for his grandmother, a precocious self-portrait.[76] At the same time, his father frequently described him as a bird or a bee.[77] Rather than biographical details, however, I will focus on the implications of the child's trans-speciation for modern art. In "The Painter of Modern Life," Baudelaire defined genius as "nothing more nor less than *childhood recovered* at will—a childhood now equipped for self-expression," an equipage that must include the white smock-dress that Sargent's children and the stereotypical artist so often wear.[78] To discover what drives this recovery project, we must ascertain what childhood's point of view or way of seeing offered the artist.

Various art historians have pointed out that *Carnation, Lily, Lily, Rose* gives to viewers "the paradoxical sensation of looking directly across to the scene, on eye level with the little girls, while hovering above them, looking at their heads framed by this indistinct field of green."[79] Many of Sargent's works, particularly his landscapes and children's portraits, offer such multiple viewpoints, as though the painter was reluctant or unable to maintain a single subject position, as though he wanted to indulge a fantasy of being simultaneously adult and child.[80] It is important to note that the elevated viewpoint in *Carnation, Lily, Lily, Rose* does not imply a position of control or privilege but rather a loss of contact with things rooted, a removal from the thick, or thicket, of things. In contrast, the lower point of view, closer to the center of the dense, unreceding green optical field, grants access to a world of visual and tactile immediacy.[81] Therein the child offers a vision of uninhibited associative responsiveness, and thus openness to potential transfiguration: becoming-flower, becoming-insect—Dolly- or Pollymorphy—which is lost through "cultivation." Though begun on a more rectangular canvas, Sargent eventually cut nearly two feet off the left-hand side—"pruned" it, in Blashfield's words. Gerdts argues that the near-square format of *Carnation, Lily, Lily, Rose* marks a departure from spatial illusionism, and, by inference, from an anthropomorphic viewpoint. The painting therefore potentially allows, or

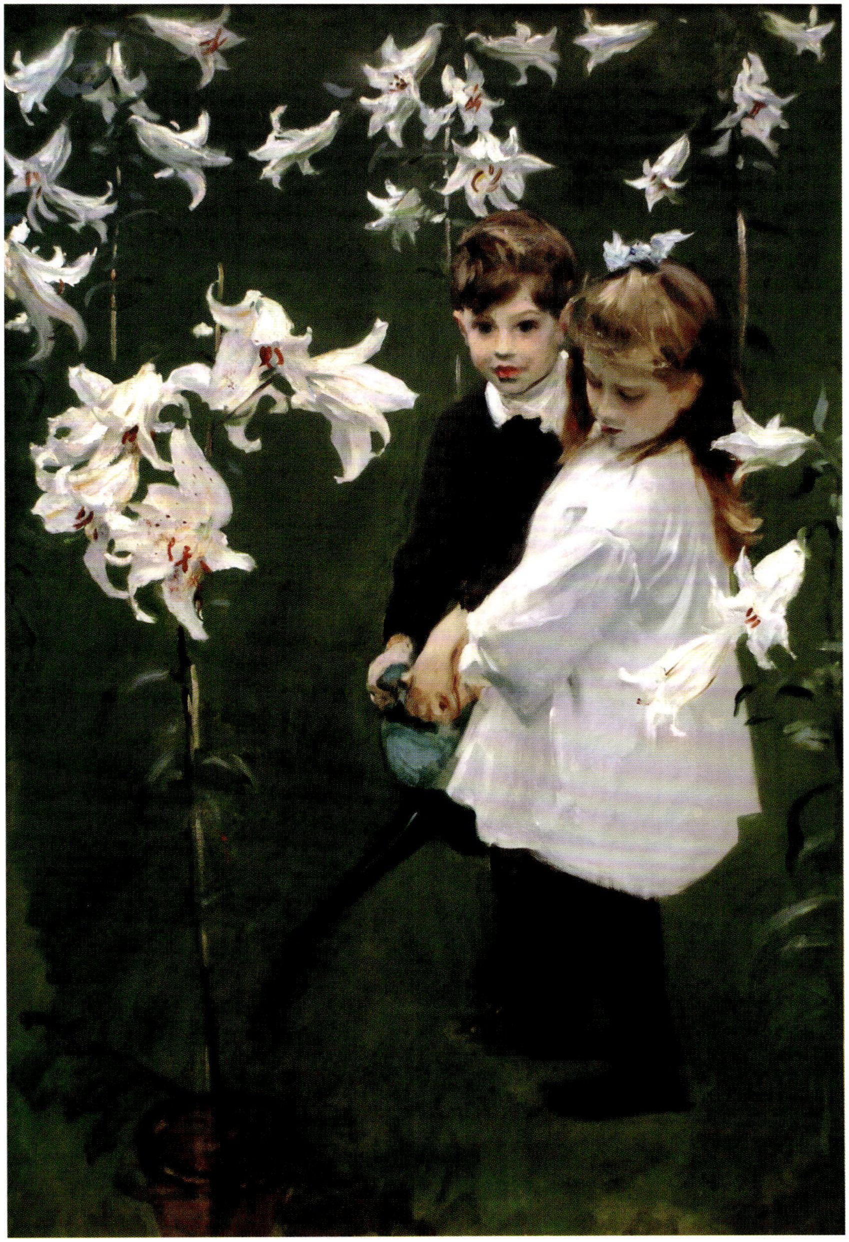

151
John Singer Sargent, *Garden Study of the Vickers Children*, 1884. Oil on canvas, 137.6 x 91.1 cm. Collection of the Flint Institute of Arts, Flint, Michigan, 1972.47.

even courts, an increase in the viewer's species identifications while it records those of its subjects and maker.[82]

To elucidate what is at the heart of this transformative vision, we may turn to another of Sargent's paintings. *Garden Study of the Vickers Children* (fig. 151) predates *Carnation, Lily, Lily, Rose* by a year but anticipates the latter with its two little children, horizonless green environment, and giant lilies. Just as there are two viewpoints in *Carnation, Lily, Lily, Rose,* there are two competing visual systems created in this painting by the overall green optical space, suggestive of immanence, and the recession indicated by the changing scale of the lilies. The two axes in the bottom-left corner created by the edge of a flowerbed or path and the black stake supporting a lily stem stage a similar confrontation. These lines emblematically map a vanishing point, where a vision (of the lily pot) is literally vanishing, yet, if read as a pentimento, is equally refusing to disappear. The mechanics of representation diagrammed here simultaneously express a knowledge of vanishing points, of mortality, and a desire for suspension of disbelief. The pot hovering between vision and dissolution, in which something is visibly *at stake,* is an object or a space

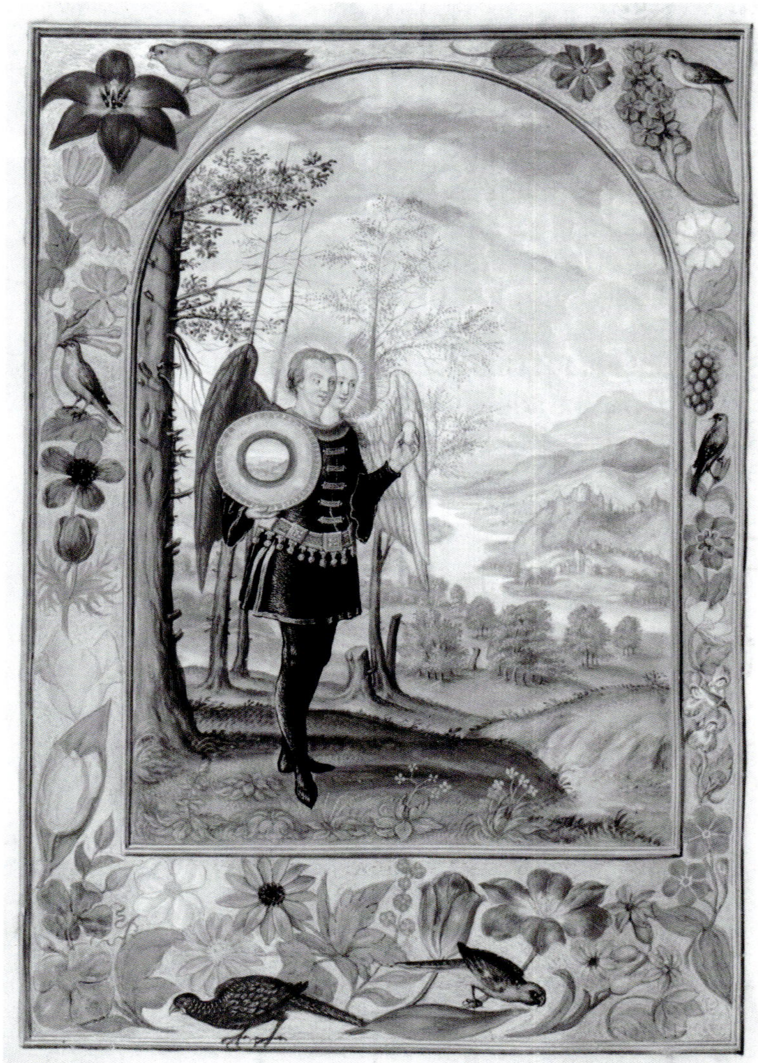

152
Salomon Trismosin, "Hermaphrodite with Egg," in *Splendor Solis*, 1582, Harley 3469 folio 19, verso. The British Library, London.

that, like childhood itself, "persists." Childhood, Jacqueline Rose writes, "is something in which we continue to be implicated and which is never simply left behind. . . . It persists as something which we endlessly rework in our attempt to build an image of our own history."[83] If the pot is a kind of place- and time-holder, it is, furthermore, a place for floral births, a vessel on which the flowering of the lilies and of Sargent's art depends. As we shall see, the adult artist had to navigate existence without autochthony, had to invest in portable—potted—roots.[84]

The pot is the focus not only of Sargent's perspectival punning but also of the painted subjects' attention. The Vickers children, Dorothy in a white smock and Billy in black, were a year apart, aged five and six. As in *Carnation, Lily, Lily, Rose,* the reciprocal relation between pollinator and plant is invoked; the red of Billy's lips picks up that of the lilies' anthers, and Dorothy's smock-dress echoes their trumpet-shaped petals. In the painting the bodies of the children are superposed, their hands intertwined, their jumble of legs shared stalks. Black and white, male and female, plant and pollinator are combined here in a composite creature with a watering can whose proboscidiform spout is making the garden grow, fertilizing it, even[85]—perhaps illustrating a Proustian situation wherein the hermaphrodite lilies "cannot be fertilised by themselves, but can by other hermaphrodites."[86] The Vickers children even recall the well-known hermaphrodite illumination in the 1582 British Library manuscript of Salomon Trismosin's *Splendor Solis* (fig. 152), which pictures a two-headed black, white, and red creature in a verdant landscape.

Sargent's representation of the fertilizing pair of Vickers children as a single hermaphrodite is effected by one of his favorite perspective-defeating compositional strategies: superposition, or a seeking out of felicitous physical conjunctions and curious constellations in the artist's field of vision. In the 1889 canvas *Two Girls on a Lawn* (fig. 153),

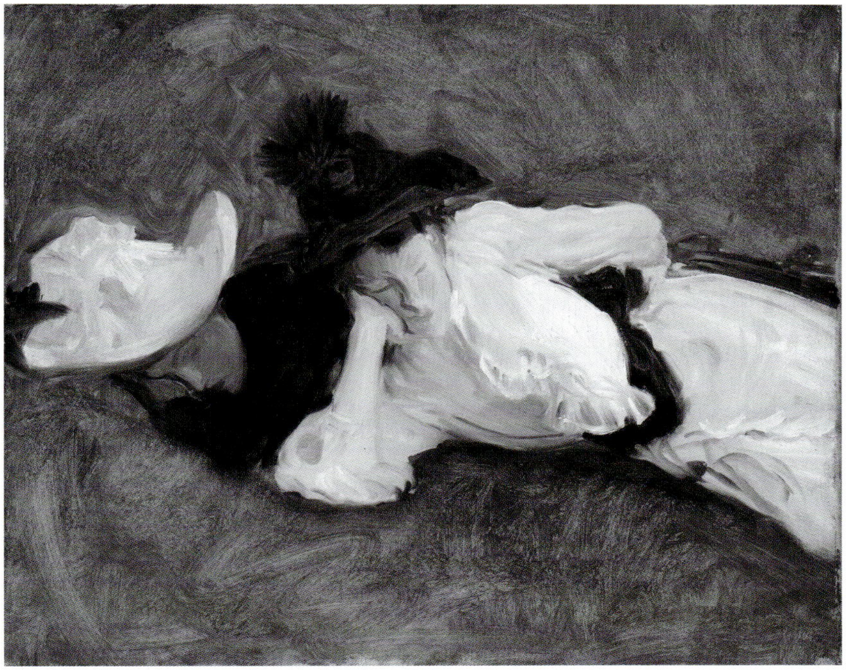

153
John Singer Sargent, *Two Girls on a Lawn*, c. 1889. Oil on canvas, 53.7 x 64.1 cm. The Metropolitan Museum of Art, gift of Mrs. Francis Ormond, 1950 (50.130.20).

for example, which depicts another Vickers child (a cousin of Dorothy and Billy) against a bright green, nonrecessive ground, the artist's perspective reveals an even more strikingly hybrid figure, something like a two-headed slug with a shared amorphous bottom half, displaying inverted black-and-white patterning.[87] This strategy is at play everywhere in Sargent's oeuvre, from the level of the individual brushstroke to that of small objects[88] to the human figure, and works to naturalize the appearance of such anomalies without abandoning their suggestiveness, as we have seen in the lantern-bellies of *Carnation, Lily, Lily, Rose*. The *Garden Study*'s resemblance to a childhood photograph of John and his sister Emily (fig. 154) has been noted.[89] Emily and John had a close relationship; Olson refers to Emily as the painter's "*soi-disant* wife" and the pair of them as "products of the same greenhouse, fed with the same nutrients." Over time, he contends, "Sargent and his sister grew into one person,"[90] an amalgam of the sort literally pictured in the painting. Sargent's identification with the Vickers children, then, may have inspired the thematics of the painting as well as its formal organization.

The synthetic figure occupying the *Garden Study* taps not only into Sargent's past but also into the allure of sexually ambiguous virgin children that underpinned the child identification of many nineteenth-century men.[91] The concept of the hermaphrodite was being illuminated by discoveries in the natural world at this moment,[92] and through its association with earlier stages of evolutionary development, it became linked with the figure of the developing child by way of an ontogeny-recapitulates-phylogeny argument. The hermaphrodite also, like the figures of the child and the androgyne, offered a model for the artist.

In the aesthetic realm, as we have seen, imaginative androgyny was considered the mark of the true artist from the Romantics onward. Popular usage may have made little distinction between the terms "androgyny" and "hermaphroditism" in the nineteenth century, as A. J. L. Busst has argued,[93] but specific disciplines were more precising,[94] and the founding myths of the two creatures—Aristophanes' spherical beings (described in Plato's *Symposium*) and Hermaphroditus (whose creation is recounted in Ovid's *Metamorphoses*)—suggest some differences important to a consideration of models of creativity. Kari Weil outlines the ideological implications of each figure. The androgyne, she writes, "suggests a spiritual or psychological state of wholeness and balance. . . . In Plato the androgyne is said to exist at the beginning, before division and before desire; s/he is a figure of origins, of fullness, and of presence." The hermaphrodite, on the other hand, results from a transformation and "calls attention to the visible and physiological fact of two differently sexed but noncomplementary bodies brought together in an unrelieved process of joining and splitting that manifests the irreparable divisions wrought by desire." Unlike the androgyne, the hermaphrodite "is a figure of the displacement of origin."[95] While these distinctions are perhaps too absolute—sexually ambiguous figures that appear in art and visual culture of the period are not so easy to categorize—they nevertheless identify key differences in the sort of artistic selves the figures could represent.

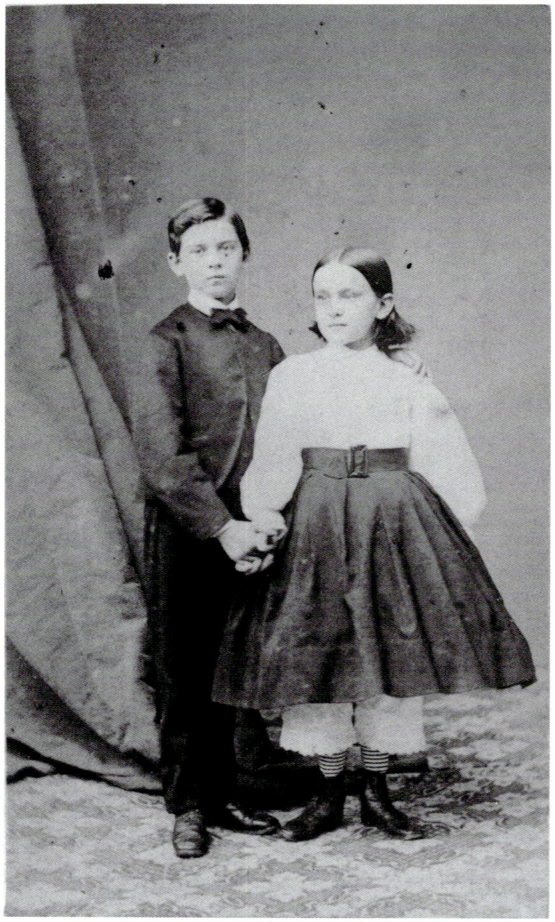

154
Photograph of John and Emily Sargent, photographer unknown, c. 1867. Private collection.

If *Garden Study of the Vickers Children* offers a vision of the artist/child as a hermaphrodite, a composite creature, an earlier work from Sargent's student days, *Nude Oriental Youth* (fig. 155), represents a fantasy of plenitudinous androgyny. The painting depicts a stereotypically eroticized Asian youth—almond-eyed, hairless, ephebish—in front of an apple tree in blossom.[96] The figure's sex is ambiguous: smooth flesh, sculpted eyebrows, and long hair feminize it, but the prepubescent genitals appear male. The body's indeterminacy does not read as hybridity, though, but rather as an ideal unity, incarnated here in a young boy in harmony with his surroundings. The gnarled trunk of a young tree snakes up behind him (a heavy branch, presumably from another, older tree, enters the frame at the top right). In addition to his physical proximity to it, a number of features suggest the boy's similarity to the young tree: the twiglike kotosticks in hair, his blossom-shaped topknot, the pair's similar heights (if one calculates the boy's from toe tip to outstretched fingers) and their parallel sexual states (just coming into bloom). The figure pulls down a branch with his left hand; he appears to be reaching up with his right to pluck a flower. It is significant that an apple *blossom*, not the fruit, hovers above his hand;

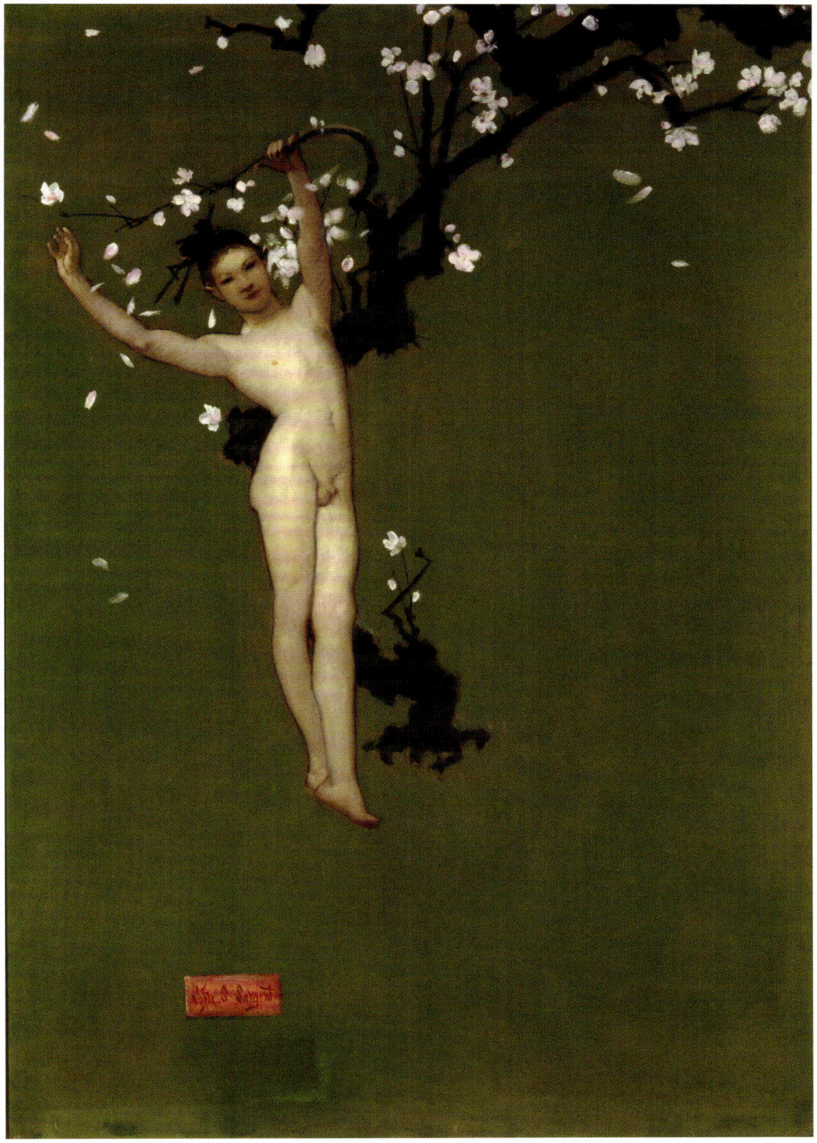

155
John Singer Sargent, *Nude Oriental Youth*, c. 1878–79. Oil on canvas, 90.3 x 64.8 cm. The Western Reserve Historical Society, Cleveland, Ohio, 76.46.5.

the painting thus offers a vision of prelapsarian harmony. Whether suggestive of the fall from paradise or a simpler defloration (and opening up to desire), however, the act of plucking has not yet taken place.

The flat green ground is impossible to read spatially, even more so than in *Carnation, Lily, Lily, Rose* and *Garden Study of the Vickers Children*. The youth, while oriented vertically, is stretched out as if reclining; his feet aren't positioned to bear weight; he effortlessly achieves a perfect balance. There are hints of sexual desire here—in the snaking branch at the top right and the insectlike base of the trunk, which extends tentacular twigs and a blossom toward the boy's penis—but the boy himself exudes serenity, and his rosy, unblemished skin and compact form emphasize his wholeness, his unfallen state.

Complete, balanced, and found in an originary garden: the figure fulfills all of Weil's criteria for the androgyne. Interestingly, Sargent could figure himself as an artist here only negatively. An unusual dabbling in *japonisme* for the painter, *Nude Oriental Youth* is signed with a mock "satsuma" stamp,[97] the fleshy pink rectangle at the lower left, in which the artist's signature appears carved out, an absence rather than a presence.[98] The artist's difficulty in inserting himself into the painting is also suggested by the pentimenti that attest to the shifting of the stamp, which in the finished painting has been moved up and to the left of its original location. The painter's desire to identify with the androgyne's body is registered in the echo of the stamp's luscious pink in the boy's nipples, but the most autographic strokes in the painting are the few single petals slowly falling.

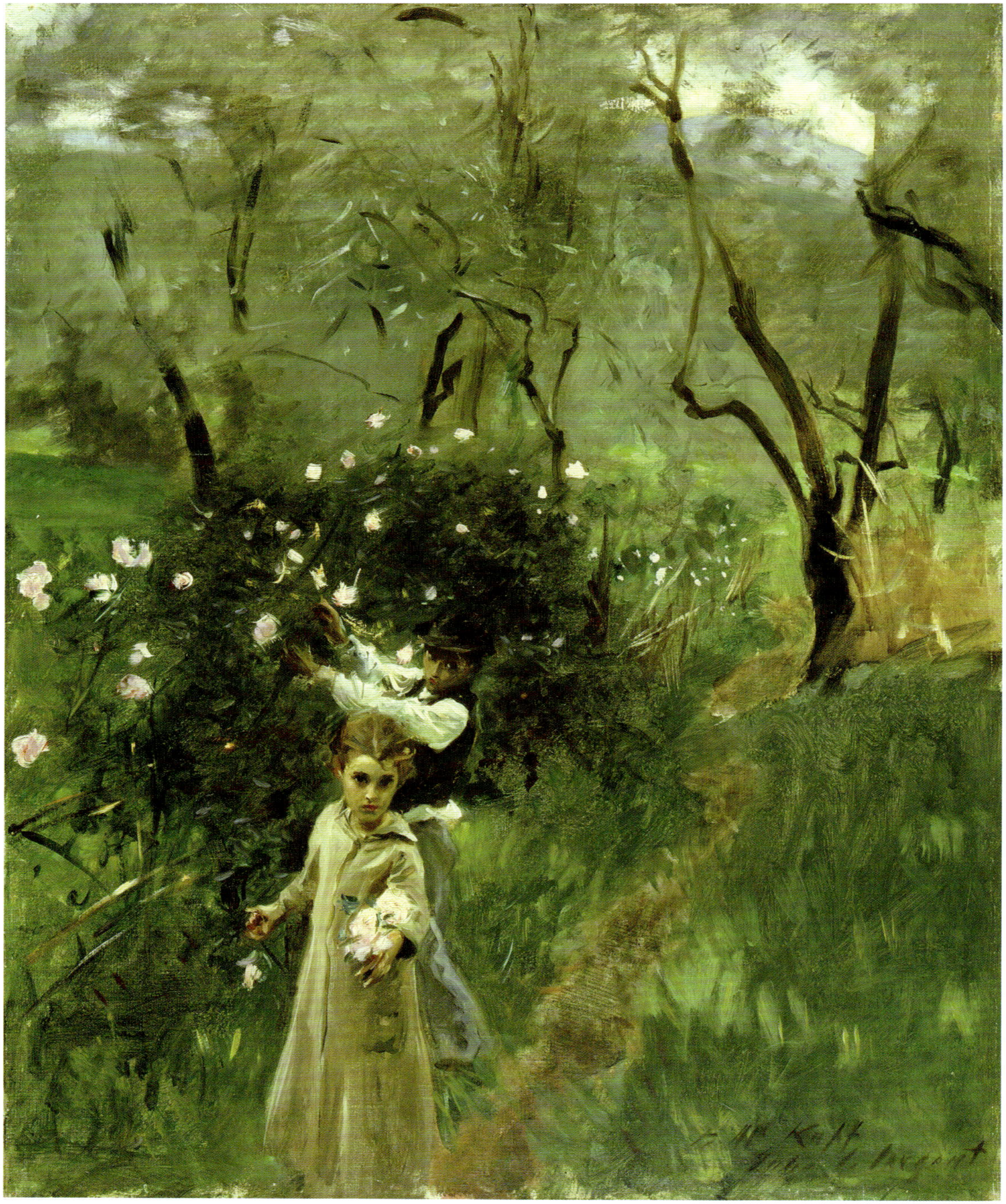

156
John Singer Sargent, *Gathering Flowers at Twilight,* c. 1883. Oil on canvas. Private collection.

Sargent never revisited this type of Orientalist fantasy, but he continued to explore the intertwining of bodies and blossoms in gardens in *Gathering Flowers at Twilight* (fig. 156), painted circa 1883. In this painting two children in rustic attire pick pale pink roses at the side of a rough path and offer them to the painter/viewer. The setting is picturesque, a little wild and ramshackle; there is a sense of natural abundance if not decorative order. The figures, though, which overlap, are not yet making the garden *grow*. From *Nude Oriental Youth* to *Garden Study of the Vickers Children* we can trace a move from a rather abstract, conceptual space of seamless unity and bodily serenity to a more "real" location and visible bodily amalgamation, from a space of natural abundance to a staked and cultivated ground. In the earlier picture the androgyne represents an unattainable ideal; in the later painting the hermaphrodite, with its slightly awkward composite figure, offers a workable solution to the invert artist's reproductive problems.

The two figures in *Carnation, Lily, Lily, Rose* evoke Aristophanes' story of the androgynes: although there are minor differences between them, they stand as mirror images of each other. If the bloody marks in the two-figure oil study suggest the pain of separation, in the final painting, rather than desperately embracing, longing "for nothing better than to be rolled into one,"[99] the two figures stand apart, each absorbed in what we have seen to be an artistic activity. Although critics referred to them as little girls, and we know that the models were female children, nothing in the painting identifies a particular sex. Short-haired and clad in common smocks, they could stand for any of Aristophanes' three sexes. As flower-children they are, of course, both hermaphroditic and comfortable in their garden setting.

Like Sargent, others saw the hermaphrodite and the invert—both composite creatures—as hopeful figures. Edward Carpenter even proposed that sexual indeterminacy might be evidence of human evolution: the increase in "these gradations of human type . . . or even the concentration of attention on them, may be the indication of some important change actually in progress." Since we don't know "what possible evolutions are to come," he argued, it is possible that "at the present time certain new types of human kind may be emerging, which will have an important part to play in the societies of the future."[100] Images of hermaphrodites, though, like images of children in gardens, can also be sentimental: they can figure a longing for a lost Platonic sexual serenity as well as hope for the future.[101]

As Eve Kosofsky Sedgwick has argued, expressions of sentimentality are figures of "vicariated desire,"[102] indices of literally improper (i.e., neither ownable nor localizable) desires forced to take substitute objects and circuitous routes. In the *Garden Study* the Vickers children are the vehicles of Sargent's vicarious identification, the dreamlike figures of a wish fulfillment. *Carnation, Lily, Lily, Rose* and *Garden Study of the Vickers Children*, then, are nostalgic for a past moment in time and for a particular relation to the world—a relation understood not just in terms of spatial position and physiology but in terms of kin. Wish fulfillment necessarily produces anxiety, however, and the latent procreative capacity of actual female child models may also have been a factor in the artist's choice of stand-in. The creation-as-procreation metaphor, as noted, was still in use, and the anxiety underpinning it could only have been exacerbated by the typical nineteenth-century invert's fear of sterility: hence Sargent's avowal of paintings as daughters, and his protestation of the naturalness and fruitfulness of his own productions.[103]

The heteronormative rhetoric of production was oppressively widespread. Victorian attitudes toward sex, as Regenia Gagnier has shown, "proffered implicit or explicit economic metaphors or theories of 'production.'" If some artists, like Oscar Wilde, insisted on art as a superior option, linking art for art's sake "with sexual pleasures contrary to bourgeois notions of productive or purposive sex,"[104] the reproductive body nevertheless returned "at the fin de siècle to haunt the consuming or pleasured body."[105] Even Remy de Gourmont (art critic, philosopher, and contemporary of Sargent), after naturalizing all the perversions in *Physique de l'amour,* concluded that "there are no living beings but those who can perpetuate life." Citing the activities of bowerbirds, he located "the origin of art" in "sexual games," and argued that although there is undoubtedly a difference between the mimetic capabilities of humans and animals, "the object and the result are the same: copulation, seed, fructification," for "on earth, life is characterized by individuals grouped into species, that is to say having the power, a male having mated with a female, to reproduce their likeness."[106]

The only real portraiture, thus, was progeny. Children's portraiture was consequently a fraught genre, unsurprisingly one in which women artists could excel with approbation. A friend of Sargent's, Harrison Morris, director of the Pennsylvania Academy of the Fine Arts and promoter of Impressionism in America, summed up the reproductive obligations of artists in an article on the genre, which begins, "The plant finds joy in its buds, the race lives anew in its offspring; and hence childhood, with its helpless little needs, its tiny mimicry, its confidence, and the simple purity of its conceptions, is the supreme human interest. All are engaged in rearing it. Even the unwilling bachelor has his economic share. The whole race exists for its preservation."[107] George de Forest Brush, another friend of Sargent's,[108] figures as Morris's ideal children's portraitist because he painted his own offspring, learning in the process that "barrenness of achievement" in this field of portraiture derives from literal barrenness "in the artist."[109] This logic of reproduction applied as rigorously to other media. James affirmed that "the terrible law of the artist" is "the law of fructification, of fertilization."[110] Sargent's position in this milieu has been aptly summarized by Sarah Burns: "rootless in the midst of family, unmated in a world largely geared for heterosexual expression," he "desired what [he] could not have," and built his career "around that lack."[111]

All of these dream-thoughts, unconscious associations, and anxieties are condensed into the strange, almost oneiric, image of *Helen Sears* from 1895 (fig. 157). A portrait of the artist as a young girl, it offers us another vision of artistic creation as Sargent conceived it, illustrating the hand-pollination the painter performs. In some sort of conservatory—a nursery in both senses—large

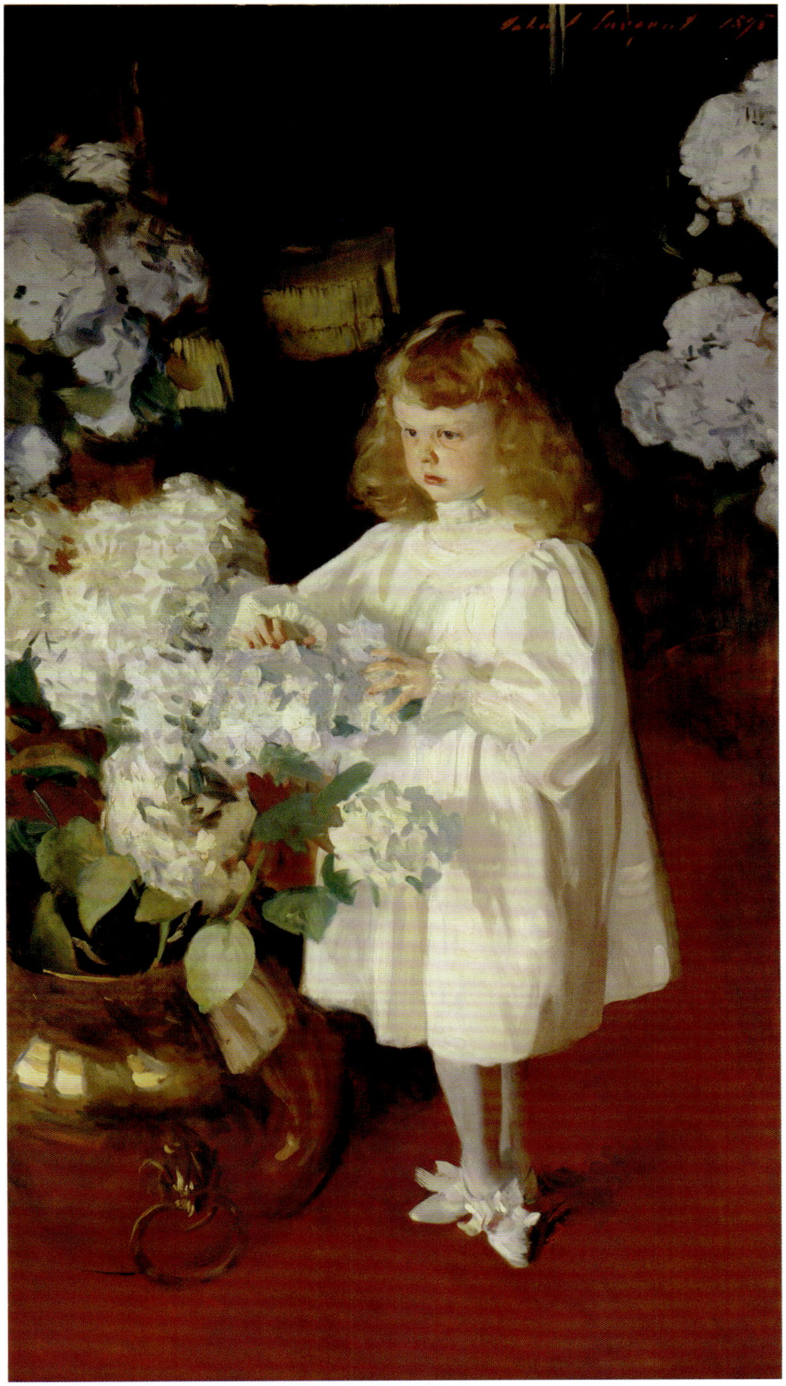

157
John Singer Sargent, *Helen Sears,* 1895. Oil on canvas, 167.3 cm x 91.4 cm. Museum of Fine Arts, Boston, gift of Mrs. J. D. Cameron Bradley, 55.1116.

brass pots holding blue and white hydrangeas stand on an earthy red-brown carpet. In a white smock-dress, Helen hovers on winged feet in the darkling evening air; upon reflection in the pot, the smock reveals its mothy wings.[112] The child-pollinator gently explores the large pale blossoms with her petal-probing, pollen-brushing fingers. Like the heat- and light-giving matches in *Carnation, Lily, Lily, Rose,* her fingers show warm and pinky-orange against the ice-blue hydrangeas (fig. 158).[113] The flowerlike lanterns in the earlier picture were metaphorical canvases; here the flowers themselves are. As Helen touches the cluster of white hydrangeas in the pot before her, they color, blush a delicate blue. Unlike the darker, opaque lumps of pigment forming the flowers in the background of the painting, these blossoms are lit up and articulated. The impasto is conspicuous on both Helen's hands and the hydrangeas, calling attention to their painterliness; it's so simple, the scene seems to declare—child's play, this finger painting—and there are touches here literally finger-painted (a practice in which Sargent occasionally engaged).[114] Like Monet, this child-artist needs no prosthesis. And in this dream performance, Sargent not only displays his technical virtuosity in accomplishing so seemingly natural a reproduction of hortensias, but also his

158
Detail of *Helen Sears*.

procreative powers, since the hortensia is a species of hydrangea named and famed for its sterility.[115]

As in *Carnation, Lily, Lily, Rose,* the hand-pollinator of the picture is subject to a flower identification as well. Turned heliotrope, her face and the frock billowing around her, full blown as the blossoms, absorb the light.[116] Green tights shoot down into the darkened ground, and at their feet these tender well-heeled stems uncurl slippery roots, creating, as in *Carnation, Lily, Lily, Rose,* a tension between rooting and floating. As if under the sway of some anamnesis of the dawn of perception in photosensitive plant surfaces, Helen's petaline fingertips—her right hand is partially unfurled, her left fully digitated—become surrogate retinas for her unseeing eyes, fingers through which her perception of the flower and *as* flower seems to dawn unconsciously. Sally Shuttleworth has argued that Victorian "psychologists were looking at child history as a map to another, alternative domain";[117] here Sargent suggests the child's ability to plumb evolutionary depths.

Something about the painting, however, is vaguely disturbing. Helen looks mesmerized, less animate even than the hortensia that reaches toward her. While critics have characterized her as unhappy, nervous, resentful, and even "dissipated,"[118] she seems to me best described as possessed of an almost vegetable vacuity. She appears to be in the grip of a "dreamy mental state," a phenomenon analyzed by Sir James Crichton-Browne, who published his findings in 1895. These states, which emerge around the age of eight, "abound amongst the educated, the refined and the neurotic classes," and are vestiges "of a previous state of existence," perhaps even traceable to "prehistoric times." While common among artists and a spur to creativity, dreamy mental states, during which the subject inhabits two places and times at once, could also lead to problems with ideas of space and "a loss of personal identity" and were thus considered indices of degeneration.[119]

Sargent knew there were multiple risks associated with floral identification for the artist. If the painter-plant wasn't flourishing, he/it was "vegetating."[120] Another hazard is illustrated by a *Punch* cartoon from 1881, entitled "Flowers of 'Culture,' or A Swinburne-Jones Cutting" (fig. 159), which portrays the painter and writer as artistic potted plants.[121] Swinburne and Burne-Jones share a root, as do their names, hence the titular elision. Ribbonlike shoots spiraling out on either side of the bulb spell "Namby-Pamby." The word derives from the name of an eighteenth-century pastoral writer, Ambrose Philips, and derides the weakness, sentimentality, and vacuity of childish pastoral fantasies; here that category includes hoping to paint or create poetry the way nature produces plants.[122] Another drawing for *Punch* by Linley Sambourne, "Our Guide to the Grosvenor Gallery" from 1878 (fig. 160), features Burne-Jones as a plant with palette-shaped leaves. A large, hovering "F. C. Bee" probes one of the blossoms with its prosthesis. The initials are those of Francis Cowley Burnand, contributor to (and later editor of) *Punch* and author of *The Colonel,* a dramatic farce mocking Aestheticism and Burne-Jones in particular in the character of the painter-plant Basil Giorgione; the play debuted in February 1881.[123] We see the author, then, gathering pollen for his parody. The label "ars sicklymen" (a pun on cyclamen) identifies Burne-Jones's colorless, waxy flowers as the product of an invalid or degenerate vegetable.[124] A caricature from 1881 of Oscar Wilde as a sunflower in a vase (fig. 161) similarly questions the vigor of the floral artist: "What's in a name? The poet is WILDE, / But his poetry's tame." The large metal urn in the background suggests the fate of poet and poetry, to become "waste" paper.[125]

The flower-poet could find it difficult to *deliver,* in more than one sense. Labor could be exhausting: Alfred Le Petit's caricature of Littré (fig. 28) acknowledges that he "gave birth to a dictionary," but mocks him because this labor "dried him out."[126] Desiccation could afflict vegetal offspring as well as artists. At a dinner party Sargent attended in 1890, a guest criticized James's fictional characters for being "just what dried specimens of plants are to growing ones. They are quite correct—only no vitality."[127] Another risk for the work of art in the age of botanical reproduction was disease. James, whose account of the floral author's pollination we examined earlier, notes that the pollen-germ of a story can be, or carry, a "virus," which may be "infused" by the "prick" of conception's "single touch." In this case, the weakened "flower" that results must be grown beneath an "umbrella and really under exquisite protection."[128] A last concern is the ephemerality of

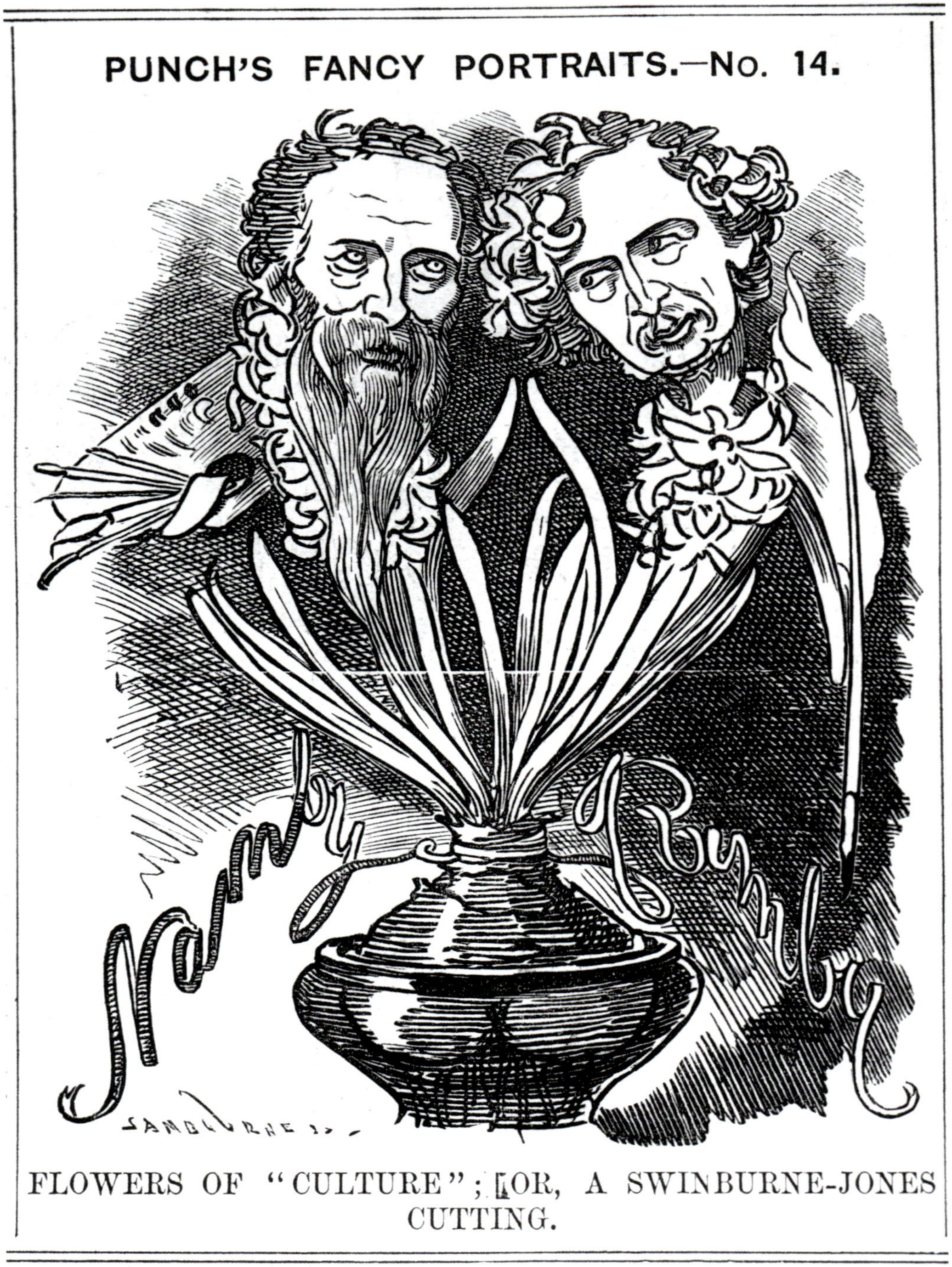

159
Linley Sambourne, "Flowers of 'Culture,' or A Swinburne-Jones Cutting," in *Punch, or the London Charivari*, 15 January 1881, 22.

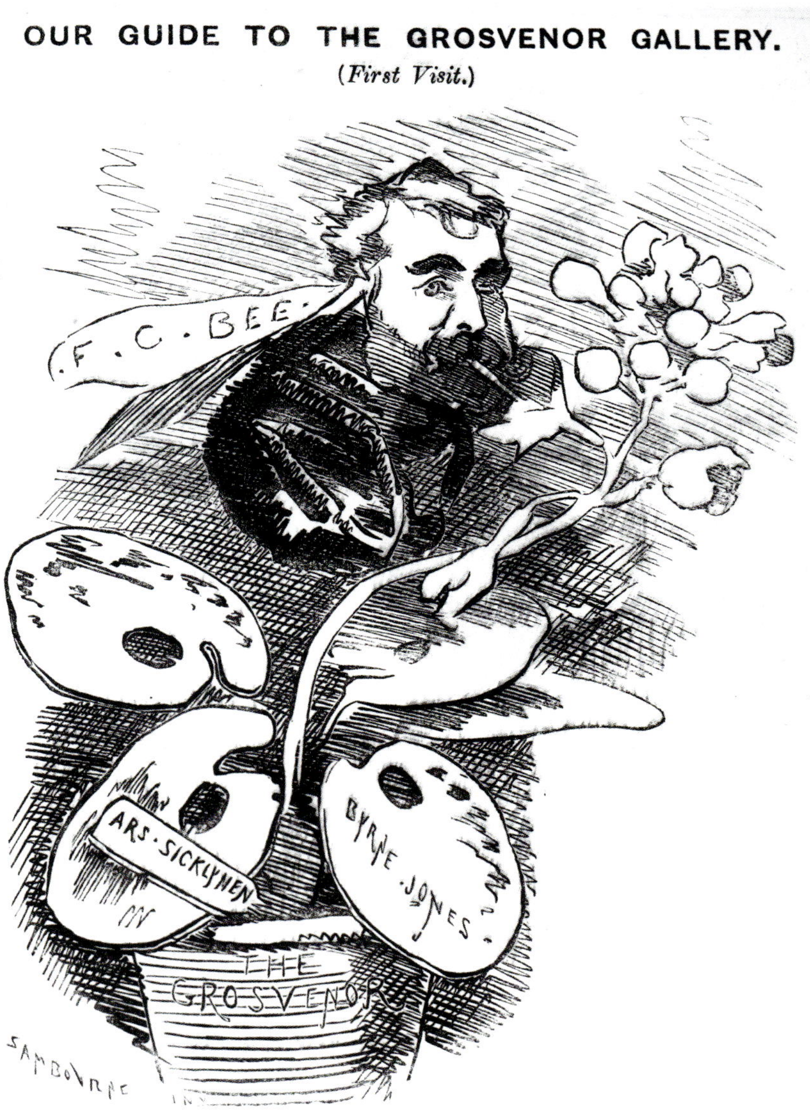

160
Linley Sambourne, “Our Guide to the Grosvenor Gallery,” in *Punch, or the London Charivari,* 22 June 1878, 285.

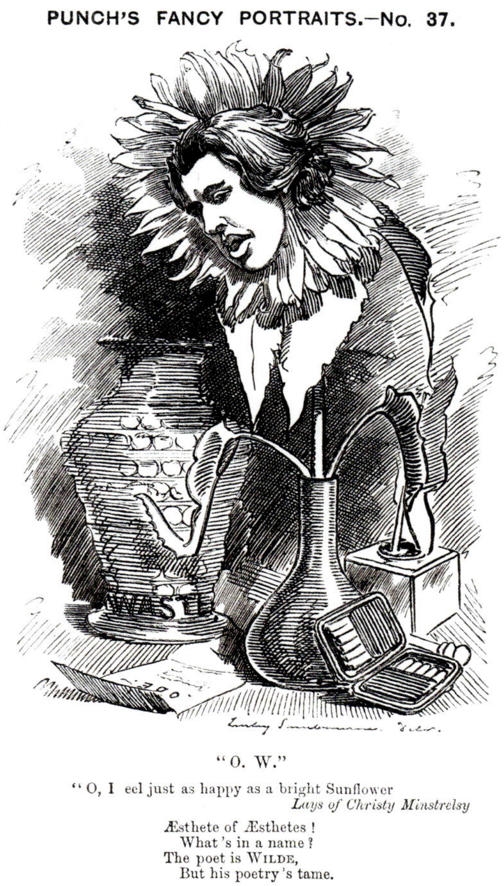

161
Linley Sambourne, “O.W.,” in *Punch, or the London Charivari,* 25 June 1881, 298.

162

Edward Lear, "Queeriflora Babyoïdes," in *Laughable Lyrics: A Fourth Book of Nonsense Poems, Songs, Botany, Music, &c.* (London: Robert John Bush, 1876).

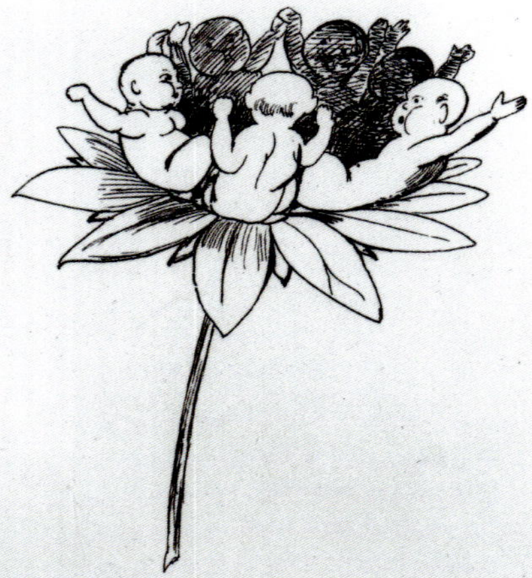

Queeriflora Babyöides.

flowers.[129] Whether imagined as thriving, namby-pamby, dried out, too fleeting, or diseased, however, Edward Lear's "Queeriflora Babyoïdes" explicitly declares that the invert's fantasy of floral births is simply "nonsense" botany (fig. 162).

Sargent may have been susceptible to a number of anxieties about floral reproduction, but none of these tells us exactly what is wrong with *Helen Sears*. Painted in Boston ten years after *Carnation, Lily, Lily, Rose,* the rhetoric and creation model employed in the later painting are the same. Transplanted to America and into a domestic interior, however, in the year that Oscar Wilde was tried and convicted, and Alfred Dreyfus exiled to Devil's Island—both cases emblematizing the impossibility of naturalization for the outsider—the fiction seems more fragile.[130] Furthermore, the flowers themselves offer a disturbing counternarrative.

Blue hydrangeas became popular in the 1880s[131] and were the Symbolist flower *par excellence* because of their oriental origin, unnatural color, and drained appearance. Comte Robert de Montesquiou—Proust's model for the baron de Charlus in *À la recherche du temps perdu,* Huysmans's for Des Esseintes, and an acquaintance of Sargent's—adopted the flowers as his personal emblem along with the bat. He drew, designed, and commissioned works of art and furniture with hortensia motifs,[132] and he owned seven pastels and paintings of hortensias by Sargent's protégé Paul Helleu, an artist and gardener who grew the flowers himself.[133] In one of Helleu's pastels, a kind of portrait of the count, bats and hydrangeas are partnered in a foggy twilight (fig. 163). While Sargent was painting *Helen Sears,* Montesquiou composed a volume of poems entitled *Les hortensias bleus.*[134]

According to Montesquiou, or "Hortensiou,"[135] as he was dubbed, the "troubling clusters" of the hortensia stood for melancholy, dissimilarity, and death[136]—hardly the stuff of *Carnation, Lily, Lily, Rose*'s pastoral fantasy—and the signification he invented for this symbol of his "unnatural," or naturally aberrant, sexuality took hold. Proust's encomium to "all those beardless men with feminine faces, garlanded with hortensias" in his essay on Gustave Moreau links the sterility of hortensias and inverts.[137] Men so garlanded do not actually appear in Moreau's oeuvre (there are blue flowers but not hortensias); Proust's understanding of the flowers came from Montesquiou. Later,

Lucie Delarue-Mardrus—wife of the translator of the first unexpurgated *Le livre des mille nuits et une nuit,* lover of Natalie Clifford Barney, and friend of Proust's and Montesquiou's—thematized their otherness in her novel *Hortensia dégénéré.*[138]

In a photograph from 1885, "Robert de Montesquiou en chasseur de chauves-souris, château de Charnizay," the count, posing as a bat hunter, holds a bat and a paper lantern while standing in a clump of hydrangea bushes at his ancestral home (fig. 164).[139] Elements of the photograph's composition recall *Carnation, Lily, Lily, Rose,* but the lantern is unlit and the bat—the pollinator, the lantern's match—is forcibly held away from it.[140] The bat hints at a vampiric underside to pollination, which does not always entail mutually beneficial encounters; indeed, it almost offers an explanation of the anemic quality of hortensias.

Sargent almost certainly chose the flowers for Helen's portrait, and he could not have been unaware of their cultural signification. In 1884 or 1885, the year he began *Carnation, Lily, Lily, Rose,* Sargent painted a portrait of the flower-painter Ernest-Ange Duez in front of a cluster of pale hydrangeas (fig. 165). Maureen O'Brien has remarked that the flowers framing "Duez's head seem a startling complement to a male portrait" of

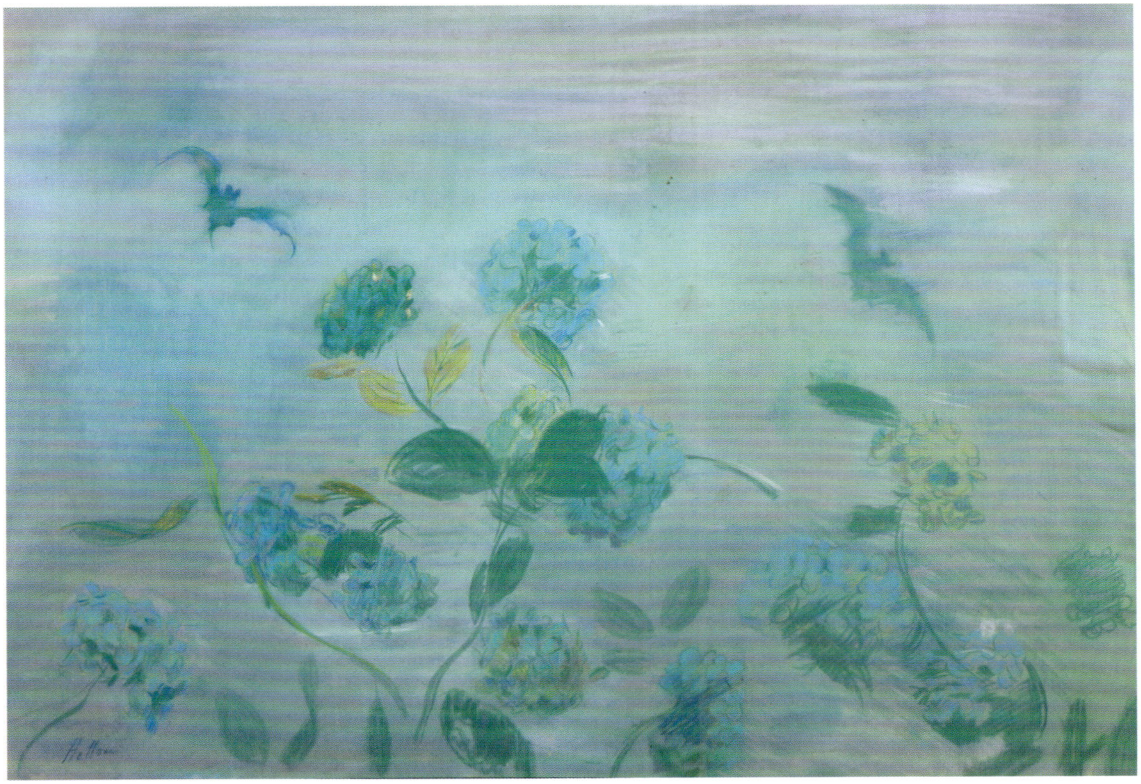

163
Paul Helleu, *Hydrangeas and Bats,* n.d. Pastel on paper laid on canvas, 69 x 99.5 cm. Private collection.

164
"Robert de Montesquiou en chasseur de chauves-souris, château de Charnizay." Manuscrits occidentaux, NAF 15039, folio 101. Bibliothèque nationale de France.

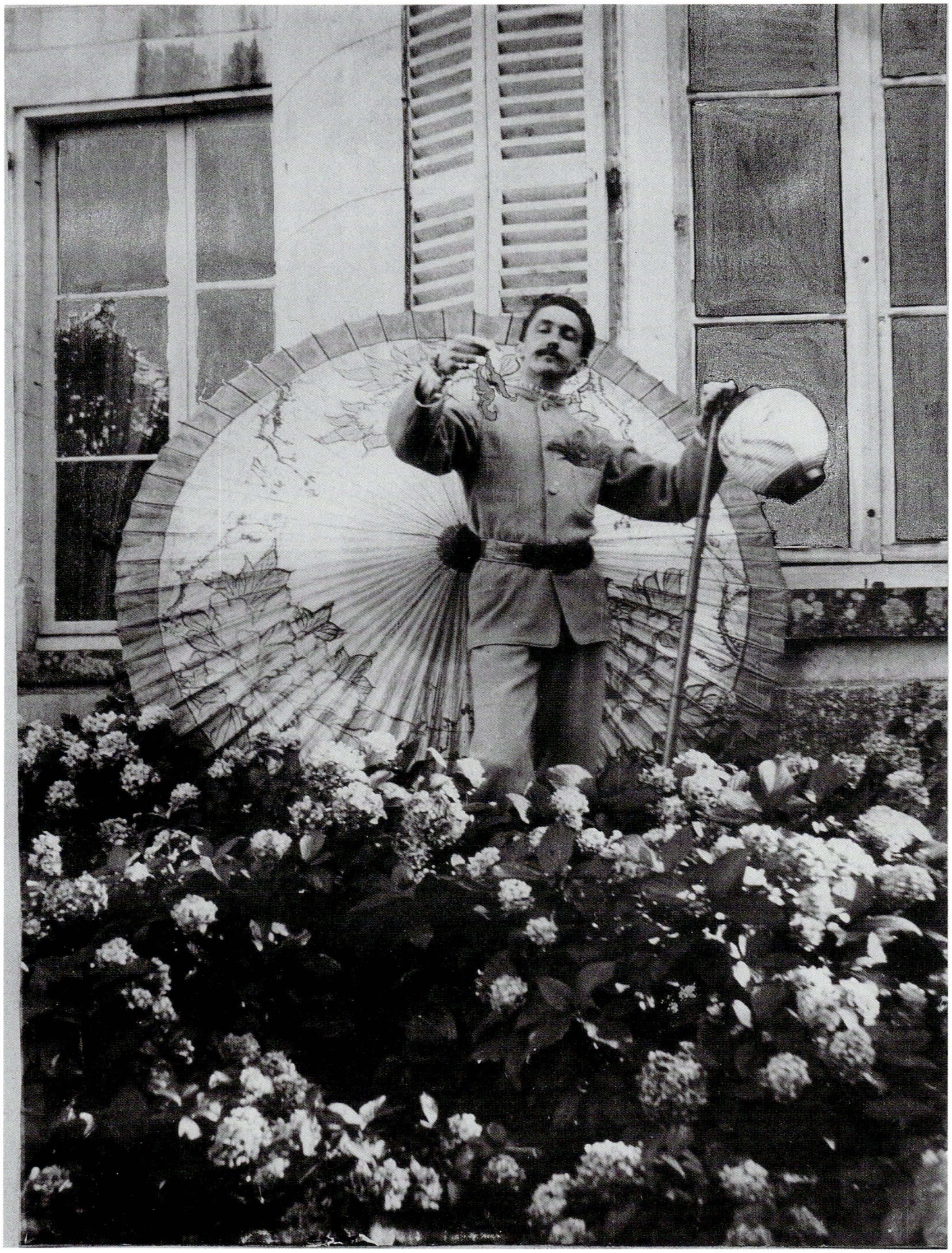

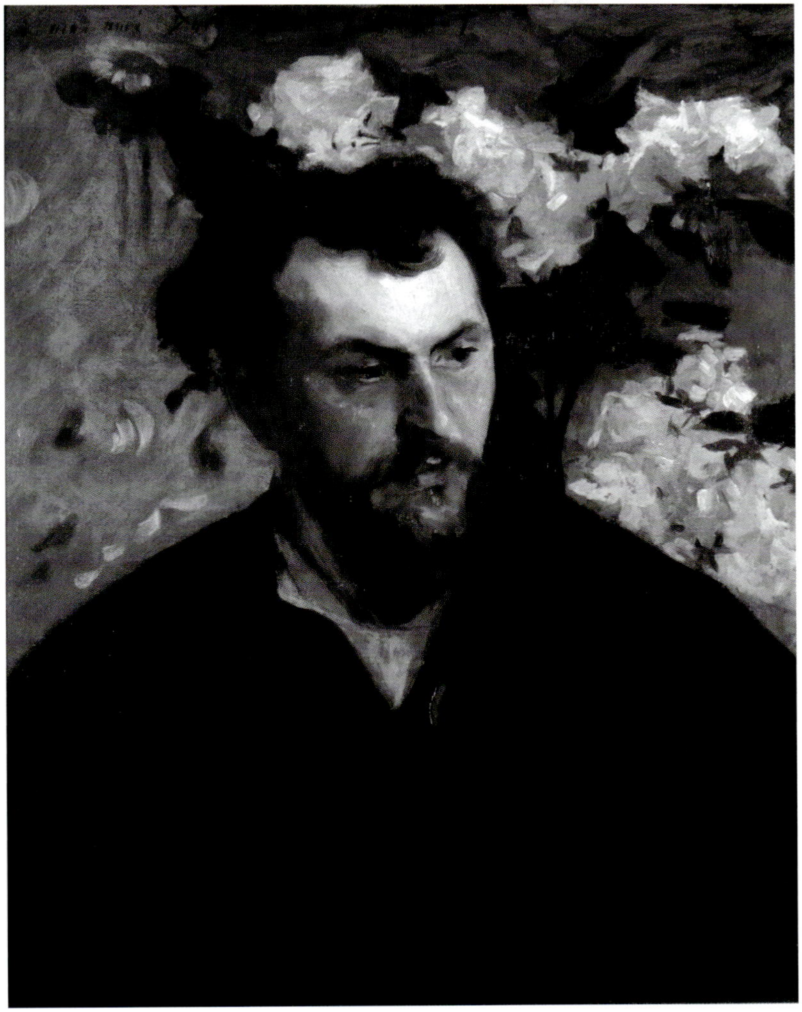

165
John Singer Sargent, *Portrait of Ernest-Ange Duez*, 1884–86. Oil on canvas, 73.7 x 60.3 cm. Montclair Art Museum, Montclair, New Jersey, gift of Dr. Arthur Hunter in memory of Ethel Parsons Hunter, 1962.32.

the period, but the hydrangeas are, in fact, the artist's offspring: one of his own paintings.[141] Sargent's portrait was probably made in exchange for another of his sitter's flower paintings, one of blue hydrangeas by the sea.[142] Favored as painterly subjects by Duez and Helleu, hydrangeas became emblems of the vegetable model of artistic production for Sargent. Five years later, in 1890, Sargent demanded blue hydrangeas for the portrait of Katharine Chase Pratt, daughter of an artist friend, in Worcester, Massachusetts (fig. 166). He sketched the flowers at a local greenhouse in preparation, though the oil study, *Hydrangeas at Lange's Greenhouses,* is untraced.[143] The painting of Pratt was considered unsuccessful and was abandoned, perhaps because of the livid, sickly pallor of both the sitter in her white dress and the flowers, but the idea behind it stayed with the painter.

In addition to the hortensia's cultural resonances, Sargent may have had in mind a specific source when he began *Helen Sears:* Jacques-Émile Blanche's 1887 painting *La petite fille aux hortensias* (fig. 167), in which the child has the same unseeing eyes as Helen and similarly grasps an umbel. The vacuity of Blanche's canvas and its stale *femme-fleur* equation seem to have tainted Sargent's conception from the start. Both Helen and the flowers in *Helen Sears* can be read as washed out or *blanched*—like Rilke's blue hydrangea, "washed out" as a "children's smock," Sargent's blossoms are as pale as Helen's smock.[144] Blanching is the process whereby vegetables are forced, artificially made to bloom out of season, and the issue of "forcing" is apt. The bloom at Helen's genitals is the most conspicuously stroked area of the canvas. And as though in response to her floral metamorphosis, the hortensia reaches, almost thrusts, toward the breast and crotch of the pear-shaped belle Hélène. But who is forcing whom? On one hand the hortensia seems to suck vampirically from Helen a supple, pleasure-giving blue nectar, which steals briefly into its blooms and across its petals; on the other, the delicate fingered flowers of the foreground are pale and wan

166

John Singer Sargent, *Portrait of Katharine Chase Pratt,* 1890. Oil on canvas, 101.6 x 76.5 cm. Worcester Art Museum.

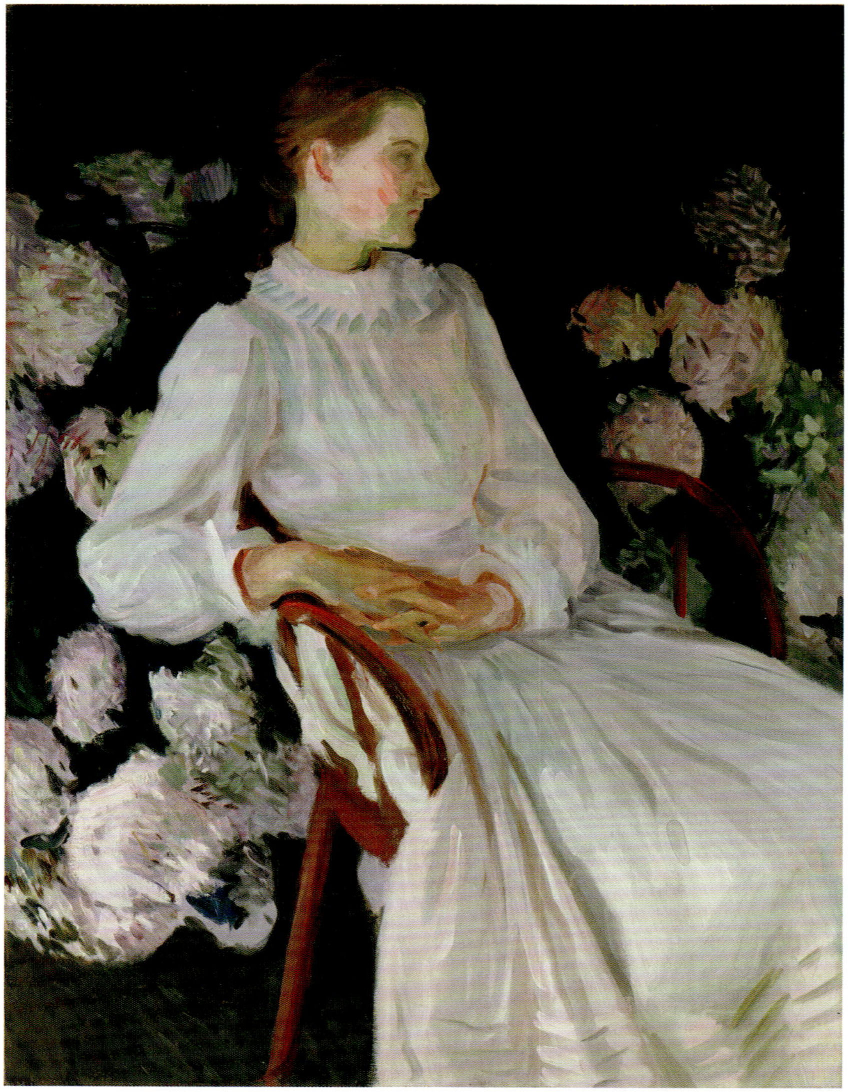

compared to the stronger coloration of the flowers in the background that Helen does not touch.

All this blanching bespeaks the emptiness of Symbolist aesthetics and the tiredness, the exhaustion, of the rhetoric of childhood, or rather, artistic innocence.[145] What if the touch of Helen *sears,* desiccates the hydrangea, the life-giving—if sterile—water vessel? Are all waspy, would-bee princesses and pollinating artists parasites? Does not the vision of Helen, archaic goddess of flowers and launcher of a thousand ships, condemn so many men and women to barren fates, the sterility of death and separation? The various ways in which this canvas poses the question of whether the vice of the artist, of the young Hellene, can really be rendered innocent(ly), without taint, attest to a profound doubt on Sargent's part, perhaps inspired by Helen's mother.

Sarah Choate Sears was a photographer, a member of the Photo-Secession and Linked Ring groups.[146] After Sargent had finished the portrait of her daughter, she sent him some pictures of Helen. He wrote back, "Many thanks for sending me the photographs. The new one of Helen has a wonderfully fine expression and makes me feel like returning to Boston and putting my umbrella through my portrait."[147] The ambiguity of the last possessive pronoun alerts us that Sargent's self-image, his conception of his artistic identity, is at stake. The impotence of the phallic prosthesis wielded here, an umbrella, is significant: it may be able to destroy successfully, but it cannot create; its whole function is to ward off contact with nature rather than facilitate its embrace, and its prosthetic arm's length emblematizes the distance separating the artist from his works of art, for all the fantasy

167
Jacques-Émile Blanche, *Little Girl with Hydrangeas,* 1887. Oil on canvas, 92 x 64.8 cm. Private collection. © 2008 Artists Rights Society (ARS), New York/ ADAGP, Paris.

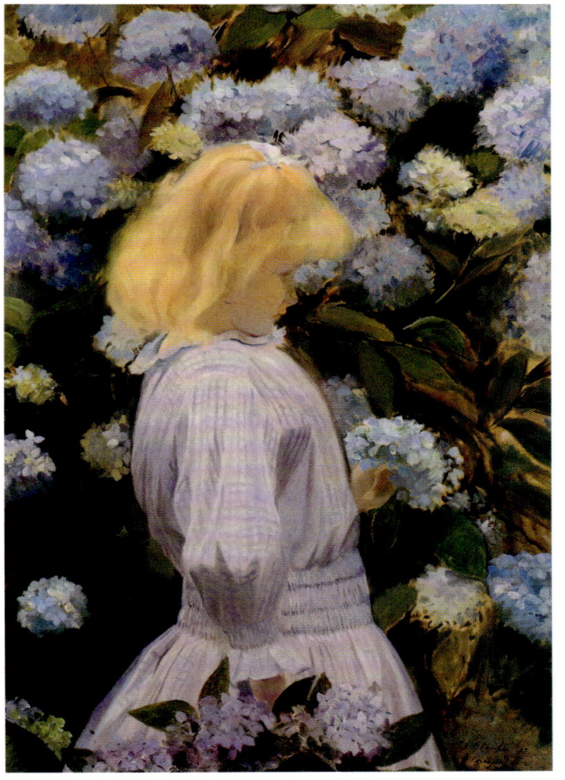

of immediacy engendered by finger painting and Impressionist impasto. Sargent's letter continues, "But how can an unfortunate painter hope to rival a photograph by a mother?" As an artist mother possessed of creative and procreative powers, prints and progeny, we can imagine that Sarah Sears may have stood as an obstacle in the way of Sargent's fructiparous fantasy, suggesting its barrenness to him.[148] Sargent's lack was all too evident beside Sarah Choate Sears's abundance: she was *choate,* complete; he, by contrast, inchoate, undeveloped, a mere beginner.[149] In one of Sarah's photographs (fig. 168), Helen is holding a Japanese lantern in front of a screen of flowers, which doubtless recalled *Carnation, Lily, Lily, Rose* to Sargent's mind and gave rise to meditation on the differences between the two paintings that rendered the allegory of hand-pollination in one successful and undermined the other.[150] If the earlier canvas represents a successful palpation by the artist, resulting in a floral birth, the latter bears, as one historian has put it, "the signs of the [surgeon's] diagnostic gaze."[151]

Unlike *Carnation, Lily, Lily, Rose, Helen Sears* was a commissioned portrait. In the former work Dolly and Polly are unconsciously absorbed in their roles, while in the latter Helen's seeming not-there-ness reads as the result of the painter's co-option of her portrait. Perhaps most important, the conservatory setting of *Helen Sears* stresses the artificiality of Sargent's allegorical equation; the winter garden seems to announce that the natural history Sargent paints for himself, his conservatory story, is a fiction. And yet critics from both the *Athenaeum* and *Portfolio* reviewing *Carnation, Lily, Lily, Rose* in 1887 thought the children were "in a conservatory."[152] The earlier painting was perhaps even more artificial a scene than *Helen Sears.* Having begun the picture late in the summer, the flowers began to die and Sargent had to "scour the cottage gardens and transplant and make shift" to keep himself in blossoms; later, hothouse or even artificial flowers had to be "tied to the withered stalks."[153] The next spring Sargent sent Lucia Millet lily bulbs to ensure a supply.[154] The garden the canvas represents is thus doubly, as Richard Ormond has pointed out, of the painter's "own creation."[155]

A taste for the pastoral, though, was in the air in England in the 1880s.[156] *Carnation, Lily, Lily, Rose*

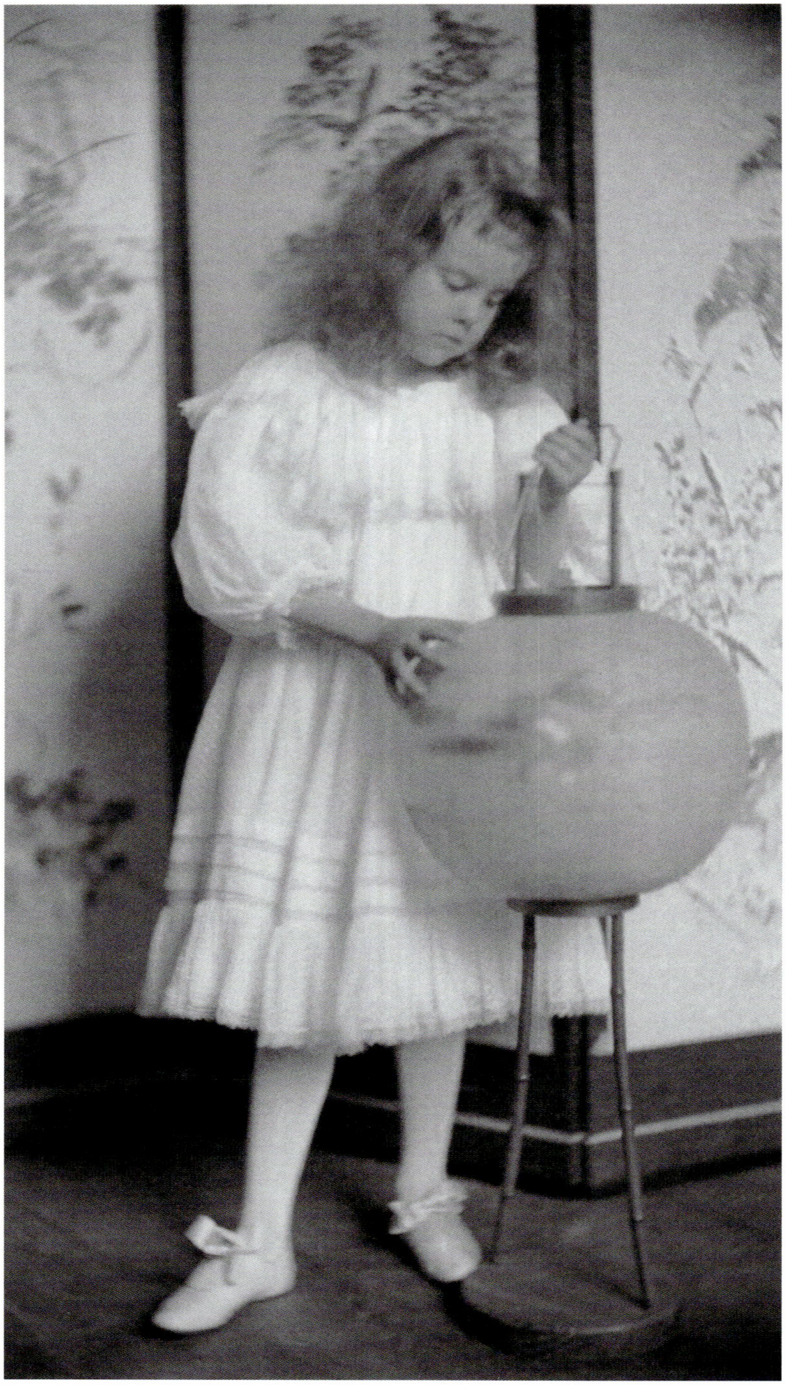

168

Sarah Choate Sears, "Helen Sears with Japanese Lantern," 1895. Platinum print, 23.7 x 13.2 cm. Descendants of Sarah Choate Sears.

would give rise to "an outburst of what might be called undergrowth painting at the Academy,"[157] and despite the uncooperativeness of Sargent's blooming subjects, Broadway really was something of an idyll, as Marc Simpson has argued.[158] At Broadway, Sargent was surrounded by expatriate and bachelor artist friends as well as a gaggle of women and children who served as convenient screens as well as seamstresses, gardeners, and painterly subjects. The little commune was always buzzing with activity—river outings, tennis parties, theatricals, musical evenings, large meals, and artistic games—and had in Edwin Austen Abbey, according to Gosse, its own "queen-bee" (Edwin Austen *Abeille*).[159] Charles Caffin called it a "fertilizing"[160] place for all its artists "busy as bees,"[161] and, predictably, Sargent had "memories of *noctes ambrosianae*" there.[162]

The rhetoric of naturalization was not out of place in an expatriate community; Broadway was, in effect, the perfect setting for the enactment of Sargent's allegorical tableau. As Anne Helmreich and others have argued, while Sargent was painting *Carnation, Lily, Lily, Rose* he "was preoccupied with his work's relation to prevailing ideas about national identity, painting schools, and techniques."[163] In the January 1886 issue of the London *Art Journal,* a critic speaking of Sargent's

American friend and fellow Broadway artist Frank Millet wrote, "it would seem to be the possible future of the best American artists to combine the superiority of English sentiment and the excellence of French technique into one rare and quite admirable whole."[164] Another, also speaking of Millet as an American, expressed a similar sentiment: "it is obviously of no small advantage to a painter to belong to a nation of which the art tendencies and peculiarities are as yet unsettled; and to be able, without violence to his origin, to gather flowers in every field of art."[165]

Carnation, Lily, Lily, Rose evinces both Impressionist technique (it has been called England's introduction to Impressionism; Fry even saw it as "a new revelation of what colour could be and what painting might attempt" before he changed his mind about Sargent)[166] and Pre-Raphaelite or Aesthetic subject matter.[167] It thus fulfilled the *Art Journal* critic's goal of Anglo-French hybridity achieved through American cross-pollination.[168] The picture's title received different punctuations in the press; one reviewer spelled it "Carnation-Lily, Lily-Rose," as if what the painting represented were hybrid flowers.[169] Hyphenation itself was a metaphor for hybrid nationality: an American critic wrote that "Sargent and Whistler developed their talents abroad and lived most of their lives abroad, and it is not too much to say that their distinguishing qualities have a European flavor." The all-American Winslow Homer, on the other hand, "was in no way hyphenated."[170]

Sargent's hybridity was a matter of fact: he was born in Florence of American parents, raised all over Europe, and as an adult settled first in Paris and then in London; his painting was compared to Velazquez, Hals, and Reynolds; and he was, as Vernon Lee put it, "an accentless mongrel."[171] Throughout John's nomadic upbringing—deemed "excellent for the budding artist"[172] by some—FitzWilliam Sargent frequently wrote to his parents of wanting to take "permanent root" somewhere the family could "thrive," though this never happened.[173] Often "made to root in strange soil,"[174] but always in a "solitary greenhouse,"[175] Sargent wound up "belong[ing] nowhere,"[176] to no one movement, nation, or exclusive orientation. Questions of sexual, national, and artistic identity were always intertwined for him: the invert and expatriate artist's longing for the rootedness of autochthony, to feel, as Genet put it, "trailing at [his] feet, the dynastic force . . . brutal, muscular, fecundating"—in other words, to be *naturalized*—left him a tangle of contradictions and competing desires.[177]

The garden world Sargent created in *Carnation, Lily, Lily, Rose* was capable of sustaining such contradictions. It was often described as a "fairy land"[178] by contemporaries—it was inspired by "fairy English gardens"[179] and filled with "fairy lights."[180] Sargent himself was even a "godchild of the fairies."[181] Already in the 1880s and '90s the term fairy connoted both enchantment and homosexuality.[182] Hans Christian Andersen, the great fairy tale writer of the nineteenth century, had described his own requited same-sex love as a "fairy tale" much earlier,[183] and in the *fin de siècle* fairy tales often promoted alternative sexualities.[184] Fairies, known to have difficulty procreating (hence their theft of children),[185] attended artistic accomplishment: James defined them as "the felicities that an artist knows when he doesn't

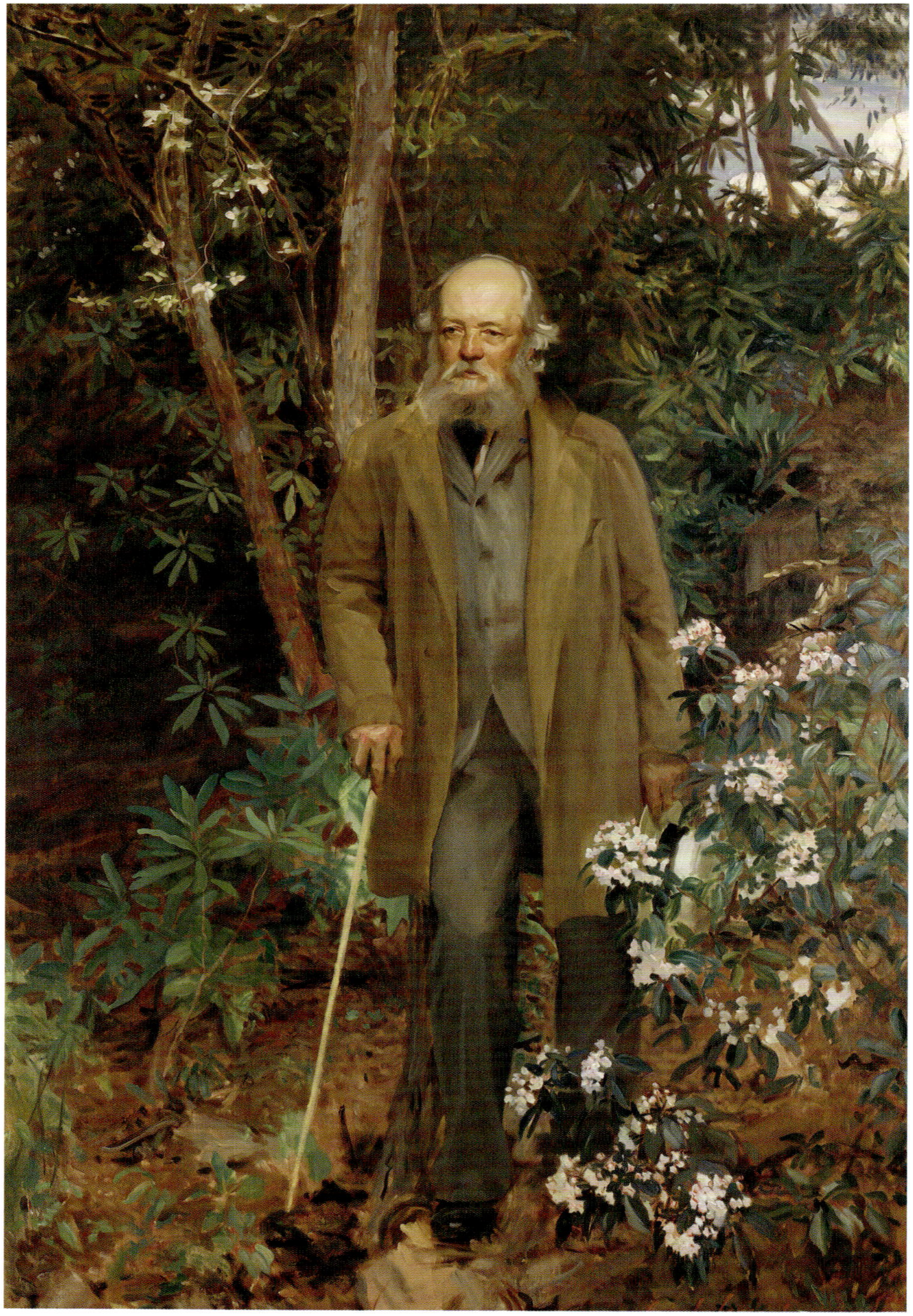

169
John Singer Sargent, *Frederick Law Olmsted,* 1895. Oil on canvas, 254 x 139.7 cm. Biltmore Estate.

know the black devils of despair."[186] In Sargent's enchanted garden, purity and voluptuousness, Impressionism and Aestheticism, parturition and sexual inversion, the fiction of roots and the freedom of flight coexisted.

The reading of *Carnation, Lily, Lily, Rose* as a fairyland was not incompatible with its "highly scientific"[187] quality, with Sargent's "biologic practice of revealing the secrets of life with a brush."[188] Biological forces were troped as "fairies" by contemporaries, for example in Arabella Buckley's popular book *The Fairy-Land of Science*.[189] Maeterlinck described Darwin's *The Various Contrivances by which Orchids are Fertilised by Insects* as a "fairylike biography,"[190] and Walter Crane and John Richard de Capel Wise dedicated their 1881 book *The First of May: A Fairy Masque* to Charles, not Erasmus, Darwin.[191] One of the novelists Rosny, who were quite "given up to scientific studies," told Vernon Lee that "the border-land of science—its still unascertained fields—will furnish that natural field for personal fancy & emotion hitherto supplied by theology: one will create one's fairyland there."[192] She immediately started reading Darwin.[193] In the 1880s fairies became the object of scientific speculation; some theorized they were "life-forms developed on a separate branch of evolution."[194] Later, various occultists took up this strand of thinking to link the material and spiritual realms, stressing fairies' connection to plants and insects. Fairies had often been represented as plant-insect hybrids in the nineteenth century;[195] in his 1913 book *The Hidden Life of Things*, Charles Leadbeater sketched evolutionary charts tracing the development of some fairies out of fungi, others out of grasses and cereals.[196] And during the Cottingley fairy frenzy, Arthur Conan Doyle (son of the fairy painter Charles Doyle) and a fellow theosophist named Edward Gardner published their observations: Gardner regarded the creatures as "allied to the *lepidoptera*" and belonging to "the same line of evolution as the *winged* insects."[197]

Lest this conflation of fairies, science, and modernity strike us as a British peculiarity, remember that Benjamin's original title for *The Arcades Project* was *Paris Arcades: A Dialectical Fairy Play* or *Fairyland*.[198] I am hardly suggesting that Sargent, like Benjamin, intended to elaborate a Marxist version of *Sleeping Beauty*, "a fairy tale of 'awakening'" from "the collective dream of the commodity phantasmagoria."[199] However, Sargent's deployment of natural historical analogies to express painting's "transformational effect"[200] articulated a yearning for an identity, a naturalization impossible to satisfy and too complicated to negotiate in existing categories of discourse.

Sargent painted a number of his adult male artist contemporaries out in nature. These figures are pictured not in the enchanted realm of the garden, however, but in more shadowy woodland settings, where they seem to embody the fragility of the fantasy of botanical reproduction. The portrait of Frederick Law Olmsted walking in the grounds of Vanderbilt's Biltmore Estate in North Carolina (fig. 169) was painted in the same year as *Helen Sears*. A renowned landscape gardener in the midst of the indigenous vegetable world[201] of his creation would seem to be a perfect figure for the botanically reproducing artist, and indeed Olmsted is vegetablized and autochthonized by the portrait,

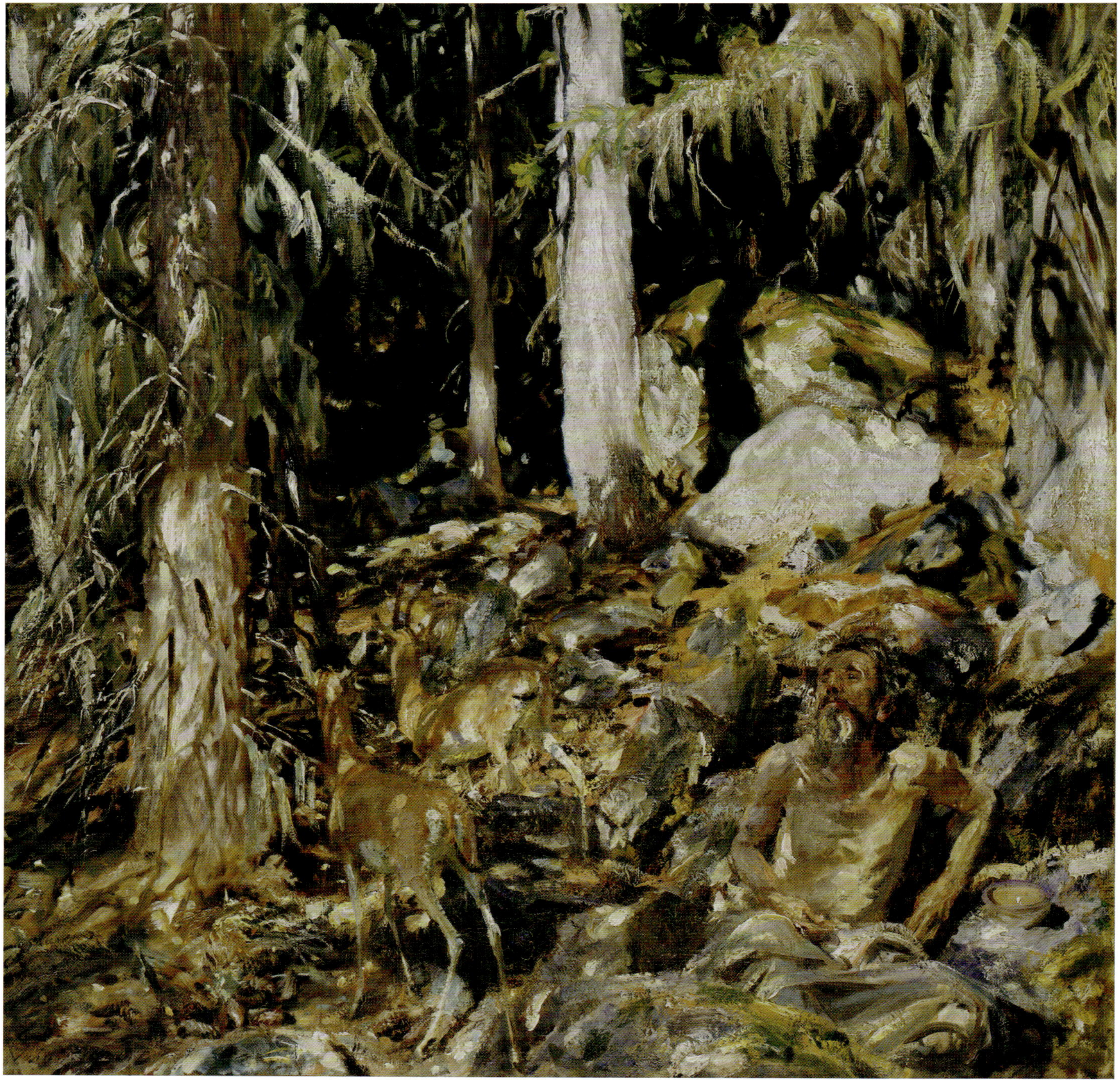

170
John Singer Sargent, *The Hermit (Il Solitario)*, 1908. Oil on canvas, 95.9 x 96.5 cm. The Metropolitan Museum of Art. Rogers Fund, 1911 (11.31).

from his feet lost in the ground to his tree-trunk-like trouser leg, earth-colored coat, up to his rhododendron-hued hat. He is also rendered vaguely insectlike by his proboscidiform yellow cane, the luminosity of which might remind us of Caffin's description of Sargent's brush as "loaded with light rather than with pigment."[202] Yet leaning against a tree he is strangely vulnerable amid all the creeping, thrusting, spraying vegetation. The seventy-three-year-old Olmsted's physical and mental condition was deteriorating in the period when the portrait was painted; this was becoming "evident not only in his forgetfulness but also in a faint bewilderment and vacancy of facial expression."[203] Fearing (with reason) this would be recorded in the portrait, the family had Frederick Jr. pose after the initial sittings.[204] The tension in the painting arises, then, from Sargent's double vision of Olmsted as, on the one hand, an artist who has lost his powers, and on the other, one nevertheless possessed of the progeny—vegetable and human—to prop him up.

Another image of an artist out in nature, *Il Solitario* from 1908 (fig. 170), depicts the painter Ambrogio Raffele naked in the woods. Sargent's portraits or genre scenes of Raffele are the only ones that have been consistently mistaken for self-portraits,[205] and their titles, including the earlier discussed *Artist in His Studio*, resist fixing their subjects' identity.[206] Max Beerbohm described *Il Solitario* as a manifestation of Sargent's yearning for a return to nature, with the sitter "*planté là* and meaning to stay so."[207] The painter's desire for union with the vegetable world is expressed through the camouflage effect of the painting, which merges figure and ground. His longing for fruitfulness, for a fertilizing encounter, is further gestured to by the hand that reads as the *solitario* indulging in the solitary vice (fig. 171).

This unusual work brings to mind two models of vegetal vigor and fertility Sargent came across in the years just prior to painting it, one involving an actual botanical specimen, the other a mythical precedent. In 1904 or 1905 Lillie Langtry, a female floral artist, sent Sargent one of the "resurrection-plants" she had acquired in America. These "dried-up withered little plants," as she described them in her accompanying letter, "preserve the germ of life even if baked in the oven, and after any lapse of time will recover their verdure in twenty-four hours, if placed in a saucer of water." Sargent's reply expresses his amazement and reports, "It is a green tree to-day. A thousand thanks for such a rarity."[208] On a rock next to the *solitario* there is a saucer; there is no water in it, however, to bring this aging, desiccated figure back to bloom and youth, to renewed life. The other vegetable paradigm the painting evokes is the grove at Nemi discussed in Frazer's *Golden Bough*.[209] We already know that Sargent was interested in Frazer's theories. Three years before this painting was made, though, Frazer gave a series of lectures at Cambridge consisting of excerpts from the forthcoming third edition of *The Golden Bough*. Dedicated to Edmund Gosse, the published version of the lectures came out the same year and would no doubt have been discussed by Sargent and his circle.[210] Frazer's *Golden Bough* and the Cambridge lectures concern the figure of the "King of the Wood," a priest of Diana in the sacred grove. The story behind this figure is that Hippolytus, who loved to hunt in the forest with Artemis, "spurned

171
Detail of *Il Solitario*.

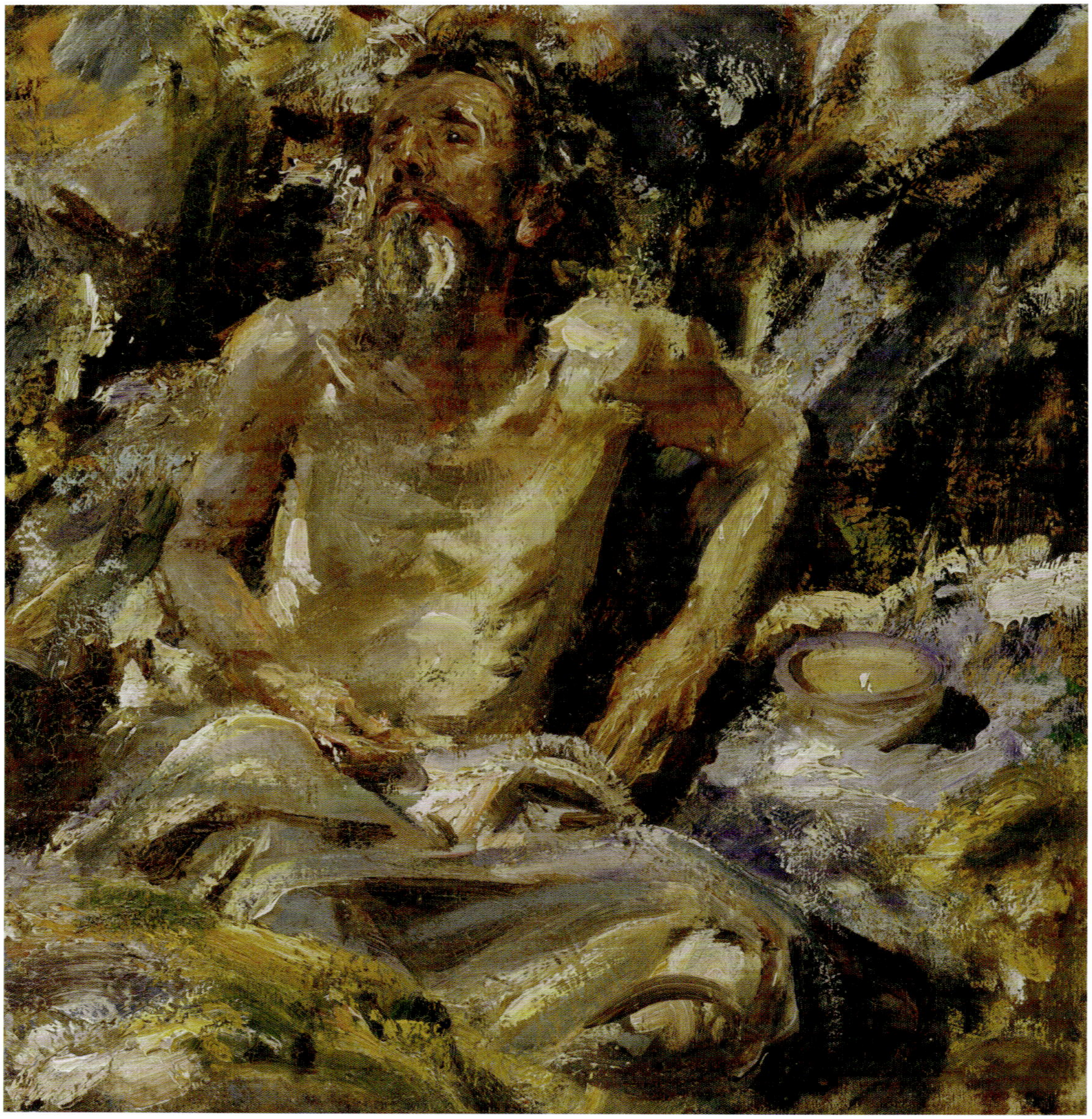

the love of women," and was punished by Aphrodite and Poseidon. Artemis rescued her companion, gave him a protective disguise "by adding years to his life," and took him to Italy, where he was to live "unknown and solitary" in the woods. Renamed Virbius, the solitary man was charged with guarding a sacred tree. Frazer hypothesized that Virbius (and later his priestly replacements, who came to be known as the Kings of the Wood) may "not only have worshipped" the tree "but embraced it as his wife"; the story thus provides another precedent for alternative, vegetable-oriented sexualities.[211] The role of the King of the Wood was to help ensure the fertility of the community. The lone figure in the woods in Sargent's painting, however, looks incapable of impregnating any of the thorny, sere trees around him.

The creative seed spilled here, then, seems doomed to dry up unused; and the melancholy autumnal setting reinforces the sense of sterility that is literally laid bare in the painter's person. This body, moreover, in its symbiosis with the vegetable world, appears diseased. The figure is not unlike Lalique's *Gynécologie* (fig. 117), with eyes raised, a hand at its genitals, naked but for a shroud and barely distinguishable from the background. The plaquey texture of the paint and the figure's skin, furthermore, suggests a syphilitic interpretation. The epidermis is almost a collage of thick, dry encrustations of paint: raw red patches alternate with grey, barklike discolorations, ostensibly colored shadows but resembling contusions or decaying flesh, especially on the subject's right arm. In *An Artist in His Studio,* Sargent pictured Raffele as an old-fashioned artist; here, the figure of the aged painter is literally superseded by the young bucks that are passing him by.[212] These two creatures are perhaps Sargent's punning comment on the danger of the Darwinian solution to the invert artist's creation problem: the withdrawal from the human species and loss of identity such mimicry demands may come at a price that is simply too dear.

5

THE STING

My life is full of scented fruits,
My garden blooms with stocks and cloves;
Yet o'er the wall my fancy shoots,
And hankers after harsher loves.

—EDMUND GOSSE, "DISCIPLINE"

Allegorizing the process of his artistic creation through portraits of little girls and old men as pollinator-plants allowed Sargent to avoid addressing the dangerous pleasures of invert sexuality explicitly.[1] In his portraits of mature artists closer to his own age, however, the floral poetics of creation helps to articulate the illicit desires it serves to screen in other paintings. In these works the underside, or perhaps more exactly the backside, of Sargent's pollinator persona—the wagtail—comes to the fore. A "wagtail," significantly, is not only a bird but was also popular slang for a harlot,[2] and was associated with inverts and art: as a nineteenth-century travel guide to London informed readers, inverts "generally congregate around the picture shops. . . . When they see what they imagine to be a chance, they place their fingers in a peculiar manner underneath the tails of their coats, and wag them about—their method of giving the office."[3]

Leaving the *hortus inclusus* of childhood behind meant moving into wilder landscapes or the potentially rather shady interiors of the cities of the plain. The "harsher loves" Gosse craved might be of the sort Proust imagined in *À la recherche du temps perdu*—for example, when the narrator enters what he thinks is a hotel in search of a drink and a bed to sleep in and instead finds himself in a male brothel where he sees Charlus "chained to a bed," bloody and "covered with bruises," receiving blows from a whip "studded with nails."[4] Or more subtly by Robert Musil in *The Confusions of Young Törless,* when the titular character's desire for the floral Basini is experienced as "something rushing and hurrying, beating against his body as though with the velvety antennae of butterflies," an anticipation of the beatings Basini will receive from the other boys at school.[5] These loves could also be found outdoors, impulsive and feral, as in the imagined tableau where an invert walking home a married friend—about whom no suspicion has begun to flower—"finds himself thrown down on to the grass without a word by the mountaineer who is shortly to become a father,"[6] or in the Bois de Boulogne, "the meeting place for all the rootless vices," in Jean Lorrain's description, where "the strong scents of garlic and rancid humanity" intermingle and shadowy forms can be glimpsed in "the wet grass."[7] These erotic encounters come in many guises, from sadomasochistic scenes of torture to quieter, if still sometimes rough and ready, tumbles in the grass.

Our wagtail was a particularly hardworking hedgewhore, painting so very many belles and beaux and would-bee flowers and *leur contant de fleurette* (charming them) with his magic wand. Yet Sargent explored many of the less bucolic aspects of botanical reproduction, from hothouse amours where pollinators are drugged with floral perfumes to "backdoor" pollination where forcible entry may be required, from vampiric pollinators to sadistically wielded painterly prostheses. He was not alone in charting this territory, and to contextualize Sargent's exploration of the various contrivances of botanically inclined artists I will first turn to the example set by his fellow expatriate James Abbott McNeill Whistler. As we saw earlier, Whistler's sitters were often flowers or pollinators; when painting and cross-fertilizing these delicate and ephemeral creatures, the artist was prone to whipping and vampirism in particular.

The sadism of the expatriate painter and etcher was widely recognized. Haldane Macfall summed up Whistler's character with the verdict that "he grasped that what the world would not see he could whip it into seeing—so he whipped it— . . . he whipped it."[8] Macfall doesn't specify with what Whistler whipped: his paintbrush, etching needle, cane of microscopic proportions ("the size of a darning needle," according to one artist peer),[9] and even his lashing tongue could be fertilizing proboscides, stingers, whips, and swords by turn. Other pollinators could arouse his ire; Whistler is reputed to have wondered whether Sargent "ought not to be taken out and whipped" for his painting, which seemed to him "almost like the work of a bad boy,"[10] and in Venice in 1880 he "impaled a scorpion on an etching needle and watched it flail."[11] It should not be surprising, then, that Whistler occasionally equipped his butterfly symbol with a long stinger (fig. 12).

In Comte Robert de Montesquiou—a fellow plant, pollinator, and artist—Whistler met, perhaps, his too perfect match, and while painting the Frenchman's portrait (fig. 172), an erotic battle for domination commenced. Whistler's passion for his sitter did not go unnoticed, and Montesquiou's biographer wondered if the emotion binding the two men was "something deeper than admiration";[12] certainly the painter's acceptance of a bed in exchange for the portrait is suggestive.[13] Whistler, in turn, may have reminded Montesquiou of his childhood companion and initiator into various vices: the Abbé Charpentier, whom the young Robert had nicknamed "The Butterfly." This insinuating creature had "introduced the child to odd practices," including spying on the staff as they used the "many-holed privies."[14]

Montesquiou, we know, identified with the blue hortensia and the bat. His friends, Whistler included, addressed him as "Bat,"[15] he kept one of the animals as a pet, and he even resembled one, "with his winged, upturned moustache and his enormous pointed ears."[16] In Whistler's portrait his arms curve like bats' wings and terminate in wrists bent clawlike, hands in gloves the color of pale skin. The association of bats and vampirism predates Bram Stoker's *Dracula*, which was not published until 1897. Sarah Bernhardt slept in a padded coffin and kept a stuffed vampire bat in her bedroom,[17] and short stories about the life-draining or blood-sucking undead abounded. Descriptions of the portrait's creation confirm the relevance of this connection. It was the butterfly painter,

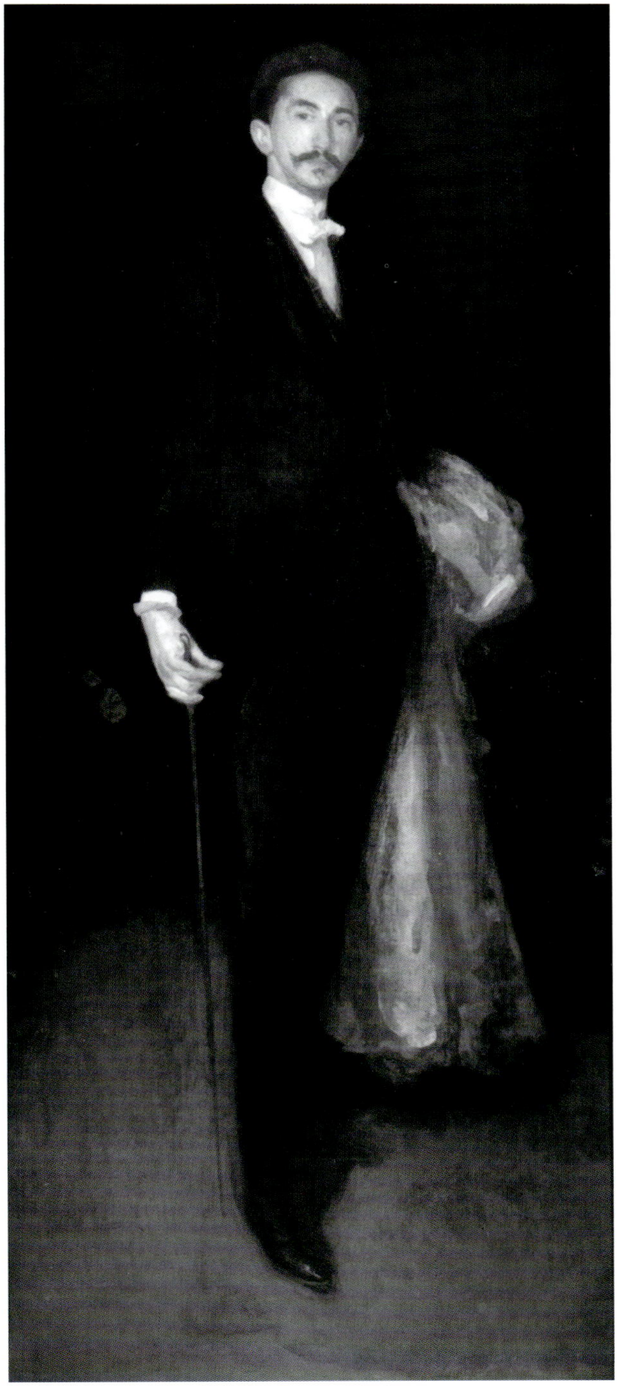

172
James McNeill Whistler, *Arrangement in Black and Gold: Comte Robert de Montesquiou-Fezensac,* 1891–92. Oil on canvas, 208.6 x 91.76 cm. The Frick Collection, New York.

however—not the bat—who played the vampire in the tale of Montesquiou's picture.

Vampirism was not the prerogative of bats alone. The fictional Dracula changed himself into flies and moths as well as a bat, and was troped as a kind of immortal bee: "the *nosferatu* do not die like the bee when he sting once," van Helsing tells us.[18] In nature other pollinators, too, were susceptible to the macabre inclination. The "butterfly pumping aroma and nectar from a rose by means of its spiral tongue" is only a variation of the "gnat piercing our hand" with its "blood-sucking tube," and lepidopters could become leechlike.[19] Such naturally nefarious pollinators had an illustrious enthusiast in the Marquis de Sade. In *Justine* a vicious blood-letter's wife is "the very image of a beautiful lily upon which the honeybee has inflicted some scratches."[20] The word used to designate her wounds, "taches" (literally spots, used to refer to patches of paint), highlights the painterliness of the tableau—vampires are aesthetic creatures—as does the label "sittings" Sade applies to the escapades of the decapitator in the novel, a "hornet" who desires to drain the pollen of his "very fresh, new-blown" flowers: "his iterated suckings" are succeeded by neck inspections during which his member throbs "with growing insistence."[21] Whistler's own painterly proboscis, as we shall see, functioned effectively as a sucker as well as a scratcher and a stinger.

In the full-length portrait Montesquiou stands in darkness, immaculately dressed in evening attire. He wears tails and flesh-colored gloves; he holds a dusty pink cloak over one arm and a thin cane in his right hand. The bat, like the butterfly that hovers beside his hand, appears ready for any nocturnal adventure. Strangely, and suggestively, the artist's sittings began at dusk: "It was [Whistler's] habit to paint when the studio was filled with gloom and lengthening shadows crept across the floor; when it was so dark the dull eye of sitter or chance visitor could scarce distinguish the figure on the canvas. This 'painting in the dark,' as some have called it, was a singular trait. He would paint with increasing force and effect as the room became darker and darker, until it seemed as if the falling of night was an inspiration."[22] And perhaps, for the vampiric painter, it was. Whistler certainly played the blood-sucking pollinator, and like many of the painter's subjects, Montesquiou felt himself thoroughly "drained" by his sittings.[23] Edmond de Goncourt "was worried; he went everywhere saying, 'Whistler is pumping the life out of his model. Were it not that he has discovered a wine made of coca that is particularly tonic, Montesquiou would not survive these sessions.'"[24]

The comte's physical draining was required by Whistler's pictorial aesthetics. All of the artist's "phantom-portraits," in Huysmans's description,[25] were a result of his sitters' subjection to a "continuous process of etherealization"; they became "eloquent disembodied souls."[26] Raitif de la Bretonne summed up this experience as "that simple sensation of being sucked out alive," exactly as "Edgar Poe described it throughout his short story called '*The [Oval] Portrait*.'"[27] In this tale the narrator chances on a disturbingly lifelike painting of a young woman in a deserted château and a manuscript relating the history of the portrait. The young wife of a painter obsessed with his art had "sat meekly for many weeks in the dark high turret-chamber where the light dripped upon the pale canvas only from overhead" from "hour to hour, and from day to day," while her image slowly emerged from out of the "vague yet deep shadow" forming the background of the painting. And as the painstaking labor

> drew nearer to its conclusion, there were admitted none into the turret; for the painter had grown wild with the ardor of his work, and turned his eyes from the canvas rarely, even to regard the countenance of his wife. And he *would* not see that the tints which he spread upon the canvas were drawn from the cheeks of her who sat beside him. And when many weeks had passed, and but little remained to do, save one brush upon the mouth and one tint upon the eye, the spirit of the lady again flickered up as the flame within the socket of the lamp. And then the brush was given, and then the tint was placed; and, for one moment, the painter stood entranced before the work which he had wrought; but in the next, while he yet gazed, he grew tremulous and very pallid, and aghast, and crying with a loud voice, "This is indeed *Life* itself!" turned suddenly to regard his beloved:—*She was dead!*[28]

Poe's artist's ability to transfer "life itself" to a painting, like Whistler's draining portraiture, is the result of willfully inflicting pain and sapping the lifeblood of the model.[29]

Over the course of Montesquiou's "innumerable and interminable" séances with the painter (he posed more than a hundred times), the count saw his portrait eerily "submit to the very laws of human growth," while he became more and more frail.[30] Whistler used to say, "it takes a man a long time to look like his portrait"; in light of the painter's draining technique, this is not mere paradox.[31] According to sitters, while painting, Whistler "would pull back, one might say, to draw a stroke from the very essence of the model, and convey it onto the canvas without giving it the time to evaporate or vanish."[32] Whistler had to press Montesquiou repeatedly to endure the endless and deleterious process of posing for the sake of his art: "the beautiful picture," already rhetorically animated, "asks of me only a few hours to come to life."[33]

Critics agreed that the uncanny animation of Whistler's black portraits, in which spirits, through the medium of the canvas, ever "seem to detach themselves bit by bit from the shadow, to come alive," could only be achieved by a "tyrannical taking possession of his models" or similar "violent seizing of character."[34] The physical and psychic submission demanded of the painter's subjects prompted the likening of Whistler's technique to that of the mesmerist in criticism and fiction. While the artist's ritual summoning of an apparition could be described in neutral terms—"it was as if the portrait were hidden within the canvas and the master by passing his wands day after day over the surface evoked the image"[35]—mesmerism was also invoked in association with Whistler for its unethical and even necrophilous connotations. The "magnetic relationship can be either good (friendly, fatherly), or evil (demoniacal)," according to E. T. A. Hoffmann, "the evil magnetiser" being "a kind of moral vampire who destroys his subject."[36] It is no coincidence that Svengali's mortal exhaustion of Trilby in Du Maurier's 1894 novel recalls descriptions of Whistler's use of his subjects: the painter was the model for his character.[37]

Whistler's painterly necromancy did not escape Montesquiou; the comte recorded in his memoirs that the artist "always seemed to say '*surge et veni!*' to the resemblance being formed."[38] In the painting that ghostly resemblance stands, or seems to hover, at the center of a pale grey-brown conjurer's circle, all around which is a black void. The sharply pointed toe of his right shoe is the most defined and solid form in the picture; smoky black leg and shadow, indistinguishable from one another, surge upward from it. The sense of the figure's *arising* out of this impossibly small vessel, like an Ifrit in *The Thousand Nights and One Night*,[39] is enhanced by his unnatural height, the absence of his left foot, the cloudy veil partially shrouding his left hand, and the cane tapering into nothingness.[40] The audience at the 1894 Salon perceived something sinister in the portrait: "the head shines forth with its unhealthy ivory flesh tints, much as some evil spirit would appear out of darkness," reported the *New York Times*, and *Le XIXe Siècle* saw the whole figure of the "apparition" seeming "to escape from the shadows and come forward toward you."[41] If half-released from its prison, however, the

173
James McNeill Whistler, *Large Bowl and Cover*, c. 1876–78. Pencil, brown ink, and wash on cream wove paper, 22.2 x 17.2 cm. Glasgow Art Gallery and Museum.

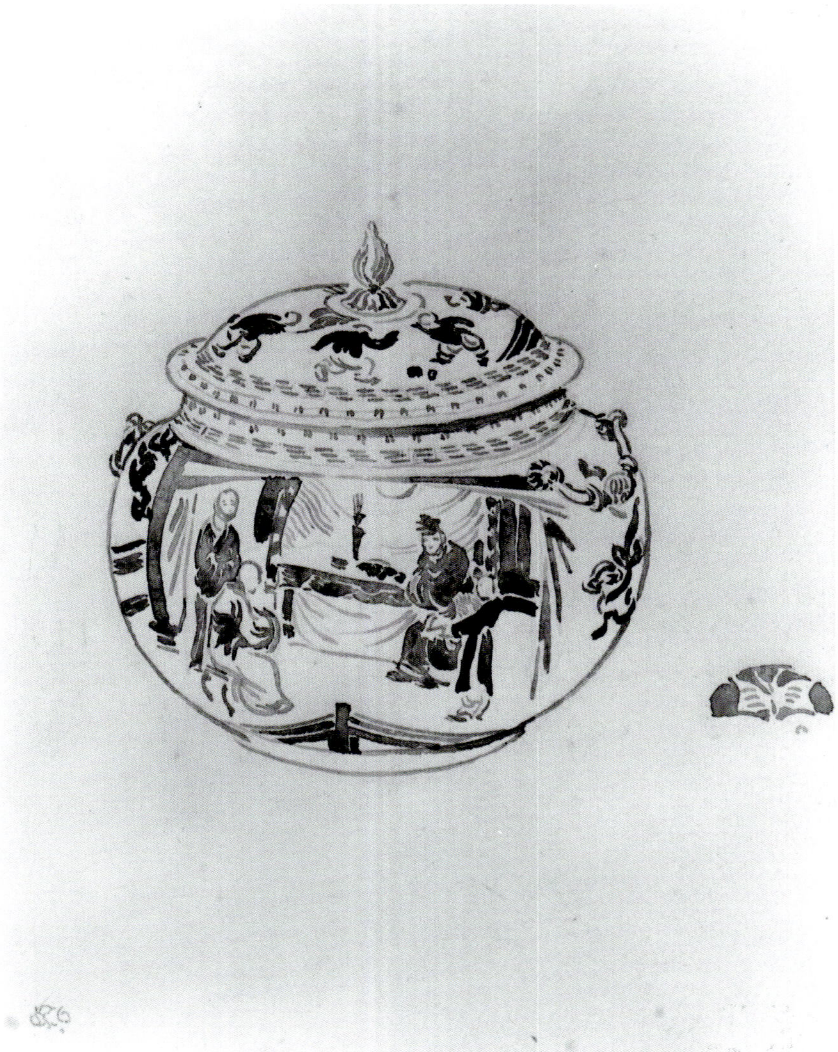

figure is still contained by the painting, the surface of which was compared to "tarnished black enamel,"[42] for "at no time" did Whistler "permit the figure to get away from or out of the background."[43]

Whistler's painting thus functions as a paradigmatic instance of the kind of controlling portraiture that, according to Vernon Lee, aims not just to "press the flower" but "to imprison and use *at pleasure* the most fleeting, intangible, and uncommunicable of all mysterious essences, a human personality."[44] The butterfly painter, then, was also a collector of sorts. Whistler's desire to snare living subjects in artistic containers might even explain his passion for china: the watercolor drawings he made for the catalogue of Sir Henry Thompson's collection of blue and white porcelain give an impression of life magically held within the confines of small vessels (fig. 173). If pots in Sargent's oeuvre ameliorate the condition of lost autochthony by rendering roots portable, for Whistler they serve to capture and hold desired and feared objects (genies, genius).

By general consensus the painter appears to have exercised total control in Montesquiou's portrait, usurping the count's spirit for himself and his painting. Whistler's own letters, however, tell a different story, of the artist terrorized, enslaved, and drained by the portrait: "this great black work that is still there" in the studio after three years is "an eternal terror and reproach until it is done!"[45] Taking on a life of its own—a pollinator's—it harassed him, demanded more: "it's all buzzing up there, the canvas isn't satisfied yet."[46]

The loss of identity Whistler experienced while painting Montesquiou was double; both the canvas and the sitter mirrored him. Over the course of the séances the painter's and the writer's powers of identification induced a kind of reciprocal anamorphosis between butterfly and bat, or a mutual inversion (W/M). Whistler's technique became more vampiric than ever in the service of the count, and as the cheiropterophilous painter became increasingly like a vampire bat,

Montesquiou "began to resemble a butterfly: he cut the ends off his moustache, grew a short imperial"; he seemed more and more like Whistler as "his wit became more biting, his taste more sure and far more simple."[47] And if the literal disappearance of the model into the dusk of his studio was, for Whistler, the condition of any resurrection on canvas, this portrait threatened the painter himself with disappearance. The moment Whistler achieved Montesquiou's perfect likeness on canvas, perceiving the uncanniness of the juxtaposition, as of "Poe's Wilson with his *doppelgänger,* of Shelley's Zoroaster with himself, or of Musset with the 'young man dressed in black, who resembled him like a brother,' such an apparition made [him] exclaim . . . '*Look at me again for a moment,* and YOÙ WILL LOOK FOREVER!'"[48] The exclamation confuses the portrait's subject, painter, and model, a consequence of the mutual and corporeal portraiture that Whistler and Montesquiou were stroking into place; Whistler, having painted himself into the portrait, realized its threat to his own life. Similarly, when Basil Hallward met Dorian Gray, he felt himself grow pale with terror: he had "come face to face with someone whose mere personality was so fascinating that, if [he] allowed it to do so, it would absorb [his] whole nature, [his] whole soul, [his] very art itself."[49] Montesquiou seems to have exercised the same power over Whistler, and subsequently the painting, in full possession of its creator's mesmeric command, as if it were an inherited aptitude, suddenly appeared capable of effecting for itself the final transfusion of life-giving nervous fluid in one metempsychotic moment—of draining its creator and becoming autonomous. There are risks, then, in creating a truly living art: painted progeny may be parricidal.

Sargent's contemporaries perceived in his art an ability to conjure the macabre and sinister akin to Whistler's. R. A. M. Stevenson (Robert Louis's cousin) described Sargent's painting of a wine cellar (fig. 174) as "a mere note of the interior of a cellar with naked figures turning a wine-press," in which, nonetheless, "the vicious tones of the wine stains on the floor and wall, the glimmer of light on bare flesh standing out sharply from the dim swimming atmosphere, and the threatening bulk of the wine-press reaching its arms into the darkness, all combined to produce the aspect of some hideous underground torture-chamber."[50] None of the figures in the painting is actually naked, though the three men turning the press shed layers of clothing as they move toward the darker interior. Stevenson's description thus speaks both to the evocative powers of Sargent's art and the sort of associations made by his peers. The opposition of the body suspended helplessly from a beam and the straining men is suggestive; one might think of Dracula, who made Mina Harker his "bountiful wine-press for a while."[51]

Like Whistler, Sargent could feel rivalry with other pollinators; butterfly collecting was a "mania" with him. Marie-Louise Pailleron recalled that the painter watched for the creatures "without flagging behind his shutter, like an assassin, and when he could seize one, he asphyxiated his victim by suffocating it under a glass with the smoke of his cigar."[52] Beyond such violence directed against competitors, Sargent also displayed an interest in the many striking instantiations of the

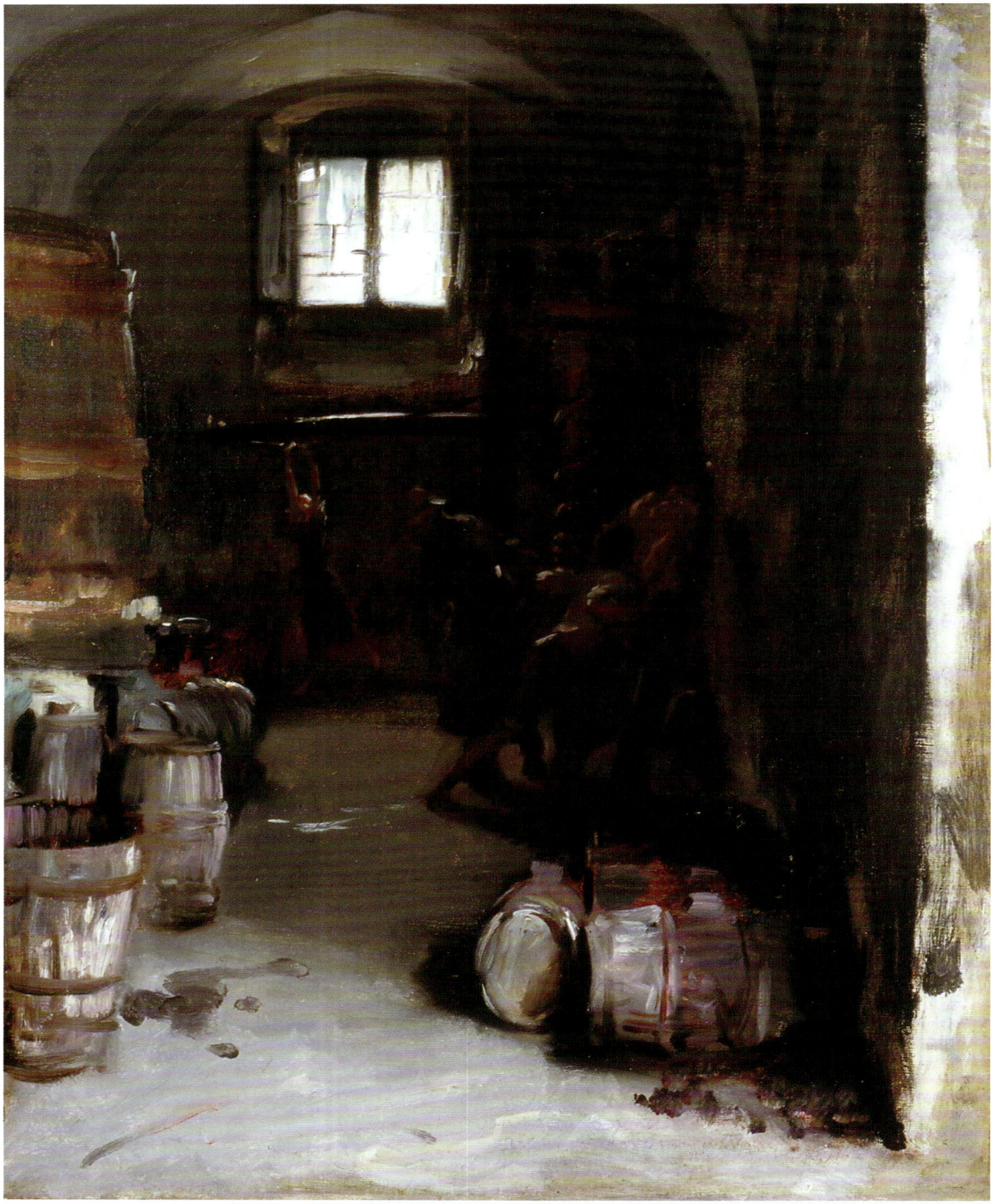

174
John Singer Sargent, *Pressing the Grapes: Florentine Wine Cellar,* c. 1882. Oil on canvas, 83.5 x 49.9 cm. The Beaverbrook Art Gallery/The Beaverbrook Foundation (in dispute, 2004), Fredericton, New Brunswick.

sadomasochistic side of floral sexuality and cross-pollinating erotics. Some oncidium orchids, for example, are fertilized by assault: they provoke attack from territorial male bees who mistake them for competitors when they vibrate in the breeze. Even the "normal" sort of pollination, however, can entail cold, hard, almost surgical violation, as a letter from one of Sargent's most peculiar sitters attests.

Frank Swettenham was the high commissioner for the Malay states and governor of the Straits Settlements as well as a writer, painter, and linguist.[53] One night in 1898, at one in the morning, he writes, a moth visited some lilies in his room:

> Once directly opposite the flower, the moth produced a delicate proboscis, which it inserted into the blossom, and then slowly pushed it right up the stamen, apparently in search of honey. When extended, this feeler was of quite abnormal length, at least two or three inches. What, however, surprised me was that, having withdrawn the probe (for that was what it looked like, a very fine steel or wire probe, such as dentists use), the instrument seemed to go back into the moth's head, or wherever it came from, to be again extended to sound the depths of another blossom.[54]

As the passage continues, the proboscis becomes explicitly an "instrument of torture": the moth "thrust that probe into the lily's nerve-centres." The retractable instrument reminds him of a man "swallowing swords."[55] Such comparisons were not unusual—Maeterlinck, for example, tells us that the tongue of the bee is "fantastic as a Chinese weapon,"[56] and Jean-Henri Fabre's popular series *Souvenirs entomologiques,* which ran to ten volumes between 1879 and 1907, describes insect prostheses as scalpels, daggers, and scissors, and includes chapters as provocatively titled as mystery novels: "Three Strokes of a Dagger," "A Skilful Killer," etc.[57]

In Sargent's portrait of Swettenham (fig. 175), the commissioner's fascination with the moth's uncannily long proboscis is echoed by the painter's for the sitter's own sharp stinger, the gleaming rapier at his side, its steel length reminiscent of the insect's probe. White wings spread, leg lifted, moustache-brushes quivering, and stinger at the ready, Swettenham becomes a man-sized moth. Naturally, this creature is drawn to fine fabrics, and he stands next to a luxurious cascading tapestry. Brushing up against its soft red and gold folds (at the heart of a red, florid, and floral interior), his white coat is streaked with yellow pollen stains. Swettenham is, moreover, like the moth he describes, nocturnal, suited to his benighted corner of the world (by comparison, India, visible to the west on the globe above, is lit and labeled).

The Swettenham portrait is typical of a number of paintings of subjects Sargent clearly admired and with whom he to some degree identified. In these paintings the "fetishistic allure"[58] of objects that resemble paintbrushes and share their aggressive potential, such as whips, canes, maulsticks, swords, and riding crops—prostheses of various sorts—is emphasized. These instruments were generally thought capable, like cigarettes, of transforming their bearers into diverse species. One early nineteenth-century

175

John Singer Sargent, *Sir Frank Swettenham*, 1904. Oil on canvas, 170.8 x 110.5 cm. National Portrait Gallery, London.

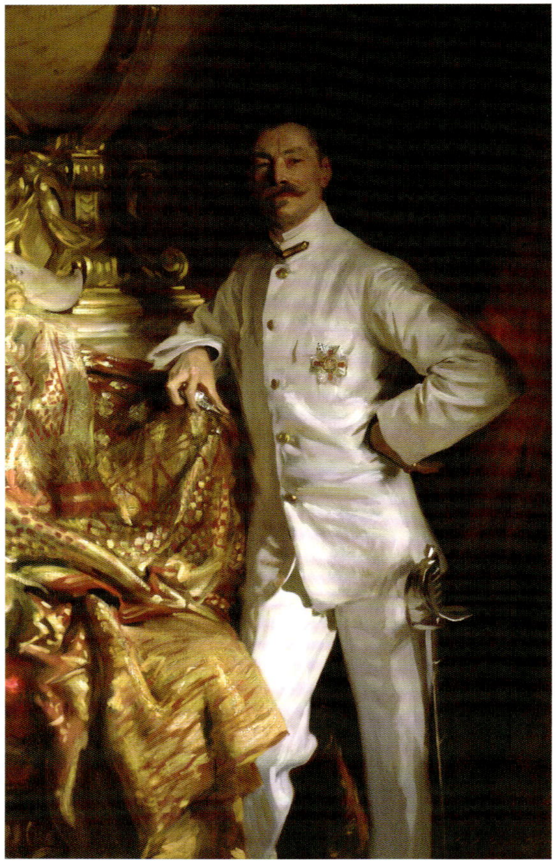

writer catalogued the metamorphoses wrought by canes in particular, which could change their owners into unicorns ("his formidable horn projects, and forces a passage through the crowd for the resolute charger"), bucks ("the stick grasped by the head, with the end advanced in the manner of a spear or bayonet, characterises the bullying buck"), dogs ("his cane, whenever any friend accosts him, is instantly agitated to and fro' like the tail of a pleased spaniel"; creatures of this breed are also known as twirlers), and of course, as they resemble proboscides, insects. They could also indicate sexual inclinations when they created "inverters," "self-tormentors," "trailers," and "whippers."[59]

In the portrait of Lord Ribblesdale in hunting costume (fig. 176), the subject, who notably sports a prominent black "butterfly" muffler,[60] is transformed by his whip into a most "sinister projection of power."[61] In an oil sketch for this portrait (fig. 177), the handle of the whip springs from the heart-shaped bulge in Ribblesdale's trousers, a phallic prosthesis the exaggerated proportions of which are echoed by the gigantic, bulging columns behind him. In the finished work the handle of the instrument is shaped like the grim reaper's scythe, and the shadow cast by this lean, black-cloaked creature in "butcher boots"[62] is that of a hooded rather than a top-hatted figure. Ribblesdale is literally as well as figuratively sinister: the stock of the whip is held in his left hand. At the end of this long, coiled proboscis, the thin line of the red popper hangs down like an unsated tongue. In the oil sketch Ribblesdale appears only as an outline, through which the background runs continuously, as though he has been caught in the process of vanishing or materializing. In the final version the malevolent pollinator, looming over the wallflower behind him (a rosette) yet posed with a ballet dancer's grace, looks as though he might spread his great black wings and disappear.

Not all of the sitters Sargent portrayed as alluringly dangerous pollinators were men of power and position. Victoria Stanley (fig. 178), dressed in red and white and sprouting feathers, holds her riding crop with equal aplomb as she casts a knowing, even impudent look at the viewer.

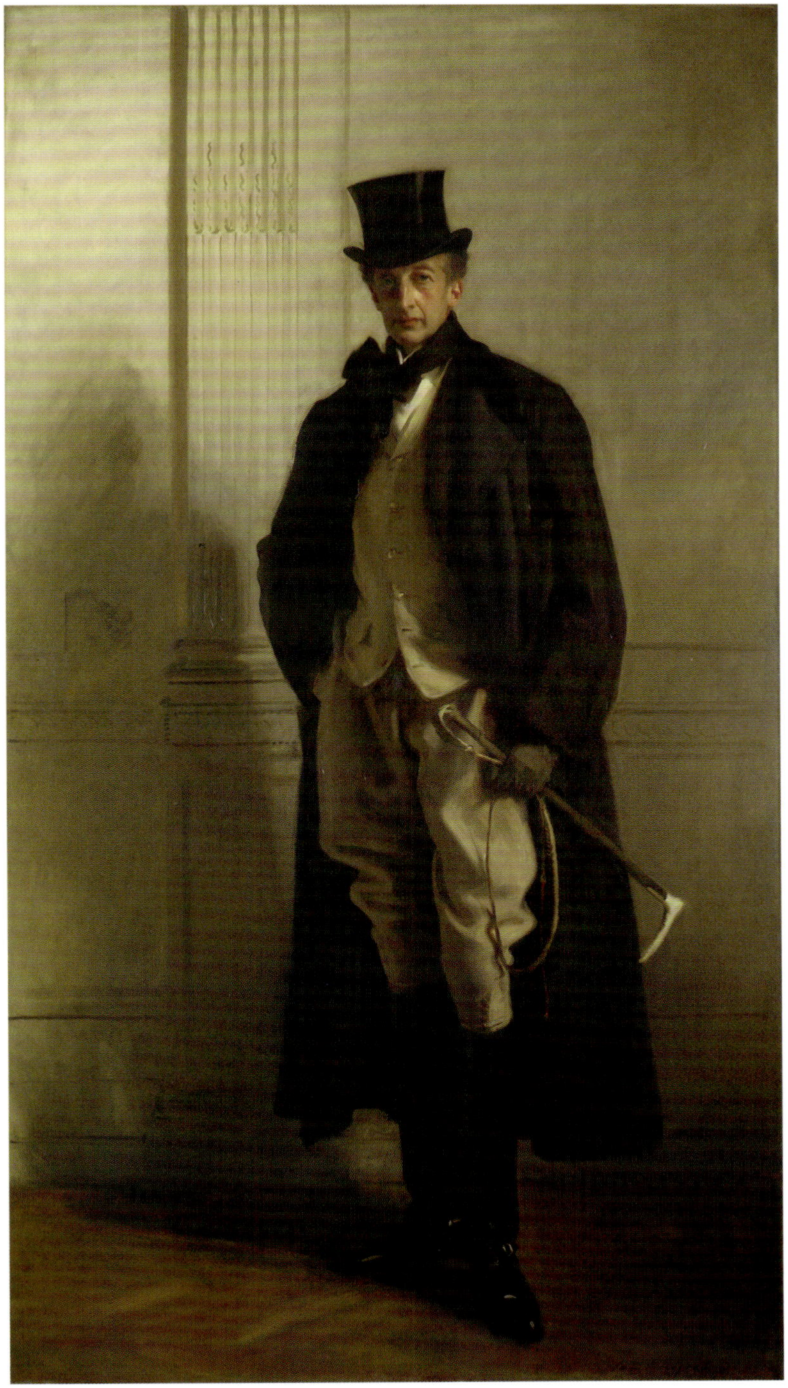

176

John Singer Sargent, *Lord Ribblesdale,* 1902. Oil on canvas, 258.5 x 143.5 cm. The National Gallery, London. Presented by Lord Ribblesdale in memory of Lady Ribblesdale and his sons, Captain the Hon. Thomas Lister and Lieutenant the Hon. Charles Lister, 1916.

"A rather savage Sargent," as D. S. MacColl put it, this precocious pollinator's command over the floral world is expressed, like Elizabeth I's over the country in the Ditchley portrait, by her feet's firm domination of the rosy Aubusson carpet.[63] The carpet is marked with rough strokes that read both as floral patterns and bruiselike marks on the buttocky or testicular orbs on which Victoria stands. Either sort of mark might have been made with the tongue of her short crop; the figure's prosthesis thus oscillates between whip and paintbrush. This is, then, another portrait of the artist as a young ladybird, in a more vicious and lascivious key than *Carnation, Lily, Lily, Rose* (recall that "ladybird" meant mistress or "wanton girl"). And, like any good, erotically cross-fertilizing artist, Victoria Stanley reads as floral as well as avian—a petally bow adorns her hat and her white skirt unfurls, corollalike, around her reproductive organs.

In addition to painting enticing portraits of budding and fully fledged sadists, Sargent sometimes painted *sadistically,* adopting for himself a Whistlerian eye and hand. "Cruelty," he would say, is "necessary to handsomeness."[64] Whether he

177
John Singer Sargent, sketch for *Lord Ribblesdale*, 1902. Oil and crayon on canvas, 87.6 x 61 cm. Private collection.

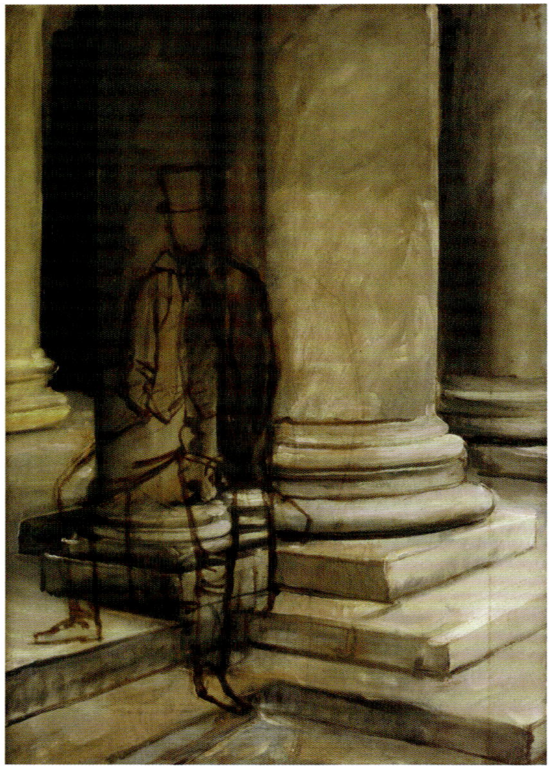

had the cruelty of the sitter or the painter in mind is unclear, but his stinging *coups de pinceau*—the whiplike licks of the brush he could use to beat or thrash his "subjects" into the often torturous shapes he desired—at times enact that dictum. "There is not in Sargent's painting courtesy of touch," admitted Martin Wood.[65] The artist himself referred to his tools as "instruments of torture."[66] His paint handling could be violent (D. S. MacColl saw his pigment as "brutalised," "imperiously driven," "torn across," "gashed and cut through"),[67] as could his finished works—in exhibitions "more or less kicking other people's pictures into bits," according to Vernon Lee.[68]

Sometimes sitters themselves inspired sexualized visual assaults. Symons described the painter's approach to his subjects in terms that seem at first a deliberate echo and rewriting of Poe's "Oval Portrait": "Sargent pours the crude light of the studio roof upon all in a man that would most escape that interrogation, crying to him roughly to speak out, stripping off some of his shyest and most honest disguises, and giving us, as the truth, whatever remains over after the soul has been frightened out of sight."[69] Although Symons refers to male subjects, female sitters most often aroused Sargent's aggression. If the portrait of the whip-wielding Victoria Stanley suggests the artist's empathy and identification, this was probably due to her age;[70] the painting's technique was still that of a "knife-thrower."[71] Mature and sexually explorative women in his oeuvre don't always fare so well in their portraits. Sargent, like Whistler, strips off their "disguises" and willfully inflicts pain, often by painting them as cut—wounded—flowers.[72]

The picture of Helen Harrison (fig. 179) is a case in point. Helen posed for Sargent during her sister's divorce trial, in the course of which it became public knowledge that "the siblings had carried on various liaisons with lovers whom on occasion they shared in the same bed."[73] Sargent was a friend of the sitter's brother-in-law, and Henry James feared that during her sittings Sargent would "stick his brush down her throat."[74] In the painting the figure looks nervous; her right thumb presses down the pinky finger of her left hand painfully, and her hair looks cropped, as though in punishment for adultery. Sargent contributes both to this decadent Helen's outer demonization and to her inner wounds: the sitter's

178

John Singer Sargent, *The Honourable Victoria Stanley*, 1899. Oil on canvas, 196.8 x 105.4 cm. Private collection.

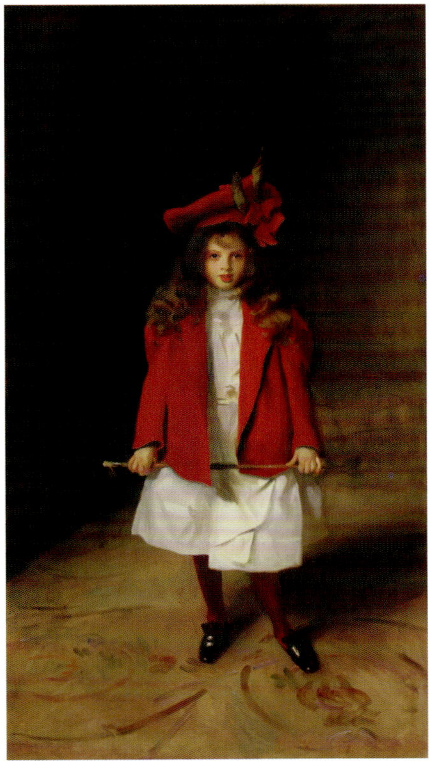

179

John Singer Sargent, *Portrait of Mrs. Robert Harrison*, 1886. Oil on canvas, 157.8 x 80.3 cm. Tate Britain.

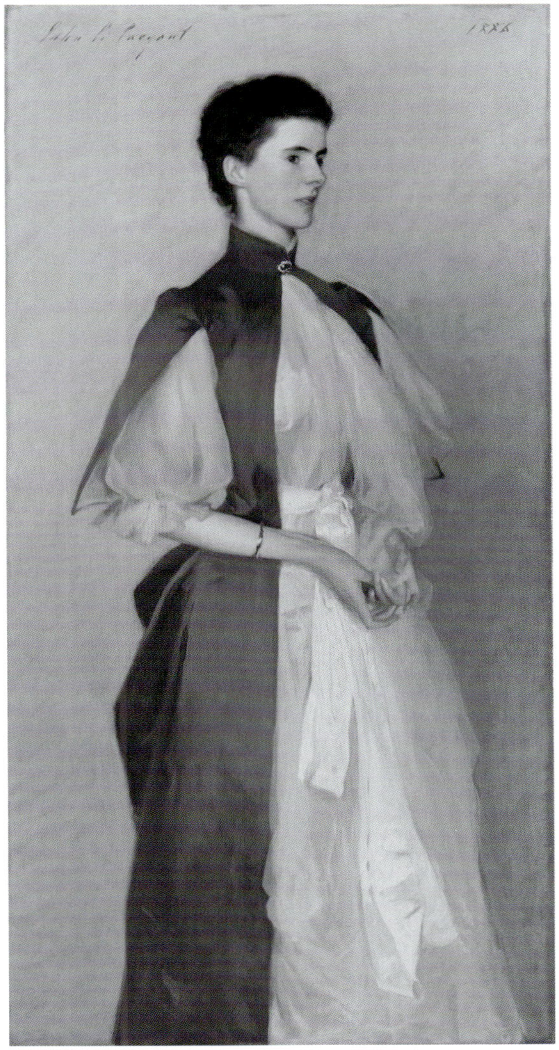

strange costume prompted a friend to write in to the *Evesham Journal and Four Shires Advertiser* that she had never seen Helen "wearing such red wing-like appendages to her costume, which look as though about to expand and convey her to the regions of Mephistopheles,"[75] and we detect, beneath the clean, perhaps even sterile gauze of her white dress, faint amaranthine stains seeping out from the shiny sword-shaped sepals of her fuchsian gown, which has begun to separate along sharp lines of dehiscence. This spreading blossom could be the floral personification of syphilis in Des Esseintes's dream, where the "sabre-shaped leaves" and petals of the Nidularium blossoms open wide to reveal their "flayed, gaping" bottoms, the "hideous wound," self-inflicted, to its bleeding interior, or rather, backside.[76] Or Sargent himself, like Baudelaire's artist of modern life who fences with his brush, may simply have reopened her wounds.[77]

Too much female independence was always an annoyance to the painter. The portrait of Elsie Palmer (fig. 180) represents an American adolescent young enough to be represented as virginally pure, but old enough to resist, like Virginie Gautreau, the painter's co-option of her portrait. She wrote in her journal that she found it "difficult to oblige" Sargent, and her portrait attests to their conflict.[78] This pure white blossom in a Tudor chapel looks, beyond quite ill, thoroughly sacrificial: a flower cut down by the painter. Elaine Kilmurray argues that the painting is a "Madonna without a child," and we might read the childlessness here as the invert artist's revenge on a nubile young woman unwilling to submit to his vision.[79] Contemporary reviewers "were startled" to find the portrait so "merciless."[80] Like Helen Harrison's, Elsie Palmer's skirt is stained scarlet, as though the sitter was bleeding, unstemmed, into the swathes of her gown. This is an image, then, of defloration, one that creates an indelible blot on the sitter's dress and on her honor.

Sargent's sadism emerges most fully, however, in portraits of women artists. Faced with their potentially twofold procreativity (exceeding the female independence, sexual or otherwise, of the subjects we have just examined), Sargent either metaphorically castrated himself, as with Sarah Choate Sears, or tried to render them unfruitful. He was not alone in this reaction, for although anxiety stemming from physiological incapacity was to some degree allayed by the vegetable model of procreation, male artists frequently attempted to make floral procreative privileges theirs exclusively. In a cartoon from around 1880, "An Aesthetic lady clutching her mahl stick" (fig. 181), a female painter in Aesthetic dress stands in front of one of her canvases. On it a stick figure of a woman in similar dress to the artist's own, with similar hair, is attempting to water a withered, flowerless plant. Women's artistic endeavors, the image insists, are

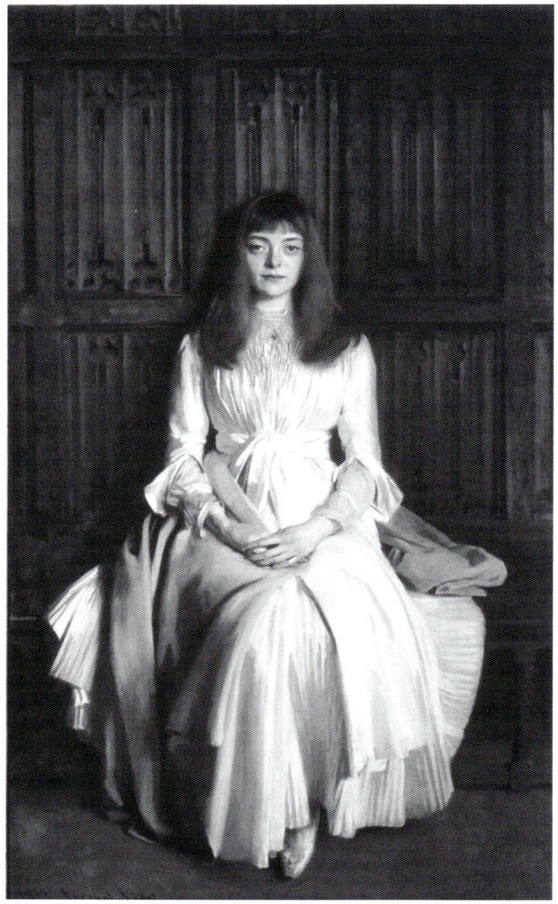

180
John Singer Sargent, *Elsie Palmer*, 1889–90. Oil on canvas, 190.5 x 114.3 cm. Colorado Springs Fine Arts Center, FA1969.3.

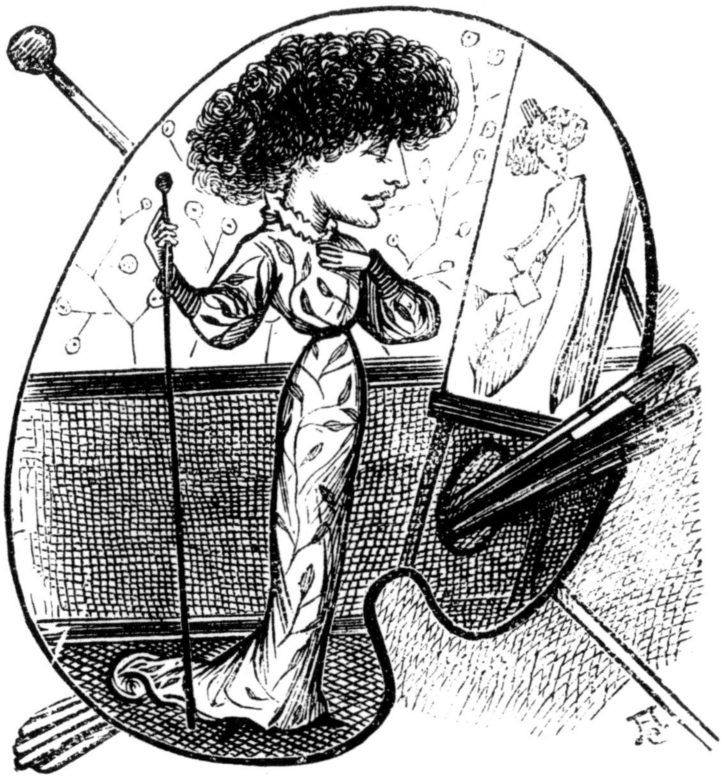

181
"An Aesthetic lady clutching her mahl stick," ***Illustrated Sporting and Dramatic News*** **(?), c. 1880.**

bound to be fruitless; the lady artist is as sere and sterile as her bare-branched offspring, denied the possibility of both horticultural and artistic creation. R. H. Ives Gammell argues that Sargent's technique was most sadistic when the painter was faced with an "artistic problem." In such instances the painter's "subconscious will to power was aroused by an opportunity to assert his superiority through his art."[81] Although Gammell is referring to technical difficulties, for Sargent most women artists fell into the category of "artistic problems."

Sargent's strangest and most strained portrait of a woman artist is one of his many paintings of Flora Priestley (fig. 182). He had met this nominally vegetal artist while they were both studying art in Paris,[82] and she became a lifelong family friend, especially close to the two Violets (Sargent and Paget, a.k.a. Vernon Lee, who dedicated a book of stories called *Hauntings* to her). Unlike most of Sargent's paintings of other artist friends, which are pervaded by a sense of empathy and dreamy atmospheric fusion, Priestley's picture seems a grotesque Boschian splicing of references and worlds: the sitter could be a broody bird or an oriental gentleman with a rice-powdered face ("whether American or Japanese it is hard to say," wrote one reviewer),[83] in a fez and ritual costume, who is performing a dance and happens to be in the middle of a deep knee bend. Her neck seemingly shackled by its high collar, waist cinched, and hair tightly bound, the sitter (or squatter) is fettered here. And perhaps as punishment for the irritatingly manifold capacity of the female artist to create, Flora's hands have been so completely inverted as to appear mutilated, rendered painfully incapacitated; they recall those hands Wilde compares to the bound feet of Chinese women and Freud's analysis of such fetishized appendages.[84] A reenactment of the castration the fetishist disavows, the painting images a literal un-*main*-ing.

Flora, who never married (her high collar suggests that she is priestly in this respect, though it is the wrong color), seems stigmatized by her flowers here as by an inherent vice. When Dorian Gray's "soul" had crept "out of its secret hiding-place," desire came "to meet it" and his natural vice "developed like a flower" into "blossoms of scarlet flame."[85] Blood-red and drooping, Flora's blossoms also read as a wound (a visible one, unlike Helen Harrison's and Elsie Palmer's). Sargent himself

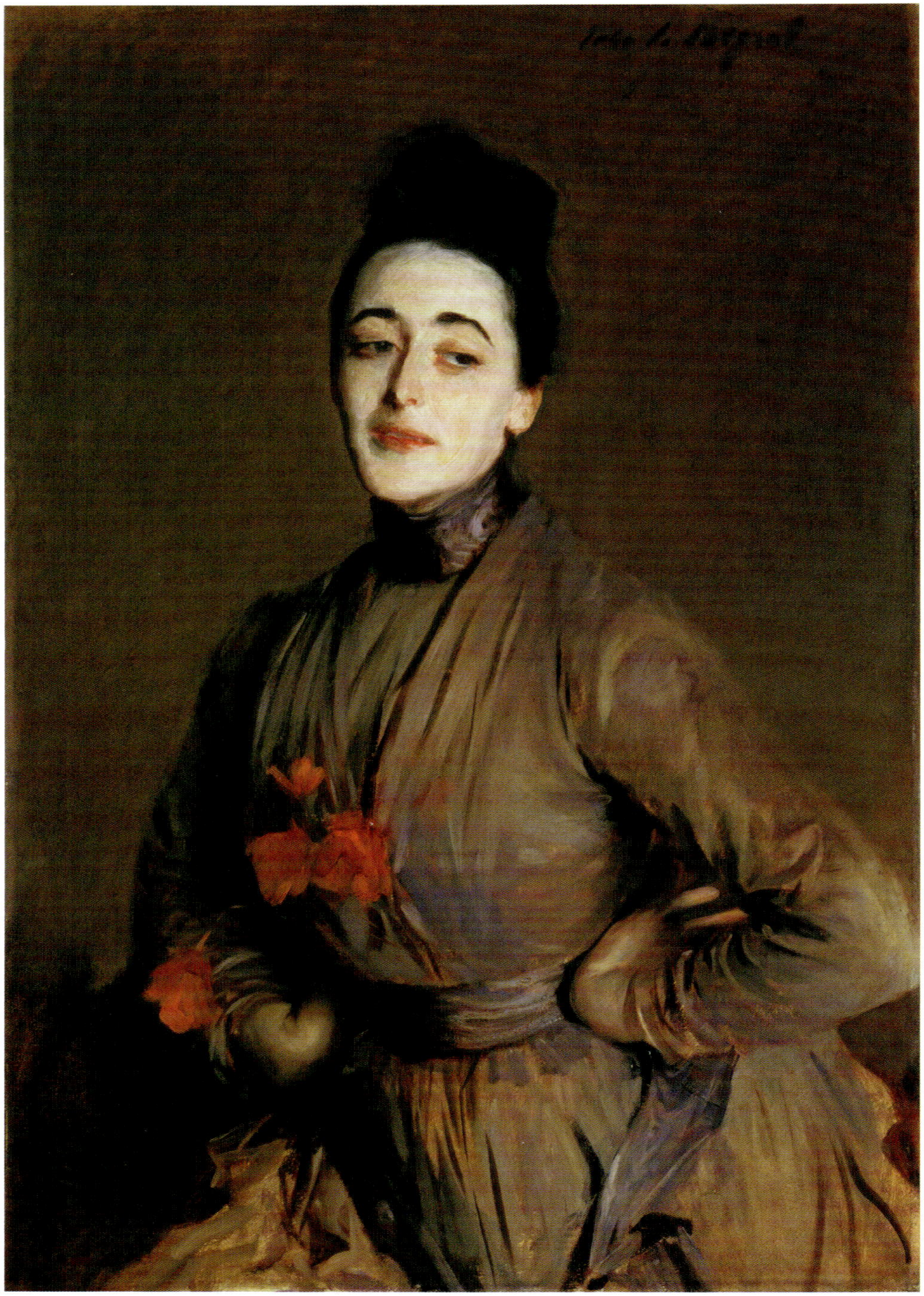

182
John Singer Sargent, *Flora Priestley*, c. 1889. Oil on canvas, 91.4 x 63.5 cm. Tate Britain.

interpreted the flowers as such—he referred to this painting as the "demoniac portrait . . . the lilac one with the ruptured heart."[86] Curiously, this chest wound is one the sitter and painter shared: Jacques-Émile Blanche described Sargent as a sort of Dimmesdale, "a being who enclosed himself in a cuirass, never undressed in front of you, as if under the shirt he carried a dressed wound,"[87] a wound that is here displaced onto and inflicted on Flora.[88]

The subject's wounded breast, combined with her rather avian air, conjures up the Proustian image of the *colombe poignardée*. These doves with red spots on their breasts look, in Reynaldo Hahn's words, "like nymphs who've stabbed themselves for love" and been transformed by the gods "into birds."[89] Having seen them at the Jardin des Plantes in the company of Hahn, Proust intended to call the parts of his novel treating the romance with the unattainable Albertine, a descendant of Gomorrha, *Les colombes poignardées*.[90] The *colombe poignardée* and the invert are, in Proust's world, figures of tainted purity and woundedness, their visual stigmata functioning like Hawthorne's scarlet letter. In her portrait, Flora Priestley appears both as a *fleur du mal* (one intimate with the painter's floral sister, Violet) and a *fleur du malheur* (misfortune, unhappiness).

Sargent's description of Flora's portrait as "demoniac" raises the idea of possession. The sitter's bent-back, tetanic hands and awkward posture suggest an almost hysterical struggle for control of her body. Her sickly pale skin, tightly covered neck, and blood-blossoms spilling forth, however, identify her adversary as a kind of vampire.[91] Like Whistler, Sargent occasionally adopted the role of vampiric pollinator—and like Whistler, when painting other artists. The deathly pallor some of Sargent's sitters exhibited did not go unnoticed. When the portrait of Paul Helleu sketching outdoors with his wife was first exhibited, it was found sinister by one American critic who saw "the absolutely colorless faces" as "the faces of the dead."[92] W. Graham Robertson's white skin suggests his delicacy (fig. 185), and we learn from him that over the course of his draining sittings he "became thinner and thinner." His wasting away was "much to the satisfaction of the artist, who used to pull and drag" his coat around him "more and more closely," as if to squeeze out every last drop.[93] Unlike Whistler, however, Sargent did not depend entirely on this possessive or vampiric mode of portraiture.

The multiplication of identities occasioned by the act of painterly cross-fertilization could be threatening, especially when other artists were sitters. In Whistler's case such potential confusion inspired his defensive painterly sadism. Sargent's reaction, however, was more often to indulge a dreamy desire for fusion—think of the heady, polleny atmospheres of the solo portraits of Paul Helleu and Robert Louis Stevenson (figs. 85–87). These paintings in particular are, as Alice Meynell put it in 1903, "essays in romantic portraiture."[94] Helleu, with just the smallest inversion, would have been the Hellen of Sargent's dreams, and Sargent described Stevenson as "the most intense creature" he had "ever met" (the writer, in return, confessed to having "lost" his heart to the painter).[95] Their portraits were thought to reveal, "as if it were, unawares, some unfamiliar aspect of [the] subjects' personality"[96]—unsurprisingly, a protean

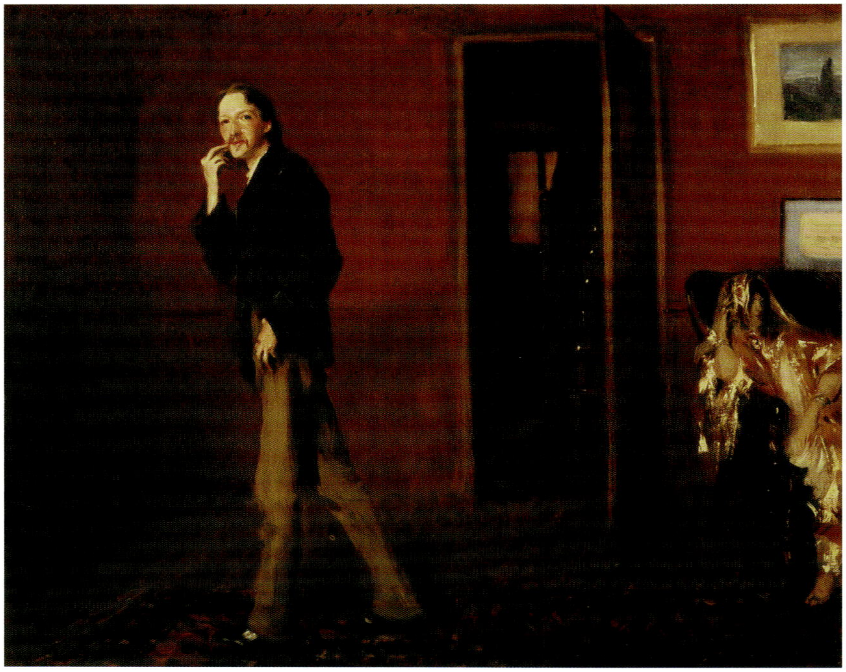

183

John Singer Sargent, *Robert Louis Stevenson and His Wife*, 1885. Oil on canvas, 52.1 x 62.2 cm. Private collection.

one: Sargent's renderings unleash the metamorphic potential of his friends, who are transmuted into flowers and trees under his touch.

Even the double portrait of Robert Louis Stevenson and his wife, Fanny (fig. 183), evokes an Ovidian fantasy of metamorphosis. The writer is rendered Daphne-like, the renowned "lissom swiftness of his movements" stilled.[97] His feet, like Ovid's nymph's, "but now so swift," are suddenly "anchored fast," and his legs, both already stiff, "wrapped in thin smooth bark."[98] His long fingers harden into gnarly twigs, and his soft, dark green velvet coat offers a luxurious cover of foliage. Sargent, famed for sprinting after his painted subjects, is implicitly figured as the loving Apollo here, Fanny offering no pursuit. A sense of captivity in the small space is palpable; in a letter, Sargent compared the room to a vivarium and its occupants to species on display, Stevenson a "caged animal" (trapped in his marriage, no doubt), and Fanny an unnamed, or unmentionable, "foreign specimen in the corner."[99] Clearly of marginal importance and interest to the painter in comparison to her husband, and notably less alive to him (she is a "specimen" rather than an "animal"), Fanny sits on a chair, her face partly shrouded by a golden shawl. In her own words, "as I said to Sargent, 'I am but a cipher under the shadow,' to which he too eagerly assented."[100] Disappearing under her shimmering wrap, the figure vanishes into a cascade of golden paint-pollen for Sargent to brush off.

Both surviving paintings of Stevenson by Sargent emphasize the writer's seductive, suggestive gaze. The importance of optical exchanges in a culture largely intolerant of sexual deviance cannot be underestimated: Henry James wrote to John Addington Symonds, the early theorist of inversion, that "victims of a common passion should sometimes exchange a look,"[101] and there are countless contemporary literary examples attesting to the importance of nonverbal communication systems among outlawed groups like thieves and sodomites. Proust demonstrates that a whole courtship may be contained in a knowing exchange of looks, and more germanely that looks, like floral hands, are subject to pollinator erotics.[102] In "the corolla of the face" the eyes are pistils or stamens;[103] "fair lily," Montesquiou wrote to a female friend, "your black pistils are your eyes."[104] The pollen of a man's "eager sensibility" too may be gathered "from the projecting stamens of his gaze."[105] In the double portrait Stevenson is smiling and touching his moustache as he stares invitingly at the painter.[106] The invitation proffered by the already plantlike writer's eyes is to pollination—they seem to register a reciprocity of desire, a recognition that the pollinating painter longs as much to brush the writer's "Velvet Coat"[107] of petals as the Velvet Coat longs to be brushed by him (Stevenson, James wrote, "arrests the eye and solicits the brush").[108] And is there not also registered, in Stevenson's eyes, a recognition that Sargent's own look carries with it a promise that the artist's hands and brush will enable the writer's reproduction? And a bemused recognition that other fannies might be worth exploring as reproductive options?[109]

Married but also male-inclined—"Mr. Stevenson does not need, as we may say, a petticoat to inflame him,"[110] according to James—like Sargent, the author was interested in probing the nature both of the invert's creations and of the creation of the invert in stories such as *Strange Case of Dr. Jekyll and Mr. Hyde*. As is the case with many of the inverts described by Wilde, Proust, and others, it is Dr. Jekyll's "almost morbid sense of shame" that impels him to conceal his pleasures and "such irregularities" as he was "guilty of," and thus commit "to a profound duplicity of life." Of his alternately dominant and recessive characters, he concludes that it is "the curse of mankind that these incongruous faggots [are] thus bound together—that in the agonised womb of consciousness, these polar twins should be continuously struggling," inverting their positions.[111] With this image Stevenson anticipates both the Merleau-Pontian model of consciousness as a fertile invagination of the flesh of the world (the "womb of consciousness") and the vegetable birth of the invert's creation (twin faggots—sticks—waiting to be born, to leaf and blossom). Hyde's is a strange birth, perhaps a Pozzian fecundating inversion: the morning after the murder of Sir Danvers Carew, when Hyde becomes truly autonomous, a "chocolate-coloured pall," dark "like the back-end of evening," hangs over London—the release of gas that must accompany the phenomenon of self-inversion, turning oneself inside out and bidding farewell to one's "bowels of mercy" writ large. This atmospheric evidence is echoed in other, more tangible images; when Hyde's rooms are ransacked the next day they reveal clothes with "their pockets inside out," and so on.[112] If Hyde literally incarnates "inversion," he is also a model of atavism. In keeping with the Darwinian explanation of sexual inversion as an atavistic phenomenon elaborated in chapter 3 (and Stevenson was a Darwin reader),[113] Hyde seems the very personification of the death drive, the "urge inherent in organic life to restore an earlier state of things," at first in his sadistic and regressive "ape-like fury," more so when he becomes "a self-destroyer," and most fully at the point of being "something not only hellish but inorganic."[114]

This death infecting life was experienced firsthand by the consumptive Stevenson, whose insidious disease was slowly reducing him to a more vegetable state, like Alphonse Daudet's syphilis, which also induced a Daphne-like metamorphosis. As he described his condition in 1887, "I have been very seedy indeed, quite a dead body."[115] Like Darwin's plants subjected to too extended self-fertilization, Stevenson and his character Mr. Hyde were both of "great apparent

184

John Singer Sargent, *Two Studies for a Portrait of Robert Louis Stevenson,* 1880–85. Charcoal on darkened off-white wove paper, 24.7 x 34.6 cm. Harvard University Art Museums, Fogg Art Museum, gift of Mrs. Francis Ormond, 1937.7.21.7.

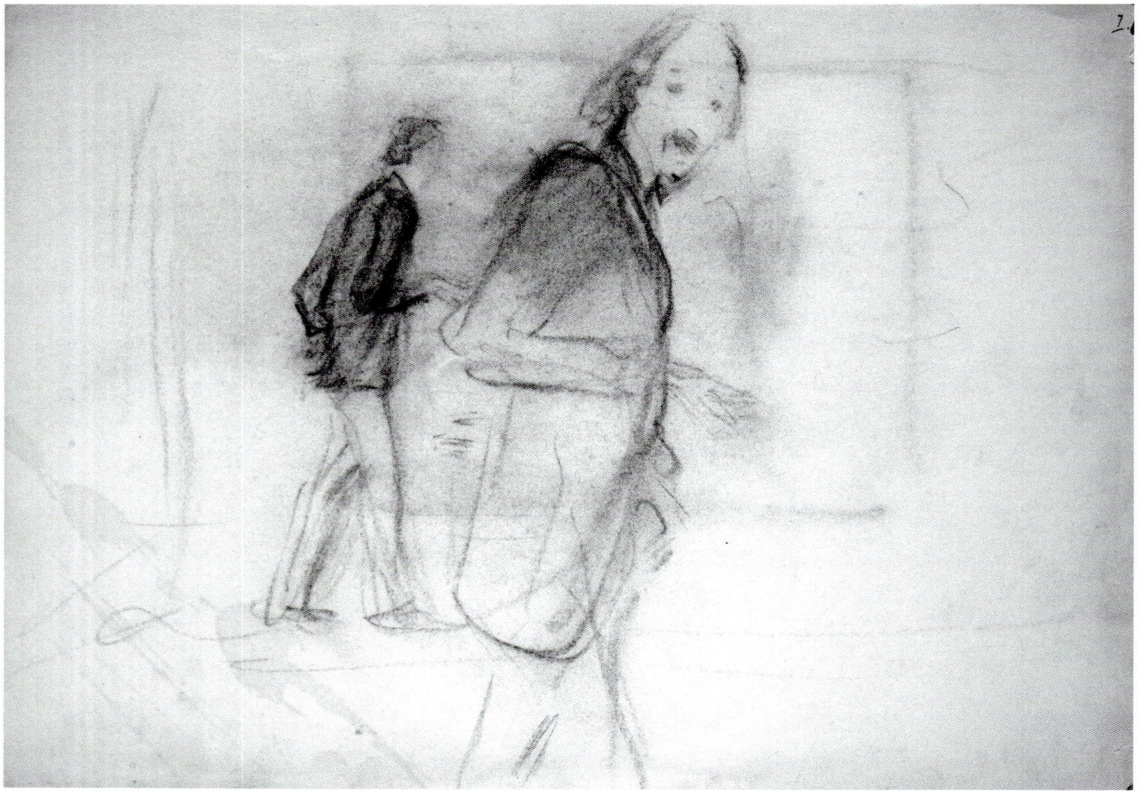

debility of constitution,"[116] the sort of weakened flowers that must live, as Henry James put it, "under exquisite protection."[117] Stevenson's rooms were aesthetic, as were Hyde's (contemporary readers often found it hard to reconcile Hyde's animality with his refined taste—his lodgings are those of a sybarite and aesthete, filled with wine, silver, good art, soft and colorful carpets),[118] and the writer, despite his obvious delicacy, had a "curious liking for queer company" and a "strange relish of rough jests."[119] Léon Daudet, the son of Alphonse, attributed the schizophrenic characters of Wilde, Stevenson, and Jekyll/Hyde to hereditary syphilis: Wilde "strangely recalled that double character of Stevenson, sometimes excellent and beneficent under the traits of Dr. Jekyll, sometimes implacable and bestial under the mask of Mr. Hyde. His was, in sum, a heredo type," i.e., one suffering from hereditary syphilis, "charged with an ancestral burden too heavy for a stump of will."[120] The slippage between author and fictional character in Daudet's comparison—is Stevenson or Jekyll/Hyde the "heredo type"?—is particularly suggestive here.

Sargent grasped the doubleness of his subject immediately. His preliminary sketch literally doubles the figure (fig. 184), and the features of Fanny's face in the double portrait, on close inspection, are remarkably like Stevenson's own.

In the solo portrait the two identical rings on Stevenson's left hand—the wedding band replicated on the second finger—are also indices of his ambivalent orientation (in the double portrait he does not wear a wedding ring). Furthermore, the hectic flush of Stevenson's cheeks in both paintings, his *bloom,* is a dubious sign. The blush suggests his physical frailty and emotional delicacy, but in the nineteenth century such coloring also signified modesty and paradoxically, as we saw in chapter 1, sexual readiness. It was open to homoerotic as well as heteronormative readings: Pater, for instance, described Winckelmann's "romantic, fervent friendships with young men" as "bringing him into contact with the pride of human form, and staining the thoughts with its bloom."[121] The blush was also suggestive of perverse knowledge and enjoyment. It was "one of James's most queerly ambiguous images," as Eric Savoy has argued;[122] and just as Saint-Loup's expressions of delight, when "the too delicate skin of his cheeks allowed a bright red glow to shine through them" and "his eyes sparkled," in "no way excluded moral duplicity,"[123] the blush in Stevenson's portrait, as physiologically accurate and sincere as it may be, functions simultaneously as a sign of sickness and health, innocence and perversion, pleasure and perhaps even pain.

If Stevenson was a queer character, so was the painter. Intriguingly, Martha Kingsbury has suggested that Sargent may in fact have inspired Stevenson to write *Dr. Jekyll and Mr. Hyde,* begun in the autumn of 1885, when Sargent was painting the author for the second time.[124] Many contemporaries knew or presumed that Sargent led a double life. His assistant Thomas Fox noted that "of his outgoings and incomings even those who knew him best and saw him most learned comparatively little."[125] Occasional slips or revelations left him flustered. Consider the following discussion of *The Picture of Dorian Gray* recorded in Lucia Fairchild's diary, in which Sargent plays a Charlusian virile hetero—who, like Proust's character, castigates effeminate men:

> Sarg., "Horrid, isn't it?" I agreed. Sarg., "Yes, dull—unnecessary—it's a book that ought to be thrown out the window." I, "Why did you once advise me to read Oscar Wilde?" He, "I? I couldn't have! He's so weak—so ridiculous, it seems to me—" "But I'm sure you did—you said he was awfully witty—" He, "Ah, well, his newspaper articles and reviews perhaps—Yes, I remember an essay on 'The Art of Lying' that was quite good, very witty."[126]

Sargent knew rather more of Wilde's art of lying than he let on; he and Wilde were more than casual acquaintances. Although we don't know if they engaged in any underworld activities together, they met at dinner parties even before Sargent moved to London,[127] and once ensconced in his Tite Street digs in Chelsea—Wilde lived on the same street—Sargent had the writer round to lunch occasionally. Wilde gave Sargent a copy of the 1882 edition of James Rennell Rodd's *Rose Leaf and Apple Leaf,* for which he had written an introduction, and his inscription refers to their friendship and their shared beliefs.[128]

If he was a public reviler of the very aesthete who happened to inscribe his books and with whom he underhandedly associated, Sargent's

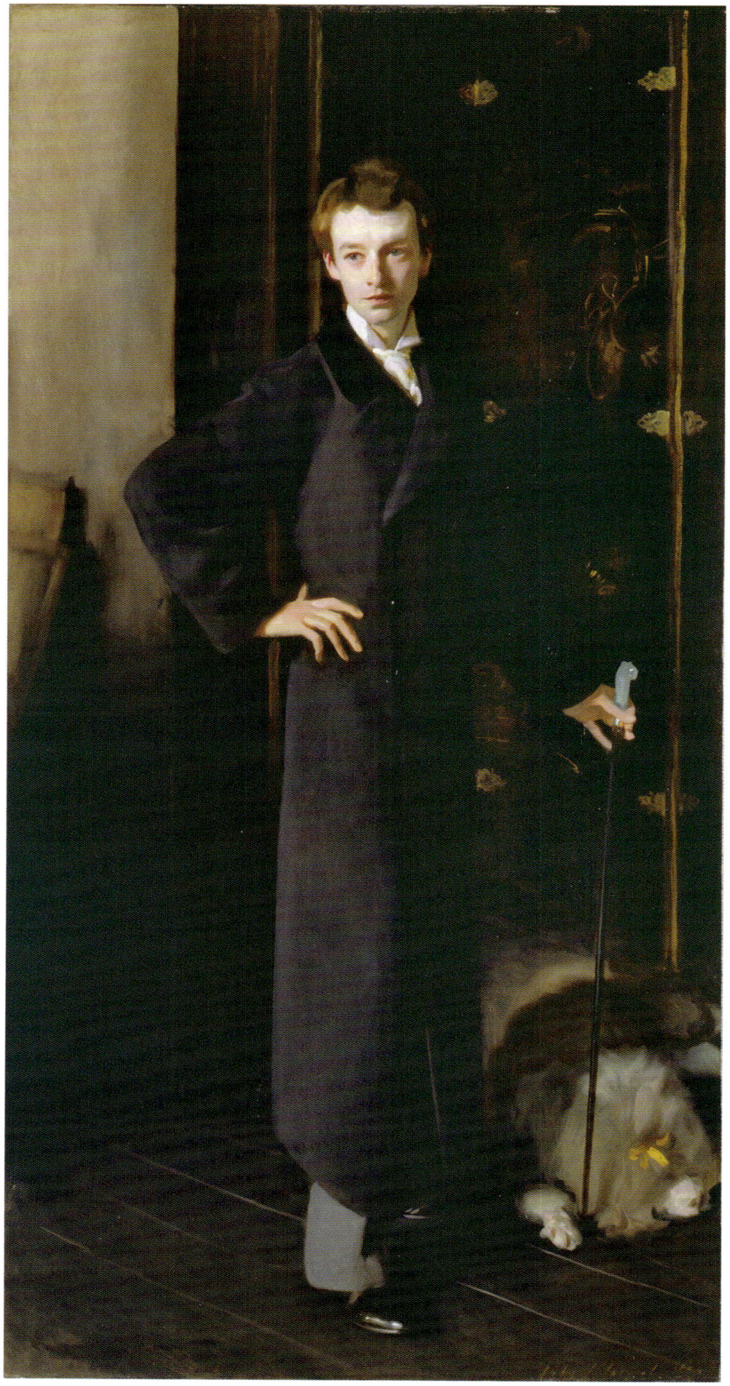

185
John Singer Sargent, *W. Graham Robertson*, 1894. Oil on canvas, 230.5 x 118.7 cm. Tate Britain.

portrait of the playwright and set designer W. Graham Robertson (fig. 185), one of Wilde's intimates, nevertheless attests to the link between the two men and their shared tastes. Trevor Fairbrother has analyzed the "subtle visual codings of effeteness and boyish beauty" in the painting and contends that the work "made the bravest public statement possible about the existence of homosexuals and their contributions to the style, manners, and thinking of the day,"[129] or, as the less identity-politic Proust would have put it, about the existence of sodomites, who "form in every land an oriental colony, cultured, musical, malicious."[130] Robertson, who had posed as Diana and other "immortals of uncertain shape and sex" for Walter Crane, whose delicate hands had been sought after and painted by Albert Moore and others,[131] who wrote a play called *Pinkie and the Fairies*, and who was the constant companion of Oscar Wilde (though, one biographer claims, "the sinister evil inherent in the older man" was "never once revealed to him" because he had a "rare mental and moral quality of pure gold"),[132] graciously agreed to model for Sargent when the older man ardently pursued him.[133]

In the painting Robertson is posed for Sargent like Jupien for Charlus, with a slight but "becoming tilt to his body" and his hand provocatively "on his hip,"[134] a pose adopted by the

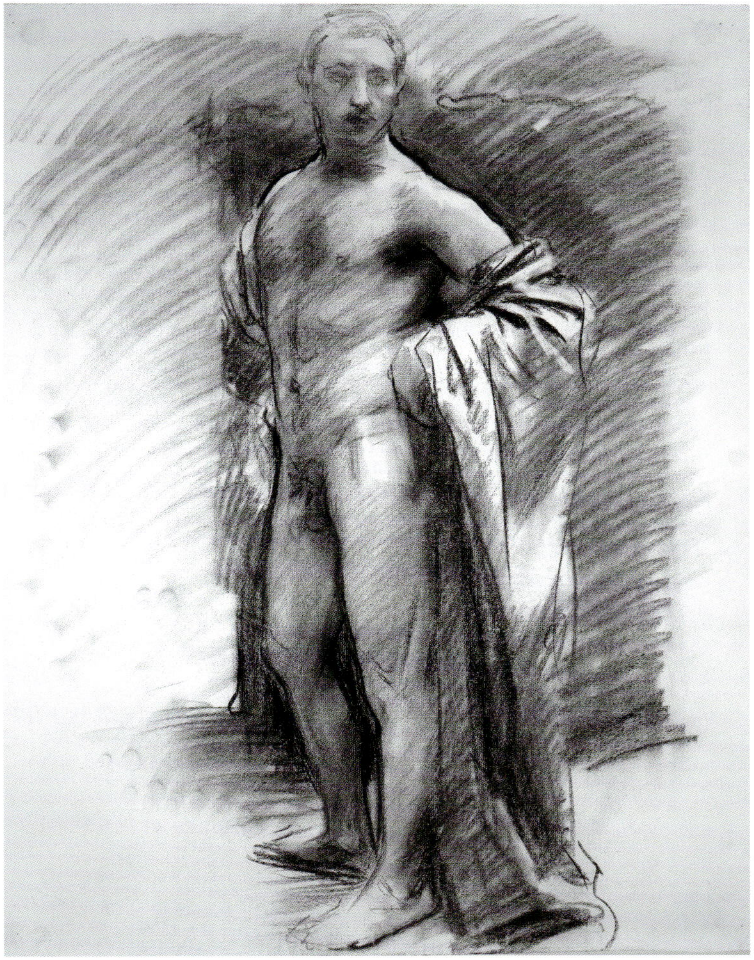

186
John Singer Sargent, *Standing Male Nude with Arms Akimbo*, c. 1890–1915. Charcoal on off-white laid paper, 63.5 x 48.7 cm. Harvard University Art Museums, Fogg Art Museum, gift of Mrs. Francis Ormond, 1937.9.9.

models for many of Sargent's highly erotic sketches of male nudes (fig. 186). Robertson holds a thin, jade-handled cane, and his aging poodle, Mouton, snoozes at his feet. The oriental screen before which this golden boy, a blond beauty like Dorian Gray, stands suggests the proclivities of his tastes. Curiously, the oil study for this work (fig. 187) seems to depict a sitter ten or twenty years older than the figure in the finished portrait, with disturbingly crude, thick fingers. Behind Robertson's boyish figure, then, somewhere in the background, perhaps even behind the screen as in Wilde's tale, is an aging picture of a far less delicate character hidden from view. And behind this vision of youth there might lurk a dangerous, vampiric predator who hunts to maintain his bloom,[135] to keep his gold from turning to grey.[136]

In the finished painting Robertson's floral hand, at once so archly and languorously splayed, invites the painter's and viewer's brushing (*effleurage*). The sitter's slightly moony face is angled to display to best advantage his mellifluous, even rather enchanting features, and most significantly to reflect the honey-colored light of the passage to his rear. His elbow indicates the vague forms of pictures within,[137] but we would have to cross the room, brush past the painting's subject, and penetrate his interior to actually conceive what they are. The portrait may be seen, then, as a metapictorial staging of the process of painting as pollination, the painter's actual and the viewer's solicited brush with the floral boy leading to a deposit of honey in the golden boy's rear chamber. If Sargent plays a vampiric pollinator here, it is because, like a vampire-bee, he desired to wring every last drop of nectar from this nocturnal, wasp-waisted pansy for his cata-honeycomb.[138] A similar structure informs Jacques-Émile Blanche's *Self-Portrait with Raphael de Ochoa* (fig. 83), where a honeycomb of golden frames behind the painter and his lover contains the by-products of their mutual brushing.

Sargent's most homoerotic works alert us to the other creation pollination facilitates in addition to vegetable reproduction: honey. "Artists' honey," Montesquiou tells us, is not of the normal sort. It

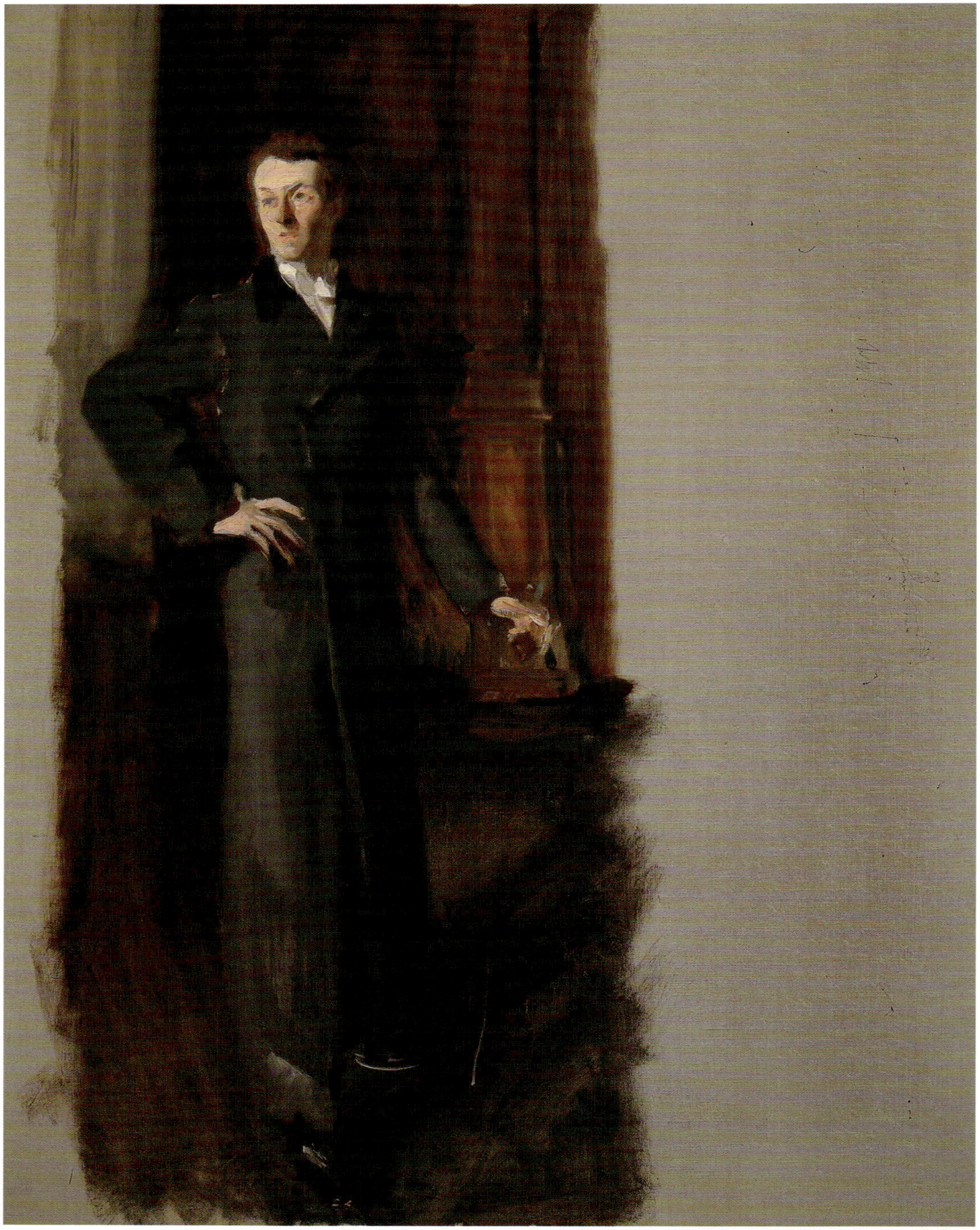

187
John Singer Sargent, *Study for W. Graham Robertson,* 1894. Oil on canvas, 91.5 x 71.2 cm. Private collection.

may be "pearl grey" and "thin," "solitary and ferocious,"[139] something like Melville's ambergris or Whitman's "albescent" kind. Distilling honey from flowers that "shame the bee and frighten the sparrows" and lure "the bats alone,"[140] like the inodorous and sterile blue hortensia, decadent *fin-de-siècle* pollinators produced a strange substance, a honey that some denounced as both noxious and foreign. Tennyson, for instance, decried art imbued "with poisonous honey stolen from the flowers of France," a phrase that could be applied to any number of Aesthetic and Decadent works.[141] Huysmans's *À rebours* was a primary source of that sickly substance; it is the book that definitively converts Dorian Gray—or, to be precise, finishes inverting him—and its honeyed words are appropriately stored in a "pearl-coloured octagonal stand" that looks like "the work of some strange Egyptian bees that wrought in silver" rather than gold.[142] In 1898 an American critic also identified "the Verlaine crop of the early eighties," including the poets Rimbaud and Gustave Kahn, as producers of toxic honey.[143] The syphilitic Verlaine himself, according to Edmund Gosse (who, you will recall, identified Sargent's wagtail tendencies), was a "giant hawk-moth" who "uncoiled his proboscis" in intoxicating "absinthe-corollas" and produced for no one or for art's sake.[144]

The strangeness and thinness of artists' honey arise from its solitary production, as most invert pollinators, like Montesquiou the bat, were bees of "silence" and "the dusk," exiled or exorbitant beings.[145] Wilde's pseudonym during his self-imposed exile in France was Sebastian Melmoth, the last name derived from the titular hero of his great-uncle's Gothic novel *Melmoth the Wanderer.* Condemned to roam the earth in solitude, Melmoth is a "brooder over the dark and unproductive nest of eternal sterility," a solitary pollinator.[146] In taking the name, Wilde identified himself as a honey moth (*miel*-moth) as well as a diseased or vicious one (*mal*-moth). Whistler, the stinging butterfly, "did not understand the spirit of the hive," according to Walter Sickert, and unlike the Impressionists, who "peopled a universe with their art," he left the world "childless."[147] Sargent's pastel portrait of Paul Helleu hints at the pollinator-artist's unfamiliar mission: the mesmeric allure of the scented smoke wafting from Helleu's hand (figs. 86, 87) seems to intoxicate the artist, evoking Proust's "bee drugged with tobacco smoke that had ceased to take any thought for preserving the accumulation of its labours and the hopes of its hive."[148] Yet if Sargent came to see that floral progeny was not so easily grown as contemporary visual culture suggested (fig. 188), or was perhaps too sentimental, honey offered another option. Although largely negatively connoted at the end of the century, it had not always been, as we know: artists' honey had also, since antiquity, been seen as the golden fruit of artistic labor, and in Sargent's paintings back rooms or private chambers filled with paintings, like W. Graham Robertson's, or books, retain this winning hue.

In calling attention to mellification, Sargent's homoerotic works also turn the spatial maneuverings pollination requires into erotic allegories. "Artists' entrance," "back door," and "back window" were all slang for the anus, and the sexual charge of such settings was widely acknowledged.[149] The anatomy of the architecture in *W. Graham Robertson* is shared by other overtly homoerotic portraits by Sargent. *Robert Louis Stevenson and His Wife,* for example (fig. 183), features a rear aperture opening onto a narrow, shadowy red passage and another door beyond. In the portrait of the painter Alberto Falchetti (fig. 189), the shady, black-clad artist stands just before a red-framed doorway, a passage onto which Sargent has inscribed his name. Bathed in a yellow light, the small space onto which Falchetti backs holds paintbrushes—instruments of art-making that here could also be read as flowers—and perhaps more significantly, "drawers," which are already open a crack and invite further exploration, prying or permitted. This typical arrangement implies the sight of the hidden backside or flipside of the sitter's public face, and suggests that the role of the artist is to literally get behind him. Like the dark, narrow streets that appear frequently in Sargent's Venetian sketches, these interior spaces are the sites of private indiscretions, of acts that can only be adumbrated preposterously.

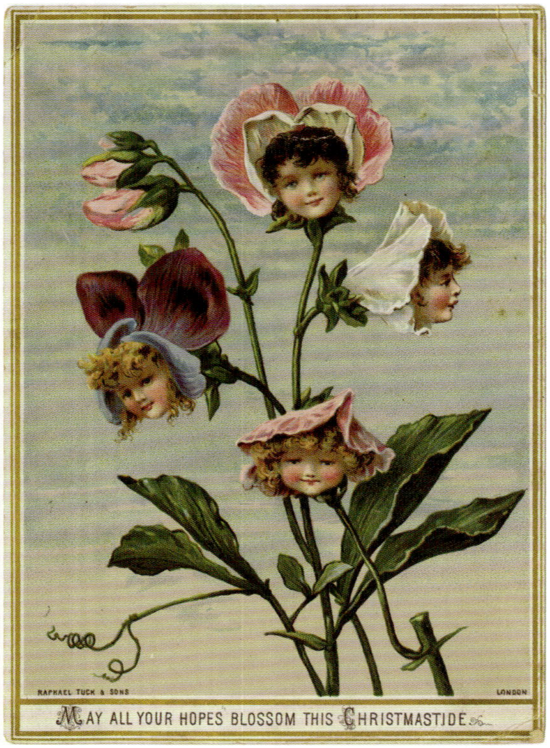

188
Raphael Tuck and Sons, "May all your hopes blossom," greeting card, 1880s, 11 x 15.4 cm.

Such rearward allusion governs the portrait of Edward Robinson (fig. 190), an archaeologist, scholar, and director of the Metropolitan Museum of Art (later of the Boston Museum of Fine Arts). The figure is wedged into a corner between two desks, behind which rises a grid of glinting, gold-stamped tomes: a private library. In a very odd stance for a portrait—"a little wanting in the sense of repose," as one reviewer put it[150]—Robinson's buttocks rest on the surface of the desk at the bottom right. He leans back, supporting himself with the provocatively splayed fingers of his hands, and archly turns his pale golden head toward the painter. An Etruscan relief depicting a drunken Dionysus supported by Eros—the backside of a mirror—can be seen between Robinson's hands and behind his bottom. To attain the promise held out by the mirror, the books, or the sitter, we would have to turn any of them over onto the desk, to probe their depths.

Sometimes the bee's insertion of its proboscis into the nectary is not through the flower's frontal opening but rather, "as it were, through a back-door," in the words of a prominent nineteenth-century botanist.[151] This sort of pollination, then, while not the norm, is not unnatural. Bees are skillful navigators of the often labyrinthine invaginations and protuberances of floral pellicles because, Darwin writes, "having long practised the

189

John Singer Sargent, *Portrait of the Artist Alberto Falchetti,* c. 1905. Oil on canvas, 74.9 x 54.6 cm. Private collection.

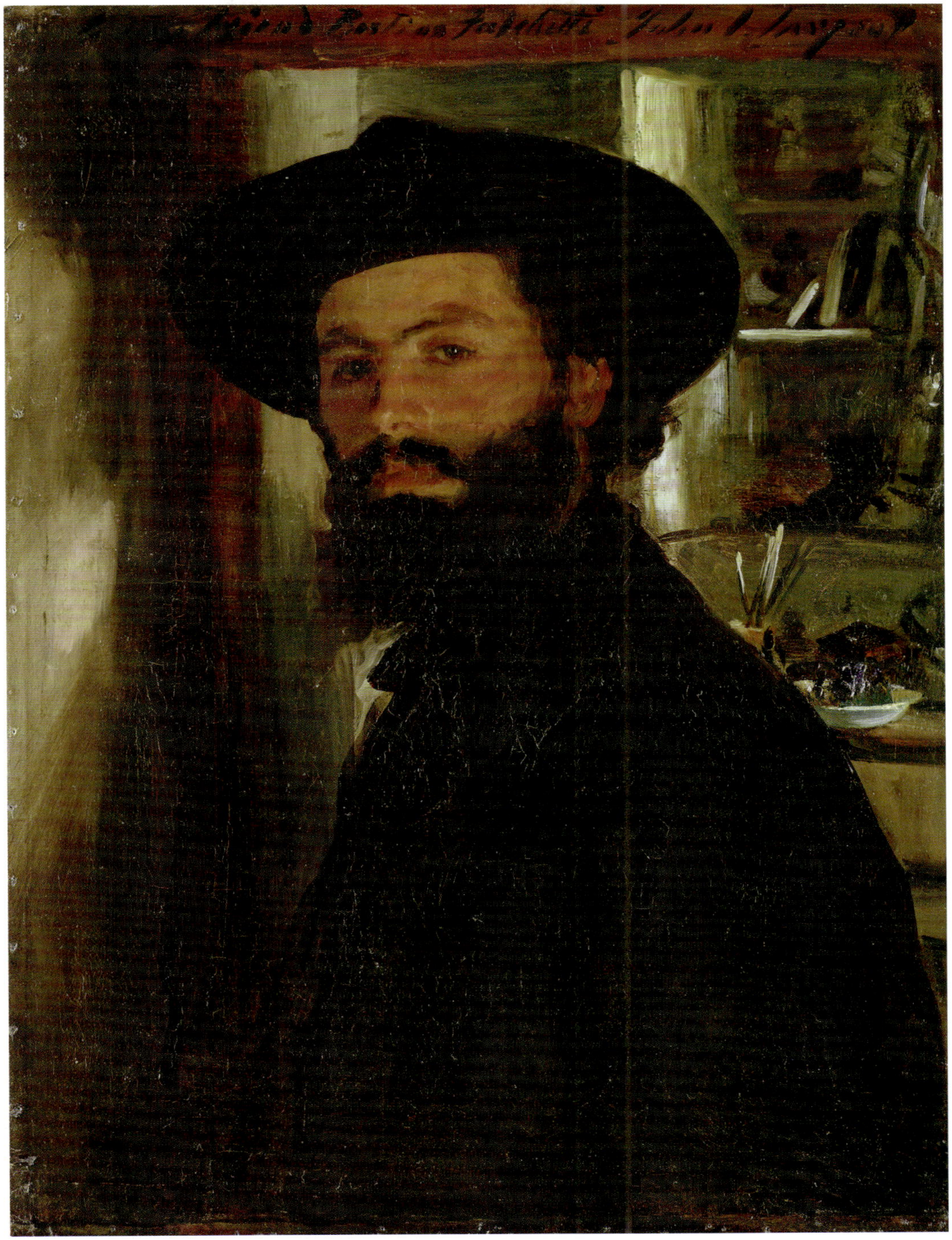

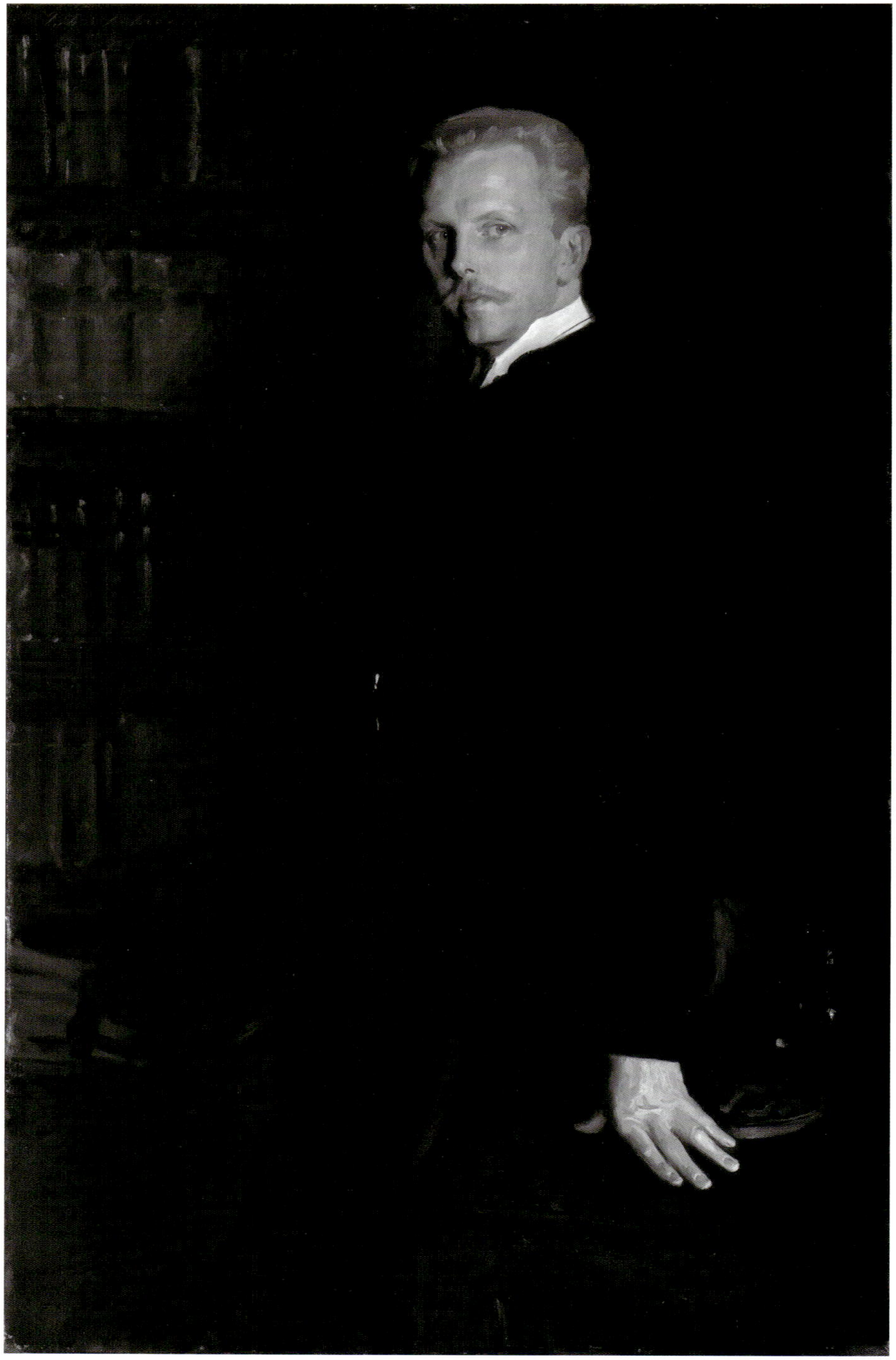

190

John Singer Sargent, *Edward Robinson,* 1903. Oil on canvas, 143.5 x 92.1 cm. The Metropolitan Museum of Art, gift of Mrs. Edward Robinson, 1931 (31.60).

instinct of moulding cells and pots of wax, or of enlarging their old cocoons with tubes of wax," they are accustomed to working both "on the inside and the outside," comfortable, like Pozzi, with the phenomenon of inversion.[152] It is generally in large, thriving flower populations that a noticeable proportion of bees seem compelled to bite or bore through the labella of flowers, the envelopes that secrete nectar and are "often modelled into variously shaped receptacles for holding this fluid." Labella, indeed, render themselves "attractive so as to be gnawed by insects"[153] who have chosen, exclusively or otherwise, this alternative lifestyle, despite the fact that "it is usually easier for the bee to get the honey in the ordinary way."[154]

In Sargent's paintings rear chambers are temptingly on display; anal penetration is hinted at without being definitively identified. But the painter's visual rhetoric drew on a well-established discourse: the spatial and conceptual-anatomical structures Sargent's paintings exhibit were employed by authors as well, and in various accounts of such private spaces—which can be flowery, honeycombed, or visceral—the issue of free admission or forcible entry is made explicit. Henry James, for example, offered the sculptor Hendrik Andersen unrestricted access to his "little cool & bowery (& even flowery) Lamb House, where a small but secure apartment opens, from tonight, to its utmost width to you, & holds itself open till you come."[155] Teleny likewise welcomes des Grieux to a "small chamber" he has specially built for him; the walls are "covered over with some warm, white, soft, quilted stuff, studded all over with frosted silver buttons," the floor carpeted with "white fleece," and the room, already like the inside of a flower, exudes an "overpowering smell of white heliotrope."[156] The chamber is Teleny's buttonhole writ large.

Dr. Jekyll and Mr. Hyde, on the other hand, begins with the "Story of the Door," a locked "cellar door" at the back of the Jekyll's property that "smells of Queer Street." Hyde goes stealthily through the back door into the laboratory, where an alternative sort of creation takes place again and again. He is pictured in Utterson's dream as Jekyll's nocturnal rapist, penetrating increasingly more intimate apertures of both house and body, the vision going dark as "the curtains of the bed plucked apart" threaten to reveal too much.[157] Stevenson's preoccupation with the secrets of bodily interiors is encapsulated in the polysemy of Hyde's very name, which suggests both that which is hidden or concealed—"If he be Mr. Hyde," Mr. Utterson proclaims, "I shall be Mr. Seek"—and the skin sheathing the inside of the body. Jekyll describes his respectable colleague Lanyon as a "hide-bound pedant," his own Hyde becoming less and less bound as, over the course of the story, Jekyll's "strange preference or bondage" increases.[158]

Sargent painted rear chambers of all types, as though responding to the particular character of his sitters: Falchetti's is blossoming, Robertson's and Robinson's honeycombed, Stevenson's anal. Sargent's portrait of Alfred Wertheimer (fig. 191), a chemist known for his "dissipated habits,"[159] could illustrate Stevenson's story. Leaning on a pile of books stacked on a table, Alfred stands in front of a tapestry or a dark doorway; to the left, two glass Kjeldahl flasks hang on the wall, the phallic shape and transparency of which promise access to interiors, even sublimation.

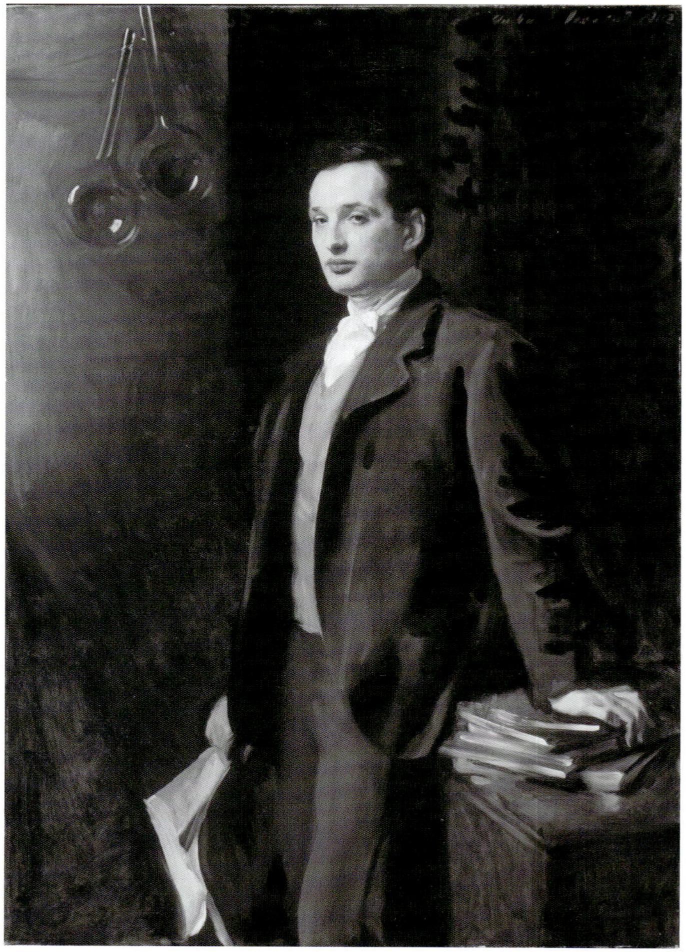

191
John Singer Sargent, *Alfred, Son of Asher Wertheimer*, c. 1901. Oil on canvas, 163 x 115 cm. Tate Britain.

Forced or welcome, backdoor pollination offered an alternative route to creation. If it could result in floral progeny or honey, it could also promise the artist a kind of perpetuity, not through offspring but through resurrection. Jekyll's self-inversion, for instance, is meant to afford him a kind of immortality—Hyde is literally both his flesh and blood and his heir.[160] Poe's "Ligeia" offers another example. After the death of the eponymous character, her widower moves to an abbey in the "wildest" and most remote area of England. In an untenanted part of his compound, beyond call of the servants, he constructs a pentagonal chamber high in a turret. This chamber is like a cell of golden honeycomb, fitted with "carpets of tufted gold" and "golden candelabra," the furniture and walls hung with "richest cloth of gold." In this "capacious" apian cell—the perverse bridal suite of the narrator's second, unloved, and doomed wife—the mannish Ligeia returns to life through "a gigantic sarcophagus of black granite" in a corner, from the ultimate back door: death.[161]

We can get a glimpse of what Sargent thought death held in store for him in the *Hell* scene in the Boston Public Library (heaven is not pictured). In the lunette (fig. 192), a giant green demon slithers among naked, writhing men. The demon, in human form but with an animal mouth, pulls a jumble of whitish bodies toward its gaping maw as if to devour them; others are locked between or under its enormous limbs. Swirling tentacles of gold flame curl around arms and thighs. They are curiously solid: sometimes ribbony, sometimes like seaweed or strong vines, and generally grasping rather than licking. Sargent described the murals, which drew in large part on Flaubert's *Salammbô,* as "in my most belly achey mood—with gold, gore, and phosphorescent Hellens. What a surprise to the community."[162] Indeed! The tableau—teeming with entwined male nudes depicted, so some would protest, for the edification of bibliophiles—was described by Andy Warhol with a phrase that could have won him Monty Python's "Proust Summarising Competition," as "a gang-bang."[163]

The decadent sexual overtones of the painting are suggested in Sargent's own description.

192
John Singer Sargent, *Hell,* installed 1916. Oil and gilded or painted Lincrusta-Walton on canvas, 254.6 x 507.4 cm. West wall lunette, north end, Special Collections Hall, Boston Public Library.

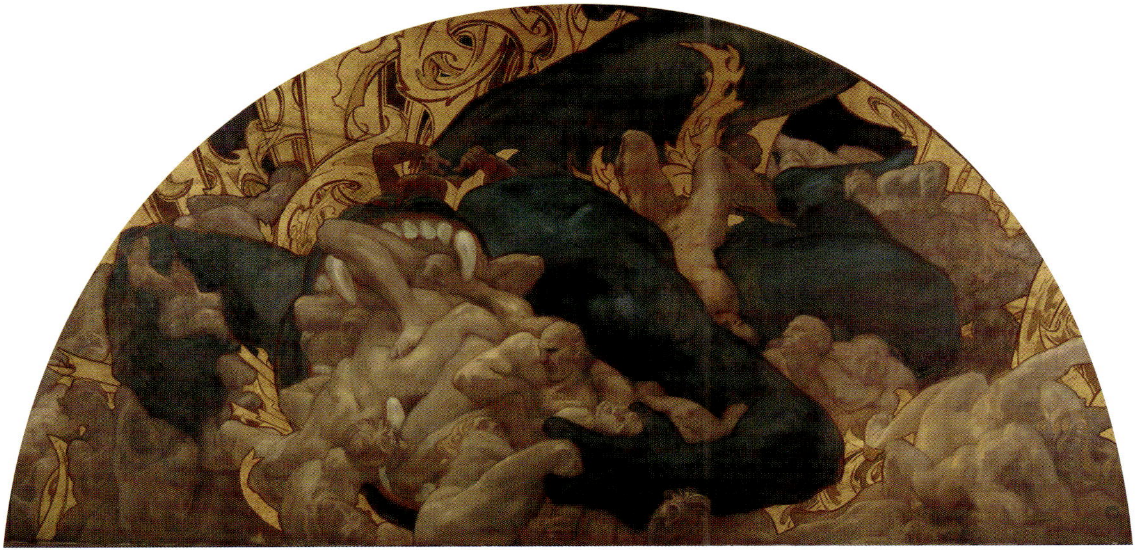

Phosphorescence is a by-product of, and became a byword for, decadence, from Baudelaire on.[164] Jean Lorrain, according to Edmond de Goncourt, employed the term as a synonym for sexual desire; a diary entry records him "jokingly complain[ing] that Pozzi's butchering in his body gave it a phosphorescence," leaving him with "a rabid need to fuck."[165] In the painting various figures seem to be engaged in sexual acts; the arrangement of the two men under the demon's right thigh, for example (the lower thigh—his torso is twisted), suggests anal intercourse. These "phosphorescent Hellens" are more Sargent's sort than Miss Sears.

Curiously, there does not appear to be any "gore" in the picture—no blood, feces, slime, or filth. In fact, the scene, for all of its jumbled bodies, is for the most part strangely sleepy and gentle. The demon's teeth penetrate nothing; if anything, he seems to be rather delicately drawing everyone into his embrace. The leg that emerges from between the demon's fangs rests gently on the shoulder of a man below it; that man's arm is curled companionably around someone else. The figure visible in the shadow above the demon's mouth can be seen as yawning and stretching more than recoiling in horror, and the man hanging down over the demon's right thigh, his legs unconcernedly splayed, appears to be dozing. If other figures do not share this tranquility, they appear to be in discomfort rather than agony. The slightly grumpy-looking bald man between the demon's left hand and shoulder is far less distressed than in the charcoal study for the figure (fig. 193), which may be, as Trevor Fairbrother describes it, Sargent's "most forceful evocation of bodily anguish."[166] Or he may just be yawning too.

Like Hamlet, Sargent transmutes death into sleep, but he does not worry, as his dramatic

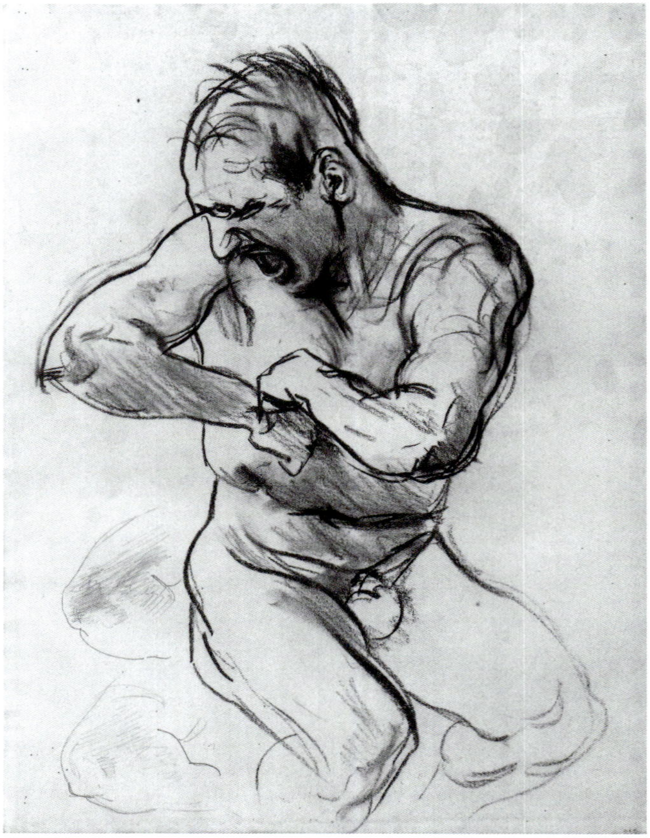

193
John Singer Sargent, *Man Screaming, Study for "Hell"* (study for mural paintings installed in three phases [1895, 1905, 1916] at the Boston Public Library, and related drawings), 1890–1916. Charcoal and stump with pencil on beige paper, 61.9 x 47.6 cm. Gift of Miss Emily Sargent and Mrs. Francis Ormond, sisters of the artist, 49.99. Collection of The Corcoran Gallery of Art, Washington, D.C.

counterpart does, about the horrors its dreams may hold; sleep for the painter offers release from social constraints. Sargent's watercolor *Tommies Bathing* (fig. 194) has an aura of fleeting liberty, perhaps even abandon. On their leaf-shaped island, the pale and succulent bodies of the two anonymous sleeping soldiers are caressed by the sun and the luxuriant grasses, and even open to penetration by the occasional blade. Set against the background of World War I and its apocalyptic destruction, this Whitmanian postcoital sprawl in the grass is a moment of respite from violent obligations, from true horrors.

Thus far we have witnessed the deployment of a number of transformative stings and brushes by artist's proboscides that attempt to ensure a kind of immortality for the invert and artist's offspring.[167] The last one we shall consider, however, is the oldest, and in fact the most familiar. In fairy tales a prick or sting is followed by the long, vegetal sleep of the beloved. The prince is free to roam an overgrown dreamworld, unfettered by ordinary considerations, as is the artist who deals with sleeping subjects. It is only when Albertine sleeps and becomes "a plant," "a long blossoming stem," that M. feels he need no longer "live on the surface of" himself and, anxieties allayed, becomes aware of recovering his own "faculty of dreaming."[168] When "no longer observed by" her, the castration anxiety she provokes in him evaporates and M. can unfold, invert himself, become permeable; he can unfurl his own feelers, all the better to beetle over her in his creative imagination. Likewise, in the images of sleepers that abound in Sargent's oeuvre, we sense that the artist, in making them, was able to pull out his tender creative tentacle without "pain of losing [his] proboscis."[169]

Under the corollas of umbrellas love lies sleeping, eternally youthful and beautiful, in Sargent's intricate webs of floral "intertwingles"[170] and dandy-long-legs (fig. 195). Such suggestive intertwinings and indiscreet exposures (with their multitudinous crotch shots) suggest the creatures' complicity in their compromising positions. Difficult to distinguish from one another and from the landscape, melded into the flesh of the world with sealed eyes and stilled hands offering no resistance, they can be modeled to match the painter's desire. The sleep of these spread- or

194
John Singer Sargent, *Tommies Bathing*, 1918. Watercolor and graphite on white wove paper, 38.9 x 52.7 cm. The Metropolitan Museum of Art, gift of Mrs. Francis Ormond, 1950 (50.130.48).

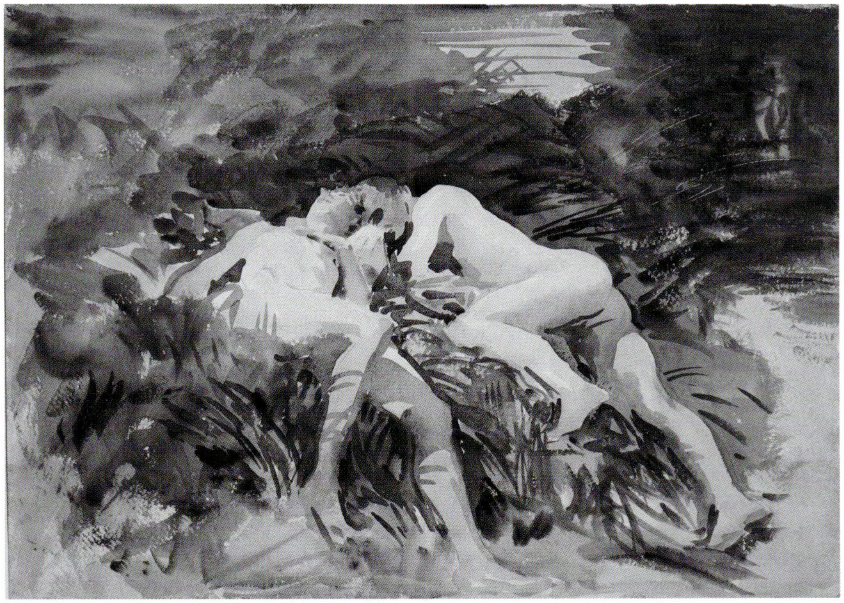

195
John Singer Sargent, *Group with Parasols*, c. 1905. Oil on canvas, 55.3 x 70.8 cm. Private collection.

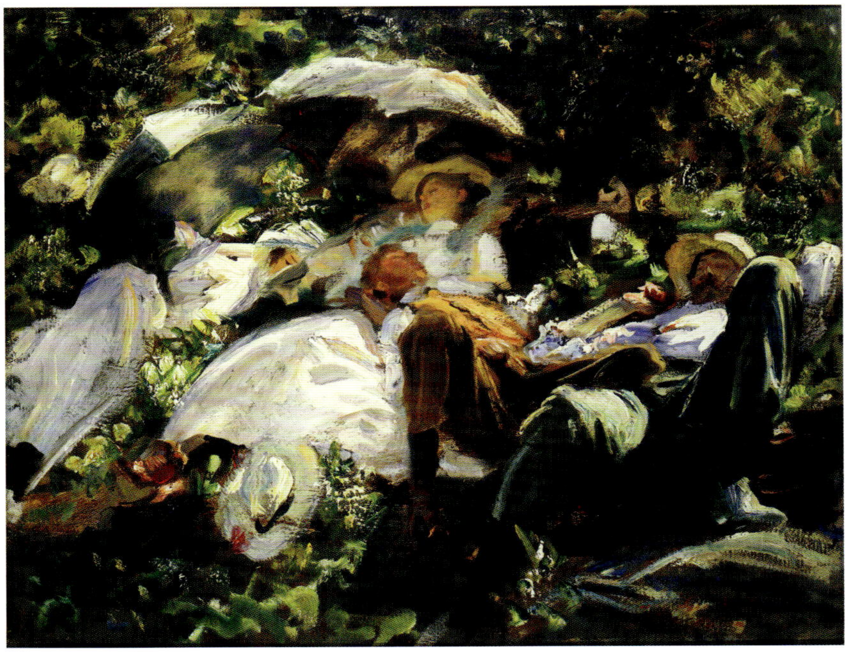

cross-legged dandies and daisies lying sprawled and overlapping in the grass images a peaceful and plenitudinous jumble of bodies and embodies the exiled invert artist's fantasy of being, in M.'s words, "fastened to a deep, invisible soil by the articulations (of which my tiredness made me conscious) of muscular and nutritious roots."[171]

CONCLUSION

LEAF TAKING

If I had wanted to make a real sex movie, I would have filmed a flower giving birth to another flower.

—ANDY WARHOL

The metaphorics and practice of botanical art-making did not disappear with Sargent or with Proust. Sargent's analogizing of hands and flowers, and human and floral sexuality, should be seen as contributing to a persisting iconographic tradition.[1] More important, the models of cross-fertilization and vegetable reproduction formulated by the artist and his peers continued to offer a compelling paradigm for art creation throughout the twentieth century. I will briefly indicate two garden paths to be followed, one leading through the field of abstract painting, the other through photography and into new media.

In the development of pictorial abstraction, the conception of the painter as a pollen-brushing cross-fertilizer helped free up the artist from mimetic concerns—the bee does not brush to create a recognizable tableau but rather in response to a somatic imperative. The floral poetics of the physical process of painting, and its techniques and materials, underpin a vein of abstraction extending from Kupka through artists like O'Keeffe to the contemporary painter-pollinator Wolfgang Laib.

František Kupka, who exhibited the first fully abstract paintings in Paris in 1912,[2] slowly evolved his early nonfigurative works from organic stimuli through a process of distillation, not unlike the Renaissance apian artist. Although this might lead us to expect the painter's withdrawal from materiality and the world, the opposite is the case. The physical act of painting was crucial to his project and his conception of artistic fertilization; this is reflected both in his facture and in the titles of his works, from the *Roman du rose* paintings to the Tales of Pistils and Stamens series from 1919 to 1923—both suggesting narratives that rely on the acts of touching, grasping, brushing, and being brushed.

The painter had from early in his career identified himself as a garden dweller through paintings like his 1894–95 *Self-Portrait Among*

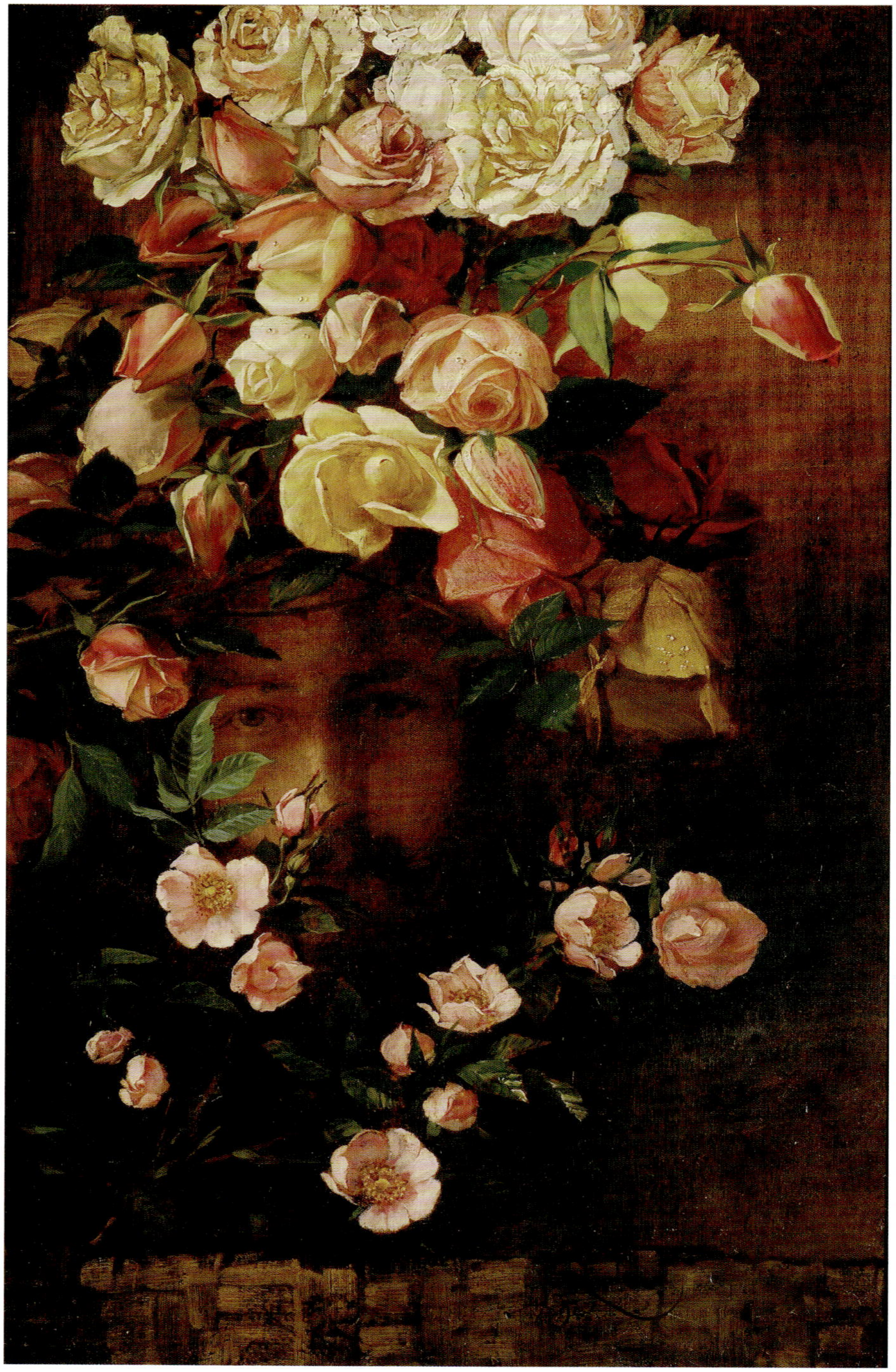

196

František Kupka, *Self-Portrait Among Roses,* 1894–95. Oil on wood, 65 x 42 cm. Museum Kampa—Jan and Meda Mládek Foundation Collection, Prague. © 2008 Artists Rights Society (ARS), New York/ADAGP, Paris.

Roses (fig. 196), where he peers out from a wreath of flowers. His prominent right eye, framed by green leaves, suggests a peculiarly floral artistic vision, the fertility of which becomes manifest as the simple climbing roses peeping over a low wall at the bottom of the canvas give rise to the fantastic flowers of the artist's brain blossoming above, wilder, multiplied. These blooms become a virtual exercise in color and form: nuances of the rosy palette, from palest pinks to deep scarlets, soft peaches to lemon yellows, are explored; every state of openness and closure is depicted; the layering of petaline surface on petaline surface becomes ever more complex. If *Self-Portrait Among Roses* links artistic production with floral reproduction, some of Kupka's other early Symbolist works also explicitly link floral and human procreation: *The Beginnings of Life,* for example, from 1900 to 1903, depicts an embryo floating above the chalice of a water lily, to which it is delicately tethered through the intermediary of a vegetal placenta. Kupka's mature abstractions often equate floral sexual encounters with the generation of form and life itself.

Of the four paintings that originally made up the Tales of Pistils and Stamens series, only three exist today, but one of Kupka's routes into full abstraction can still be traced through them. While the title of the paintings indicates that the subject of the works is floral fecundation by the encounter of pistil and stamen, the center of one of the first versions is filled with writhing, orgiastic forms identifiable as small male and female nudes.[3] Human and floral sexualities are thus linked. Baroque patterns of color and shape spiral out from the luminous white core of the composition, further suggesting that the production of art, the creation of images, is analogous to organic reproduction. In the later *Tale of Pistils and Stamens* in the Centre Georges Pompidou, the same arcing, striated forms surge forth from a molten core. The compositional vortex of this primeval scene channels our vision down to a microscopic scale, perhaps giving us the perspective of a bee. Significantly, there are no recognizable figures here. The exfoliating patches of paint might evoke feathers, scales, petals, bones—each stroke is full of plastic potential—but nothing has yet coalesced into an identifiable substance.

The artful contrivances of floral reproduction by pollination are invoked in another set of paintings, the titles of which also point to the natural romance behind Kupka's abstractions. In the 1910s Kupka painted a group of nonfigurative works exploring specific colors. The majority of the canvases are titled *Form of* the particular hue (*Form of Vermilion, Form of Orangy,* etc.), but the color pink, or *rose* in French, is the exception in this loose series; its studies are called *Roman du rose,* evoking the medieval floral romance (the title is *du rose* rather than *de la rose,* since the color, unlike the flower, is a masculine noun).

In a study for *Roman du rose* (fig. 197), butterfly wings at the lower center, made up of parallel brushstrokes of white, lavender, red, and black, gently sweep over a floral form, whose pinky-white outer petals give way to a series of vulval red and orange surfaces that meet at the center in a delicate tuft. As though in response to the act of brushing, and following the arc of the wings, small red kernels bud out from this warm blossom at the top right. Irregularly shaped and

197

František Kupka, *Study for Roman du rose*, 1911–24. Oil on canvas, 80 x 67 cm. Museum Kampa—Jan and Meda Mládek Foundation Collection, Prague. © 2008 Artists Rights Society (ARS), New York/ADAGP, Paris.

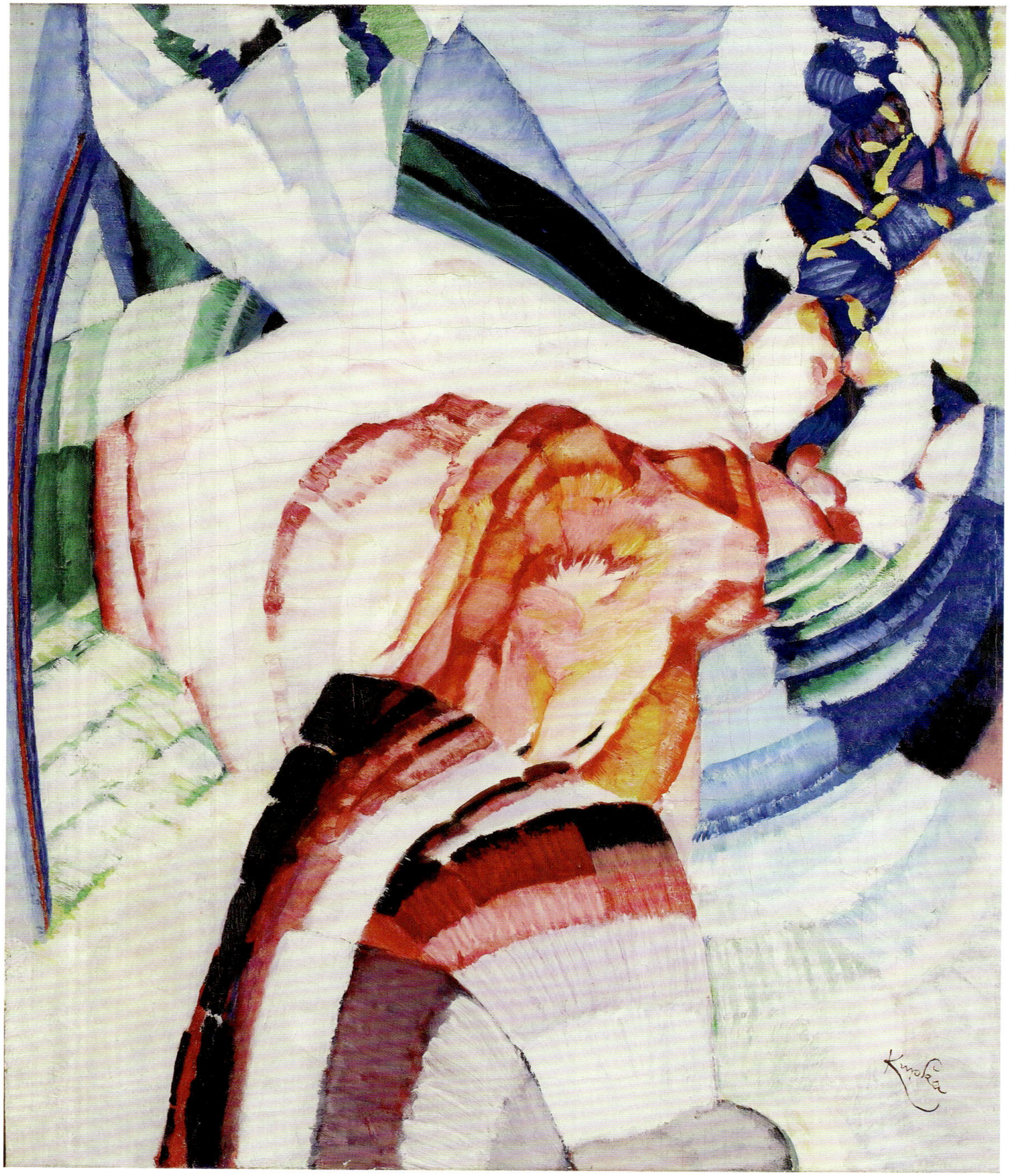

colored forms issue from these buds, fragmented and indefinable. Spots of yellow, red, blue, and white accrete like primordial building blocks, shifting direction and twisting into shape under the white light of an iridian sun. Undergirded by the shoots of blue and green lines, these confused forms make up a segment of the dazzling arabesque of colors and forms inaugurated by the brushing of wing and flower, an arabesque that circles up and, imaginarily, over the top of the canvas, sweeping down again on the left side in a blue and red arc. On this side of the painting lines and planes extend in cool greens and blues. Chromatically, then, the work moves from fiery generation to cooling form; compositionally, though, a cycle is established that will endlessly repeat in the movement around the canvas. Significantly, we only see fractions of recognizable objects—wings and petals but neither a whole butterfly nor an entire flower. The emphasis rests on part-objects and their interactions, on the moment of contact and brushing of surfaces.

The white space at the bottom of the painting, out of which everything springs, is covered with minute, delicately tinted parallel brushstrokes, and thus reads not as empty but as full, as though teeming with potential. The parallel brushstroke Kupka employs conspicuously here is key to his poetics: it functions as a basic unit, not only of form but also of time and space, measuring their incremental development and mutual implication. The painter's process of laying down brushstrokes and generating pictorial form, moreover, is identified with the painterly process of pollen brushing. The very material approach to abstraction that Kupka's works inaugurate, then, begins with floral reproduction.

Recently the idea of abstract painting as a process of pollination has been realized in the work of Wolfgang Laib. A literal instantiation of the pollen-gathering artist, Laib collects pollen from plants and uses it as his pigment to create color field paintings on the floors of galleries; his 1992 *Pollen from Hazelnut* was the product of months of meticulous harvesting. The end results of his activity are purely abstract, luminous, shimmering, enveloping-à-la-Rothko planes, sprinkled, à la Pollock, on the ground, their sheer size luring the viewer into an apian experience, a floral romance.

Georgia O'Keeffe's erotic floral forms belong within this genealogy, despite the persistent reductive equation of her abstractions with an essentialist conception of female sexuality. Anne Wagner has argued compellingly that O'Keeffe "staked her claim as a modernist painter on her effort to rewrite the terms in which the body could be experienced in representation," and that the artist assumed different gender identifications in her works—feminine in the vaginal petals of *Black Iris III,* for example, masculine in the phallic Jack-in-the-Pulpit series.[4] The reading of O'Keeffe's art as masquerade seems to me exactly right, but I would argue that Wagner does not go far enough. The dominant focus in almost all of O'Keeffe's floral works is the petals: parts that can fold and unfold over time, invert themselves, be firm or pliant. Surrounding the sexual organs of the flower, petals can take on "masculine" or "feminine" forms without *being* either; they are the primary instruments of visual floral mimicry. Wagner sees the Jack-in-the-Pulpit series as masculine masquerade on O'Keeffe's part—"never, in short, has the phallic symbol been more decisively colonized in paint"[5]—based on the prominent spadix. The spadix is not the masculine organ, though (it is a spike of tiny flowers), and this particular plant (*Arisaema triphyllum*) changes sex from year to year depending on its level of nourishment.[6] Botanical accuracy isn't the issue, however, so much as the painter's phantasmatic identification with a *plant*, rather than human, part, and particularly its surface.

If Paul Rosenfeld was one of the most reductivist critics, linking O'Keeffe's floral abstractions with the "essence of very womanhood," he nevertheless noted that "there are canvases of O'Keeffe's that make one to feel life in the dim regions where human and animal and plant are one, indistinguishable, and where the state of existence is blind pressure and dumb unfolding."[7] This inchoate, primordial possibility in her works suggests that we ought to read the artist's painterly performances in terms not only of gender but also of species (an idea similarly suggested by Stieglitz's photographs of O'Keeffe's strange, creatural hands) and primary processes of organic development. Wagner is no doubt right that a phallic symbol was prominently employed in the Jack-in-the-Pulpit series to taunt O'Keeffe's critics, but there is more to her forms than meets the eye; they persistently invoke the sense of touch through the contact of petals and other parts. The "near hallucinatory myopia" many of her works entail for the viewer, who is confronted with "a world to be lost in,"[8] encourages a nonanthropomorphic vision and other sensory responses.

In the National Gallery's *Jack-in-the-Pulpit #4* from 1930, the incandescent flame created by a large blue-black spadix surges upward through its surrounding spathe to what look like green planes and valleys and blue sky beyond. This world-making image is not without a procreative charge, the match functioning as an erotic metaphor. Nor is it without theatrics: the dark form of the spadix can read as a footlight in this drama of creation as easily as a match. If O'Keeffe phantasmatically assumes the role of the spadix, at the burning core of this floral world, "her" phallic form penetrates nothing: the light flaring up in this primordial grotto illuminates an interior that is not one. Nothing is revealed to the eye except a surface, but what we do get a sense of is the effect of heat and light on that pellicle, registered on the black sheath as a chromatic response to such primary stimuli: black becomes blue, grey, and modulated, evoking the space and sky visible in the four corners of the painting; undifferentiated ground blushes into life. The folding of the spathe to reveal its pale reverse above the flame also registers response to its generative flaring. Color and form are thus depicted as tactile responses to heat and light, artistic creation as something felt as much as seen. Metamorphically motile, the white glow emanating from the spadix can also be read as the stylized bristles of a brush stroking on color, the curling inward of the spathe at the top perhaps a response to its smooth but pointed touch. Floral and artistic creation are thus intimately linked. Readable, then, as a seed germinating in earth, the sex of a flower or a woman or a man, a geological chasm, or the touch of a paintbrush on a smooth surface, *Jack-in-the-Pulpit #4* invokes the ambiguous time, space, and species of Kupka's abstractions and naturalizes the very appearance of colors and forms, however abstract.

Photographers as well as painters felt the lure of flora. Huysmans's dictum that the only real artists are the horticulturalists found a proponent in Edward Steichen, who grew poppies and developed new varieties of flowers, creating a "pure blue petunia" and a perennial delphinium. In 1936 he even exhibited his own delphinium hybrids at the Museum of Modern Art, where he would later become director of the Department of Photography.[9] Photography and flowers had been linked, though, since Fox Talbot started making his photogenic drawings. A fleeting contact of surfaces produced these negative images, joining botanical and mechanical reproduction, and even a type of readymade inversion (tonal) that would be explored by later Rrose lovers like Man Ray.

Fantasies of promiscuously intermingling human and floral sexualities have, moreover, occupied a significant place in the iconography of invert-oriented photographers, from von Gloeden to Mapplethorpe and beyond. Wilhelm von Gloeden's homoerotic photographs of flower-bedecked nude youths made him popular with inverts throughout the Western world; Oscar Wilde

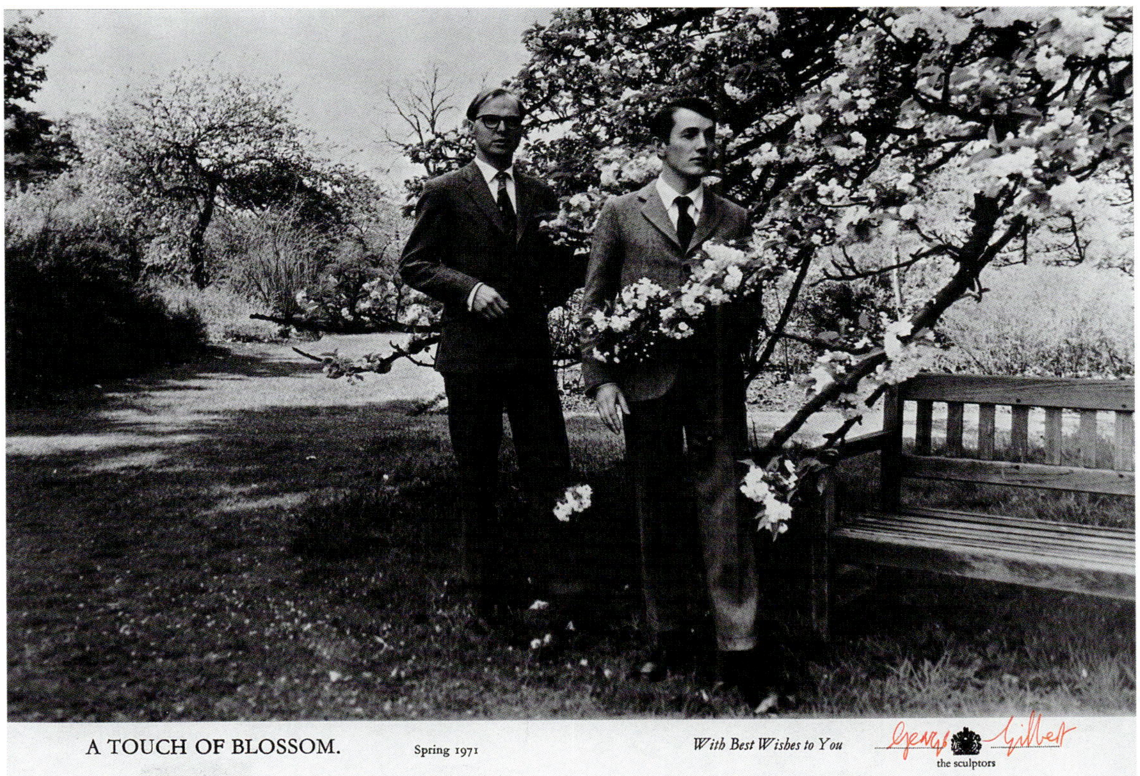

198
Gilbert and George, *A Touch of Blossom*, 1971. Offset lithograph, 30 x 40 cm. Postal sculpture.

even visited him in Sicily in the 1890s.[10] Carrying on this tradition, an untitled print from George Platt Lynes's Narcissus series, circa 1939, links floral, human, and photographic reproduction: he print pictures the arousal of Narcissus on perceiving his double on the surface of the water—his becoming reproduction—an arousal that peaks in an ejaculation of frothy white flowers. Spurting all over his chest, arms, and face, these literal and figurative sprays of blossom are captured as white stains on the surface of the photograph, forming a jubilant pattern that conveys the ecstatic fusion, the cross-fertilization of self and world, through which the image comes into being.

A Touch of Blossom from 1971 by Gilbert and George (fig. 198) mobilizes various of the tropes examined in this study in a kind of photo-synthesis: flower-brushing as productive of art; vegetable romance, with the chase of Apollo and Daphne enacted by the figures,[11] the foremost one (Gilbert) imprisoned by blossom-loaded branches; and imminent man-to-man contact suggested by the expectantly extended hands, and facilitated by the flowery sprays. The postcard format of the

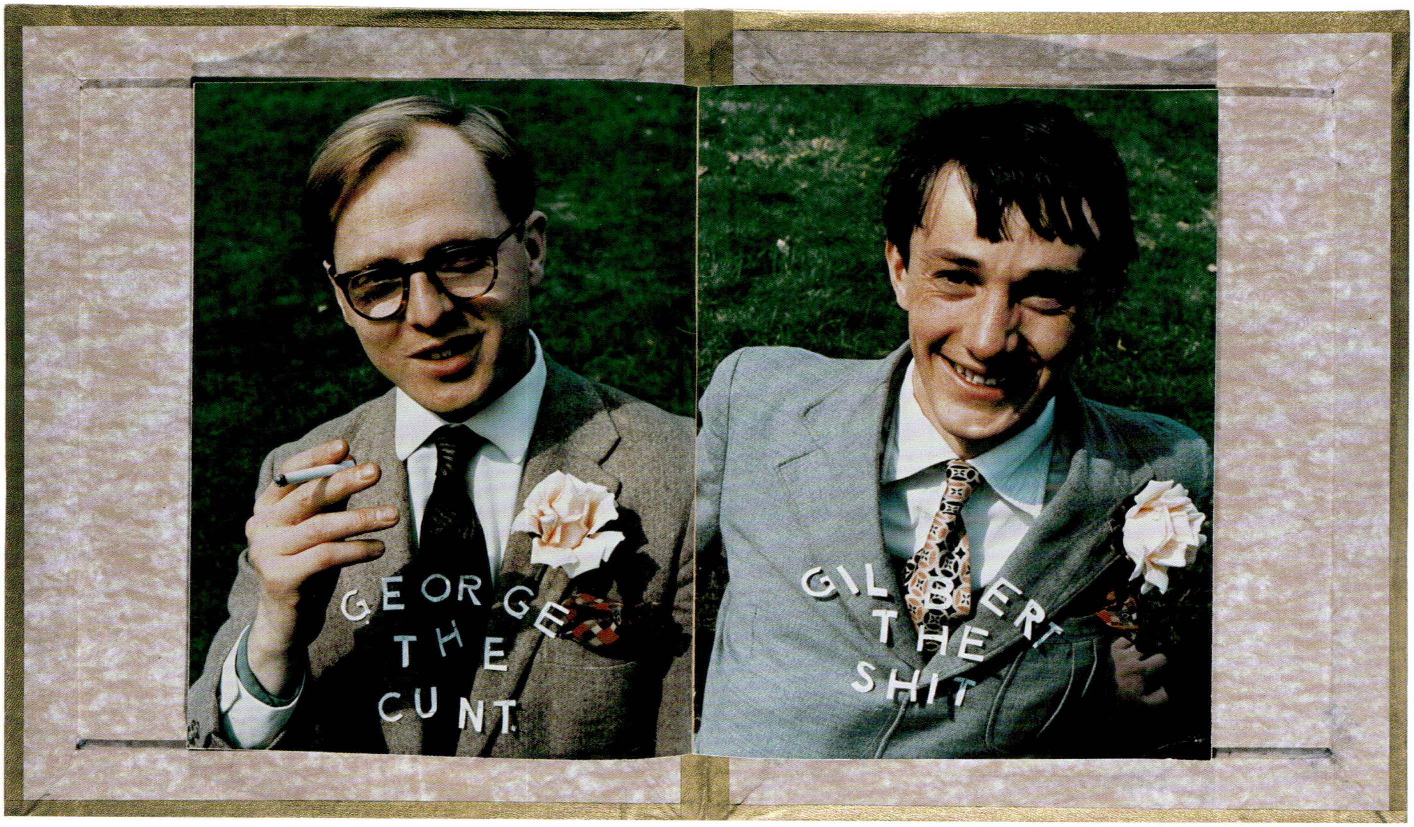

199
Gilbert and George, *George the Cunt and Gilbert the Shit*, 1969. Magazine sculpture.

hands, one of which is visible beneath the rose at the lower right. Closing the magazine, pressing the two leaves against each other, enfolds the two within a foliate realm wherein horizontal and vertical become parallel planes, and George's cigarette winds up in Gilbert's flower, his buttonhole.

A grainy photograph of Sargent at age thirty (fig. 200) captures the painter as pollinator and plant: provocatively sprawled, holding a cigarette between the tendrilly fingers of his elegant (and very Sargent) left hand, stretching his right wing out against the sofa. My exploration of the evolution of pollinator erotics, floral models of reproduction (artistic and otherwise), and a period poetics of homosexuality has sketched a loose but flourishing community of turn-of-the-century inverts and artists with Sargent at its heart. I have tried to restore pansies, flowers, wagtails, and fairies to their original garden setting, and accord the bug its rightful place in the history of buggery, sod in sodomy, and so on. At the same time, I have tried to show the contribution of these fantastic creatures to the development of artistic modernism, specifically through their elaboration of a particular poetic understanding of materials and the creative process. Regarding critical appraisals of Sargent's quality relative to Cézanne's, Forbes Watson commented that "occasionally the modernist forgets, as a Wells character remarked, that 'there is something about a rose which a lily has not got.'"[12] Without claiming to have fully deciphered this enigmatic statement, I hope *A Touch of Blossom* has illuminated the diverse aptitudes of some of the artists who flowered in the gardens of modernism.

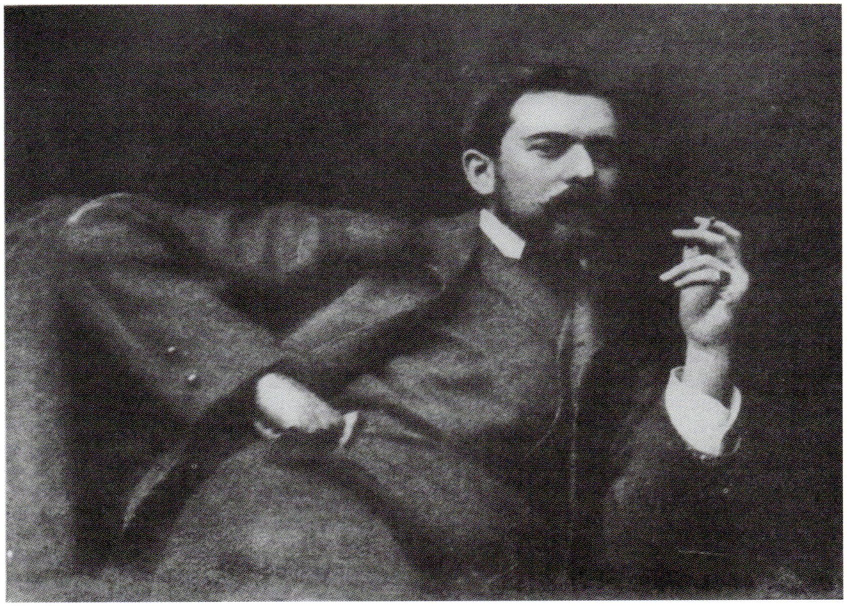

200
Photograph of John Singer Sargent, c. 1884. Reproduced in Evan Charteris, *John Sargent* (New York: Charles Scribner's Sons, 1927), facing page 78.

work, like the calling cards left by Proust's social bees on their rounds of visits, allows viewers to experience a touch of blossom, a material dissemination, without getting stains on their shirts.

Gilbert and George also utilize the trope of pollination by nonnormative means. In a "magazine sculpture" (a decidedly nonnormative medium) published in *Studio International* in 1970 (fig. 199), photographs of the pair against green backgrounds, in similar though not identical poses and dress (both wear suits and ties, pale pink roses in their buttonholes, and handkerchiefs in their pockets), appear on facing pages. George, on the left, holds a cigarette nonchalantly in his right hand; pinned to his chest are paper letters that spell out "GEORGE THE CUNT." He appears to be standing or sitting upright in front of a neatly clipped lawn. On the right, "GILBERT THE SHIT" is leaning back on the grass, supported by his

NOTES

Abbreviations

CP 1 = Ormond and Kilmurray, *Complete Paintings*, vol. 1

CP 2 = Ormond and Kilmurray, *Complete Paintings*, vol. 2

CP 3 = Ormond and Kilmurray, *Complete Paintings*, vol. 3

CP 4 = Ormond and Kilmurray, *Complete Paintings*, vol. 4

OED = Oxford English Dictionary

RGA = Simpson, "Reconstructing the Golden Age: American Artists in Broadway, Worcestershire, 1885 to 1889"

RTPE = Proust, *Remembrance of Things Past*

RTPF = Proust, *À la recherche du temps perdu*

Introduction

1. Plato, "Ion," *The Collected Dialogues of Plato, Including the Letters,* ed. Edith Hamilton and Huntington Cairns, trans. Lane Cooper (Princeton: Princeton University Press, 1989), 220.

2. Petrarch to Boccaccio, 28 October 1365, in *Petrarch, the first modern scholar and man of letters; a selection from his correspondence with Boccaccio and other friends, designed to illustrate the beginnings of the Renaissance,* ed. and trans. James Harvey Robinson (New York: G. P. Putnam's Sons, 1898), 291.

3. Carlo Cesare Malvasia, *The Life of Guido Reni* (1678), trans. Catherine and Robert Enggass (University Park: Pennsylvania State University Press, 1980), 35.

4. Ibid., 133. Guido Reni, for instance, took his "figures that are so well proportioned and correct" from Raphael, "his sense of purity in the attitudes and propriety in the poses" from Correggio, etc.

5. Seneca, "On Gathering Ideas," epistle 84, in *Ad lucilium epistulae morales,* trans. Richard M. Gummere, 3 vols. (London: William Heinemann, 1930), 2:279.

6. Winckelmann extolled the artists of antiquity for their judicious aesthetic sampling: "as a bee gathers from many flowers, so their concept of beauty was not limited to the individual attributes of a single beauty." Johann Joachim Winckelmann, *History of the Art of Antiquity* (1764), trans. Harry Francis Mallgrave (Los Angeles: Getty Research Institute, 2006), 198. Anton Rafael Mengs and other contemporaries concurred. For more on the apian artist in the eighteenth century, see Elizabeth C. Mansfield, *Too Beautiful to Picture: Zeuxis, Myth, and Mimesis* (Minneapolis: University of Minnesota Press, 2007), 67–74.

7. Il "a dû porter, dès le collège, les signes inquiétants du pur artiste . . . avec ses antennes, ses yeux d'abeille aux multiples facettes." Jacques-Émile Blanche, *Mes modèles* (1928) (Paris: Stock, 1984), 100. All translations are my own unless otherwise indicated.

8. Philip Knight, *Flower Poetics in Nineteenth-Century France* (Oxford: Clarendon Press, 1986), vi, 241.

9. "J'ai fait une nouvelle *fleur.*" Baudelaire to Auguste Poulet-Malassis, 4 February 1859, in Charles Baudelaire, *Correspondance,* ed. Claude Pichois, 2 vols. (Paris: Gallimard, 1973), 1:546, quoted in Barbara Johnson, *Mother Tongues: Sexuality, Trials, Motherhood, Translation* (Cambridge: Harvard University Press, 2003), 168.

10. "J'ai essayé d'inventer de nouvelles fleurs . . . de nouvelles langues." Arthur Rimbaud, "Une saison en enfer" (1873), in *A Season in Hell and The Drunken Boat,* trans. Louise Varèse (New York: New Directions, 1961), 86–87. "Langue" can be translated as "tongue" or "language."

11. "Je dis: une fleur!" Stéphane Mallarmé, "Crise de vers" (1895), in *Oeuvres complètes,* ed. Henri Mondor and G. Jean-Aubry (Paris: Gallimard, 1945), 368.

12. Historians of all varieties agree on this. See, for example, James H. Rubin, *Manet's Silence and the Poetics of Bouquets* (Cambridge: Harvard University Press, 1994); William H. Gerdts, *Down Garden Paths: The Floral Environment in American Art* (London: Associated University Presses, 1983); and Jean Clay, "Ointments, Makeup, Pollen," *October* 27 (Winter 1983): 3–44.

13. "Cet homme [Whistler] disait, de l'objet d'art digne de ce nom, qu'il devait être exécuté comme la nature produit les plantes." Comte Robert de Montesquiou, *Les pas effacés: Mémoires,* 3 vols. (Paris: Émile-Paul Frères, 1923), 2:261.

14. "Il produit ses oeuvres tout aussi simplement qu'un pommier produit des pommes." Max Buchon made this observation in "Recueil de dissertations sur le réalisme" (Neuchâtel, 1856), quoted and translated in Linda Nochlin, *Realism and Tradition in Art, 1848–1900* (Englewood Cliffs, N.J.: Prentice Hall, 1966), 46.

15. "Les horticulteurs sont les seuls et les vrais artistes." Joris-Karl Huysmans, *À rebours* (1884) (Paris: Fasquelle, 1907), 126.

16. "Mon plus beau chef-d'oeuvre c'est mon jardin." Quoted in Jean-Pierre Hoschedé, *Claude Monet, ce mal connu: Intimité familiale d'un demi-siècle à Giverny de 1883 à 1926,* 2 vols. (Geneva: Pierre Cailler, 1960), 1:70.

17. Mary Hale, "The Sargent I Knew," *World Today,* November 1927, 566.

18. Michael Fried, *Courbet's Realism* (Chicago: University of Chicago Press, 1990), 343n47.

19. Oskar Bätschman reads the illustration as a commentary on the art market: "art works in the galleries have become as

aggressive as the advertisements in the fight for buyers." The still life is the only work being actively consumed, if not by the intended buyer. Oskar Bätschman, *The Artist in the Modern World: The Conflict Between Market and Self-Expression* (Cologne: Dumont, 1997), 131.

20. Effie Millais's notebook, quoted in John Guille Millais, *The Life and Letters of Sir John Everett Millais*, vol. 1 (Toronto: Morang, 1900), 324.

21. Carl Hendrik Andreas Bjerregaard, "A Philosopher and a Butterfly," *Monthly Illustrator and Home and Country* 3, no. 11 (1895): 382. Nine engravings of butterflies alighting on flowers illustrate the article and reinforce the argument made in the text.

22. The Prix de Rome was a scholarship given out by the French Academy of Painting and Sculpture, which, by the second half of the nineteenth century, had become a bastion of conservatism. A sign on the wall behind Didier declares him an *homme-canon* as much as an *homme-papillon,* and announces his 1857 award with the mocking phrase "Prix des places"; the tethered painter does not seem likely to get far. For a different interpretation of this caricature, see Daniel Wildenstein, *Claude Monet: Biographie et catalogue raisonné,* 5 vols. (Lausanne: Bibliothèque des Arts, 1974–91), 5:151, cat. no. D515. Wildenstein reads the centaur as a woman trying to trap Didier into marriage.

23. Charles Darwin, *On the Origin of Species* (1859) (Cambridge: Harvard University Press, 1998), 97. In this case, by "pencil" Darwin means a paintbrush (OED, s.v. "pencil," 1.a., 1.b.); elsewhere, however, he includes an illustration of the pollinia of *orchis mascula* attached to a graphite pencil. Charles Darwin, *The Various Contrivances by which Orchids are Fertilised by Insects,* 2nd ed. (London: John Murray, 1877), 12, fig. 2.

24. Oscar Wilde, "The Decay of Lying" (1889), in *De Profundis and Other Writings* (Harmondsworth: Penguin, 1954), 74, 87.

25. RTPE 1:764; "la précieuse fleurette dont certains maîtres anciens signaient leurs tableaux," RTPF 1:711.

26. Moore's anthemion was inspired by that of the painter Simeon Solomon, who had crowned his monogram with one since 1860. Robyn Asleson, *Albert Moore* (London: Phaidon, 2000), 86. Anthemia usually resemble honeysuckle clusters or palmettes. In one version of Moore's 1877 painting *End of the Story* (see ibid., 138), a small cluster of pink honeysuckle is placed just above the anthemion, echoing its arrangement. The blossomy rather than leafy character of his symbol is unsurprising given Moore's flowery oeuvre.

27. Georges Rivière, *Renoir et ses amis* (Paris: H. Floury, 1921), 44, quoted in Clare Willsdon, *In the Gardens of Impressionism* (New York: Vendome Press, 2004), 15, 256n15. That this name was inspired by the location of what would later be known as the first Impressionist exhibition, Nadar's studio on the Boulevard des Capucines, does not diminish its floral significance.

28. Lillie Langtry attributed Wilde's flower worship to the influence of the gardening artist Frank Miles. Lillie Langtry, *The Days I Knew* (London: Hutchinson, 1925), 61–62.

29. Pierre Loti was the pseudonym of the French writer Louis Viaud. "Loti" is "an Indian flower which loves to blush unseen," though "island flower" is also a common translation. Edmund Gosse, "Biographical Note," in Pierre Loti, *An Iceland Fisherman* (1886), trans. Jules Cambon (New York: P. F. Collier and Son, 1902), xxiii.

30. Elaine Scarry, "Imagining Flowers: Perceptual Mimesis (Particularly Delphinium)," *Representations* 75 (Winter 1997): 108. Darwin makes the case most forcefully in *The Power of Movement in Plants,* first published in 1880 and translated into French in 1882.

31. "[N]e cultiver que la fleur de leur propre jardin, fleur unique." Odilon Redon, *À soi-même: Journal, 1867–1915* (Paris: J. Corti, 1989), 118.

32. Walter Pater, *The Renaissance* (1873) (Oxford: Oxford University Press, 1998), 49.

33. Max Nordau, *Degeneration* (1st German ed. 1892; 1st English trans. 1895) (Lincoln: University of Nebraska Press, 1993), 192.

34. The author of Beltraffio, for example, is a "rose," and a number of James's artists have floral names, like Florimund Daintry in "A New England Winter." Henry James, "The Author of Beltraffio," in *In the Cage and Other Tales,* ed. Morton Dauwen Zabel (London: Rupert Hart-Davis, 1958), 30. James does not neglect pollinator-artists: the ever industrious painter Sam Singleton in *Roderick Hudson* resembles "some curious little insect with a remarkable mechanical instinct in its *antennae.*" Henry James, *Roderick Hudson* (1875) (Harmondsworth: Penguin, 1986), 360.

35. Henry James, preface to the 1908 New York edition of *The Spoils of Poynton* (first published in the *Atlantic Monthly* in 1896), reprinted in *The Spoils of Poynton* (Oxford: Oxford University Press, 2000), xxxix, xl, xli, xliv.

36. Asleson, *Albert Moore,* 179, 226n170.

37. In another painting purchased by Connal, *Reading Aloud,* Moore included a bee "as a parallel to his own anthemion signature." Ibid., 179, 168.

38. Quoted in Giorgio Vasari, *The Lives of the Artists* (1550, 1568), trans. Julia Conaway Bondanella and Peter Bondanella (Oxford: Oxford University Press, 1998), 479. On this trope, see Frederika H. Jacobs, *The Living Image in Renaissance Art* (Cambridge: Cambridge University Press, 2005), 23–24.

39. Sir Philip Sidney, *Astrophil and Stella* (1591), in *Selected Poems,* ed. Catherine Bates (Harmondsworth: Penguin, 1994), 99, sonnet 1, line 12.

40. "It seemed like the secret labor of a germination, a being who was growing inside him," which produced that "exaltation" and "nausea" familiar to women. "On aurait dit le travail sourd d'une germination, un être qui naissait en lui, cette exaltation et cette nausée que les femmes connaissent." Émile Zola, *L'oeuvre* (1886), vol. 15 of *Oeuvres complètes,* ed. Maurice Le Blond (Paris: Bernouard, 1928), 234.

41. As Susan Stanford Friedman has pointed out, women's procreative as well as creative potential makes the metaphor easier to collude in when employed by women artists; with men, there is "a perpetual tension in the metaphor, one that threatens to overwhelm the movement toward resolution." Friedman, "Creativity

and the Childbirth Metaphor: Gender Difference in Literary Discourse," in *Speaking of Gender*, ed. Elaine Showalter (New York: Routledge, 1989), 78. The tension increases when the trope is employed by male "inverts." Thus, as one sexual "deviant" speaking of another, Jean Lorrain snubbed Proust by describing him as "one of those pretty little society boys who've managed to get themselves pregnant with literature." Lorrain, review of *Les hortensias bleus* (1 July 1896), quoted in George Painter, *Marcel Proust: A Biography*, 2 vols. (New York: Vintage, 1978), 1:209.

42. Hans Christian Andersen, *The Complete Fairy Tales and Stories*, trans. Erik Christian Haugaard (New York: Anchor, 1983), 29. "Tommelise," translated as "Thumbelina" or "Inchelina," was first published in 1835.

43. The image of Cupid emerging from the center of a rose was a common feature of nineteenth-century valentines. Ruth Web Lee, *A History of Valentines* (New York: Studio Publications, 1952), 146.

44. Quoted in Terry Castle, "Lab'ring Bards: Birth *Topoi* and English Poetics, 1660–1820," *Journal of English and Germanic Philology* 78 (April 1979): 201.

45. Raymond Stephanson has recently offered a corrective to Castle's too absolute account, analyzing a "hierarchy of *kinds* of poetic birthing" in the eighteenth century. He notes that over the course of the century the model of writing as a sort of head birth (based on the myth of Athena springing from Zeus's skull) gave way to one of body birth. The fantasy of head birth, which required neither mother nor body, "encouraged fictions of authorial power, autonomy, or removal from material contingencies"; at the top of the hierarchy, it was "the supreme act of male genius." The rise of physiological metaphors (of gestation, labor, etc.) reflected the conditions of the new literary marketplace and the often "difficult entry of author and his book-child into the public scrutiny of readers and critics." An "emergent literary capitalism and the professionalization of authors" necessitated the abandonment of the abstraction of the ideal model proposed by Pope. Raymond Stephanson, *The Yard of Wit: Male Creativity and Sexuality, 1650–1750* (Philadelphia: University of Pennsylvania Press, 2004), 97, 103–5, 98, 107.

46. There is an extensive literature on the androgyne imagination in the modern period. See, for example, Nancy Topping Bazin and Alma Freeman, "The Androgynous Vision," *Women's Studies* 2 (1974): 185–215; Catriona MacLeod, *Embodying Ambiguity: Androgyny and Aesthetics from Winckelmann to Keller* (Detroit: Wayne State University Press, 1998); Mario Praz, *The Romantic Agony*, trans. Angus Davidson (London: Oxford University Press, 1933), especially the section entitled "Byzantium"; Sue Ann Prince, "Figuring Sexual Difference: Paul Cézanne's Bathers (1870–1906) and Pablo Picasso's Nudes (1906–1907)" (PhD diss., University of Pennsylvania, 1997), esp. 173n93; Lisa Rado, *The Modern Androgyne Imagination: A Failed Sublime* (Charlottesville: University Press of Virginia, 2000); and Kari Weil, *Androgyny and the Denial of Difference* (Charlottesville: University Press of Virginia, 1992). As various feminist critics have pointed out, the Romantic model of creative androgyny appropriates and incorporates "feminine qualities such as sensibility into a singularly male consciousness" rather than actually representing a desire for the fusion of two equal sexes. Rado, *Modern Androgyne Imagination*, 12. I return to this subject in chapter 4.

47. Castle, "Lab'ring Bards," 205.

48. Meyer Howard Abrams, *The Mirror and the Lamp: Romantic Theory and the Critical Tradition* (New York: W. W. Norton, 1958), 169.

49. MacLeod, *Embodying Ambiguity*, 80. As Bazin and Freeman explain, "in its most basic sense, the experience of androgyny is the experience of wholeness." "Androgynous Vision," 185.

50. "Das erste Genie, das sich selbst durchdrang, fand hier den typischen Keim einer unermeßlichen Welt." Novalis, "Blüthenstaub," *Athenäum* 1 (1798): 100. Translation taken from Novalis, "Miscellaneous Remarks (Original Version of Pollen)," trans. Alexander Gelley, *New Literary History* 22 (Spring 1991): 399.

51. "Auch die Philosophie hat ihre Blüthen. Das sind die Gedanken"; "Die Kunst Bücher zu schreiben ist noch nicht erfunden. Sie ist aber auf dem Punkt erfunden zu werden. Fragmente dieser Art sind litterarische Sämereien. Es mag freilich manches taube Körnchen darunter sein: indessen, wenn nur einiges aufgeht!" (The art of writing books has not yet been invented. But it is on the point of being invented. Fragments of this sort are literary seeds. There may well be barren grain among them—Yet if only some sprout). Novalis, "Blüthenstaub," 73, 106 (trans. Gelley, 400).

52. Coleridge wrote that for the artist, "events and images, the lively and spirit-stirring machinery of the external world, are like light, and air, and moisture, to the seed of the Mind." Samuel Taylor Coleridge, *Treatise on Method*, ed. A. D. Snyder (London: Constable and Co., 1934), 7, quoted in Abrams, *Mirror and the Lamp*, 172. Nordau's earlier-cited comparison of healthy (i.e., not modern) poets to chlorophyllic plants is nostalgic for a model of the poet-plant who "dives into the soil, and, by the honest labour of its own roots, procures for itself the nutritive materials out of which it constructs its blossoms and fruit." Nordau's emphasis on labor reflects his concern with contemporary decadence and luxury. Nordau, *Degeneration*, 192.

53. "[U]ne sorte de botanique appliquée, non aux plantes, mais aux oeuvres humaines"; "les productions de l'esprit humain, comme celles de la nature vivante, ne s'expliquent que par leur milieu"; "chaque zone a sa culture et sa végétation propres." Fortunately for Taine, works of art were already "rangées par familles dans les musées et les bibliothèques, comme les plantes dans un herbier" (arranged according to families in museums and libraries, like plants in a herbarium). Hippolyte Taine, *Philosophie de l'art* (1865) (Paris: Hachette, 1893), 15, 11, 10, 16.

54. John Addington Symonds, "On Some Principles of Criticism," in *Essays Speculative and Suggestive* (London: Chapman and Hall, 1893), 66.

55. Havelock Ellis and John Addington Symonds, *Sexual Inversion* (1897) (New York: Arno Press, 1975), 132. An article by

James Weir called "Viraginity and Effemination" devotes a third of its length to natural historical precedents. "In following up the chain of evolution in animal life from its inception in primordial protoplasm to its end, as we now find it, we discover that the interlinking organisms are . . . either asexual or hermaphroditic." Weir, "Viraginity and Effemination," *Medical Record* (16 September 1893): 359. Krafft-Ebing also cites various medical studies on this theme. Richard von Krafft-Ebing, *Psychopathia Sexualis* (1886), 12th ed. 1903, ed. Brian King, trans. F. J. Rebman (Burbank, Calif.: Bloat, 1999), 284, 670n88.

56. Havelock Ellis's terms are typically amorphous. After defining "simple inversion" as homosexuality and "psychosexual hermaphroditism" as bisexuality in *Sexual Inversion,* he acknowledges the impossibility of maintaining any stable categories: "psychosexual hermaphroditism merges imperceptibly into simple inversion." Ellis and Symonds, *Sexual Inversion,* 42, 73, 115. In 1903 Otto Weininger retrospectively narrowed the scope of sexual phenomena stemming from hermaphroditic biology and inheritance to homosexuality alone: "Reference has been made, and in recent years has increasingly been made, to the relation between homo-sexuality and the presence of bisexual rudiments in the embryonic stages of animals and plants." Weininger's definition of homosexuality, however, still suggests the fluidity of *fin-de-siècle* inversion: "Homo-sexuality is merely the sexual condition of these intermediate sexual forms that stretch from one ideally [*sic*] sexual condition to the other sexual condition. In my view all actual organisms have both homo-sexuality and hetero-sexuality." Otto Weininger, *Sex and Character* (1903) (London: W. Heinemann, 1906), 48, excerpted in Lucy Bland and Laura Doan, *Sexology Uncensored: The Documents of Sexual Science* (Chicago: University of Chicago Press, 1998), 58.

57. As Jay Prosser has convincingly argued, "what sexologists sought to describe through sexual inversion was not homosexuality but differing degrees of *gender* inversion . . . while homosexuals certainly number among inverts, the category described a much larger gender-inverted condition of which homosexuality was only one aspect." Jay Prosser, "Transsexuals and Transsexologists: Inversion and the Emergence of Transsexual Subjectivity," in *Sexology in Culture: Labelling Bodies and Desires,* ed. Lucy Bland and Laura Doan (Chicago: University of Chicago Press, 1998), 117.

58. Nicole Albert, "From Myth to Pathology: Perversions of Gender-Types in Late Nineteenth-Century Literature and Clinical Medicine," *Diogenes* 52 (November 2005): 121–22.

59. Isidore Geoffroy Saint-Hilaire, son of the teratologist Étienne, argued in 1832 that hermaphrodites were simply variant organizations of individuals within the species and thus could not be considered monsters. "[C]es conditions ne sont [pas] étrangères . . . à l'espèce à laquelle appartient l'individu qui les présente . . . elles ne sont anomales que relativement à son sexe" (The features of hermaphroditism are not foreign . . . to the species to which the individual who presents them belongs . . . they are only anomalous relative to his or her sex). Isidore Geoffroy Saint-Hilaire, *Histoire générale et particulière des anomalies de l'organisation chez l'homme et les animaux,* 3 vols. (Paris: J.-B. Baillière, 1832–37), 2:32. As Julia Epstein glosses this passage, "none of the formations in themselves are *mal*formations; rather, it is the combination and association of certain characteristics that normally exist only separately." Julia Epstein, *Altered Conditions: Disease, Medicine, and Storytelling* (New York: Routledge, 1995), 100. For Geoffroy Saint-Hilaire, the laws governing anomalies, or "déviation de type spécifique" (deviation from species type) in general, "ne sont eux-mêmes que des corollaries des lois les plus générales de l'organisation" (are in themselves none other than corollaries of more general laws of organization). *Histoire générale et particulière,* 1:30, xi, quoted and translated in Epstein, *Altered Conditions,* 100, 223n52, 99, 223n51. Just as Geoffroy Saint-Hilaire argued that hermaphroditism was not pathological, Weininger cautioned "those who regard sexual inversion as pathological" to remember that all sorts of "transitional stages" exist, "reaching from the most masculine male to the most effeminate male and so on to the sexual invert, the false and true hermaphrodite; and then, on the other side, successively through the sapphist to the virago and so on until the most feminine virgin is reached." Weininger, *Sex and Character,* 48, excerpted in Bland and Doan, *Sexology Uncensored,* 57.

60. Arthur Schopenhauer, *The World as Will and Idea* (1859), trans. R. B. Haldane and J. Kemp, 3 vols. (London: Routledge and Kegan Paul, 1957), 3:350. "Nämlich der die Auswahl zur Geschlechtsbefriedigung instinktiv leitende Schönheitssinn wird irre geführt, wenn er in Hang zur Päderastie ausartet; Dem analog, wie die Schmeißfliege (*Musca vomitoria*), statt ihre Eier, ihrem Instinkt gemäß, in faulendes Fleisch zu legen, sie in die Blüthe des *Arum dracunculus* legt, verleitet durch den kadaverosen Geruch dieser Pflanze." Arthur Schopenhauer, *Die Welt als Wille und Vorstellung,* 2 vols. (Leipzig: F. A. Brockhaus, 1859), 2:618. The *Arum Dracunculus* is a plant of rather forbidding appearance: its tonguelike crimson or purple spathe is pointed, with wrinkled edges; the long, dark purple-black spadix can overshoot the spathe. The plant is not carnivorous; it uses flies to disperse its pollen.

61. "[E]nraciné comme une plante." RTPF 2:604.

62. RTPE 2:626–27; "prenait des poses avec la coquetterie qu'aurait pu avoir l'orchidée pour le bourdon providentiellement survenu," RTPF 2:604.

63. RTPE 2:628; "sifflant comme un gros bourdon," RTPF 2:606.

64. RTPE 2:627; "d'un naturel, dont la beauté allait croissant," RTPF 2:605.

65. "[F]écondation." RTPF 2:627.

66. These "accidents" should not lead us to doubt Maurice's bat identification. His gaydar is quite accurate, and he proves himself a good bat at the cricket match later in the novel.

67. E. M. Forster, *Maurice* (written 1913–14) (New York: W. W. Norton, 1987), 185, 184, 253, 187, 190, 188.

68. Oscar Wilde, *The Picture of Dorian Gray* (1891) (Harmondsworth: Penguin, 1985), 42, 47.

69. Honoré de Balzac, *Lost Illusions* (Harmondsworth: Penguin, 1971), 634–35; Balzac, *Illusions perdues* (1837–43) (Paris: Garnier, 1961), 702–3.

70. George Bauer, "Gay Incipit: Botanical Connections, Nosegays, and Bouquets," in *Articulations of Difference: Gender Studies and Writing in French*, ed. Dominique Fisher and Lawrence Schehr (Stanford: Stanford University Press, 1997), 73.

71. RTPE 2:625; "j'avais déjà tiré de la ruse apparente des fleurs une conséquence sur toute une partie inconsciente de l'oeuvre littéraire," RTPF 2:603.

72. Abrams, *Mirror and the Lamp*, 201–13.

73. Ibid., 187. As Schlegel put it, "the more divine a man or a work of a man is, the more it resembles a plant . . . the highest, most perfect mode of life would actually be nothing more than *pure vegetating*." Friedrich Schlegel, *Lucinde and the Fragments* (1799), trans. Peter Firchow (Minneapolis: University of Minnesota Press, 1971), 66; "Je göttlicher ein Mensch oder ein Werk des Menschen ist, je ähnlicher werden sie der Pflanze . . . das höchste vollendetste Leben [wäre] nichts als ein *reines Vegetieren*," Schlegel, *Kritische Friedrich-Schlegel-Ausgabe*, vol. 5, ed. Hans Eichner (Munich: F. Schöningh, 1962), 27, quoted in MacLeod, *Embodying Ambiguity*, 81.

74. "[S]orte de phénomène d'atavisme"; "Il en est de l'art comme des jardins, où l'on voit parfois jaillir d'une plate-bande des fleurs qui en ont été arrachées depuis des années. Il suffit qu'un ferment, enfoui au plus profond, remonte petit à petit à la surface et jette sa pousse à la barbe du jardinier, pour causer le miracle." The author continues, "Certains artistes, se détachant de l'époque et du milieu où ils s'épanouissent, remontent le cours des années et produisent des oeuvres qui semblent empruntées à une période reculée" (Certain artists, detaching themselves from the epoch and milieu where they are flowering, scale the course of the years and produce works which seem borrowed from a remote period). "Remonter" means to flower again as well as to go backward or forward. "Le modernisme de Frans Hals," *L'Art Moderne* (Brussels) 3 (23 September 1883): 301.

75. Julia Kristeva, *Time and Sense: Proust and the Experience of Literature*, trans. Ross Guberman (New York: Columbia University Press, 1996), 87.

76. Gilles Deleuze and Félix Guattari, *A Thousand Plateaus: Capitalism and Schizophrenia*, trans. Brian Massumi (Minneapolis: University of Minnesota Press, 1987), 10.

77. "La botanique . . . n'est autre chose jusqu'ici que la science de l'amour. Science peu séparée de la zoologie. Elles se traduisent l'une l'autre. L'amour est le terrain, à peu près indistinct, où l'animal est fleur, la fleur est animée." Jules Michelet, *La montagne* (Paris: A. Lacroix, Verboeckhoven, 1868), 249.

78. Cyril Beaumont, *Complete Book of Ballets: A Guide to the Principal Ballets of the Nineteenth and Twentieth Centuries* (Garden City, N.Y.: Garden City Publishing, 1941), 19–23.

79. Arthur Symons, "Violet," in *The Collected Works of Arthur Symons*, vol. 1, *Poems* (London: Martin Secker, 1924), 185, section 11, "Kisses," written 21 January 1893, lines 5–6, 3, 4.

80. "Tandis qu'il effeuille en fuyant brins à brins / Des roses, comme un lys entr'ouvrant ses pétales, / Sa tunique s'écarte aux rondeurs de ses reins." Jean Lorrain, *L'ombre ardente* (Paris: E. Fasquelle, 1897), 235.

81. Franco Borsi and Ezio Godoli, *Paris 1900* (New York: Rizzoli, 1977), 196.

82. Fuller "managed to be openly lesbian while evoking virtually no titillation or disapproval in her public." Rhonda Garelick, *Electric Salome: Loie Fuller's Performance of Modernism* (Princeton: Princeton University Press, 2007), 4. Significantly, she "never relied upon the motif of heterosexual seduction traditionally present in cabaret dance, especially in the popular North African and 'Oriental' style dances that dominated cabaret stages in Paris at the turn of the century." Rhonda Garelick, "Electric Salomé: Loie Fuller at the Exposition Universelle of 1900," in *Imperialism and Theatre: Essays on World Theatre, Drama, and Performance*, ed. J. Ellen Gainor (New York: Routledge, 1995), 85.

83. Isadora Duncan, *My Life* (1927) (New York: Liveright, 1955), 94, 96.

84. While innovative in many ways, Fuller's dances had roots in the Romantic ballet, a fantasy world full of flowers, butterflies, birds, and other ethereal creatures. Garelick, *Electric Salomé*, 130, 127.

85. Duncan, *My Life*, 95.

86. "L'exquis fantôme . . . effleure le sol avec la légèreté de la libellule, sautille, se pose à peine comme un oiseau, glisse, en secouant ses ailes tremblantes, à la façon d'une chauve-souris." Roger Marx, "Loïe Fuller," *Les Arts de la Vie* 3 (May 1905): 269. In a third part of the article he emphasized the "végétation fantastique et passionnelle" (fantastic and passionate vegetation) Fuller became while dancing. Marx, "Loïe Fuller," *Les Arts de la Vie* 3 (June 1905): 354.

87. Paul van Capelleveen, "Butterfly, Abstraction, or Flower? Another Look at Charles Ricketts' Design for *The Picture of Dorian Gray*," *Private Library* 1 (Winter 1998): 191.

88. Pater, *Renaissance*, 49.

89. Edmund Gosse, "J. A. S.," in *The Collected Poems of Edmund Gosse* (London: Heinemann, 1911), 321, line 6. The initials of the title are Symonds's.

90. Carl Little, *The Watercolors of John Singer Sargent* (Berkeley and Los Angeles: University of California Press, 1998), 13.

91. Quoted in Evan Charteris, *John Sargent* (New York: Charles Scribner's Sons, 1927), 4–5.

92. See Rachel Field, "John Sargent's Boyhood Sketches," *St. Nicholas* 53 (June 1926): 777.

93. Charles Loeffler to Evan Charteris, 30 January 1926, quoted in Charteris, *John Sargent*, 148. See also Martin Birnbaum on this passion. Birnbaum, *John Singer Sargent: A Conversation Piece* (New York: William E. Rudge's Sons, 1941), 5.

94. Marie-Louise Pailleron, *Le paradis perdu: Souvenirs d'enfance* (Paris: Éditions Albin Michel, 1947), 157–58.

95. Bunker to Isabella Stewart Gardner, 2 September 1888, quoted in Erica Hirshler, *Dennis Miller Bunker: American Impressionist,* exh. cat. (Boston: Museum of Fine Arts, 1994), 56.

96. William Gerdts has suggested that Bunker's academic training in Gérôme's studio, which had ill equipped him for working *en plein air,* was responsible for the ruin of those vegetable works. See Gerdts, "The Arch-Apostle of the Dab-and-Spot School: John Singer Sargent as an Impressionist," in *John Singer Sargent,* exh. cat., ed. Patricia Hills (New York: Whitney Museum of American Art and Harry N. Abrams, 1986), 123.

97. Trevor Fairbrother has done pioneering work on Sargent's sexuality and its role in the construction of his images. See in particular Fairbrother, "A Private Album: John Singer Sargent's Drawings of Nude Male Models," *Arts Magazine* (December 1981): 70–79; "Warhol Meets Sargent at Whitney," *Arts Magazine* (February 1987): 64–72; "Sargent's Genre Paintings and the Issues of Suppression and Privacy," in *American Art Around 1900: Lectures in Memory of Daniel Fraad,* ed. Doreen Bolger and Nicolai Cikovsky Jr. (Washington, D.C.: National Gallery of Art, 1990), 29–49; *John Singer Sargent* (New York: Harry N. Abrams, 1994); and *John Singer Sargent: The Sensualist* (New Haven: Yale University Press; Seattle: Seattle Art Museum, 2000).

98. Most of the works I examine are by male artists. This is in part due to the fact that the artistic discourse around female sexuality comes into its own slightly later in the twentieth century, in the works of writers like Colette, Djuna Barnes, and others. It is also in part an effect of my focus on what was considered progressive painting at the time. In 1896 the Union centrale des arts décoratifs organized an exhibition called The Arts of Woman; the catalogue cover depicts a butterfly alighting on a flower (reproduced in Debora L. Silverman, *Art Nouveau in Fin-de-Siècle France: Politics, Psychology, and Style* [Berkeley and Los Angeles: University of California Press, 1989], 205). This might be considered an example of cross-fertilized art—the artists were certainly inspired by nature in various ways—but the artists in the show were not well enough known and did not generate enough critical discourse to sustain the kinds of readings I perform here, for which I rely on evidence drawn from a broad range of documentary sources, including contemporary reviews, letters, and diaries.

99. Henry James, "John S. Sargent," *Harper's New Monthly Magazine,* October 1887, 684.

Chapter 1

1. Knight, *Flower Poetics in Nineteenth-Century France,* 63.

2. "[T]rop de pervenche, trop de lierre, trop de vigne vierge, trop de saules et trop de truffes." Rachilde [Marguerite Eymery], preface to *À mort* (Paris: Monnier, 1886), reprinted in *Rachilde–Maurice Barrès: Correspondance inédite, 1885–1914,* ed. Michael R. Finn (Brest: Centre d'Étude des Correspondances et Journaux Intimes des XIXe et XXe Siècles, 2002), 159–60.

3. Edmund Gosse, a poet and critic, was the son of the famous Victorian naturalist Philip Gosse, whose books, including *Popular Ornithology* (1849), *A Naturalist's Rambles on the Devonshire Coast* (1853), *The Aquarium: An Unveiling of Wonders of the Deep* (1854), *Evenings at the Microscope* (1859), *The Romance of Natural History* (1860), and *Actinologia Britannica: A History of the British Sea Anemones and Corals* (1860), contributed greatly to the popularization of natural history in the Victorian period. As a child, Edmund helped his father with his specimen collecting and illustrations. Edmund Gosse, *Father and Son: A Study of Two Temperaments* (1907), ed. Peter Abbs (Harmondsworth: Penguin, 1989), 124–25.

4. Otto de Joux cited the Uranians' (Karl Heinrich Ulrichs's term for inverts) "love of flowers," and Ambroise Tardieu remarked that pederasts like to hold flowers in their hands. In general this identifying trait was probably easier to spot than the characteristic "infundibuliform anus." Otto de Joux, *Die Enterbten des Liebesglückes* (Leipzig: M. Spohr, 1893), quoted in Edward Carpenter, *The Intermediate Sex* (1908) (London: Allen and Unwin, 1918), 33; "[L]a déformation infundibuliforme de l'anus," Ambroise Tardieu, *Étude médico-légale sur les attentats aux moeurs* (1857), 3rd ed. (Paris: J.-B. Baillière, 1859), 138, 142–43. François Carlier, head of the vice squad (the Service des moeurs) with the Paris police, noted in his study of prostitution that inverts ("pederasts," in his account) were drawn to trades like floristry, and that they liked to dress as women and deck themselves with "artificial flowers, chaplets, and garlands" on special occasions ("ouvriers en fleurs fines"; "c'est avec des fleurs artificielles, des couronnes et des guirlandes qu'ils se parent"). F. Carlier, *Études de pathologie sociale: Les deux prostitutions* (Paris: E. Dentu, 1887), 325. For an analysis of Carlier's study, see William Peniston, *Pederasts and Others: Urban Culture and Sexual Identity in Nineteenth-Century Paris* (New York: Harrington Park Press, 2004). In Rachilde's *Monsieur Venus,* the sexually ambiguous Jacques Silvert is a type of florist (he makes artificial flowers). He first appears overrun by blossoms: "a garland of roses spiraled around his body and over his loose-hanging smock; fat satin roses, velvet-skinned and garnet-red, passing between his legs, running up to his shoulders and wound about his neck." Rachilde, *Monsieur Venus,* trans. Liz Heron (Sawtry, Cambridgeshire: Dedalus, 1992), 11; "Autour de son torse, sur la blouse flottante, courait en spirale une guirlande de roses, des roses fort larges de satin chair velouté de grenat, qui lui passaient entre les jambes, filaient jusqu'aux épaules et venaient s'enrouler au col," Rachilde, *Monsieur Vénus* (1884) (Paris: Flammarion, 1977), 24.

5. Jules Michelet, *The Insect* (1857) (London: T. Nelson and Sons, 1875), 305; "Celles-ci vivent, et elles sentent, elles désirent," Michelet, *L'insecte* (Paris: Hachette, 1858), 321.

6. John Ruskin, *The Queen of the Air* (1869) (Kent: George Allen, 1906), 88–89.

7. "L'autre jour, disait la Jonquille, mon maître, en me montrant à un de ses amis, s'est écrié: Voyez cette jolie fleur! C'est le Désir.—Moi, répondit la Tubéreuse, je suis la Volupté.—J'aime bien mieux être le Désir.—Cela vous plaît à dire, mais tout le monde n'est pas de votre avis.—Vous ne venez qu'après moi.—Mais je vous fais

oublier" (The other day, said the Jonquil, my master, as he showed me to one of his friends, exclaimed: See this pretty flower! It is Desire.—Me, replied the Tuberose, I am Pleasure.—I am much happier being Desire.—It pleases you to say so, but not everyone is of your opinion.—You only come after me.—But with me you are forgotten). And so on. The text is by Taxile Delord. J. J. Grandville, *Les fleurs animées* (1847), 2 vols. (Paris: Garnier Frères, 1867), 2:138–39.

8. Liebig, a manufacturer of "meat extract," started producing chromolithographic trade cards in 1872. The cards usually came in sets of six and were printed in multiple European languages; they are still avidly collected today.

9. On the vernacularization of Linnaean botany and its terminology, see Amy King, *Bloom: The Botanical Vernacular in the English Novel* (Oxford: Oxford University Press, 2003).

10. Erasmus Darwin, *The Temple of Nature, or The Origin of Society* (London: J. Johnson, 1803), 63, canto 2, lines 265–66.

11. Ibid., line 263.

12. Erasmus Darwin, *The Loves of the Plants* (1789) (Oxford: Woodstock Books, 1991), 2, canto 1, lines 19–20.

13. OED, s.v. "orkhis." Vanilla comes from the Spanish *vainilla,* a diminutive of *vaina,* sheath, from the Latin *vagina*. OED, s.v. "vanilla."

14. A text from 1682 notes that function is "hinted from the *Shape* of the *Parts.* For in the *Florid Attire,* the *Blade* doth not unaptly resemble a small *Penis,* with the *Sheath* upon it, as its *Praeputium*. And in the *Seed-like Attire,* the several *Thecae,* are like so many little *Testicles.* And the *Globulets* and other small *Particles* upon the *Blade* or *Penis,* and in the *Thecae,* are as the *Vegetable Sperme*. Which, so soon as the *Penis* is exerted, or the *Testicles* come to break, falls down upon the *Seed-Case* or *Womb,* and so Touches it with a *Prolifick* Virtue." Nehemiah Grew, *The Anatomy of Plants* (London: W. Rawlins, 1682), 172. See also Londa Schiebinger, *Nature's Body: Gender in the Making of Modern Science* (Boston: Beacon Press, 1993), chapter 1, "The Private Lives of Plants," for a discussion of gendered modes of botanical classification.

15. Philogynes Clitorides, *The Natural History of the Frutex Vulvaria, or Flowering Shrub* (London: W. James, 1732), 6. On this botany of the boudoir, see Julie Peakman, *Mighty Lewd Books: The Development of Pornography in Eighteenth-Century England* (Houndmills, Basingstoke: Palgrave Macmillan, 2003), 67–78.

16. Quoted in Ronald Pearsall, *The Worm in the Bud: The World of Victorian Sexuality* (London: Weidenfeld and Nicolson, 1969), 235–36.

17. Peakman, *Mighty Lewd Books,* 72.

18. Wilfred Blunt, *The Compleat Naturalist: A Life of Linnaeus* (London: William Collins Sons, 1971), 165–66, quoted in Schiebinger, *Nature's Body,* 26.

19. The falling blossoms in Greuze's *Broken Pitcher* might symbolize deflowering or the onset of menstruation. For a brief discussion of the rose in Rococo art, see Philip Stewart, "Representations of Love in the French Eighteenth Century," *Studies in Iconography* 14 (1978): 131, 147n14. Bowers, garlands, floral crowns, and bouquets were more than the accoutrements of courtship; they could represent its stages. Wreaths could signal different phases in amatory progress, *fête galante* activities, or simply emblematize "genuine passion." Anne Betty Weinshenker, "'The Lover Crowned' in Eighteenth-Century French Art," *Studies in Eighteenth-Century Culture* 16 (1986): 290.

20. On the Ordre des Chevaliers et Nymphes de la Rose and other "sociétés d'amour," see Olivier Blanc, "Visibilité du libertinage féminin sous Louis XVI," in *Femmes et libertinage au xviiie siècle ou Les caprices de Cythère,* ed. Anne Richardot (Rennes: Presses Universitaires de Rennes, 2003), 49.

21. King, *Bloom,* passim.

22. Edmund Gosse, *On Viol and Flute* (1873) (New York: Henry Holt, 1883), 167, lines 37–42.

23. Honoré de Balzac, *The Lily of the Valley,* trans. Lucienne Hill (New York: Carroll and Graf, 1989), 183; "les grâces de son corsage fleurissaient déjà," Balzac, *Le lys dans la vallée* (1836), ed. Moise Le Yaouanc (Paris: Garnier, 1966), 243.

24. Octave Mirbeau, *Torture Garden,* trans. Michael Richardson (Sawtry, Cambridgeshire: Dedalus, 1997), 76; "ce buste gonflé, comme une capsule de fleur saoule de pollen," Mirbeau, *Le jardin des supplices* (1899) (Paris: Fasquelle, 1957), 63.

25. "[B]outon de sein près d'éclore." Théophile Gautier, "À une robe rose," in Gautier, *Émaux et camées* (1852) (Paris: Charpentier, 1872), 88, line 19.

26. Christina Bradstreet, "'Wicked with Roses': Floral Femininity and the Erotics of Scent," *Nineteenth-Century Art Worldwide* 6 (Spring 2007), http://www.19thc-artworldwide.org/spring_07/articles/brad.shtml.

27. This card was part of a series of erotic *femmes-fleurs* depicting the same woman, in slightly different poses, emerging from a variety of flowers.

28. "[G]azon," Louis de Landes, *Glossaire érotique de la langue française* (1861) (Paris: Éditions de Paris, 2004), 196; "forêt humide," Alfred Delvau, *Dictionnaire érotique moderne* (1864) (Geneva: Slatkine Reprints, 1968), 191; "green-grove," "sweet-briar," John S. Farmer, *Slang and Its Analogues Past and Present,* 7 vols. (London: John S. Farmer, 1890–1904), 3:206, 7:42.

29. Delvau, *Dictionnaire érotique moderne,* 226; de Landes, *Glossaire érotique,* 212; Farmer, *Slang and Its Analogues,* 3:119. In addition to "garden," Farmer lists "flower," "hortus," "lady-flower," and "moss-rose" (3:33, 3:361, 4:141, 4:359).

30. Joris-Karl Huysmans, *Against Nature,* trans. Margaret Mauldon (Oxford: Oxford University Press, 1998), 81; "il vit s'épanouir sous les cuisses à l'air, le farouche Nidularium," Huysmans, *À rebours,* 131.

31. Albert Barrère and Charles Leland, eds., *A Dictionary of Slang, Jargon, and Cant,* 2 vols. (London: Ballantyne Press, 1889–90), 2:302.

32. "[P]étales," "violettes." Pierre Guiraud, *Dictionnaire historique, stylistique, rhétorique, étymologique, de la littérature érotique* (Paris: Payot, 1978), 41.

33. In the same text, another is "a red hawthorn berry." Oscar Wilde et al., *Teleny, or The Reversal of the Medal* (1893), ed. John McRae (London: GMP, 1986), 44, 78.

34. Paula Bennett, "Critical Clitoridectomy: Female Sexual Imagery and Feminist Psychoanalytic Theory," *Signs* (Winter 1993): 243. See also Bennett's "'Pomegranate-Flowers': The Phantasmic Productions of Late-Nineteenth-Century Anglo-American Women Poets," in *Solitary Pleasures: The Historical, Literary, and Artistic Discourses of Autoeroticism*, ed. Paula Bennett and Vernon Rosario (New York: Routledge, 1995), 189–213.

35. Poem 1377, *The Complete Poems of Emily Dickinson*, ed. Thomas H. Johnson (Boston: Little, Brown, 1961), 592.

36. Guiraud, *Dictionnaire historique*, 42.

37. "En mangeant une banane, on dirait qu'on mange le clitoris d'un fruit." Edmond de Goncourt and Jules de Goncourt, *Journal: Mémoires de la vie littéraire*, ed. Robert Ricatte, 3 vols. (Paris: Robert Laffont, 1989), entry for 30 September 1864, 1:1103, quoted in Stéphanie Champeau, "Le végétal dans le *Journal* des Goncourt," in *Le végétal*, ed. Jean-Pierre Cléro and Alain Niderst (Rouen: Publications de l'Université de Rouen, 1999), 208.

38. Mirbeau, *Torture Garden*, 123; "durs bourgeons prêts à fleurir," *Le jardin des supplices*, 135.

39. Symons, "Violet," 185, section 11, line 11.

40. George Du Maurier, *Trilby* (1894) (Oxford: Oxford University Press, 1998), 51.

41. Mary Putnam Jacobi, *The Question of Rest for Women During Menstruation* (1877) (New York: Putnam, 1886), 167, quoted in Thomas Laqueur, *Making Sex: Body and Gender from the Greeks to Freud* (Cambridge: Harvard University Press, 1990), 224.

42. OED, s.v. "flower," 2.b.; Pierre Larousse, *Grand dictionnaire universel du dix-neuvième siècle*, 15 vols. (Paris: Administration du Grand Dictionnaire Universel, 1866–76), s.v. "fleurs."

43. "For twenty-five days in every month the camellias were white, and for five they were red." Alexandre Dumas *fils, La Dame aux Camélias*, trans. David Coward (Oxford: Oxford University Press, 2000), 9; "Pendant vingt-cinq jours du mois, les camélias étaient blancs, et pendant cinq ils étaient rouges," Dumas *fils, La dame aux camélias* (1848) (Paris: Calmann-Lévy, 1931), 13.

44. OED, s.v. "deflower," 1; "déflorer," Delvau, *Dictionnaire érotique moderne*, 132; "cueillir la fraise, la noisette, la fleur, un bouton de rose," etc., Delvau, 124; see also de Landes, *Glossaire érotique*, 99; "to pluck a rose," Farmer, *Slang and Its Analogues*, 6:53; "effeuiller" (female masturbation), de Landes, 124, Delvau, 146.

45. Farmer, *Slang and Its Analogues*, 1:245. Fruity compliments also abounded: attractive females were dubbed "peaches" and "plums." Barrère and Leland, *Dictionary of Slang*, 2:120.

46. Neville Williams, *Powder and Paint: A History of the Englishwoman's Toilet, Elizabeth I–Elizabeth II* (London: Longmans, Green, 1957), 105–6.

47. Alain Corbin, *The Foul and the Fragrant: Odor and the French Social Imagination* (Cambridge: Harvard University Press, 1986), 186.

48. Ibid., 193. A century earlier in London, fashionable coiffures even included fruits and vegetables. Ann B. Shteir, *Cultivating Women, Cultivating Science: Flora's Daughters and Botany in England, 1760–1860* (Baltimore: Johns Hopkins University Press, 1996), 43.

49. See C. Wilhelm, "Flowers and Fancies: From the Garden to the Stage, Part 1," *Magazine of Art* (November 1898): 172–79, and "Flowers and Fancies: From the Garden to the Stage, Part 2," *Magazine of Art* (January 1899): 121–28. To give one example, the ballet *Thea, ou La fée aux fleurs* opened in London in 1847 and consisted largely of dances of animated flowers. Beaumont, *Complete Book of Ballets*, 299–301.

50. Ouida [Mary Louise Ramé], *Moths* (1880) (Peterborough, Ont.: Broadview, 2005), 299.

51. She identifies Sargent, Whistler, and Chase as "a few of the better-known artists who painted portraits of women as flowers." Annette Stott, "Floral Femininity: A Pictorial Definition," *American Art* 6 (Spring 1992): 72.

52. H. Barbara Weinberg and Stephanie L. Herdrich, "John Singer Sargent in the Metropolitan Museum of Art," *Metropolitan Museum of Art Bulletin* (Spring 2000): 23.

53. Stott points out the analogous arrangement of figures and bouquets, but she identifies the flowers as peonies rather than magnolias. Stott, "Floral Femininity," 72.

54. Visiting the Academy galleries in 1902, Rodin remarked in front of the picture, "Voilà le van Dyck de l'époque! Sargent n'a jamais rien fait de mieux que cela. C'est un bouquet de fleurs!" Quoted in William Howe Downes, *John S. Sargent, His Life and Work* (Boston: Little, Brown, 1925), 58.

55. Joséphin Péladan, "L'esthétique au Salon de 1883," *L'Artiste* 53 (May 1883): 385, quoted in Barbara Dayer Gallati, "From Souvenir to High Art: Childhood on Display," in Barbara Dayer Gallati, Erica Hirshler, and Richard Ormond, *Great Expectations: John Singer Sargent Painting Children*, exh. cat. (New York: Brooklyn Museum and Bullfinch Press, 2004), 81.

56. Downes, *John S. Sargent*, 179.

57. "Society of American Artists," *New York Times*, 26 April 1891, quoted in Gallati, "From Souvenir to High Art," 115.

58. Paul Bourget, *Outre-Mer: Impressions of America* (London: T. Fisher Unwin, 1895), 108.

59. "[D]ans une prairie, au grand abaissement des tulipes tendant vers elle leur corolles curieuses. D'une saveur un peu étrange, ce dernier morceau m'a infiniment charmé: il fait penser à la poésie de Charles Baudelaire." Armand Sylvestre, "Le monde des arts," *La Vie Moderne* (29 May 1880): 340, quoted in CP 1:47n7.

60. Claude Phillips, "The Royal Academy II," *Academy* 993 (16 May 1891): 471.

61. Stott, "Floral Femininity," 66.

62. It is important to note that the picture's success rested at least in part on the critics' understanding of it as a triumph of *plein-air* painting. Rochegrosse "has come out of the *Bibliothèque Nationale* into the full blaze of sunlight and set up his canvas in a hillside meadow," wrote the reviewer for the *Times* ("The Salon of the Champs Elysées," *Times* [London], 1 May 1894, 13). In *Le Rappel* Charles Frémine called the work "le plus frais tableau" (the freshest) to be seen at the Salon, commenting, "On nous parle tout le temps de la peinture en plein air; eh bien! en voilà" (people always talk to us about painting *en plein air;* well! Here you go). Charles Frémine, "Le Salon des Champs-Élysées," *Le Rappel,* 1 May 1894, 1.

63. Stanley Appelbaum, ed., *Bizarreries and Fantasies of Grandville* (New York: Dover, 1974), xii.

64. "Die Mädchen: Komm'! Komm'! / Holder Knabe, / lass mich dir blühen! / Dir zu wonniger Labe / gilt mein minniges Mühe . . . Parsifal: Ihr wild holdes Blumengedränge, / soll ich mit euch spielen, entlasst mich der Enge!" Richard Wagner, *Parsifal,* trans. H. L. Corder and F. Corder (Philadelphia: Oliver Ditson, 1903), 28. While the translation gives Parsifal himself a vegetable identity (he is a stripling), the German original specifies his humanness (Knabe).

65. "Le char des *femmes-fleurs,* pendant un quart d'heure arrêté, sert de pissotière aux guerriers antiques, tandis que les *femmes-fleurs* mangent des œufs durs." Albert Flament, *Le bal du Pré-Catelan* (Paris: Librairie Arthème Fayard, 1946), 57, entry for 18 February 1896.

66. Robert Machray, *The Night Side of London* (London: Macqueen, 1902), 6, quoted in Kristina Huneault, "Flower-Girls and Fictions: Selling on the Streets," in Huneault, *Difficult Subjects: Working Women and Visual Culture, Britain, 1880–1914* (Aldershot: Ashgate, 2002), 80.

67. "[G]arden-whore," Barrère and Leland, *Dictionary of Slang,* 1:397; "columbine," ibid., 1:265, Farmer *Slang and Its Analogues,* 2:158; "artichoke," Barrère and Leland, 1:49, Farmer, 1:74; "hedge-whore," Farmer, 3:296.

68. "Violette" and "Camélia" were common prostitutes' pseudonyms in Paris. Alain Corbin, *Women for Hire: Prostitution and Sexuality in France After 1850,* trans. Alan Sheridan (Cambridge: Harvard University Press, 1990), 77. Unsurprisingly, a "whore-master" could be called a "flower-fancier" and a brothel a "hot-house." Farmer, *Slang and Its Analogues,* 3:33, 364.

69. Thomas Wentworth Higginson, "Water-Lilies," *Atlantic Monthly,* September 1858, 473. See Christopher Looby's analysis of Higginson's flower poetics in "Flowers of Manhood: Race, Sex, and Floriculture from Thomas Wentworth Higginson to Robert Mapplethorpe," *Criticism* 37 (Winter 1995): 111–33.

70. Stott, "Floral Femininity," 73.

71. Iota [Katherine Mannington], *A Yellow Aster* (1894) (New York: Optimus Printing, n.d.), 109.

72. See Gretchen Wilson, *With All Her Might: The Life of Gertrude Harding, Militant Suffragette* (New York: Holmes and Meier, 1998), 99–104; and Luigi Berliocchi, *The Orchid in Lore and Legend,* trans. Lenore Rosenberg and Anita Weston, ed. Mark Griffiths (Portland, Ore.: Timber Press, 2000), 11.

73. Henry James, *Watch and Ward* (1871) (New York: Grove Press, 1960), 46, 81. The image of a young girl's tender heart opening up to manly rays also appears in Louisa May Alcott's 1876 novel *Rose in Bloom.* The heroine Rose eventually marries her much older cousin Mac. He woos her slowly, until "he felt that now the last leaf had folded back, the golden heart lay open to the light, and his Rose had bloomed." Louisa May Alcott, *Rose in Bloom* (N.p.: Dodo Press, 2005), 263.

74. Denys Hinton, "John Singer Sargent R.A.: A Re-evaluation?" *Connoisseur* 157 (December 1964): 250.

75. Émile Zola, *The Kill,* trans. Alexander Teixeira de Mattos (London: Paul Elek, 1958), 267, translation slightly modified. "Il devenait fleur. Ses membres verdissaient, s'allongeaient, dans son costume collant de satin vert; la tige flexible, les jambes légèrement recourbées, allaient s'enfoncer en terre, prendre racine, pendant que le buste, orné de larges pans de satin blanc, s'épanouissait en une corolle merveilleuse. La chevelure blonde de Maxime complétait l'illusion, mettait, avec ses longues frisures, des pistils jaunes au milieu de la blancheur des pétales. Et la grande fleur naissante, humaine encore, penchait la tête vers la source, les yeux noyés, le visage souriant d'une extase voluptueuse, comme si le beau Narcisse eût enfin contenté dans la mort les désirs qu'il s'était inspirés à lui-même." Zola, *La curée* (1872), in *Les Rougon-Macquart,* ed. A. Lanoux and H. Mitterand, 5 vols. (Paris: Gallimard, 1963), 1:553.

76. "[O]lives," Guiraud, *Dictionnaire historique,* 469; "marrons," "prunes de Monsieur," and "gourdes," Delvau, *Dictionnaire érotique moderne,* 253, 320, and 212, respectively.

77. "[N]ectar," de Landes, *Glossaire érotique,* 259. Semen's milkiness is "like the sap of the fig tree or the euphorbia." Wilde et al., *Teleny,* 138.

78. Homer Bostwick, *A Treatise on the Nature and Treatment of Seminal Diseases, Impotency, and Other Kindred Affections* (New York: Burgess, Stringer, 1847), 23, quoted in Christopher Looby, "'The Roots of the Orchis, the Iuli of Chestnuts': The Odor of Male Solitude," in Bennett and Rosario, *Solitary Pleasures,* 165.

79. John Cleland, *The Memoirs of a Woman of Pleasure, Fanny Hill* (1748–49) (Ware, Hertfordshire: Wordsworth Editions, 1996), 152. A French slang equivalent of the swelling and vegetating mimosa was "l'herbe qui croît dans la main." Delvau, *Dictionnaire érotique moderne,* 255. Shelley saw the sensitive plant as autoerotically self-sustaining: "But none ever trembled and panted with bliss / In the garden, the field, or the wilderness, / Like a doe in the noontide with love's sweet want / As the companionless Sensitive Plant." Percy Bysshe Shelley, *The Sensitive Plant* (1820) (London: William Heinemann, 1911), 21, lines 9–12.

80. "[C]arotte" and "champignon," Delvau, *Dictionnaire érotique moderne,* 82 and 89, respectively; "aspergès," de Landes, *Glossaire érotique,* 20, and Delvau, 29; "radish," Farmer, *Slang and Its Analogues,* 5:289; "tige," Guiraud, *Dictionnaire historique,* 597; "branche," de Landes, 50;

"racine," de Landes, 310; "man-root," Farmer, 5:289; "épine," de Landes, 138.

81. Wilde et al., *Teleny*, 97.

82. "[A]rroser," de Landes, *Glossaire érotique*, 19; "arroser" and "arroser le bouton," Delvau, *Dictionnaire érotique moderne*, 28. Guiraud also lists a nineteenth-century source using "irrigateur" as slang for penis. *Dictionnaire historique*, 397.

83. "[P]lanter des hommes," "planter le cresson," "planter le mai," de Landes, *Glossaire érotique*, 285–86.

84. Alfred Le Petit, *Fleurs, fruits et légumes du jour* (Paris: Au Bureau de l'Éclipse, 1871).

85. Claudette Sartiliot, *Herbarium Verbarium: The Discourse of Flowers* (Lincoln: University of Nebraska Press, 1993); Stanislas Aquarone, *The Life and Works of Émile Littré, 1801–1881* (Leyden: A. W. Sythoff, 1958), 12.

86. The American doctor and author Oliver Wendell Holmes affirmed that the hothouse of city life could cultivate specimens as delectable as those found in a country garden: "these beauties that grow and ripen against the city-walls, these young fellows with cheeks like peaches and young girls with cheeks like nectarines, show that the most perfect forms of artificial life can do as much for the human product as garden-culture for strawberries and blackberries." Oliver Wendell Holmes, *Elsie Venner: A Romance of Destiny* (1861) (Dublin: Nonsuch, 2007), 283.

87. Michael Booth, *Victorian Spectacular Theatre, 1850–1910* (Boston: Routledge and Kegan Paul, 1981), 78.

88. Willsdon, *In the Gardens of Impressionism*, 179.

89. The libretto was by Robert de Flers and the music by Reynaldo Hahn, both intimate friends of Proust.

90. Jules Champfleury [Jules Fleury-Husson], *La reine des carottes*, in *Pantomimes*, ed. Isabelle Baugé (Quercy: Cicero, 1995), 67–80.

91. Scarry, "Imagining Flowers," 105.

92. Guillaume de Lorris and Jean de Meun, *The Romance of the Rose*, trans. Charles Dahlberg (Princeton: Princeton University Press, 1995), 53; "Des fueilles i a quatre paire, / que Nature par grant maistire / i ot asisses tire a tire," de Lorris and de Meun, *Le roman de la rose* (written 1225–40, 1275), ed. Félix Lecoy, 3 vols. (Paris: Honoré Champion, 1965), 1:51, lines 1660–62.

93. Andrew Marvell, "The Garden," in *The Complete Poems*, ed. Elizabeth Story Donno (Harmondsworth: Penguin, 1996), 100, lines 29–30.

94. Andrew Marvell, "Upon Appleton House," ibid., 93, lines 589–90, and 94, lines 615–16.

95. Marvell, "The Garden," 100, lines 17–18.

96. Nathaniel Hawthorne, *The Marble Faun, or The Romance of Monte Beni* (1860) (Harmondsworth: Penguin, 1990), 74. Vernon Lee and Sargent read the novel together as children and to some degree lived it; as they played, Mrs. Sargent entertained the living models for Hawthorne's characters—"Harriet Hosmer, Randolph Rogers and W. W. Story, and so forth," whom "we children in 1868–69 were perpetually reading about—in Hawthorne's 'Marble Faun.'" Vernon Lee, "J. S. S.: In Memoriam," in Charteris, *John Sargent*, 243.

97. Walt Whitman, "Faces," in *Leaves of Grass and Other Writings*, ed. Michael Moon (New York: W. W. Norton, 2002), 391, lines 62–67.

98. Xavier-Boniface Saintine, *Picciola, or The Prison Flower* (1836) (Chicago: A. C. McClurg, 1890). On Proust's childhood reading, see Jean-Yves Tadié, *Marcel Proust*, trans. Euan Cameron (Harmondsworth: Viking, 2000), 53.

99. "Je vois, dit la jeune fille, éclatant en sanglots, je vois que vous aimez tant les tulipes, qu'il n'y a plus place dans votre coeur pour une autre affection"; "Nous l'avouons à la honte de notre héros et de l'horticulture, de ses deux amours, celui que Cornélius se sentit le plus enclin à regretter, ce fut l'amour de Rosa." Alexandre Dumas *père*, *La tulipe noire* (1850) (Paris: Calmann-Lévy, 1938), 218, 219.

100. "Au sommet du vieux donjon croissait une Giroflée. Un prisonnier la voyait de sa fenêtre. C'était sa joie, sa consolation, son unique espérance. Il l'aimait comme on aime une femme." Grandville, *Les fleurs animées*, 2:119.

101. Women botanized for pleasure, edification, and sometimes professionally. See Shteir, *Cultivating Women, Cultivating Science*.

102. Richard Polwhele, *The Unsex'd Females* (New York: William Cobbett, 1800), 10–11, lines 29–34, quoted in King, *Bloom*, 52.

103. In *Proserpina* he proposed that "Ophryds" should designate the whole family of orchids. Since he could not bring himself to utter the offending word, however, several confusing pages of indirect allusion (he identifies the genera Ophrys and Satyrium as part of the family) are required to make his point. Ruskin also objected to the "recent phrenzy for the investigation of digestive and reproductive operations in plants," which "may by this time have furnished the microscopic malice of botanists with providentially disgusting reasons, or demoniacally nasty necessities, for every possible spur, spike, jag, sting, rent, blotch, flaw, freckle, filth, or venom." The "gentle and happy scholar of flowers has nothing whatever to do" with these "obscene processes and prurient apparitions." John Ruskin, *Proserpina: Studies of Wayside Flowers*, 2 vols. (Sunnyside, Orpington, Kent: George Allen, 1879, 1886), 1:199–203, 2:6.

104. Grant Allen, *Physiological Aesthetics* (1877) (London: Elibron Classics, 2005), 214.

105. Honoré de Balzac, *A Harlot High and Low (Splendeurs et misères des courtisanes)*, trans. Rayner Heppenstall (Harmondsworth: Penguin, 1970), 56; "elle coquetait avec les arbres," Balzac, *Splendeurs et misères des courtisanes* (1847), ed. A. Adam (Paris: Garnier, 1958), 57.

106. Rachilde, *Pleasure (Volupté)*, in *Madame La Mort and Other Plays*, trans. and ed. Kiki Gounaridou and Frazer Lively (Baltimore: Johns Hopkins University Press, 1998), 87, translation slightly modified. "Je vais m'étendre par terre tout contre une grosse jacinthe rose qui a poussé au bas du jardin, près d'une charmille. On est dans l'ombre comme ici. Je jette ma robe par-dessus ma tête et

j'entoure la fleur de mes bras pour que le parfum me monte tout entier dans le nez, et je respire . . . je respire. . . . Il me semble que je mange du miel pendant que des abeilles en s'envolant me frôlent les paupières de leurs ailes de sucre! (*Elle se pâme.*) Tu ne peux rien y comprendre! Mais c'est si délicieux que je t'en oublie!" Rachilde, "Volupté," *Mercure de France* 9 (September 1893): 29. Ellipses in original.

107. *Punch, or the London Charivari,* 9 April 1881, 161.

108. Edward Lear, *The Complete Verse and Other Nonsense,* ed. Vivien Noakes (Harmondsworth: Penguin, 2001), 112.

109. W. S. Gilbert and Arthur Sullivan, *Patience,* in *The Complete Annotated Gilbert and Sullivan,* ed. Ian Bradley (Oxford: Oxford University Press, 1996), 293, lines 413–14, 419–20.

110. Nordau, *Degeneration,* 62.

111. "He was tremendously interested in the speculations and evidence of 'The Golden Bough.'" G. P. Jacomb-Hood, "John Sargent," *Cornhill Magazine,* September 1925, 281. "On one occasion he reported his pleasure at having 'horrified' a guest of [his sister] Emily's by offering Frazer's interpretation of the 'primitive' origins of contemporary mourning rituals." From a letter from Sargent to Charles Martin Loeffler, 26 December 1921, box 3, Loeffler Papers, Library of Congress, summarized by Sally M. Promey, *Painting Religion in Public: John Singer Sargent's "Triumph of Religion" at the Boston Public Library* (Princeton: Princeton University Press, 1999), 29.

112. James George Frazer, *The Golden Bough: A Study in Magic and Religion* (1890), 2nd ed., 3 vols. (New York: Macmillan, 1900), 2:206. Erasmus Darwin also cited anthropological precedent in his work, likening the "Adonis" flower, which contains many male and female parts, to "the Society called the Areoi in the island of Otaheite [that] consists of about 100 males and 100 females who form one promiscuous marriage." *Loves of the Plants,* 164n.

113. "Dès qu'il s'abandonne, il flotte; il se sent alors devenir tout petit, léger, bizarre, végétal; il lui pousse des feuilles par tout le corps; et bientôt l'eau de la rivière peut coucher sur la rive le délicat rameau de chêne que notre ami Gribouille est devenu. . . . [S]ans doute la grand'mère de Nohant ne pensait guère écrire là quelque chose de débauchant; mais je témoigne que nulle page d'*Aphrodite* ne put troubler nul écolier autant que cette métamorphose de Gribouille en végétal le petit ignorant que j'étais." André Gide, *Si le grain ne meurt* (1924) (Paris: Gallimard, 1928), 62.

114. Several articles by George Bauer and Christopher Looby are notable exceptions. See Bauer, "Gay Incipit"; Looby, "Flowers of Manhood"; and Looby, "'Roots of the Orchis.'"

115. Gide, *Si le grain ne meurt,* 33, 129. For Wilde, too, flowers had always been "part of desire." Wilde, *De Profundis* (1897), in *De Profundis and Other Writings,* 207.

116. Joanna Luft, "Unfixing the Rose as a Fixture of the Female Sex in Guillaume de Lorris's and Jean de Meun's Thirteenth-Century *The Romance of the Rose,*" paper presented at the Fixed Ideas conference, Harvard University, 1 March 2002.

117. De Lorris and de Meun, *Romance of the Rose,* 53; "la tige ere droite con jons, / et par desus siet li boutons / si qu'i ne cline ne ne pent," *Roman de la rose,* 1:51–52, lines 1663–65.

118. *Romance of the Rose,* 353; "tretout soavet, san moi poindre, / le bouton pris a elloichier," "A la parfin, tant vos an di, / un po de greine i espandi, / quant j'oi le bouton elloichié," *Roman de la rose,* 3:152, lines 21678–79, 21689–91.

119. Marvell, "The Garden," 101, lines 58–59.

120. For easy reference, see Darwin's "Catalogue of the Poetic Exhibition," *Loves of the Plants,* 174–75.

121. Ibid., 149–50.

122. In the nineteenth century most popular manifestations of romantic botany, as we have seen, ignored the hermaphroditic sexuality of flowers, despite its being the very basis of Linnaean nomenclature, which reflects the number of masculine and feminine elements in the flower. Flowers' potential for self-fertilization was itself a common enough notion; after all, if one instance of blossoming romance were to emblematize the nineteenth century, it might be the one that, after Ovid, so fascinated Freud and Havelock Ellis: narcissism. Yet Edmund Gosse's eroticized account of the pistils and stamens of a single man-flower in his poem "Manes: The Heretic" reads almost as a deliberate mockery of Erasmus Darwin's descriptions:

> The pistils trembled, and the stamens flew
>
> Straight to the centre, where their god they knew,
>
> Clung quivering there, enkindled and aglow,
>
> Sank, big with blessing, on the leaves below.

The Manichaean heresy here (Manes is a variant spelling of Mani) seems to be the recognition of the dualism of the floral body and thus its capacity for autoerotic fulfillment, which is imaged as a kind of manual stimulation, a self-touching. Edmund Gosse, "Manes, the Heretic" (1894), in *Collected Poems,* 255, lines 51–54.

123. Louis Crompton, *Byron and Greek Love: Homophobia in Nineteenth-Century England* (Berkeley and Los Angeles: University of California Press, 1985), 128.

124. Byron to Matthews, 22 June 1809, in *Byron's Letters and Journals,* ed. Leslie A. Marchand, 12 vols. (Cambridge: Harvard University Press, 1973–82): 1:206–7, quoted in ibid., 127. Byron's letter supports Christopher Looby's argument that "if sexual displacement or substitution is a factor in human fascination with flowers," then "exotic flowers in particular may frequently be place-holders for certain impulses of racial xenophilia." Looby, "Flowers of Manhood," 125.

125. Quoted in Crompton, *Byron and Greek Love,* 129.

126. Neil Bartlett, *Who Was That Man? A Present for Mr. Oscar Wilde* (London: Serpent's Tail, 1988), 56.

127. Wilde, *Picture of Dorian Gray,* 46, 25.

128. George Chauncey, *Gay New York: Gender, Urban Culture, and the Making of the Gay Male World, 1890–1940* (New York: Basic Books, 1994), 15.

129. These dyed flowers were worn by Wilde and his young male friends at the opening night of *Lady Windermere's Fan* in 1892. Although Wilde claimed that the flower didn't symbolize anything, he apparently "knew" it was "the emblem worn by homosexuals in Paris." Regenia Gagnier, *Idylls of the Marketplace: Oscar Wilde and the Victorian Public* (Stanford: Stanford University Press, 1986), 164. Robert Hichens's 1894 novel *The Green Carnation* satirized the witty, homoerotic culture of Aestheticism and the romance of Wilde and Alfred Douglas (Esmé Amarinth and Reggie Hastings in the book). In 1895, the year of Wilde's trial, the book was withdrawn from circulation. Robert Hichens, *The Green Carnation* (New York: Dover, 1970). George Du Maurier criticized such floral cults in *The Martian:* the painter Barty belongs to a "little clique" of Greek-loving men who "worshipped the tuberose"; for the narrator they are not "men of my own sex" but "inspired epicenes" with "perverted passions." George Du Maurier, *The Martian* (New York: Harper and Brothers, 1897), 327, 328. Ludwig Dilsner's 1899 play *Jasminblüthe* (Jasmine Blossoms) offers a medicalized version of the invert's floral symbol: the flowers "of the title are the headmaster's simile for homosexuality as a transitional stage, their crippled stamens half turned to petals." Laurence Senelick, "The Homosexual as Villain and Victim in Fin-de-Siècle Drama," *Journal of the History of Sexuality* 4 (October 1993): 213.

130. The calamus poems are among Whitman's most homoerotic works. Knowing that his European readers might not be familiar with the plant, Whitman described calamus as "the biggest and hardiest kind of spears of grass," and as having a "fresh, acquatic, pungent bouquet." Whitman to Moncure D. Conway, 1 November 1867, in Whitman, *The Correspondence,* ed. Edwin Haviland Miller, 6 vols. (New York: New York University Press, 1961–77), 1:347, quoted by Michael Moon in Whitman, *Leaves of Grass and Other Writings,* 96–97n1. Slightly later, the American art connoisseur Arthur Lyon Raile used the wild rose as a symbol for "the comrade-love of ancient Greece." His volume of poems *The Wild Rose* (1909) was an enlarged edition of *Itamos: A Volume of Poems,* first published in 1903. Timothy d'Arch Smith, *Love in Earnest: Some Notes on the Lives and Writings of English "Uranian" Poets from 1889 to 1930* (London: Routledge and Kegan Paul, 1970), 115.

131. Herman Melville, "Billy Budd, Sailor" (written 1885–91, 1st ed. 1924), in *Billy Budd and Other Stories* (Harmondsworth: Penguin, 1986), 345.

132. RTPE 3:862. "Si jamais vous étiez curieux, un soir, de voir, je ne dis pas quarante, mais une dizaine de voleurs, vous n'avez qu'à venir ici; pour savoir si je suis là vous n'avez qu'à regarder la fenêtre de là-haut, je laisse ma petite fenêtre ouverte et éclairée, cela veut dire que je suis venu, qu'on peut entrer; c'est mon Sésame à moi. Je dis seulement Sésame. Car pour les lys, si c'est eux que vous voulez, je vous conseille d'aller les chercher ailleurs." RTPF 3:833.

133. The idea of sending messages through flowers was imported to Europe in the eighteenth century by Lady Mary Wortley Montagu and Seigneur Aubry de la Mottraye. It caught on quickly, and hundreds of floral dictionaries were published over the course of the next century. On the language of flowers and its reception in the West, see Beverly Seaton, *The Language of Flowers: A History* (Charlottesville: University Press of Virginia, 1995), chapter 4; and Jack Goody, *The Culture of Flowers* (Cambridge: Cambridge University Press, 1993), chapter 8.

134. "[I]nventé dans le loisir de leur vie solitaire"; "comme d'un chiffre pour des déclarations lesbiennes." Joseph Hammer-Purgstall, "Sur le langage des fleurs," *Fundgruben des Orients* 1 (1809): 34.

135. Baudelaire "wanted at first to call the whole volume not *Les Fleurs du mal* but *Les Lesbiennes*"; these were interchangeable terms in his aesthetic and moral conception of things. Baudelaire in fact advertised his forthcoming volume as *Les lesbiennes* between 1845 and 1847. Marcel Proust, "Concerning Baudelaire" (1921), in *Against Sainte-Beuve and Other Essays,* trans. John Sturrock (Harmondsworth: Penguin, 1994), 301; see also Claude Pichois's introduction to *Les fleurs du mal* (1861) (Paris: Gallimard, 1996), 14. The term "fleur du mal" was quickly vernacularized; see Delvau, *Dictionnaire érotique moderne,* 189.

136. Oscar Wilde to Lord Alfred Douglas, January 1893 (?), in *The Complete Letters of Oscar Wilde,* ed. Merlin Holland and Rupert Hart-Davis (London: Fourth Estate, 2000), 544. In Joseph Méry's 1859 novel *Monsieur Auguste,* the dandy classicist of the title, who is in love with the painter Octave, is "an animal that Buffon hasn't classified." Early in the text, M. Auguste has the *blood* of a water lily in his veins, later, the juice; he becomes more plantlike as the novel progresses. "[U]n animal que Buffon n'a pas classé"; "sang de nénufar"; "[il] a dans les veines du jus de nénufar." Joseph Méry, *Monsieur Auguste* (Paris: Librairie Nouvelle, 1859), 51, 28, 193.

137. Alexander Michaelson [Marc-André Raffalovich], "Verglas and Pollen," *New Blackfriars* 9 (May 1928): 281.

138. "Laisse errer mes doigts dans la mousse / Où le bouton de rose brille, / Laisse-moi, parmi l'herbe claire, / Boire les gouttes de rosée / Dont la fleur tendre est arrosée." Paul Verlaine, "Printemps," in *One Hundred and One Poems by Paul Verlaine,* trans. Norman R. Shapiro (Chicago: University of Chicago Press, 1999), 150–51. Translation modified.

139. Marquis Donatien Alphonse François de Sade, *The 120 Days of Sodom and Other Writings,* trans. Austryn Wainhouse and Richard Seaver (New York: Grove Press, 1987), 214. "Entrouvrait-on ce cul délicieux, un bouton de rose s'offrait alors à vous, et c'était dans toute sa fraîcheur et dans l'incarnat le plus tendre que la nature voulait vous le présenter." Sade, *Les 120 journées de Sodome,* 2 vols. (Paris: Union Générale d'Éditions, 1975), 1:48. This image appears frequently in Sade's writings. In *Justine* we learn that some content themselves "with the rose, others more fastidious or more depraved (for the question has not yet been decided) wish to bring to full flower the bud that grows adjacently." Sade, *Justine, Philosophy in the Bedroom, and Other Writings,* trans. Richard Seaver and Austryn Wainhouse (New York: Grove Press, 1990), 464; "les uns se contentent de la rose, d'autres plus

délicats ou plus dépravés (car la question n'est pas résolue) veulent épanouir le bouton qui fleurit à côté," Sade, *Justine, ou Les malheurs de la virtu* (1791) (Paris: Jean-Jacques Pauvert, 1966), 17.

140. "À *Domo d'Ossola* les lieux . . . présentaient un aspect enchanteur et féérique; ils étaient peints à fresque et représentaient des bouquets de roses qui s'épanouissaient comme des trous du cul de blondes, avec une touche de pourpre au milieu. Il est fort agréable de s'accroupir, ayant l'oeil sur ces anus fleuris, ou sur ces fleurs anales, dépliant leur pétales: les fronçures d'un sphincter, prêt à boire une pine, ou à vomir un étron." Théophile Gautier, *Lettre à la présidente: Voyage en Italie, 1850* (Naples: Imprimerie du Musée Secret du Roi de Naples, 1890), 13.

141. "[R]osette," Delvau, *Dictionnaire érotique moderne*, 333; a sodomite could also be termed a "chevalier de la rosette," ibid., 94, 334.

142. Wilde et al., *Teleny*, 65. See also Guiraud, *Dictionnaire historique*, 333.

143. "Obscur et froncé comme un oeillet violet / Il respire, humblement tapi parmi la mousse / Humide encor d'amour." "L'idole: Sonnet du trou du cul," from *Album zutique*, in Arthur Rimbaud, *Oeuvre-vie*, ed. Alain Borer and Andrée Montègre (Paris: Arléa, 1991), 264, lines 1–3.

144. Joris-Karl Huysmans, *Becalmed*, trans. Terry Hale (London: Atlas, 1992), 128, translation modified. "[P]lantée de potirons. Tous palpitaient, se soulevaient enfiévrés, tiraient sur les tiges qui les attachaient au sol. Jacques eut l'immédiate perception qu'il voyait un champ de fesses mongoles, un potager de derrières appartenant à la race jaune. Il examinait les rainures profondes, bien arquées, qui s'enfonçaient dans ces sphères aux épidermes rebondis, d'un orange vif. Puis, une curiosité infâme lui vint." Huysmans, *En rade* (1887) (Paris: Plon-Nourrit, 1928), 201–2.

145. The word "'infâme' had been synonymous with 'sodomite' in the French language since the Reformation." Senelick, "Homosexual as Villain and Victim," 227.

146. Frederick Rolfe to "C.," 28 November 1909, quoted in Chris White, *Nineteenth-Century Writings on Homosexuality: A Sourcebook* (New York: Routledge, 1999), 338.

147. Wilde et al., *Teleny*, 119.

148. For more on Ricketts's book designs, see Maureen Watry, *The Vale Press: Charles Ricketts, a Publisher in Earnest* (New Castle, Del.: Oak Knoll Press; London: British Library, 2004).

149. John Addington Symonds, *New and Old: A Volume of Verse* (London: Smith, Elder, 1880), 232, line 38. In addition to collaborating with Havelock Ellis on *Sexual Inversion*, Symonds published *A Problem in Greek Ethics* in 1883 and *A Problem in Modern Ethics* in 1891, translated Michelangelo's sonnets, and wrote studies of the lives and work of famous inverts like Michelangelo, Cellini, and Whitman.

150. Frank Harris, *Oscar Wilde, His Life and Confessions*, 2 vols. (N.p.: privately printed, 1916), 1:190.

151. "[D]ie goldenen Früchte des Lebens"; "straf ihn von hinten / Mit dem Pfahle, der dir rot von den Hüften entspringt." Johann Wolfgang von Goethe, "Prologue," *Roman Elegies* (1795), in *Erotic Poems*, trans. David Luke (Oxford: Oxford University Press, 1997), 2–3, lines 3, 9–10. Translation modified.

152. "[F]alx," "palus," "pessulus," "raster," "vomer." Richard Burton and L. C. Smithers, *Priapeia, sive, Diversorum poetarum in Priapum lusus, or Sportive epigrams on Priapus* (1890) (Ware, Hertfordshire: Wordsworth Editions, 1995). All sorts of vegetable terms for the genitals are also listed.

153. RTPE 2:645; "à un homme, comme le volubilis jette ses vrilles là où se trouve une pioche ou un râteau," RTPF 2:621.

154. Graham Robb, *Strangers: Homosexual Love in the Nineteenth Century* (New York: W. W. Norton, 2004), 151; Bauer, "Gay Incipit," 72–78.

155. Bauer, "Gay Incipit," 77.

156. In French, as Bauer points out, the word for eyelet is *oeillet*, which also means carnation. The conflation of floral opening and fleshly orifice finds its apotheosis in Rimbaud and Verlaine's purple carnation/asshole. Ibid., 75.

157. Wilde et al., *Teleny*, 28–29, 39. Teleny gives des Grieux his buttonhole "that its smell may remind you of me to-night, and perhaps make you dream of me." While I focus on visual floral signifiers, the importance of scent to *fin-de-siècle* queer identity and community has been explored by Looby and others. In his analysis of Raffalovich's floral tropes, Ed Madden also argues that "floral scent and memory (cultural or personal) are imbricated in the production of homosexual identity." See Madden, "Say It with Flowers: The Poetry of Marc-André Raffalovich," *College Literature* 24 (February 1997): 16.

158. The portrait is reproduced in color in David Colvin, *Aubrey Beardsley: A Slave to Beauty* (London: Orion, 1998), 66.

159. William Hamilton Gibson, *Blossom Hosts and Insect Guests: How the Heath Family, The Bluets, The Figworts, The Orchids, and Similar Wild Flowers Welcome The Bee, The Fly, The Wasp, The Moth, and Other Faithful Insects* (1880), ed. Eleanor E. Davie (New York: Newson, 1901), xix. The word "contrivances," as Robert Fraser glosses it, suggests "difficulty, guile, a stealthy outdoing of numerical odds . . . cunning, complexity, the perplexed and perplexing confounding of the obvious in favour of the various, the splendiferous, the efficiently motley and odd." Fraser, *Proust and the Victorians: The Lamp of Memory* (New York: St. Martin's Press, 1994), 204.

160. On unwelcome visitors, see Anton Joseph Kerner von Marilaun, *Flowers and Their Unbidden Guests* (1876), trans. W. Ogle (London: C. Kegan Paul, 1878). Our subsequent lapse in awareness of and interest in pollinators is the subject of Stephen Buchmann and Gary Nabham's *The Forgotten Pollinators* (Washington D.C.: Island Press; Covelo, Calif.: Shearwater Books, 1996). I thank Robin Kelsey for this reference.

161. Philip Henry Gosse, *Evenings at the Microscope, or Researches Among the Minuter Organs and Forms of Animal Life* (1859) (New York: D. Appleton, 1889), 177.

162. The second series includes images of butterfly performers at the Théatre italien, Théatre français, and Ballet champêtre. Charles-Germain de Saint-Aubin (1721–

1786) was the brother of Gabriel (1725–1780), creator of *La comparaison du bouton de rose,* described above (fig. 18).

163. RTPE 2:536; "mettre des cartes dans la maison," RTPF 2:516. Proust's insistence on *reading* pollen was perhaps inspired by the translation he read of Darwin's *The Different Forms of Flowers on Plants of the Same Species;* in the preface by Amedée Coutance, bees gather "messages" from the stamens they brush against. Charles Darwin, *Des différantes formes de fleurs dans les plantes de la même espèce,* trans. Edouard Heckel, preface by Amedée Coutance (Paris: Reinwald, 1878), quoted and translated in Fraser, *Proust and the Victorians,* 206.

164. "N'approche pas, répondit la Rose dédaigneuse. . . . Tu es présomptueux et tu sais flatter; tu es un trompeur, n'approche pas!" P.-J. Stahl, *Scènes de la vie privée et publique des animaux,* 2 vols. (Paris: J. Hetzel et Paulin, 1842), 1:120.

165. George Du Maurier, "Entomological Studies," *Punch, or the London Charivari,* 19 March 1870, 107.

166. The poem appeared in the 1856 edition of *Leaves of Grass* under this title; in 1867 Whitman changed it to "Spontaneous Me." *Leaves of Grass and Other Writings,* 89, line 17.

167. Comte de Caylus, "Rosanella," in *The Green Fairy Book,* ed. Andrew Lang (London: Longmans, Green, 1892), 48–55. *The Green Fairy Book* was part of a wildly successful series of books of fairy tales from around the world, edited by Lang, each with a different color in the title, that ran to twelve volumes between 1889 and 1910. "Rosanie" first appeared in Caylus's *Féeries nouvelles* (Paris: La Haye, 1741).

168. RTPE 1:255, 254, translation modified; "faire catleya," "il y a un peu . . . je pense que c'est du pollen qui s'est répandu sur vous; vous permettez que je l'essuie avec ma main? Je ne vais pas trop fort, je ne suis pas trop brutal?" RTPF 1:234, 232.

169. James Fenimore Cooper, *The Oak Openings, or The Bee-Hunter* (1848) (New York: Co-operative Publication Society, n.d.). I thank Michael Ziser for bringing this book to my attention.

170. Alcott, *Rose in Bloom,* 5.

171. Ouida, *Moths,* 533, 539.

172. Marquis de Sade, *Justine,* 488; "dès que l'abeille en a pompé le suc, le calice de la rose se referme; on n'imaginerait pas qu'il ait jamais pu s'entrouvrir," *Justine,* 52.

173. "[O]iseau," de Landes, *Glossaire érotique,* 241; "bird," Farmer, *Slang and Its Analogues,* 5:288; "antenne" and "aiguillon," de Landes, 16 and 12, Delvau, *Dictionnaire érotique moderne,* 12; "beak," Farmer, 5:288. Fragonard, as Mary Sheriff has shown, employed the long-established trope of bird as penis in numerous canvases. Mary D. Sheriff, *Fragonard: Art and Eroticism* (Chicago: University of Chicago Press, 1990), 109.

174. "[P]apillon de l'amour," Delvau, *Dictionnaire érotique moderne,* 286; "papillonner," Guiraud, *Dictionnaire historique,* 476.

175. Beaumont, *Complete Book of Ballets,* 83.

176. Quoted in Lionel Lambourne, "Fairies and the Stage," in Jeremy Maas et al., *Victorian Fairy Painting,* exh. cat. (London: Merrell Holberton, 1997), 49.

177. *La source* was first performed in Paris in 1866 and featured dancing butterflies, insects, and animate flowers. Beaumont, *Complete Book of Ballets,* 354–55. The ballet may be familiar to some readers through Degas's painting *Mme. Eugénie Fiocre in the Ballet "La source"* (1867–68), which is now in the Brooklyn Museum.

178. Ibid., 291.

179. Henry James, *The Sacred Fount* (1901) (New York: New Directions, 1983), 54.

180. Symons, "Bianca," part 5 ("Escalade"), in *Collected Works of Arthur Symons,* 1:259, lines 1–2.

181. Cleland, *Fanny Hill,* 153.

182. Barrère and Leland, *Dictionary of Slang,* 2:1.

183. Carle Armory, *Le monsieur aux chrysanthèmes: Pièce en trois actes* (Paris: Librairie Molière, 1908), quoted in Senelick, "Homosexual as Villain and Victim," 223.

184. "Belles-Dames-Nacrées qui vous accueillent avec une aimable aisance"; "il y a des Guêpes à fines tailles qui sont d'un commerce dangereux." Stahl, *Scènes de la vie privée et publique des animaux,* 2:140, 117.

185. See Michelle Tolini, "'Beetle Abominations' and Birds on Bonnets: Zoological Fantasy in Late-Nineteenth-Century Dress," *Nineteenth-Century Art Worldwide* 1 (Spring 2002), http://www.19thc-artworldwide.org/spring_02/articles/toli.shtml.

186. Langtry, *Days I Knew,* 138.

187. Ouida, *Moths,* 48.

188. Mary (Mrs. H. R.) Haweis, *The Art of Beauty* (London: Chatto and Windus, 1878), 98.

189. *Punch, or the London Charivari,* 14 May 1892, 231. Other cartoons depicting ladies as pollinators include "The Butterfly Was A 'Lady,'" *Punch,* 17 June 1871, 245, and "Suffrage for Both Sexes," *Punch,* 2 April 1870, 128.

190. Charles Merrill Mount, "Carolus-Duran and the Development of Sargent," *Art Quarterly* 26 (Winter 1963): 390.

191. Delmar Harmood Banner, "John Singer Sargent," *Nineteenth Century and After,* February 1926, 301.

192. CP 1:188.

193. Sargent's sketches of Peter Harrison reveal a "fetish for the knee." Richard Ormond, "In the Alps," in Warren Adelson et al., *Sargent Abroad: Figures and Landscapes* (New York: Abbeville Press, 1997), 83.

194. Gillian Beer, *Darwin's Plots: Evolutionary Narrative in Darwin, George Eliot, and Nineteenth-Century Fiction* (Cambridge: Cambridge University Press, 2000), 37.

195. Quoted in Michaelson, "Verglas and Pollen," 281.

196. Jacques-Émile Blanche, *Portraits of a Lifetime: The Late Victorian Era, the Edwardian Pageant, 1870–1914,* trans. and ed. Walter Clement (New York: Coward-McCann, 1938), 37.

197. "We probably, he and I, were attached to flowers and to flowering shrubs in the same earnest and affectionate spirit. . . .

Slender carnations, round balls of acacia, irregular ranunculus, auroral azaleas, pale hydrangeas, saxifrages called by him ghost flowers, king-cups, yellow marguerites, marigolds, red berries, we cherished with peculiar and parallel intensity." Michaelson, "Verglas and Pollen," 280–81.

198. Stéphane Mallarmé, "The Impressionists and Edouard Manet," first published in the *Art Monthly Review and Photographic Portfolio* 1 (30 September 1876): 117–22, reprinted in *Documents Stéphane Mallarmé,* ed. Carl Barbier, 7 vols. (Paris: Nizet, 1968), 1:74. Redon concurred; he described his experience of Hals portraits in Harlem: "Il y a autour de moi des figures épanouies d'une fraicheur de carnation fine qui donnent désir de peindre" (There are around me figures blooming with a freshness of delicate carnation that makes one want to paint). Redon, *À soi-même,* 78.

199. *La métamorphose d'Ovide figurée* (Lyon, 1583); *Catalogue of the Library of the Late John Singer Sargent, R.A., D.C.L., L.L.D* (London: Messrs. Christie, Manson and Wood, 29 July 1925), 11, cat. no. 85. Sargent painted two watercolors of Daphne's transformation, based on a statue in the garden of the Villa Garzoni at Collodi. See Bruce Robertson, ed., *Sargent and Italy,* exh. cat. (Princeton: Princeton University Press and Los Angeles County Museum of Art, 2003), 112.

200. See Jane Dini, "Public Bodies: Form and Identity in the Work of John Singer Sargent" (PhD diss., University of California, Santa Barbara, 1998), 77. Michael Booth discusses the costumes, scenery, and technology of these plays, pantomimes, etc., in *Victorian Spectacular Theatre,* 36–39, 78–84.

201. Charteris, *John Sargent,* 78.

202. Arthur Symons, "Whistler," in Symons, *Studies in Seven Arts* (New York: E. P. Dutton, 1907), 146, 143.

203. Dennis Miller Bunker to Joseph Evans, 18 June 1888, roll 1201, frames 145–49, Dennis Miller Bunker Papers, Archives of American Art, Smithsonian Institution, quoted in RGA, 554.

204. Laurie Schneider, "Butterfly or Scorpion: A Note on the Iconography of Whistler's Signatures," *Source: Notes in the History of Art* 2 (Winter 1983): 26.

205. "[S]on air de pédéraste fantastique et macabre." Edmond and Jules de Goncourt, *Journal,* entry for 23 April 1881, 2:892.

206. Whistler's pictorial insignia were, like Albert Moore's, inspired by Simeon Solomon's anthemion-topped monogram; Solomon was arrested in 1873 for attempted sodomy.

Chapter 2

1. There are many others: dead men's fingers and red-handed orchis, for example. Foxglove has a number of dactylic synonyms, including digitalis, bloody fingers, fairy thimble, and witches', fairy's, folk's, and Virgin's gloves.

2. OED, s.v. "phalanx," 4.a., 5; "unguis," 2, 4; "carpo-1" and "carpo-2."

3. Émile Zola, *The Abbé Mouret's Sin,* trans. Alec Brown (London: Elek Books, 1957), 160. "Il la crut d'abord cachée derrière un grenadier; mais c'étaient deux fleurs en bouton qu'il avait pris pour les deux noeuds roses de ses poignets." Zola, *La faute de l'abbé Mouret* (1875), in *Les Rougon-Macquart,* 1:1367.

4. Andrew Wynter, "A Garden Scene," in Wynter, *Subtle Brains and Lissom Fingers, Being Some of the Chisel-Marks of Our Industrial and Scientific Progress* (London: Robert Hardwicke, 1863), 421, lines 13–14.

5. Francis William Bain, *A Digit of the Moon: A Hindoo Love Story* (1899), 5th ed. (London: James Parker, 1906), 23. Sargent owned at least six of Bain's books, including a first edition of *A Digit of the Moon.* He found it a "curious and charming book" and sent it to a friend so they could discuss it. *Catalogue of the Library of Sargent,* 16, cat. nos. 136, 137; Sargent to Daniel Curtis, 20 October 1902, box 1, folder 4, John Singer Sargent–Thomas A. Fox Papers, Boston Athenaeum.

6. "[M]es mains, les lys de mon âme." Maurice Maeterlinck, *Hothouses: Poems, 1889,* trans. Richard Howard (Princeton: Princeton University Press, 2003), 34, line 3.

7. *Modern Etiquette in Public and Private* (London: Frederick Warne, 1887), 177. The text indicates some of the hand's potential revelations: "Selfishness shows itself directly in the hands in the prominent thumb and grasping fingers. . . . Long fingers are a sign of refinement. A short stubby hand argues a lack of sensibility. A small thin thumb denotes weakness of character; the thumb should stand up boldly in front of the fingers, like a general commanding an army. . . . The self-indulgent hand is to be seen both in man and in woman—who does not know its characteristic points, its brilliant purity of colour, untouched by any sign of labour, its perfect shape, only spoilt by the fingers being a shade too short?" Short and stubby or boldly erect, commanding or self-indulgent—as such suggestive descriptions make clear, shape and size revealed character. Similar lists or catalogues of hand types and manual errors from the period abound.

8. Arthur Conan Doyle, "A Case of Identity" (1892), in *The Adventures of Sherlock Holmes,* ed. Richard Lancelyn Green (Oxford: Oxford University Press, 1994), 40.

9. Robert Louis Stevenson, *Strange Case of Dr. Jekyll and Mr. Hyde* (London: Longmans, Green, 1886), 121, 52.

10. Giovanni Morelli [Ivan Lermolieff], *Italian Painters: Critical Studies of Their Works,* trans. Constance Jocelyn Ffoulkes, 2 vols. (London: John Murray, 1892–93), 1:76–77.

11. Morelli was the most renowned proponent of this method, but its principles can be traced back to the seventeenth century. See Carlo Ginzburg, "Clues: Roots of an Evidential Paradigm," in Ginzburg, *Clues, Myths, and the Historical Method,* trans. John Tedeschi and Anne C. Tedeschi (Baltimore: Johns Hopkins University Press, 1989), esp. 108–18.

12. Morelli, *Italian Painters,* 1:37, 47, 80, 81.

13. Sigmund Freud, "The Moses of Michelangelo" (1914), in *Art and Literature,* PFL 14, trans. James Strachey, ed. Albert Dickson (Harmondsworth: Penguin, 1985), 265.

14. RTPE 1:980. "[C]elles d'Andrée, maigres et bien plus fines, avaient comme une vie particulière, docile au commandement de la jeune fille, mais indépendante, et elles s'allongeaient souvent devant elle comme de nobles lévriers, avec des paresses, de longs rêves, de brusques étirements d'une phalange, à cause desquels Elstir avait fait plusieurs études de ces mains. Et dans l'une où on voyait Andrée les chauffer devant le feu, elles avaient sous l'éclairage la diaphanéité dorée de deux feuilles d'automne." RTPF 1:919.

15. For an illustration, see Carter Ratcliff, *John Singer Sargent* (New York: Artabras, 1982), 122, no. 172. Arthur Symons, who also visited the Exposition, made the same dancers' "spectral hands that thrill" the subject of a poem. Symons, "Javanese Dancers" (written October–December 1889), *Collected Works of Arthur Symons,* 1:125, line 16.

16. Sargent owned a copy of Muybridge's *Animal Locomotion, an Electro-Photographic Investigation of Consecutive Phases of Animal Movements, 1872–1885* (Philadelphia: J. B. Lippincott, 1887) and a selection of plates. See *Catalogue of the Library of Sargent,* cat. no. 217. For more on the dancers and the fair, see D. Dodge Thompson, "John Singer Sargent's Javanese Dancers," *Antiques* 138 (July 1990): 124–33.

17. John Donne, "Elegie VIII: The Comparison" (1633), in *The Complete English Poems,* ed. C. A. Patrides (London: J. M. Dent, 1994), 93, lines 27–28, 33–34. Sargent once painted out the hands in a portrait of a young female relative because they had turned out "like radishes." Cara Burch to David McKibbin, 21 February 1958, quoted in CP 1:215.

18. RTPF 1:919.

19. If they do not offer to do so of their own accord, hands can be made to indicate social position: "hard, unremitting labour" effects the change of the already "gaunt, horny, clay-coloured, begrimed workman's hand," for example, "into a kind of hoof." Wilde et al., *Teleny,* 33.

20. Ruskin, *Queen of the Air,* 114.

21. In European folklore the onion reputedly excited "sensuality" and was considered an aliment "symbolic of generation." Angelo de Gubernatis, *La mythologie des plantes* (1878–82), 2 vols. (New York: Arno Press, 1978), 2:256.

22. Edwin Blashfield, "John Singer Sargent—Recollections," *North American Review* (June–August 1925): 652.

23. Margaretta Lovell, *Venice: The American View, 1860–1920,* exh. cat. (San Francisco: Fine Arts Museums of San Francisco, 1984), 106.

24. Sargent to Nancy Astor, 5 February 1913, quoted in Stanley Olson, *John Singer Sargent, His Portrait* (New York: St. Martin's Press, 1986), 228.

25. Symons, "Whistler," 124.

26. Langtry, *Days I Knew,* 63.

27. Helen Rochefort, "A Portrait of the Greatest Portrait Painter," *Boston Evening Transcript,* 16 September 1916, 4. "One is not long in Sargent's company without noticing [them]."

28. "[L]ongues mains adroites et fines." Pailleron, *Paradis perdu,* 157.

29. Downes, *John S. Sargent,* 7. Curiously, in *The Art of Beauty* Mary Haweis described the practice of substituting more elegant hands for those of the portrait's sitter as akin to stripping "a flower of its natural leaves." She considered badly painted hands, moreover, like artificial leaves. Haweis, *Art of Beauty,* 168, 169.

30. Mary-Kate O'Hare, "Fashioning Manhood: John Singer Sargent and the Construction of Masculinity at the Turn of the Century" (PhD diss., Rutgers University, 2004), 32; Robert de Montesquiou, "Le pavé rouge: Quelques réflexions sur 'l'oeuvre' de M. Sargent," *Les Arts de la Vie* 3 (June 1905): 348.

31. David Fraser Jenkins, "Pre-Raphaelite and Impressionist: Sargent's *Carnation, Lily, Lily, Rose,*" in *British Art in Focus: Patrons' Papers I* (London: Tate, 1998), 2, 3.

32. "*Mrs. Vickers,* the first major portrait Sargent created after the *Madame X* scandal, serves in his oeuvre as the counterpoint, similar in scale and palette, utterly different in purpose. The Gautreau portrait shocked; the Vickers portrait reassured." David Park Curry, "Portrait of Mrs. Albert Vickers," *American Art Review* 5 (Winter 1993): 166.

33. Theodore Child, "The Paris Salon," *Art Amateur* 13 (July 1885): 24, quoted in Marc Simpson, Richard Ormond, and H. Barbara Weinberg, *Uncanny Spectacle: The Public Career of the Young John Singer Sargent,* exh. cat. (New Haven: Yale University Press, 1997), 144.

34. "Il n'est pas orthodoxe." Paul Mantz, "Le Salon: V," *Le Temps,* 7 June 1885, 1, quoted in Simpson, Ormond, and Weinberg, *Uncanny Spectacle,* 144.

35. "[T]out le caractère irréel et mystique des héroïnes de son compatriote Edgard Poë." Charles Ponsonailhe, "Salon de 1885: La peinture (Suite) III," *L'Artiste* 55 (June 1885): 452.

36. Edgar Allan Poe, "Ligeia" (1838), in *The Complete Tales and Poems of Edgar Allan Poe* (New York: Vintage, 1975), 655.

37. Ibid., 654.

38. "[F]antômateuse." Ponsonailhe, "Salon de 1885," 452. The sitter's skin suggests a position somewhere between death and life, in places a "leaden hue of total absence of blood," in others a "high colour . . . suggestive of dyspepsia." Montezuma [Montague Marks], "My Note Book," *Art Amateur* 15 (July 1886): 24, quoted in Andrew Stephenson, "'A Keen Sight for the Sign of the Races': John Singer Sargent, Whiteness, and the Fashioning of Anglo-Performativity," *Visual Culture in Britain* 6 (Winter 2005): 219.

39. Sargent's English friend Edmund Gosse described the magnolia flower as both "vast" and oriented toward night and the unknown when it rends "its roseate husk" open "to the dusk." Gosse, "The Whitethroat," in *On Viol and Flute,* 9, lines 41–42.

40. Sabine Baring-Gould's story "A Dead Finger" describes such an uncannily animate appendage: a vampiric finger, after feeding, becomes a hand that pulls a half-materialized arm after it "just as a silkworm might pull after it the tangle of its web." Sabine Baring-Gould, "A Dead Finger," in *A Book of Ghosts* (1904) (Penyffordd, Chester: Ash-Tree Press, 1996), 217.

41. Poe, "Ligeia," 657, 656.

42. While "jette sa pousse" (sprays or propels its shoot) is used by the author of "Le modernisme de Frans Hals" to refer to plant life (301), Lacan uses the word "pousse" to describe the erotic, ejaculatory quality of the "seer's 'shoot.'" Jacques Lacan, *The Four Fundamental Concepts of Psycho-Analysis*, ed. Jacques-Alain Miller, trans. Alan Sheridan (New York: W. W. Norton, 1981), 72.

43. Poe's "Ligeias" are strangely "insexuelles" and have "poitrines garçonnières." Huysmans, *À rebours*, 254.

44. Théophile Gautier, *Mademoiselle de Maupin* (New York: Pierre Louÿs Society, 1927), 205, 197, 384. "[C]es belles mains sortant de leurs manchettes comme les pistils d'une fleur du milieu de leurs pétales,—il avait l'air non du plus beau des hommes, mais de la plus belle des femmes"; "comme l'oeil de ses ongles se dessine nettement! Quel poli et quel éclat! on dirait des feuilles intérieures d'une rose"; "Je suis d'un troisième sexe à part." Gautier, *Mademoiselle de Maupin* (1836) (Fribourg: Éditions du Lac, 1973), 225, 217, 420.

45. William Cohen, "Manual Conduct in *Great Expectations*," *ELH* 60 (Spring 1993): 221–22.

46. "Fine Arts: Third Annual Exhibition of the Society of American Artists," *New York Herald*, 22 March 1880, 8, quoted in Lacey Taylor Jordan, "John Singer Sargent's Images of Artists in an International Context" (PhD diss., Emory University, 1999), 62.

47. Proust, Montesquiou, and Sargent all competed for Delafosse's favors. Tadié, *Marcel Proust*, 181n.

48. "The Sargent Exhibition at Copley Hall, Boston," *Artist* 24 (March 1899): xlvi, quoted in CP 2:106.

49. Bourget, *Outre-Mer*, 108.

50. Henry Adams to Mabel Hooper La Farge, 8 July 1894, in Harold Cater, ed., *Henry Adams and His Friends: A Collection of His Unpublished Letters* (Boston: Houghton Mifflin, 1947), 325.

51. The same word designates glans and acorn in French (*gland*) and German (*Eichel*).

52. Clementina Anstruther-Thomson to Vernon Lee, n.d., Vernon Lee Papers, Special Collections, Miller Library, Colby College, Waterville, Maine, quoted in CP 2:146.

53. Charteris, *John Sargent*, 157.

54. Archibald's wife, Isabelle, née Parrott, was a pollinator. In a companion portrait by Sargent, her clawlike hands and ostrich-feather frippery attest to her avian inheritance. For an illustration, see CP 1:146, cat. no. 143.

55. Algernon Charles Swinburne, "Hermaphroditus," *Poems and Ballads* (written 1863, published 1866), quoted in Brian Reade, ed., *Sexual Heretics: Male Homosexuality in English Literature from 1850 to 1900* (New York: Coward-McCann, 1971), 72, line 31. Hermaphrodites were often troped as "double flowers." See, for example, Paul Morand's *La fleur double* (Paris: Émile-Paul Frères, 1924). While this doubleness might appear redundant, given the hermaphroditic condition of most flowers, it is predicated on Darwin's studies of the mutations that result from multiple generations of floral self-fertilization. Vegetal mutations are taken up in chapter 3.

56. Whitman, "Spontaneous Me," in *Leaves of Grass and Other Writings*, 89, line 13.

57. On the *bergère*, see CP 2:67. Sargent's furniture preferences were for the Louis Quinze and Louis Seize styles. CP 2:xxii–xxiii.

58. Huysmans, *Against Nature*, 54, translation modified. "Le xviiie siècle a su envelopper la femme d'une atmosphère vicieuse, contournant les meubles selon la forme de ses charmes, imitant les contractions de ses plaisirs, les volutes de ses spasmes, avec les ondulations, les tortillements du bois." *À rebours*, 87. Martin Maloney notes that her left arm, the one grasping the chair, is "highly charged." Maloney, "Girls, Girls, Girls," *Tate* 16 (Winter 1998): 28.

59. Maloney, "Girls, Girls, Girls," 28.

60. Huysmans, *Against Nature*, 149; "fécondantes pluies de pollen," *À rebours*, 243.

61. As was typical of female masturbators (according to nineteenth-century medicine), Lady Agnew, née Gertrude Vernon (1865–1932), suffered from ill health and neurasthenia, and died of pernicious anemia. She had no children and lived for years at a time apart from her husband. Julia Rayer Rolfe, "Sargent and Lady Agnew," in Rolfe et al., *The Portrait of a Lady: Sargent and Lady Agnew*, exh. cat. (Edinburgh: National Gallery of Scotland, 1997), 17, 32.

62. "Une mémoire prodigieuse, stupéfiante, que cette mémoire de Mirbeau, nommant, retrouvant sans jamais broncher, énumérant tous les noms de fleurs, et avec des notes amoureuses dans la voix, quand il les décrit, et des gestes caressants, qui ont l'air de craindre de les faner sous ses doigts." Edmond and Jules de Goncourt, *Journal*, entry for 29 December 1894, 3:1065.

63. Gosse found it "odd" that Shelley *didn't* "dwell upon the stimulus of touch upon the exquisite pinnate leaflets of the [sensitive] plant" in his study. Gosse, "Introduction," in Shelley, *Sensitive Plant*, xiii.

64. Maurice Maeterlinck, *The Double Garden*, trans. Alexander Teixeira de Mattos (New York: Dodd, Mead, 1904), 265; "ceux qui l'aperçoivent pour la première fois, n'en croyant pas leurs yeux, veulent toucher du doigt la bleuâtre merveille," Maeterlinck, *Le double jardin* (Paris: Charpentier, 1904), 219.

65. Higginson, "Water-Lilies," 469.

66. Journal entry for 9 May 1871, in Gerard Manley Hopkins, *Poems and Prose*, ed. W. H. Gardner (Harmondsworth: Penguin, 1985), 123–24.

67. William Wordsworth, "A Poet's Epitaph" (1800, written 1798–99), in *William Wordsworth*, ed. Stephen Gill (Oxford: Oxford University Press, 1984), 151–52, lines 18–20.

68. Maurice Maeterlinck, *Pelleas and Mélisanda and The Sightless*, trans. Laurence Alma-Tadema (London: Walter Scott Publishing, 1895), act 2, scene 2, p. 53; "Oh! ces petites mains que je pourrais écraser comme des fleurs," Maeterlinck, *Pelléas et Mélisande* (1892) (Brussels: P. Lacomblez, 1908), 29.

69. "Le guerrier convulsif brûle ses mains de caresses, de caresses incestueuses et brutales . . . des baisers sacrilèges." "Ô ma fille, vos mains sont des corolles fines; / Vos mains sont un bouquet de jeune aubépines; / L'haleine du printemps souffle de votre chair: / Je ne moissonne pas les fleurs avec le fer." Pierre Quillard, *La fille aux mains coupées,* in *Deux pièces symbolistes: Pierre Quillard, "La fille aux mains coupées," et Charles van Lerberghe, "Les flaireurs,"* ed. Jeremy Whistle (Exeter: University of Exeter, 1976), 5.

70. "La volontaire martyre pose sans trembler ses mains jaillissant des manches sur une table"; "Brandissant dans la pénombre les deux torches jumelles des bras mutilés"; "épouvantable et radieux, un double nénuphar aux tiges d'écarlate flotte dans une écume rose." Quillard, *La fille aux mains coupées,* 6.

71. Berliocchi, *Orchid in Lore and Legend,* 18.

72. Algernon Charles Swinburne, "Étude réaliste" (1883), in *The Springtide of Life: Poems of Childhood,* ed. Edmund Gosse (Philadelphia: J. B. Lippincott, 1918), 9, lines 12–13. In "A Clasp of Hands" (1904), they are "Soft, small and sweet as sunniest flowers" that open "bud by bud." Ibid., 27, lines 1, 3.

73. Swinburne, "After a Reading" (1884), in ibid., 78, line 14.

74. Jean Genet, *Our Lady of the Flowers,* trans. Bernard Frechtman (New York: Grove Press, 1963), 112; "de leurs mains . . . s'échappaient des fusées de fleurs au milieu desquelles elles se tenaient de la façon la plus simple du monde," Genet, *Notre-Dame-des-Fleurs* (1943) (Saint-Amand: Bussière, 1976), 96–97.

75. Martin Kemp, "The Handy Worke of the Incomprehensible Creator," in *Writing on Hands: Memory and Knowledge in Early Modern Europe,* exh. cat., ed. Claire Richter Sherman (Carlisle, Pa.: Trout Gallery, 2000), 22.

76. "What can be more curious," Darwin asks, "than that the hand of a man, formed for grasping, that of a mole for digging, the leg of the horse, the paddle of the porpoise, and the wing of the bat, should all be constructed on the same pattern, and should include the same bones, in the same relative positions?" He argued that descent was "the only certainly known cause of similarity in organic beings" and that shared morphology across such divergent species was evidence for evolution. Darwin, *On the Origin of Species,* 434, 456. The analysis of hands in fact enabled the debunking of both the theory of preformation and of various claims concerning the significance of *man*'s role in the world with respect to other creatures, including women. Anomalies in hand structure contributed to the study of inheritance; when he became head of the German Academy of Sciences in 1740, the French scientist Pierre Maupertuis found in the polydactylous Ruhe family proof both of genetic inheritance and, since he found that this trait "could be handed on by either sex for many generations," of the fact that both parents contribute to generation. Elizabeth Gasking, *Investigations into Generation, 1651–1828* (London: Hutchinson, 1967), 80.

77. As the title suggests and the charming illustrations confirm, alternate appendages of all sorts function as artfully as hands. If human superiority could no longer be guaranteed by a distinctive feeler, though, the familiar hierarchy of species remains in place in this text. "A definite order has been preserved," Wood assures the reader in the preface, "the Mammalia having precedence," and man, of course, being "the chief of all the mammalia." Human creations, he argued, differ from those of other organisms, which are "planned, not by reason, but by instinct." Mental rather than physical apprehension, then, was the principal sign of humans' evolutionary progress and divine favor for Wood. John George Wood, *Homes Without Hands* (London: Longmans, Green, 1884), iii, 3. *Homes Without Hands* went through at least sixteen reprintings between 1865 and 1898, not including abridged versions.

78. Charles Darwin, *The Descent of Man, and Selection in Relation to Sex* (2nd ed., 1879) (Harmondsworth: Penguin, 2004), 69.

79. Remy de Gourmont, *The Physiology of Love* (1903) (New York: Rarity Press, 1932), 168. "Les mains de l'homme, si merveilleuses qu'elles nous paraissent, n'ajoutent à peu près rien à son intelligence. . . . L'organe extérieur est secondaire: n'importe quoi, bec, queue prenante, dents, trompe, pattes, eût fait l'office de mains." Gourmont, *Physique de l'amour* (Paris: Mercure de France, 1903), 262.

80. Gourmont, *Physiology of Love,* 169, 50; "Les exercices merveilleux de la main" ne sont que "luxure"; "La main . . . des quadrumanes n'a qu'un rôle sexuel, la masturbation," *Physique de l'amour,* 264, 84.

81. Max Beerbohm, *A Book of Caricatures* (London: Methuen, 1907), plate XI.

82. Whitman, "Song of Myself," section 27, *Leaves of Grass and Other Writings,* 50.

83. Wilde et al., *Teleny,* 33.

84. RTPE 1:980–81, translation slightly modified. "Mais, plus grasses, les mains d'Albertine cédaient un instant, puis résistaient à la pression de la main qui les serrait, donnant une sensation toute particulière. La pression de la main d'Albertine avait une douceur sensuelle qui était comme en harmonie avec la coloration rose, légèrement mauve, de sa peau. Cette pression semblait vous faire pénétrer dans la jeune fille, dans la profondeur de ses sens, comme la sonorité de son rire, indécent à la façon d'un roucoulement ou de certains cris. Elle était de ces femmes à qui c'est un si grand plaisir de serrer la main qu'on est reconnaissant à la civilisation d'avoir fait du shake-hand un acte permis entre jeunes gens et jeunes filles qui s'abordent." RTPF 1:919.

85. John Kasson cites various writers' cautions in "Venturing Forth: Bodily Management in Public," chapter 4 of his *Rudeness and Civility: Manners in Nineteenth-Century Urban America* (New York: Hill and Wang, 1990), 142.

86. Frederick Maryatt, *A Diary in America, with Remarks on Its Institutions,* ed. Sydney Jackman (New York: William Colyer, 1839), 39. Vernon Lee reported to her mother with horror that "Mrs. Stanley sometimes had to shake hands with 2000 people in one afternoon there!!" Lee to her mother, 18 July 1891, *Vernon Lee's Letters,* ed. Irene Cooper Willis (London: privately printed, 1937), 336.

87. *The Illustrated Manners Book: A Manual of Good Behavior and Polite Accomplishments* (New York: Leland, Clay, 1855), 103.

88. *The Laws of Etiquette* (Philadelphia: Carey, Lea, and Blanchard, 1836), 38.

89. Wilde et al., *Teleny,* 33–34.

90. Ibid., 115.

91. Robert Tomes, *Bazar Book of Decorum: The Care of the Person, Manners, Etiquette, and Ceremonials* (New York: Harper and Brothers, 1870), 65.

92. "Vrai, Planchemol, tu devrais te faire drainer les mains . . . tu as là dedans une humidité dont tu gratifies les autres!" Edmond de Goncourt, *La Faustin* (Paris: Charpentier, 1882), 19.

93. On the significance of "tea" to queer circles, see Emily Eells, *Proust's Cup of Tea: Homoeroticism and Victorian Culture* (Aldershot: Ashgate, 2002); and Kristeva, *Time and Sense.* Kristeva gives the following brief summary of tea's underworld connotations: "According to the Larousse dictionary, French slang uses *teacup* for a public urinal, *have some tea* for sex, especially homosexual sex, and *teapot* for the sort of public urinal frequented by homosexuals. One writer notes that 'homosexuals paid regular visits to their teapots and teacups, the nickname they gave to urinals, and this created a lovely nocturnal ballet.' . . . [Another explains that] '[Encounters] happened most often at "bay." The "bays" were "teacups." We referred to them in this way because we had a coded language. Because of their snobbishness the queers of the aristocratic sixteenth district refused to say "urinal" [*pissotière*].' . . . Proust was familiar with these terms." *Time and Sense,* 347n60. Jarrod Hayes examines the significance of tea and tea parties to Proustian memory and creativity as well as sexual identity in "Proust in the Tearoom," *PMLA* 110 (October 1995): 992–1005.

94. Farmer, *Slang and Its Analogues,* 7:88.

95. Moll gives the example of "Lisa the handle," who stands in this way. "[D]as Henkellieschen, weil er die Arme oft wie Henkel in die Hüften stemmt." Albert Moll, *Die conträre Sexualempfindung* (Berlin: Fischer, 1891), 86.

96. Joseph W. Howe, *Excessive Venery, Masturbation, and Continence* (New York: E. B. Treat, 1887), 72–73n1, quoted in Cohen, "Manual Conduct in *Great Expectations,*" 223.

97. Wilde et al., *Teleny,* 33.

98. Philippe-Auguste, comte de Villiers de l'Isle-Adam, *Eve of the Future Eden,* trans. Marilyn Gaddis Rose (Lawrence, Kans.: Coronado Press, 1981), 73. "Je doterai cette Ombre . . . de toutes les mysticités passionnées des *Ligéias* [*sic*] d'Edgar Poe." Villiers de l'Isle-Adam, *L'Ève future* (Paris: M. de Brunhoff, 1886), 103. By using the plural of the name, Villiers may be referring to the many death-defying, sexually ambiguous characters in Poe's oeuvre.

99. Villiers de l'Isle-Adam, *Eve of the Future Eden,* 70, 243; "Vous étiez imbu, pénétré, vous, du fluide nerveux . . . [quand] Hadaly *vous a pris par la main.* . . . À l'instant même il s'envola sur ces invisibles réseaux demeurés," *L'Ève future,* 98, 360–61.

100. *Eve of the Future Eden,* 92; "Le baume subtil se répandra de là dans l'organisme magnéto-métallique de Hadaly," *L'Ève future,* 133.

101. Maurice Merleau-Ponty, *The Visible and the Invisible* (1948), trans. Alphonso Lingis, ed. Claude Lefort (Evanston: Northwestern University Press, 1968), 268.

102. Delacroix to Mme. de Forget, 4 August 1848, quoted in T. J. Clark, *The Absolute Bourgeois: Artists and Politics in France, 1848–1851* (Berkeley and Los Angeles: University of California Press, 1999), 125. Translation slightly modified. "J'ai rêvé, cette nuit même, qu'elle me faisait cadeau d'une petite main en une substance charmante, ressemblant à de la chair, mais qui n'en était point. C'était une sorte de joujou, de babiole à mettre dans un *Petit Dunkerque;* de plus cette main répondait à la pression, était douée en un mot de sensibilité et j'en suis resté là avec la main de mon rêve. En me réveillant, je me suis dit qu'un bijou comme cela serait bien commode en voyage." Eugène Delacroix, *Correspondance générale,* ed. André Joubin, 5 vols. (Paris: Plon, 1935–38), 2:354–55.

103. James Delbourgo, *A Most Amazing Scene of Wonders: Electricity and Enlightenment in Early America* (Cambridge: Harvard University Press, 2005), chapter 4, "Electrical Politics and Political Electricity," 135.

104. Paul Gilmore examines the electric visions of Coleridge, Shelley, Byron, Whitman, and Poe, and their construction of an imagined community of British and American poets. Gilmore, "Romantic Electricity, or The Materiality of Aesthetics," *American Literature* 76 (September 2004): 473–74.

105. Paola Bertucci, "The Electrical Body of Knowledge: Medical Electricity and Experimental Philosophy in the Mid-Eighteenth Century," in *Electric Bodies: Episodes in the History of Medical Electricity,* ed. Paola Bertucci and Giuliano Pancaldi (Bologna: Università di Bologna, 2001). On electricity as a metaphor for "emotional connection" and "bodily excitement," see Gilmore, "Romantic Electricity," 474.

106. Christoph Asendorf, *Batteries of Life: On the History of Things and Their Perception in Modernity,* trans. Don Reneau (Berkeley and Los Angeles: University of California Press, 1993), 155.

107. Whitman, "Song of Myself," 50, lines 619–20.

108. Marco Piccolino, *The Taming of the Ray: Electric Fish Research in the Enlightenment from John Walsh to Alessandro Volta* (Florence: Leo S. Olschki, 2003), 16–17.

109. Carl Hendrik Andreas Bjerregaard, "The Elementals, the Elementary Spirits, and the Relationship Between Them and Human Beings," *The Path* 1 (January 1887): 329.

110. Although torpedoes and eels are distinct species, the terms were used interchangeably in the eighteenth century and throughout most of the nineteenth. Delbourgo, *Most Amazing Scene of Wonders,* 192.

111. Eels can "make long journeys through wet grass" and survive in wet foliage and moss for some time; there are many accounts of them foraying into fields in search of food. An early nineteenth-century account describes "the eels from Lake Hedenlunda" that "wandered into the fields

at night and ate pea pods, making a smacking sound with their lips." Comfortable in water and air, eels also burrow into mud, producing "large quantities of mucus which serves to lubricate them." Christopher Moriarty, *Eels: A Natural and Unnatural History* (New York: Universe, 1978), 11, 24, 11. See also Elizabeth Pennisi, "Much Ado About Eels," *Bioscience* 39 (October 1989): 595, 596.

112. Oppian, *Oppian's Halieuticks of the nature of fishes and fishing of the ancients in V. books. Translated from the Greek, with an account of Oppian's life and writings, and a catalogue of his fishes,* trans. William Diaper (London: Rob. Shippen, 1722), 37, lines 849–58.

113. "[D]ie seit Jahrhunderten schwebende Frage." Sigmund Freud, "Beobachtungen über Gestaltung und feineren Bau der als Hoden beschriebenen Lappenorgane des Aals," *Sitzungsberichte der Kaiserlichen Akademie der Wissenschaften, Wien* (April 1877): 420. Freud's paper was written partly in response to a slightly earlier study by a scientist named Syrski, who claimed to have uncovered "den lange gesuchten Hoden der Aale" (the long sought-after testicles of the eel) in a long, frilly, flaplike organ consisting of two rows that appeared where the ovary is usually found. Syrski's "Lappenorgan," or flap-organ, was rather less distinctive than the scientific world would have liked. To begin with, it so closely resembled an immature ovary ("das Aussehen des unentwickelten Lappenorgans" approaches that of the "ganz unreifen Ovariums") that it was impossible to discern a structural difference other than the doubling. Second, as "niemand ein reifes Lappenorgan und ein reifes Aalmännchen gesehen [hat]" (no one had ever seen a mature flap-organ or a mature male eel), it was impossible to tell whether the flap-organ was, in fact, the testicle of the eel or not. Freud's dissections could not determine the function of the organ, nor could he prove whether it developed out of the ovary or was an independently developing organ (ibid., 420, 423, 429). Darwin skirted the issue by describing eels as "androgynous." Darwin, *Descent of Man,* 189n28.

114. Lucretia Lovejoy, *An Elegy on the Lamented Death of the Electrical Eel, or Gymnotus Electricus* (London: T. Hookham, 1779), 23, 21, 27, 24.

115. Jean Lorrain, "Glaucous Eyes," in *Nightmares of an Ether-Drinker,* trans. Brian Stableford (Carlton-in-Coverdale: Tartarus, 2002), 52–53. "Fontenay, ne dites point de mal des anguilles, leur souplesse a du bon." "Vous êtes un vicieux, Chauchat, nous le savons; mais, pour les gens honnêtes, à gouts avouables et reconnus par la loi, le mérite d'une femme est d'être vraiment femme et non de rappeler je ne sais quelle équivoque de sexe ou de poisson." Jean Lorrain, "Les yeux glauques," in *Buveurs d'âmes* (1893) (Paris: Charpentier, 1929), 84.

116. Farmer, *Slang and Its Analogues,* 2:401.

117. As the *New York Times* put it, "art, music, and the drama, literature, popular entertaining, flowers, and fishes will assemble themselves together in the great building and help us to be gay together." "The Westminster Aquarium," *New York Times,* 24 October 1875, 4.

118. Melville, *Moby-Dick* (1851), ed. Charles Child Walcutt (New York: Bantam Books, 1981), 384–85. At the time of Melville's writing, "eel" was a nickname for a New Englander. OED, s.v. "eel," 4.

119. Jean Lorrain, *Monsieur de Phocas,* trans. Francis Amery (Sawtry, Cambridgeshire: Dedalus, 1994), 31; "la douceur fondante de ses mains . . . leur glissement entre les doigts, telle une fuite de couleuvre . . . une étreinte insaisissable de doigts fluides et froids!" Lorrain, *Monsieur de Phocas* (1901) (Paris: Albin Michel, 1929), 19.

120. "[C]omme d'un nombre infini de serpentins cabalistiques." Flament, *Le bal du Pré-Catelan,* 41.

121. Recorded by Lucia Fairchild in her diary entry for 2 October 1890, Boston, quoted in Lucia Miller, "John Singer Sargent in the Diaries of Lucia Fairchild, 1890 and 1891," *Archives of American Art Journal* 26, no. 4 (1986): 11, 7.

122. "Held between the fingers, it filaments and slowly strains away. Chemical analyses do not explain this character." "L'eau de mer, même la plus pure . . . est légèrement blanchâtre et un peu visqueuse. Retenue entre les doigts, elle *file* et passe lentement. Les analyses chimiques n'expliquent pas ce caractère. . . . Les plantes, les animaux marins, sont vêtus de cette substance, dont la mucosité, consolidée autour d'eux, a un effet de gélatine, parfois fixe et parfois tremblante." Jules Michelet, *La mer* (1861) (Paris: Michel Lévy Frères, 1875), 111.

123. Entry in Matilda Gay's diary, quoted in William Rieder, *A Charmed Couple: The Art and Life of Walter and Matilda Gay* (New York: Harry N. Abrams, 2000), 141.

124. RTPE 2:230; "une main à demi dépliée flottant, comme l'aileron d'un requin, à côté de sa poitrine, et qu'il laissait presser indistinctement par ses vieux amis et par les inconnus qu'on lui présentait," RTPF 2:224. As I indicated above, the transgression of this mode of touching partly consists in its crossing of class boundaries. The unrestrained, even lewd homosociality of the duc de Guermantes, who is perversely open to contact with his social inferiors, is demonstrated at his meeting with M.'s bourgeois father, whose hand he held "in his own, stroking it even, to prove to him, with the shamelessness of a courtesan, that he did not begrudge him the privilege of contact with the ducal flesh," RTPE 2:28; "lui prenait la main, et la retenant dans la sienne, la lui caressant même pour lui prouver, avec une impudeur de courtisane, qu'il ne lui marchandait pas le contact de sa chair précieuse," RTPF 2:33.

125. Gustave Flaubert, *The Temptation of Saint Anthony* (1874 ed.), trans. Kitty Mrosovky (Harmondsworth: Penguin, 1983), 171–72. "J'ai habité le monde *informe* où sommeillaient des bêtes hermaphrodites, sous le poids d'une atmosphère opaque, dans la profondeur des ondes ténébreuses,—quand les doigts, les nageoires et les ailes étaient confondus." Flaubert, *La tentation de Saint-Antoine—1874,* vol. 4 of *Oeuvres complètes,* 18 vols. (Paris: Conard, 1910–33), 128–29 (emphasis added).

126. Rachilde, *The Juggler,* trans. Melanie Hawthorne (New Brunswick: Rutgers

University Press, 1990), 10; une "paume voluptueuse, collante comme une ventouse," Rachilde, *La jongleuse* (1900) (Paris: Des Femmes, 1982), 34.

127. Ruskin's diary, entry for 9 March 1868, quoted in Pearsall, *Worm in the Bud,* 441.

128. Villiers de l'Isle-Adam, *Eve of the Future Eden,* 92; "l'huile de roses," *L'Ève future,* 119.

129. "[D]oigt," Delvau, *Dictionnaire érotique moderne,* 139; de Landes, *Glossaire érotique,* 115.

130. Darwin, *Various Contrivances,* 2nd ed., 276.

131. Ruskin, *Queen of the Air,* 112.

132. Wilde et al., *Teleny,* 144.

133. The thumb that completes the sphincter's circle, straightening the snake, seems to confirm Gourmont's theory that the digit evolved primarily for sexual stimulation.

134. RTPE 2:650. "Méduse! Orchidée! Quand je ne suivais que mon instinct, la méduse me répugnait à Balbec; mais si je savais la regarder, comme Michelet, du point de vue de l'histoire naturelle et de l'esthétique, je voyais une délicieuse girandole d'azur. Ne sont-elles pas, avec le velours transparent de leurs pétales, comme les mauves orchidées de la mer?" RTPF 2:626–27. A similar conflation of floral and marine realms can be found in a gold and enamel Lalique pendant, c. 1899, in the Victoria and Albert Museum (M.521-1924). Perhaps best described as florally octopoid, the pendant consists of an androgynous head that sprouts enormous, coiling, Medusean blue locks of hair covered with small white flowers; these flowers resemble suckers and emphasize the tentacular quality of the hair.

135. Similarly, in John Addington Symonds's poem "An Allegory," the narrator plants a queer, colorless flower, a symbol of his invert desire, in a wine-dark agate vase decorated with nereids, a submarine base for his flower. Symonds, "An Allegory," in *New and Old,* 24–26.

136. Georges Bataille, "The Language of Flowers," trans. Roger Banner and Allan Stoekl, in *Karl Blossfeldt: Art Forms in Nature, the Complete Edition* (Munich: Schirmer Art Books, 1999), 10–11. "[I]l ne faudrait pas moins, pour détruire l'impression favorable, que la vision fantastique et impossible des racines qui grouillent, sous la surface du sol"; "alors que celles-ci s'élèvent noblement, celles-là, ignobles et gluantes, se vautrent dans l'intérieur du sol"; "il semble d'ailleurs impossible d'éliminer une opposition aussi flagrante que celle qui différencie la tige et la racine." Bataille, "Le langage des fleurs" (1929), *Oeuvres complètes,* 12 vols. (Paris: Gallimard, 1970), 1:177, 178.

137. Thiébaut interprets the ambiguity of the hand—which may be masculine or feminine, human or hybrid, rising from the sea or about to be engulfed by it—as an index of Gallé's increasing debility due to an amorphous disease ("pernicious anemia," from which he would die in 1904). Gallé read Darwin, and perhaps viewed his bodily regression as a return to the sea. Philippe Thiébaut, "*La main aux algues et aux coquillages:* Le testament artistique d'Émile Gallé," *48/14: La Revue du Musée d'Orsay* 18 (Spring 2004): 20.

138. "[G]ens avec des étoiles de mer en guise de mains." Montesquiou, "Le pavé rouge," 337.

139. Blanche was a renowned though closeted invert. Léon Daudet described him as a "grande fifille zézayante" (big lisping girly girl). Léon Daudet, *Salons et journaux,* part four of *Souvenirs des milieux littéraires, politiques, artistiques et médicaux,* 2 vols. (Paris: Nouvelle Librairie Nationale, 1920), 1:601. After Wilde's trial in 1895, Blanche married Rose Lemoinne, but the union remained unconsummated. Henri Loyrette, *Degas* (Paris: Fayard, 1991), 532.

140. "[D]e mêmes tendances que les miennes." Jacques-Émile Blanche, *La pêche aux souvenirs* (Paris: Flammarion, 1949), 166. Raphael de Ochoa was a cousin by marriage of Reynaldo Hahn. In 1884 Blanche and Ochoa traveled together to Madrid, visiting the Prado, where Raphael's uncle was a curator.

141. Louise d'Argencourt, *European Paintings of the Nineteenth Century,* vol. 1, *Aligny–Gros* (Cleveland: Cleveland Museum of Art, 1974), cat. no. 11.

142. Sargent painted Blanche in exchange for a Louis XV armchair and inlaid desk rather than one of the latter's paintings, as was his custom with friends. Trevor Fairbrother, "Possessions and Props: The Collection of John Singer Sargent," *Antiques* (February 2001): 318. On Sargent's exchanges of works with artist friends, see Jordan, "John Singer Sargent's Images of Artists."

143. Bell also notes that Blanche told Ethel Sands at a luncheon that Sargent was a "homosexual." Clive Bell (quoting Blanche) to Mary Hutchinson, 31 July 1927, Harry Ransom Humanities Research Center, University of Texas at Austin, quoted in Trevor Fairbrother, "The Complications of Being Sargent," *John Singer Sargent: Portraits of the Wertheimer Family,* exh. cat., ed. Norman L. Kleeblatt (New York: Jewish Museum, 1999), 39, 42n17.

144. Jules Michelet, *The Mountain,* trans. W. H. Davenport Adams (London: T. Nelson and Sons, 1875), 168; "une petite mer ondoyante," *La montagne,* 189.

145. Loeffler to Charteris, 30 January 1926, quoted in Charteris, *John Sargent,* 148.

146. Sargent's affinity with these aesthetic birds inspired Charles Martin Loeffler to dedicate his song "Les Paons," based on a poem by Gustave Kahn, to the painter. The poem was first published under the title "Intermède IX" in *La Vogue* 2 (19–26 July 1886): 158–59. For more on the relationship between Loeffler, Kahn, and Sargent, see C. Stanley Eby, "Stylistic Analysis of Selected Songs for Voice and Piano by Charles Martin Loeffler: *Quatre mélodies,* Poésies de Gustave Kahn, Op. 10; and *Four Poems,* Op. 15" (PhD diss., Boston University, 1997).

147. Once there were two. Fanny Stevenson is said to have destroyed the first, painted in 1884, because she didn't like it. CP 1:141, cat. no. 139.

148. Greta, "Art in Boston," *Art Amateur* 18 (December 1887): 3.

149. Stevenson to Sidney Colvin, 19–20 March 1891, in Robert Louis Stevenson, *Selected Letters of Robert Louis Stevenson,* ed. Ernest Mehew (New Haven: Yale University Press, 1997), 453.

150. Marcel Proust, *Jean Santeuil,* trans. Gerard Hopkins (Harmondsworth: Penguin, 1985), 271, translation modified.

"[J]oie en attendant répandue comme une poussière de germes dans le cabinet de toilette," Proust, *Jean Santeuil* (written c. 1895–1900), 3 vols. (Paris: Gallimard, 1952), 2:53.

151. This is not to say that women didn't smoke, of course. Vernon Lee described leaning out of the Paters' attic window to smoke a "smuggled cigarette," the odor of which still made her fear discovery. Lee to her mother, 18 June 1884, *Vernon Lee's Letters,* 146. Sargent's 1885 painting *The Candelabrum* plays on the suggestion of such female impropriety: a wisp of smoke rises from an empty candelabrum wreathed by white roses; behind these a woman is holding something, presumably but not definitely a cigarette. For an illustration, see CP 1:169, cat. no. 163.

152. OED, s.v. "fag," *n.* 4.b.

153. Curiously, in the trade card for tobacco we saw in chapter 1 (fig. 29), the personified plant appears to be a cross-dresser. It wears a top hat, smokes a cigar, and sports a moustache and sideburns, and is thus presumably male. The leaves of the figure's attire, however, take the form of a dress.

154. Delvau, *Dictionnaire érotique moderne,* 97–98. See also Guiraud, *Dictionnaire historique,* 227.

155. Signed "Σ," "A Love Song," *Kottabos* 3 (1881), quoted in Reade, *Sexual Heretics,* 154–55, lines 9–16. Louise d'Argencourt is quite right to suggest that the cigar held by the creeping fingers of Raphael de Ochoa in Jacques-Émile Blanche's *Self-Portrait with Raphael de Ochoa* (fig. 83) is meant to evoke "the perfume of a lover." *European Paintings of the Nineteenth Century,* cat. no. 11. In *L'Ève future* Edison provocatively lights his cigar in the heart of a pink camellia, showing Lord Ewald what can be his if he so chooses. Villiers de l'Isle-Adam, *Eve of the Future Eden,* 111; *L'Ève future,* 159.

156. Olson, *John Singer Sargent,* 106.

157. Gautier, *Mademoiselle de Maupin,* 204; "le soir, quand au moment de nous quitter il me tend son adorable main si douce et se satinée, toute ma vie se porte à la place qu'il a touchée"; "mon âme se suspend à ses lèvres, comme une abeille à une fleur, pour y boire le miel de ses paroles," Gautier, *Mademoiselle de Maupin,* 224.

158. Marvell, "The Garden," 100, line 18.

159. On Sargent's experimentation with the characteristic mediums and techniques of his artist subjects, see Jordan, "John Singer Sargent's Images of Artists," 96.

Chapter 3

1. Marcel Proust, "Chardin and Rembrandt," in *Against Sainte-Beuve and Other Essays,* 131. "Ainsi le gynécologiste pourrait étonner une femme qui vient d'accoucher en lui expliquant ce qui s'est passé dans son corps, en lui décrivant le processus physiologique de l'acte qu'elle a eu la force mystérieuse d'accomplir," "serait-il [Chardin] fort étonné." Proust, "Chardin," *Contre Sainte-Beuve* (Paris: Gallimard, 1954), 376. The chapter epigraph is from Mirbeau, *Torture Garden,* 28; "l'art! . . . le beau! . . . sais-tu ce que c'est? . . . Eh bien, mon garçon, le beau, c'est un ventre de femme, ouvert, tout sanglant, avec des pinces dedans!" *Le jardin des supplices,* xvi.

2. Proust, "Chardin and Rembrandt," 131; "Une femme n'a pas besoin de savoir la médicine pour enfanter," "Chardin," 376.

3. RTPE 2:636. "Jusqu'ici je m'étais trouvé, en face de M. de Charlus, de la même façon qu'un homme distrait, lequel, devant une femme enceinte dont il n'a pas remarqué la taille alourdie, s'obstine, tandis qu'elle lui répète en souriant: 'oui, je suis un peu fatiguée en ce moment,' à lui demander indiscrètement: 'qu'avez-vous donc?' Mais que quelqu'un lui dise: 'elle est grosse,' soudain il aperçoit le ventre et ne verra plus que lui." RTPF 2:613.

4. RTPE 3:738; "un chirurgien qui, sous le poli d'un ventre de femme, verrait le mal interne qui le ronge," RTPF 3:718.

5. Robert Proust, *Chirurgie de l'appareil génital de la femme* (Paris: Masson, 1904).

6. Melville, *Moby-Dick,* 129.

7. Merleau-Ponty, "The Intertwining—The Chiasm," in *Visible and the Invisible,* 154, 152, 146.

8. Merleau-Ponty, "Working Notes," in ibid., 209, 218.

9. "There is double and crossed situating of the visible in the tangible and of the tangible in the visible." "Intertwining," 134.

10. Der Maler malt "durch Auflegen der Hand"; der Operateur "dringt vielmehr operativ in ihn ein." Walter Benjamin, *Das Kunstwerk im Zeitalter seiner technischen Reproduzierbarkeit* (1936) (Frankfurt am Main: Suhrkamp Verlag, 1977), 31, 32.

11. Under Pozzi's influence, Robert Proust, who had started his career as a proctologist, switched over to gynecology. Lawrence Joseph, "Marcel Proust et 'Docteur Dieu': Lettres inédites à Samuel Pozzi," *Bulletin Marcel Proust* 51 (2001): 13.

12. Claude Vanderpooten, *Samuel Pozzi: Chirurgien et ami des femmes* (Paris: Éditions in Fine, 1992), 261.

13. They "met at the Godebskis and, following Pozzi's rules, played a kind of orgiastic truth game." Arthur Gold and Robert Fizdale, *Misia: The Life of Misia Sert* (New York: Vintage, 1992), 24.

14. Samuel Pozzi, *Traité de gynécologie clinique et opératoire* (Paris: G. Masson, 1890; 2nd ed., Paris: G. Masson, 1892; 3rd ed., Paris: Masson et Cie, 1897). Unless otherwise indicated, I will cite from the first edition.

15. The phrase appears throughout "The Intertwining" and is discussed in Merleau-Ponty's "Working Notes," 248–51.

16. "Jamais homme, mieux que lui, n'a su exprimer le plaisir aigu et comme frissonnant, des doigts, au contact d'une belle faïence sans défaut." Henry Marcel, preface to *Catalogue des objets d'art et d'ameublement: Composant la première partie de la collection de feu le Prof. S. Pozzi* (Paris: Galerie Georges Petit, 23–24 June 1919), 7.

17. "Toucher l'utérus, les ovaires, pincer doucement les trompes entre les doigts, sentir l'estomac, l'intestin, chauds, pleins, vivants . . . palper les artères, l'aorte battantes, la petite vésicule biliaire tendue, souple ou bourrée de calculs comme un épi de maïs . . . Quelle griserie!" (To touch the uterus, the ovaries, to softly pinch the fallopian tubes between the fingers, to feel the stomach, the intestine, hot, full, living . . . to palpate the arteries, the beating

aorta, the small gall-bladder, taut, supple, or stuffed with stones like an ear of corn). Vanderpooten, *Samuel Pozzi,* 87.

18. "L'index est le doigt le plus commode pour le toucher." "On se rendra compte alors successivement de la direction du col, de son volume, de sa forme, de sa consistance, de son degré de béance, de l'état de ses commissures." Pozzi, *Traité de gynécologie,* 99.

19. Bimanual exploration was first employed by the French doctor Puzos in 1753, but Pozzi was responsible for its popularization. Ibid., 101, viii.

20. "Tandis que l'index de la main droite pratique le toucher [vaginal], la main gauche est posée au-dessus du pubis transversalement, et les doigts s'enfoncent par une pression douce que refoule les organes génitaux internes vers le doigt vaginal." Ibid., 102–3.

21. "On essaye tout d'abord de se rendre un compte exact de la position de l'utérus dans la région hypogastrique, puis on passe aux parties latérales, et tandis que la main abdominale déprime les flancs, le doigt vaginal va à sa rencontre en s'enfonçant dans les culs-de-sac vaginaux; on explore ainsi très bien la base des ligaments larges et les annexes de l'utérus anormalement augmentés de volume. Il faut noter en même temps la sensibilité des parties; à l'état de santé, la pression du côté des annexes, le soulèvement ou ballottement de l'utérus ne sont pas douloureux." Ibid., 103.

22. Various of the *Traité*'s illustrations were drawn by Charles Nicolet under Pozzi's direction, and others were reproduced from earlier published works on gynecology. I will indicate wherever those I include are not Nicolet's. Ibid., xii.

23. Merleau-Ponty, "Intertwining," 130–31.

24. Ibid., 136.

25. "Invitations to watch the worldly surgeon perform his revolutionary gynecological operations were much sought after by his society friends." Gold and Fizdale, *Misia,* 24.

26. "Ouvrir des ventres humains, y travailler sans douleur et les refermer sans que mort s'ensuive. . . . Voir enfin, vivante, la chair intérieure promise de tous leurs anciens, jusqu'alors inaccessible. . . . Quelle chance extraordinaire!" Vanderpooten, *Samuel Pozzi,* 87.

27. Merleau-Ponty, "Working Notes," 218.

28. These could be fungous, polypoid, rhiziform, and so on; "fongueuses," "polypeuses," "rhiziformes." *Traité,* 1014, 187, 703. French slang for such growths included "mushrooms" and "flowers of Venus." "[C]hampignon," "fleurons de Vénus," Delvau, *Dictionnaire érotique moderne,* 89, 189.

29. "Ludwig Pernice a donné la description et le dessin d'un myosarcome strio-cellulaire de l'utérus, en forme de grappe, observé sur une nullipare, sujette depuis six mois à des hémorrhagies. La tumeur, née du museau de tanche, avait le volume d'un poing et demi et ressemblait fort à une grappe de raisin à grains violacés contenant un liquide gélatineux." *Traité,* 377.

30. "Imperfect Fertilization," *Vick's Magazine,* December 1883, 381.

31. The following references are a partial sample of his vegetal vocabulary throughout the *Traité:* "un grain de chènevis," 716, 731; "un grain de millet," 626; "pois," 188, 715; "gros pois," 677; "noisette," 188, 580, 731; "grappe," 377; "cerise," 715; "noix," 693, 716; "haricot," 1019; "champignon," 372; "petite poire," 641; "pomme," 287; "orange," 693; "noix de coco," 641; "chou-fleur," 372, 770, 1014.

32. Ibid., 767.

33. In "kystes parovariens," the liquid "est clair comme l'eau de roche." Ibid., 703.

34. "[S]ucre d'orge," ibid., 703, 731; "citrin," 728; "du vert pomme, à la teinte du café ou du chocolat," 703. Such extraordinarily colorful visions of the human interior were not uncommon. Charles Blanc, in his *Grammaire des arts du dessin,* imagined it to be as rainbow-hued as Pozzi suggests—"À l'intérieur, le corps humain est diapré, comme la nature, de mille couleurs, ainsi que l'annoncent déjà le vermeillon de ses lèvres, l'ivoire de ses dents, les tons bleu, brun, jaune et orangé de sa prunelle" (iridescent, like nature, with a thousand colors, as the vermilion of its lips, the ivory of its teeth, the blue, brown, yellow, and orangy tones of its eyes already announce). Charles Blanc, *Grammaire des arts du dessin* (1867) (Paris: Librairie Renouard, 1880), 31.

35. "[L]actescent," *Traité,* 770; "rayon de miel," 694.

36. Herman Melville, *Mardi, and a Voyage Thither* (1849), ed. Harrison Hayford et al. (Evanston: Northwestern University Press, 1998), 375; *Moby-Dick,* 378. Pozzi owned Sargent's watercolor sketch for *Fumée d'ambre gris;* it is listed in the auction catalogue as "Mauresque." *Catalogue des objets d'art et d'ameublement,* 17, no. 24. For an illustration, see CP 4:305, cat. no. 791. When the oil painting was first exhibited at the 1880 Salon, its white-draped figure, who holds the cloth of her hood out to catch the ambergris smoke, was caricatured as a mushroom in *Le Charivari*'s "Salon pour rire." The "Champignon du Sérail" was potentially venereal; the caption reads, "Espérons qu'il n'est pas vénéneux" (Let's hope it's not poisonous). *Le Charivari,* 12 May 1880, 3; see CP 4:303.

37. Melville, *Moby-Dick,* 384.

38. The relation between whales and healthy women was explored earlier by Michelet in *La mer;* he focuses on the sea mammal's milk, the nine months it carries its young, etc. Whales and women had since the late sixteenth century also been intimately connected through whalebone corsets. Michelet, *La mer,* 240.

39. "[C]omme disparaissent les baleines dans l'Océan, ou, au centre d'Afrique, les gigantesques éléphants et les hippopotames démesurés, traqués par de trop nombreux chasseurs." Quoted in Vanderpooten, *Samuel Pozzi,* 87. Images of hunting are found elsewhere in the treatise—for example, when Pozzi advises the use of "deux gros drains accolés en manière de canons de fusil et poussés doucement jusqu'au fond de foyer" (two large draining tubes, fastened together like the barrels of a gun, and pushed carefully into the depths of the abcess). *Traité,* 687.

40. Vanderpooten, *Samuel Pozzi,* 87.

41. Melville, *Moby-Dick,* 107.

42. See *Collection S. Pozzi (deuxième partie): Art antique* (Paris: Galerie Georges Petit, 25–27 June 1919).

43. "On peut donc dire que non seulement tous les kystes de l'ovaire ont une origine congénitale, mais que beaucoup d'entre eux ont une date congénitale, et peuvent ou rester stationnaires, ou se développer ultérieurement." *Traité,* 701.

44. The lining of the uterus, which is described as "une pyramide posée sur sa pointe" (an upside-down pyramid—that of an Egyptian mother), is similarly puissant: "la muqueuse utérine n'est pas comparable aux autres muqueuses; elle jouit d'un pouvoir de régénération tout spécial" (the uterine mucosa is not comparable to other mucous membranes; it enjoys a special power of regeneration). *Traité,* 447, 210.

45. The billowing fabriclike ends of the fallopian tubes are called pavilions (*pavillons*).

46. "[L]e doigt se promène"; "glissant"; "le doigt explorera." *Traité,* 99, 98, 99.

47. Bland-Sutton occasionally approaches this georgic sensibility, for example, when he writes that the fimbriae of the fallopian tubes "float freely like water-weeds in a pond." J. Bland-Sutton, *Surgical Diseases of the Ovaries and Fallopian Tubes, Including Tubal Pregnancy* (London: Cassell, 1896), 4. A popular Dr. Warner's Coraline Corsets ad (fig. 19, right) likewise figures the female interior as a watery floral realm. Water lilies and pads fill the torso, the plant's looping stems suggesting various kinds of organic plumbing. Dr. Warner's corsets were made, as their advertisements announce, from the fibers of the coraline plant rather than whalebone; their name nevertheless conjures up the sea.

48. "[H]étérotopie plastique, employé par Lebert." *Traité,* 709. Pozzi's surgically opened female abdomens are also heterotopias in the Foucauldian sense: these sites are not freely accessible but can be entered by a few who have undergone the necessary purifications; they juxtapose (at least in Pozzi's descriptions) several emblematic heterotopic sites: the garden, the brothel, the colony, and the museum; and they function as heterotopias of crisis, deviation, and compensation. See Michel Foucault, "Of Other Spaces," trans. Jay Miskowiec, *Diacritics* 16 (Spring 1986): 22–27.

49. "Il demeure isolé, comme un pistil au centre du calice d'une fleur, et on peut autour de lui accumuler des topiques destinés à le maintenir aseptique et à le momifier." *Traité,* 310. In the 1892 and 1897 editions of the *Traité,* the pedicle's location becomes "the calyx of *the* flower," though whether that *fleur* is the woman or the womb is unclear. "[D]u calice de la fleur," *Traité* (1892), 313; *Traité* (1897), 337.

50. Mutability has always been seen as their key characteristic: "The Polipe chaunge themselues into the likenesse of euerie obiect." Robert Greene, *Mamillia, the Second Part of the Triumph of Pallas* (London: Th. C. for William Ponsonbie, 1593), sig. L2v. The *Encyclopédie* curiously describes the polyp as an "insecte singulier et merveilleux" (singular and marvellous insect). "Polype," *Encyclopédie, ou Dictionnaire raisonné des sciences, des arts et des metiers* (1751–65), 12:945, quoted in Aram Vartanian, "Trembley's Polyp, La Mettrie, and Eighteenth-Century French Materialism," *Journal of the History of Ideas* 11 (June 1950): 260.

51. James Elkins, "On Visual Desperation and the Bodies of Protozoa," *Representations* 40 (Autumn 1992): 43.

52. Shirley Roe, "John Turberville Needham and the Generation of Living Organisms," *Isis* 74 (June 1983): 159, 177.

53. Michelet, *Insect,* 82; "imperceptibles constructeurs du globe," Michelet, *L'insecte,* 33.

54. Philip Henry Gosse, *A Naturalist's Rambles on the Devonshire Coast* (London: John van Voorst, 1853), 19.

55. Gosse, *Father and Son,* 124–25.

56. Other "habitat" plates include depictions of coral, ferns, and hummingbirds.

57. Gosse, *Father and Son,* 125.

58. Barnes's work was translated into French in 1874; Pozzi used some of Barnes's illustrations in his own treatise. Robert Barnes, *Traité clinique des maladies des femmes,* trans. A. Cordes (Paris: G. Masson, 1876).

59. Gosse, *Father and Son,* 124.

60. Merleau-Ponty, "Intertwining," 147.

61. The same structural logic determines the relationship between the eye and the gaze as that between the hand as probe or palpator and flesh of the world, for as the hand is formed in the heart of the tangible, the eye is "formed in the heart of the visible." Ibid., 130.

62. "Working Notes," 263, 264, 255.

63. Melville, *Mardi,* 507.

64. Ibid., 506, 508.

65. Merleau-Ponty, "Intertwining," 148; "Working Notes," 272.

66. Melville, *Mardi,* 506.

67. Quoted in Elizabeth Foster, "Historical Note," in ibid., 657.

68. Michelet, *Insect,* 203–4. "Parfois, comme pour nous contredire et confondre nos idées, elle [la nature] fait ravissants de formes des organes qui, selon nous, accomplissent de basses fonctions. Je pense à l'extrême beauté, à la tendre délicatesse de cet arbre de corail qui pompe incessamment le chyle de nos intestins." *L'insecte,* 193.

69. Wilde et al., *Teleny,* 168.

70. They also evoke Redon's polyp lithograph in the *Origins* series, *The misshapen polyp floated on the shores, a sort of smiling and hideous Cyclops,* with its hair-ringed, spherical eyeball bulging out of the flesh of its face.

71. Wilde et al., *Teleny,* 35.

72. Zola, *The Kill,* 262. "La comtesse Vanska donnait son ardeur sombre au corail, allongée, les bras levés, chargés de pendeloques rouges, pareille à un polype monstrueux et adorable, qui montrait des chairs de femme dans des nacres roses et entrebâillées de coquillages." *La curée,* 1:548.

73. Charles Kingsley, *Glaucus, or The Wonders of the Shore* (1855) (London: Macmillan, 1890), 232.

74. Lacan, *Four Fundamental Concepts of Psycho-Analysis,* 82. Pozzi uses the same metaphor to describe uterine inversion; the uterus inverts itself "en doigt de gant" (like the finger of a glove). *Traité* (1897), 597–98.

75. This was not a commissioned portrait. Sargent showed it at the exhibition of *Les Vingt* in Brussels in 1884.

76. "[F]leur de sang." Michelet, *La mer,* 147.

77. A letter from Robert Proust's daughter describing a visit to the doctor, then age sixty-eight, records the following scene: Pozzi "s'est tourné vers moi et m'a demandé: 'Voulez-vous voir Pozzi au temps de sa splendeur?' et d'un geste théâtral il a tiré un rideau rouge qui cachait son portrait en pied par Sargent" (turned toward [her] and asked: "Would you like to see Pozzi in the time of his splendour?" and with a theatrical gesture he pulled aside a red curtain which hid his full-length portrait by Sargent). Claude Francis and Fernande Garnier, *Marcel Proust et les siens: Suivi des souvenirs de Suzy Mante-Proust* (Paris: Plon, 1981), 157, quoted in Joseph, "Marcel Proust et 'Docteur Dieu,'" 19n27. On the green curtain concealing *L'origine du monde,* see Maxime du Camp, *Les convulsions de Paris,* 2nd ed., vol. 2 (Paris: Hachette, 1879), 263.

78. "Je suis infiniment plus femme que vous." Pozzi to Augustine Bulteau, 16 January 1903, quoted in Vanderpooten, *Samuel Pozzi,* 280.

79. "[E]nvie (de femme enceinte)." Pozzi to Bulteau, 8 April 1902, ibid., 268.

80. "J'aurais envie d'être disséqué, votre scalpel me manque"; "nous nous anatomiserons plus aisément en tête-à-tête, scalpel à scalpel." Pozzi to Bulteau, 10 March and 29 April 1900, ibid., 244, 246. The knife wielding involved in this intimate exploration suggests the Goncourts' thoughts on their own self-investigative procedures: "[L]'observation, au lieu d'émousser en moi la sensibilité, l'a étendue, raffinée, développée, mise à nu. . . . On devient, à force de s'étudier, au lieu de s'endurcir, une sorte d'écorché moral et sensitif, blessé à la moindre impression, sans défense, sans enveloppe, tout saignant" (observation, instead of blunting my sensibilities, has extended them, developed them; it has left me laid bare. . . . One becomes, as a result of this constant self-analysis, the opposite of inured: a kind of skinned and chafed moral and sensitive being, wounded by the minutest impression, defenseless, without a casing, all bloody and raw). Edmond de Goncourt and Jules de Goncourt, *Idées et sensations* (Paris: Librairie Internationale, 1866), 187–88, quoted and translated in Silverman, *Art Nouveau in Fin-de-Siècle France,* 37–38.

81. "Pardonnez cette encre cruotique. C'est goût professionel." Pozzi to Montesquiou, 27 March 1884, quoted in Vanderpooten, *Samuel Pozzi,* 148.

82. Jules Michelet, *Love,* trans. J. W. Palmer (New York: Rudd and Carleton, 1860), 140. "Peu d'hommes (qui ont les deux sexes, et pourtant sont les plus puissant mâles) ont le don d'incubation." Michelet, *L'amour* (1858) (Paris: Calmann-Lévy, 1882), 184. This might explain Pozzi's legendary womanizing.

83. Juliet Bellow, "Engendering the Masculine Interior: John Singer Sargent's *Dr. Pozzi at Home,*" master's thesis (abridged) (University of Pennsylvania, 2000), 2, 8, 10, 11. I am very grateful to the author for sending me a copy of her paper.

84. The word Proust uses is "écoeurent," which literally means to remove the core from the interior of the body, and which colloquially signifies nausea and indigestion. "Chardin," 365.

85. "La nature morte deviendra surtout la nature vivante." Ibid., 366. The viewer, then, experiences a kind of perceptual morning sickness before recognizing Chardin's living offspring. Proust's word "surtout," which means overcoat as well as "above all," is an odd echo of Melville's narratives of literal human inversion, which must, the Mardians argue, account "for that otherwise nonsensical term—'the coat of the stomach'; for originally it must have been a surtout, instead of an inner garment." *Mardi,* 507.

86. "[U]ne poire est aussi vivante qu'une femme." Proust, "Chardin," 374.

87. "Ils sont si inseparables"; "vous l'avez reconnu et pour la première fois vous l'avez goûté." Ibid., 365.

88. "Je nourris mes malades avec de la beauté." Quoted in "Le professeur Pozzi tombe victime d'un dément," *Le Siècle* (Paris), 15 June 1918, 1.

89. "[P]our rendre la maladie belle et la souffrance presque heureuse." Montesquiou, *Les pas effacés,* 3:220. Dr. Trépan's idea of beauty (see epigraph), however, was surely as operative in the clinic as any painterly one.

90. Zola, *Abbé Mouret's Sin,* 61; "la mère naturelle, laissant tomber de ses doigts . . . une sueur d'engendrement," *La faute de l'abbé Mouret,* 1:1264.

91. Pozzi's investment in his patients could be perfectly illustrated by Georges Rochegrosse's *Knight Among the Flower-Maidens* (fig. 25), where the blossoming *femmes-fleurs* assume the positions to best reflect the knight's flowery vision of himself, his mirrorlike steel equipment operating solely in the service of his narcissism.

92. "Chevelure immense de verdure, piquée d'une pluie de fleurs, dont les mèches débordaient de toutes parts, s'échappaient en un échevèlement fou, faisaient songer à quelque fille géante, pâmée au loin sur les reins, renversant la tête dans un spasme de passion, dans un ruissellement de crins superbes, étalés comme une mare de parfums." Zola, *La faute de l'abbé Mouret,* 1:1347.

93. "[C]ave," "caverne," "crevasse," Delvau, *Dictionnaire érotique moderne,* 272, s.v. "nature de la femme (la)"; "crevasse," de Landes, *Glossaire érotique,* 98; "grotte," Guiraud, *Dictionnaire historique,* 378; "grotto," Farmer, *Slang and Its Analogues,* 3:221, "crevice," 4:337 (under "monosyllable").

94. "On a longtemps considéré les végétations comme l'indice constant d'une infection vénérienne, soit blennorrhagique, soit syphilitique." *Traité,* 1015.

95. The affinity of botany and venereology doubtless contributed to Linnaeus's successful transformation from a "clap doctor in Stockholm . . . specializing in syphilis" to Sweden's foremost naturalist. Lisbet Koerner, *Linnaeus: Nature and Nation* (Cambridge: Harvard University Press, 2001), 56.

96. Jean-Louis Alibert, *Clinique de l'hôpital Saint-Louis ou traité complet des maladies de la peau* (Paris: Cormon et Blanc, 1833), 260, 261.

97. Barbara Maria Stafford, *Body Criticism: Imaging the Unseen in Enlightenment Art and Medicine* (Cambridge: MIT Press, 1991), 300, 304.

98. Barrère and Leland, *Dictionary of Slang,* 1:397.

99. "[T]out le jardin des affections cutanées"; "un assez joli bouquet d'infirmités." Michel Corday, *Vénus* (Paris: Bibliothèque Charpentier, 1901), 251, 252, quoted in Claude Quétel, *History of Syphilis,* trans. Judith Braddock and Brian Pike (Baltimore: Johns Hopkins University Press, 1990), 147.

100. See Linda Hutcheon and Michael Hutcheon, *Opera: Desire, Disease, Death* (Lincoln: University of Nebraska Press, 1996), chapter 3, "Syphilis, Suffering, and the Social Order: Richard Wagner's *Parsifal.*"

101. Huysmans, *Against Nature,* 81; *À rebours,* 130.

102. André Couvreur, *Les mancenilles* (Paris: Plon, 1900), quoted in Corbin, *Women for Hire,* 270.

103. "Les tibias s'exfolient en exostoses, comme des colonnes de vert antique dans une ruine romaine." Gautier, *Lettre à la présidente,* 39–40.

104. Alphonse Daudet, *In the Land of Pain,* ed. and trans. Julian Barnes (New York: Knopf, 2002), 43. "La sensation mythologique, l'insensibilisation et le durcissement du torse étreint dans une gaine de bois ou de pierre, et le malade, à mesure que la paralysie monte, se changeant peu à peu en arbre, en rocher, comme une nymphe des *Métamorphoses.*" Daudet, *La doulou: Aux pays de la douleur* (1929) (Paris: Arléa, 1995), 37–38. Barnes's translation of the passage enacts the author's changing identity by shifting from the first person to the third, though the original is entirely in the third person.

105. In Lorrain's 1893 short story "Ophelius" the narrator's best friend, Claudius, falls in love with a young man who resembles the flowery figure in Botticelli's *Primavera.* Obsessed from the moment he first sees the painting's personification of spring (described as studded with anemones and corals, as both flowery and marine), when he meets the living male embodiment of this figure the romance leads to illness and death. "La Primavera était là, vaguement animée sous la vacillante lueur du flambeau, droite sous ses longs cheveux piqués d'anémones et de branchages aux ramifications de madrépores." Jean Lorrain, "Ophelius," in *Buveurs d'âmes,* 157. Lorrain, whose syphilitic condition was exacerbated by the deleterious effects of his ether drinking, was drawn to such pre-Raphaelite figures. In a letter of 16 January 1885 to Oscar Méténier, Lorrain described an infected blond sailor he had picked up and poetically renamed Ophelius: "un matelot blond que j'appelle *Ophélius.* . . . Il lui manque deux dents, et il est couvert de taches de rousseur, mais avec cela il ressemble à la Primavera de Botticelli" (He is missing two teeth, and he is covered in red patches, but with these he ressembles Botticelli's *Primavera*). Jean Lorrain, *Correspondance: Lettres à Barbey d'Aurevilly, François Coppée, Oscar Méténier, Catulle Mendès, Edmond Deschaumes, Mecislas Golberg, etc.* (Paris: Éditions Baudinière, 1929), 70. See also Philippe Jullian, *Jean Lorrain, ou Le Satiricon 1900* (Paris: Fayard, 1974), 35. In Lorrain's case, the treatment of his syphilitic metamorphosis (his anal fistulae had created "a rose of orifices" around his anus) also partook of a garden poetics. He irrigated his fistula-flower, according to Edmond de Goncourt, with "tuyaux semblables à des tuyaux d'arrosage de jardin" (tubes like watering hoses in the garden). Edmond and Jules de Goncourt, *Journal,* entry for 7 July 1895, 3:1152. (With fistulae "une *pomme d'arrosoir* d'orifices peut se constituer, menant à un *clapier* purulent et fétide" [a rose of orifices can form, leading to a purulent and fetid warren]. Vanderpooten, *Samuel Pozzi,* 69. The English equivalent for "pomme d'arrosoir," the multiholed sprinkling nozzle of a watering can, is a "rose." OED, s.v. "rose," 17. Pozzi's 1871 thesis was on fistulae. Samuel-Jean Pozzi, *Étude sur les fistules de l'espace pelvi-rectale supérieur ou fistules pelvi-rectales supérieures* [Paris: G. Masson, 1873]).

106. Huysmans, *Against Nature,* 74; "comme rongées par des syphilis"; "tendaient des chairs livides, marbrées de roséoles, damassées de dartres"; "montraient des épidermes poilus, creusés par des ulcères et repoussés par des chancres," *À rebours,* 120.

107. Zola, *Abbé Mouret's Sin,* 181, 180; "les cereus plantaient des végétations honteuses, des polypiers énormes"; "les mamillaria entassaient des pustules vivantes," *La faute de l'abbé Mouret,* 1:1389.

108. Maurice Maeterlinck, *The Intelligence of the Flowers,* trans. Alexander Teixeira de Mattos (Honolulu: University Press of the Pacific, 2001), 114. "Le pétale inférieur orné à sa naissance de caroncules bronzées, de moustaches mérovingiennes, et de bubons lilas de mauvais augure, s'allonge interminablement, follement, invraisemblablement, en forme de ruban tire-bouchonné, de la couleur que prennent les noyés après un mois de séjour dans la rivière." Maeterlinck, *L'intelligence des fleurs* (Paris: Bibliothèque Charpentier, 1907), 67.

109. Mirbeau, *Torture Garden,* 168, 166; "cornets évasés d'un violet foncé de pourriture à l'intérieur, à l'extérieur d'un jaune verdâtre de décomposition, et semblables à des thorax overts de bêtes mortes"; "des tigridias ouvrant des gorges mutilées"; "les anthuriums, pareils à des plèvres saignantes," *Jardin des supplices,* 202, 199.

110. Huysmans, *Against Nature,* 75, 74; "des Echinopsis, sortant de compresses en ouate des fleurs d'un rose de moignon ignoble"; "d'autres avaient le ton rose vif des cicatrices qui se ferment ou la teinte brune des croûtes qui se forment; d'autres étaient bouillonnées par des cautères, soulevées par des brûlures," *À rebours,* 121, 120.

111. *Against Nature,* 75; "ressemblaient à un sabot, à un vide-poche, au-dessus duquel se retrousserait une langue humaine, au filet tendu, telle qu'on en voit dessinées sur les planches des ouvrages traitant des affections de la gorge et de la bouche," *À rebours,* 121.

112. *Against Nature,* 77; "l'inusable héritage, l'éternelle maladie"; "depuis le commencement du monde, de pères en fils, toutes les créatures se transmettaient l'inusable heritage"; "tout n'est que syphilis," *À rebours,* 125, 124.

113. Quoted in Deborah Hayden, *Pox: Genius, Madness, and the Mysteries of Syphilis* (New York: Basic Books, 2003), 30.

114. Gustave Flaubert, *Bouvard and Pécuchet, with The Dictionary of Received Ideas,* trans. A. J. Krailsheimer (Harmondsworth: Penguin, 1976), 327. "Syphilis. Plus ou moins, tout le monde en est affecté." "Syphilis" entry in Flaubert, *Dictionnaire des idées reçues* (Paris: Jean Aubier, 1951).

115. "[M]aladie contagieuse qui se gagne de tant de manières, qui se présente sous des formes si variées et si multipliées, qu'elle n'est pas susceptible de définition philosophique." "Syphilis" entry in the *Dictionnaire des sciences médicales,* 60 vols. (Paris: C.-L.-F. Panckoucke, 1812–22), quoted in Quétel, *History of Syphilis,* 109. Alibert was also astonished by the "infinite variety" of cutaneous symptoms the disease produced. "On est frappé d'étonnement quand on songe à cette variété infinie d'excroissances et de végétations que produit le virus syphilitique." *Clinique de l'hôpital Saint-Louis,* 261. Jonathan Hutchinson recognized the syphilis spirochete's mimetic powers in an address he gave to the British Medical Society, published in 1879 as "An Address on Syphilis as an Imitator." At the *Congrès international de dermatologie et de syphiligraphie,* held in Paris in 1889 (Pozzi was one of the organizers), Dr. James White of Harvard University gave a talk on hereditary dermatoses, focusing on the links between leprosy, tuberculosis, and syphilis and enumerating their "many properties in common." The protean quality of syphilis and its connections with leprosy and tuberculosis explain why manifestations of our perverse legacy are also occasionally troped as infection by those other diseases. The young doctor Léon Reille diagnoses Eliante Donalger's onanism as leprosy in *La jongleuse;* the senses of Villiers's "callous, carefree playboys" are "leprous, vitiated"; and the symptoms of leprosy are vegetations as well. Leprosy and syphilis had for a long time been considered related diseases: when the army of Charles VIII of France captured the Kingdom of Naples in 1494, an outbreak of syphilis was put down to the Spanish "mixing lepers' blood with Greek wine." Lorrain described his "charming" tubercular symptoms in floral terms—his mucous was "starred with little white camellias." James Hutchinson, "An Address on Syphilis as an Imitator," *British Medical Journal* (1879): 541–42, quoted in Hayden, *Pox,* 33; *Congrès international de dermatologie et de syphiligraphie: Comptes rendus* (Paris: G. Masson, 1890), 363–85; Rachilde, *Juggler,* 17 ("vous avez la lèpre," *La jongleuse,* 44); Villiers de l'Isle-Adam, *Eve of the Future Eden,* 124 ("sens leprosés et plus qu'avilis de ces gais et froids viveurs," *L'Ève future,* 151); Quétel, *History of Syphilis,* 33; "C'est charmant. . . . toutes les muqueuses s'étoilent de petits . . . camélias blancs," Lorrain to Mme. Flore Bergeys, 18 February 1906, quoted in Georges Normandy, *Jean Lorrain intime* (Paris: Albin Michel, 1928), 165.

116. *Against Nature,* 77; "splendeur première," *À rebours,* 125.

117. Sigmund Freud, "Three Essays on the Theory of Sexuality" (1905), in *On Sexuality,* PFL 7, trans. James Strachey, ed. Angela Richards (Harmondsworth: Penguin, 1977), 161.

118. Charles Darwin, *The Effects of Cross and Self Fertilisation in the Vegetable Kingdom* (London: John Murray, 1876), 40.

119. RTPE 2:625; "l'autofécondation, la fécondation de la fleur par elle-même, comme les mariages répétés dans une même famille, amènerait la dégénérescence et la stérilité," RTPF 2:603.

120. Darwin, *Effects of Cross and Self Fertilisation,* 40. As mentioned in chapter 1, Ludwig Dilsner's 1899 play *Jasminblüthe* (Jasmine Blossoms) likens the invert to a blossom with "crippled stamens."

121. Sigmund Freud, *Beyond the Pleasure Principle* (1920), trans. James Strachey (New York: W. W. Norton, 1961), 30.

122. Flaubert to Louis Bouilhet, August 1854, *The Letters of Gustave Flaubert, 1830–1857,* ed. and trans. Francis Steegmuller (Cambridge: Harvard University Press, 1980), 239–40, quoted in Hayden, *Pox,* 140.

123. Hawthorne, *Marble Faun,* 234. Nordau discusses Mallarmé's "long pointed faun-like ears": "After Darwin, who was the first to point out the apish character of this peculiarity, Hartmann, Frigerio, and Lombroso, have firmly established the connection between immoderately long and pointed external ears and atavism and degeneration; and they have shown that this peculiarity is of especially frequent occurrence among criminals and lunatics." Nordau, *Degeneration,* 131. Nordau's description of Mallarmé's ears is taken from Jules Huret, *Enquête sur l'évolution littéraire* (Paris: Charpentier, 1891), 55.

124. "This was the shocking thing" about Mr. Hyde, that in him "the slime of the pit seemed to utter cries and voices; that the amorphous dust gesticulated and sinned . . . that that insurgent horror was knit to him closer than a wife, closer than an eye; lay caged in his flesh." Stevenson, *Dr. Jekyll and Mr. Hyde,* 138.

125. Nordau, *Degeneration,* 556.

126. Charles Blanc, for example, refers to the theory in his *Grammaire des arts du dessin:* "L'homme, avons-nous dit, est une résumé de toutes les créations antérieures. La science moderne nous enseigne que l'embryon humain passe, dans le cours de son développement, par la forme des animaux inférieurs" (31). Haeckel's theory was first published in 1866.

127. RTPE 2:643. Quand "la femme n'est pas seulement intérieurement unie à l'homme, mais hideusement visible, agités qu'ils sont dans un spasme d'hystérique, par un rire aigu qui convulse leurs genoux et leurs mains, [ils] ne ressemblant pas plus au commun des hommes que ces singes à l'oeil mélancolique et cerné, aux pieds prenants, qui revêtent le smoking et portent une cravate noire." RTPF 2:620.

128. Rachilde, *Juggler,* 43–44; "un jovial rire de garçon obtus"; "ce grand singe, modérément femelle," *La jongleuse,* 71, 72. When exhibited in 1865, Manet's *Olympia* was seen as a "sorte de gorille femelle" (sort of female gorilla). Amédée Cantaloube, "Le Salon de 1865," *Le Grand Journal,* 21 May 1865, quoted in Charles Bernheimer, "Manet's *Olympia:* The Figuration of Scandal," *Poetics Today* 10 (Summer 1989): 263. Bernheimer has argued that the scandal surrounding the painting is connected to the figure's sexual

ambiguity: her phallic fingers and the indeterminacy of her "manual gesture" raise the question of sexual difference (269).

129. "[C]himpanzés en redingote." Montesquiou, "Le pavé rouge," 337.

130. RTPE 2:653. "[L]es invertis, qui se rattachent volontiers à l'antique Orient ou à l'âge d'or de la Grèce, remonteraient plus haut encore, à ces époques d'essai où n'existaient ni les fleurs dioïques ni les animaux unisexués, à cet hermaphroditisme initial dont quelques rudiments d'organes mâles dans l'anatomie de la femme et d'organes femelles dans l'anatomie de l'homme semblent conserver la trace." RTPF 2:629.

131. Between 1904 and 1914 at the Hôpital Broca, Robert Proust was Pozzi's assistant and collaborator in his studies of sexual anomalies; Pozzi's own investigations began much earlier, as his publications attest. Joseph, "Marcel Proust et 'Docteur Dieu,'" 13. See also Alice Domurat Dreger, *Hermaphrodites and the Medical Invention of Sex* (Cambridge: Harvard University Press, 1998), 63–65.

132. "Il n'existe pas actuellement un seul example avéré d'hermaphrodisme vrai, de coexistence des ovaires et des testicules chez l'homme. Cette anomalie ne semble pas toutefois impossible, à priori. Cette disposition est fréquente chez les poissons et les batraciens. Elle constitue même la règle chez le crapaud." *Traité*, 1080. Pozzi's own identification with such figures is suggested in his portrait by the outstretched middle finger and the hole created by the thumb and index of his right hand; together these form a hermaphroditic object. I thank Boris Steipe for this observation and for discussing medical metaphors with me.

133. "[U]ne cause supérieure, l'atavisme, reproduisant sporadiquement dans une espèce les formes d'une autre espèce, par l'effet de ce que Darwin a appelé une phénomène de réversion." Ibid., 1088.

134. See the Introduction. Krafft-Ebing quotes a range of sources grappling with the question of sexuation: "The research of Frank Lydston (*Philadelphia Med. And Surg. Recorder*, September 1888) and Kiernan (*Medical Standard*, November 1888) suggests [1] that bisexual organization is still found in the lower animal kingdom, and [2] that monosexuality gradually develops from bisexuality"; "Klaus' research in zoology (*Zoologie*, 1891, p. 490) shows that not only does hermaphroditism exist in the lower grades of the animal world, sexual exchange (physiological?) within one individual may also occur"; "recent explanations of the homosexual instinct have come from philosophers, psychologists and natural scientists." Krafft-Ebing, *Psychopathia Sexualis*, 284, 670n88, 281.

135. Flaubert, *Temptation of Saint Anthony*, 232. "Enfin, il aperçoit de petites masses globuleuses, grosses comme des têtes d'épingles et garnies de cils tout autour. Une vibration les agite." "Ô bonheur! Bonheur! J'ai vu naître la vie, j'ai vu le mouvement commencer." *La tentation de Saint-Antoine—1874*, 4:200.

136. On the perceived connection between syphilis and artistic genius, see Quétel, *History of Syphilis*, 172–75, and Hayden, *Pox*, passim. An early twentieth-century doctor writing under the pseudonym Marc la Marche asked in his *Observations sur les maladies vénériennes*, "N'est-tu pas, Syphilis, le grand entremetteur / Poussant l'homme vers son génie, / Le ferment des pensers puissants, le transmetteur / Des germes de science et d'art?" (Are you not, Syphilis, the great go-between / Working to put man in touch with his genius, / The source of powerful thoughts, the transmitter / Of the seeds of art and science?). Marc la Marche, *Observations sur les maladies vénériennes . . . En vrac* (Paris: Jouve, 1929), quoted in Claude Quétel, *Le mal de Naples: Histoire de la syphilis* (Paris: Seghers, 1986), 220.

137. Quoted in Quétel, *History of Syphilis*, 173. "[A]ussi bien le fouet du génie et du talent, de l'heroïsme et de l'esprit, que celui de la paralysie générale, du tabes et de presque toutes les dégénérescences." Daudet, *Devant la douleur*, part 2 of *Souvenirs des milieux littéraires, politiques, artistiques et médicaux*, 1:208.

138. "It is the yeast which raises the somewhat heavy dough of the peasantry and refines it in two generations. It makes a great poet of a maid's son." Quoted in Quétel, *History of Syphilis*, 174. "Il est le ferment qui fait lever la pâte un peu lourde du sang paysan et l'affine en deux générations. Du fils d'une bonne il fait un grand poète." Daudet, *Devant la douleur*, 1:208–9.

139. "Analogue pour l'élan et l'acrobatisme au propagateur de la vie, associé à lui dans mainte conception par la transmission héréditaire, le tréponème propage à la fois l'intensité dramatique de la vie, la stérilité qui est son contraire." Daudet, *Devant la douleur*, 1:209.

140. Darwin, *Effects of Cross and Self Fertilisation*, 386.

141. RTPE 2:625. "Cependant cet essor peut être excessif, l'espèce se developer démesurément; alors, comme une antitoxine défend contre la maladie, comme le corps thyroïde règle notre embonpoint, comme la défaite vient punir l'orgueil, la fatigue le plaisir, et comme le sommeil repose à son tour de la fatigue, ainsi un acte exceptionnel d'autofécondation vient à point nommé donner son tour de vis, son coup de frein, fait rentrer dans la norme la fleur qui en était exagérément sortie." RTPF 2:603. Proust's sentence enacts the curbing of its own disproportionate florid growth.

142. Frédérique Desbuissons suggests the influence of Gautier d'Agoty's colored anatomical prints in particular. Desbuissons, "Voir le mal en face: 'L'origine du monde' de Gustave Courbet," *Ligeia* 19–20 (October 1996–June 1997): 16.

143. "Ce corps tronqué . . . n'était pas autre chose pour les contemporains que l'agent infernal de la terrible épidémie" (This truncated body . . . was for contemporaries nothing other than the infernal agent of the terrible epidemic). Michèle Haddad, "Courbet, entre présence et absence, dans deux toiles limites: *Les baigneuses* et *L'origine du monde*," in *Courbet, l'amour . . .*, exh. cat. (Ornans: Musée Gustave Courbet, 1996), 135.

144. "Ne soulève pas le rideau / Qui cache à tes yeux cette toile; / Le formidable objet qu'il voile . . . C'est lui qui te courbe avant l'âge, / Et, de noirs, fait tes cheveux blancs / C'est lui qui t'ébrèche les dents, / Lui qui te plombe le visage . . . / Il tient une

vengeance prête, / Pour quiconque lui fait affront." Jules-Antoine Castagnary, "Sur un tableau de la collection Khalil Bey," in *Courbet, raconté par lui-même et par ses amis,* ed. Pierre Courthion, 2 vols. (Geneva: Pierre Cailler, 1948, 1950), 1:194–96. Françoise Gaillard's description of Courbet's painted vagina as "la lésion par laquelle peut surgir à la surface une primitivité" (the *lesion* through which may surge to the surface a primitivity) (emphasis added) anticipates and complements Haddad's reading. Gaillard, "Allégorie d'un fantasme fin de siècle, Courbet: *L'Origine du monde,*" in *Mimesis et semiosis: Littérature et représentation; Miscellanées offertes à Henri Mitterand,* ed. Philippe Hamon and Jean-Pierre Leduc-Adine (Paris: Nathan, 1992), 433.

145. "[L]a fée de nos prairies: elle est belle et gracieuse"; "l'hiver, la Dame verte habite dans ces grottes de rochers." Xavier Marmier, *Souvenirs de voyages et traditions populaires* (Paris: P. Masgana, 1841), 76, 77.

146. "[T]rou poilu"; "ce trou qui ouvre sur les abîmes de l'animalité"; "la femme, par sa fonction de génitrice, est, dans l'imaginaire du temps, l'agent de transmission de cette animalité originelle." Gaillard, "Allégorie d'un fantasme fin de siècle," 431, 432.

147. "[S]on sexe est un gouffre sans fond qui communique avec l'origine"; "l'histoire de notre terrifiant ancêtre, l'hominien de l'époque glaciaire. Freud l'a rendu responsable de nos névroses et de nos perversions sexuelles." Ibid., 432, 433.

148. "[L]a charcuterie de Pozzi"; "cette maison de la mort, où sont décédées quatre femmes pendant le temps qu'il y a passé." Quoted in Edmond and Jules de Goncourt, *Journal,* entries for 29 April 1896 and 7 July 1895, 3:1273, 1152. Interestingly, Maxime du Camp asserted that the word for an artist who would create a work like *L'origine du monde* was "only used in butcher shops": "[Il] n'est guère usité qu'en charcuterie." *Les convulsions de Paris,* 264.

149. "J'ai voulu figurer la maladie . . . dans toute sa tragique horreur. R. Lalique, 16 mars 1901." Quoted in Vanderpooten, *Samuel Pozzi,* 263.

150. Emily Apter, "The Garden of Scopic Perversion from Monet to Mirbeau," *October* 47 (Winter 1988): 104.

151. "[U]ne énorme et fragile poche de pus éclate sous l'ongle qui cherche le plan de clivage." Vanderpooten, *Samuel Pozzi,* 87.

152. Pozzi's biographer gives a fairly graphic description of the body fluids with which the doctor's "naked hands" were in daily contact: they were "mains poisseuses des litres de liquide jaune giclant des énormes kystes percés à force d'un gros 'trocard,' pleines de sang des fibromes saignants, des 'eaux rousses' des cancers bourgeonnants du col de l'utérus qu'il faut aller reconnaître dans leur profondeur, trempées de l'urine incessante suintant des cloaques et fistules des mères déchirées par l'accouchement" (made sticky from liters of yellow liquid from enormous spurting cysts forcibly pierced by a large 'trocar,' full of blood from bleeding fibromas, of 'red waters' of burgeoning cancers of the cervix that must be reconnoitred in their depth, drenched with the urine incessantly oozing from the cloacas and fistulae of mothers torn open during labor), and so on. Vanderpooten, *Samuel Pozzi,* 37.

153. "Les ongles seront nettoyés avec un soin *très grand,* à l'aide d'une lime pointue." Pozzi, *Traité,* 2.

154. Pozzi recommends Lawson Tait's practice of "scraping with his nails" to produce "fraying of the serosa and of the fibrous texture of the ligament" in cases where tissues adhere. "[E]n grattant avec les ongles, il produit en ce point des éraillures de la séreuse et de la charpente fibreuses du ligament." *Traité,* 655.

155. "[D]oigts d'acier." Vanderpooten, *Samuel Pozzi,* 86.

156. "Pozzi assis entre les cuisses ouvertes incise sans trembler l'abdomen bombant. . . . Les têtes chauves ou blanches des spectateurs se penchent pour mieux voir, les barbes balaient les épaules de l'opérateur, reculant parfois devant un jet brutal de liquide jaune ou verdâtre qui bientôt trempe les tabliers et les pieds des chirurgiens lorsqu'un volumineux kyste de l'ovaire . . . se rompt et lance son contenu louche" (Pozzi sat between the open thighs to incise the swollen abdomen without trembling. . . . The bald or white heads of the spectators stooping to see better, the beards brushing the shoulders of the operator, recoiled now and then before a brutal jet of yellow or greenish liquid which easily soaked the aprons and the feet of the surgeons when a voluminous ovarian cyst . . . burst and shot forth its murky contents). Vanderpooten, *Samuel Pozzi,* 87.

157. Their color can be "d'un blanc nacré ou bleuâtre, marbrée par les vaisseaux veineux," their contents "clair comme l'eau de roche." *Traité,* 692, 703. Pozzi referred to a cyst on Robert de Montesquiou's eyelid as "la petite perle blanche sur le bord libre de la paupière inférieure" (the little white pearl on the clear edge of the lower lid). Pozzi to Montesquiou, 25 December 1901, quoted in Vanderpooten, *Samuel Pozzi,* 266.

158. "Son KYSTE n'était pas vulgaire: élégant, tenace en ces racines multiples, enfoui, SERTI dans le 'ligament large' où j'ai été le *désenchâsser* (excusez le vocable)." Pozzi to Montesquiou, 20 February 1898, quoted in Vanderpooten, *Samuel Pozzi,* 224 (emphasis added).

159. In Stevenson's story the Rajah's Diamond is a fetish "so imposing" that anyone would "prostrate himself in adoration" before it, because it would set free whosoever possessed it "from the primal curse," and allow him to forever "follow his own inclinations without concern or hurry, without let or hindrance." Robert Louis Stevenson, "The Rajah's Diamond" (1882), part of "The New Arabian Nights" series, *The Complete Short Stories of Robert Louis Stevenson,* ed. Charles Neider (New York: Da Capo Press, 1998), 139–40.

160. For example, in a letter of 1898: "À mon Docteur Dieu." Quoted in Vanderpooten, *Samuel Pozzi,* 227. In *À la recherche du temps perdu,* Mme. Verdurin calls Dr. Cottard, who is modeled in part on Pozzi, "Dr. Dieu." RTPE 2:995; RTPF 2:963.

161. "No doubt the life of certain inverts appears at times to change, their vice (as it is called) is no longer apparent in their habits; but nothing is ever lost: a missing jewel turns up again." RTPE 2:648. "Sans

doute la vie de certains invertis paraît quelquefois changer, leur vice (comme on dit) n'apparaît plus dans leurs habitudes; mais rien ne se perd: un bijou caché se retrouve." RTPF 2:625.

162. Delacroix's "bijou" (see chapter 2) functions as such a fetish, imaginarily involving the literal un-*main*-ing of a woman. Even when women are actually "castrated" in Pozzi's treatise, through oophorectomy or salpingo-oophorectomy, menstruation may continue, he claims, since the genital apparatus shares "the nervous system of vegetative life," which can "reproduce, so to speak, automatically and under the influence of an old stimulus, acts such as congestion of the genital apparatus." "On comprend fort bien que le système nerveux de la vie végétative, toute comme celui de la vie de relation, puisse reproduire, pour ainsi dire automatiquement et sous l'influence d'une incitation ancienne, des actes tels que la congestion de l'appareil génital." *Traité*, 589.

163. Vanderpooten's hyperbolic description of Pozzi's relations with women seems justified by contemporary accounts of Pozzi's philandering: "Il les aime passionnément, comme tout ce qui est beau. Brunes ou blondes, d'une minceur de liane comme Sarah, ou pulpeuses comme Judith Gautier, la fille du bon Théo, douces ou violentes. . . . Il mène 'double femmes,' le plus normalement du monde, évoluant entre celles du matin, femmes de salles gisantes, souffrantes, exposées, déformées, grimaçantes, pitoyables sur des lits de toile rude, ou soignantes asexuées, noires ou blanches, voilées, silencieuses, déférentes, et celles du soir ou de la nuit, de salon ou de théâtre, sur canapés, fauteuils ou bergères ou sous les feux de la rampe, chatoyantes, parées, souriantes, odoriférantes, ensorcelantes, grisantes, tentantes. . . . Il les aime toutes, à leurs heures, à ses heures. . . . il sera souvent aimé, certaines, fortes ou volages, s'envoleront, le laissant désespéré; il en aimera et quittera plusieurs, le coeur en morceaux—car il est sincère, chaque fois. Il n'a pas long chemin à faire, beaucoup seront les premières à lui dire, répéter, écrire qu'elles l'aiment. . . . Il garde leur lettres" (He loved them passionately, like all that was beautiful. Brunette or blond, of a liana slenderness like Sarah, or pulpy like Judith Gautier, daughter of the good Theo, sweet or violent. . . . He managed "double women" the most normally in the world, moving between those of the morning, women of lying-in rooms, suffering, exposed, deformed, grimacing, pitiable on beds of rough canvas, or convalescent asexuals, black or white, muffled, silent, deferential, and those of the evening or the night, of the salon or the theater, on sofas, arm-chairs or wing-chairs or under the flames of the footlights, iridescent, adorned, smiling, odoriferous, spellbinding, intoxicating, tempting. . . . He loved them all, as it suited them, as it suited him. . . . he would often be loved . . . [yet] certain, strong, or flighty, they would vanish, leaving him in despair; he would love and leave many, his heart in pieces—because he was sincere, each time. He didn't have much difficulty, many would be the first to speak to him, to repeat, to write that they loved him. . . . He kept their letters). Vanderpooten, *Samuel Pozzi*, 91–92. Pozzi is reputed to have said to his wife, "I don't deceive you, my dear, I supplement you." Painter, *Marcel Proust*, 1:101.

164. As Villiers de l'Isle-Adam's Dr. Ryllh says, "in my hands the sick may lose their lives, but never their hope." *Eve of the Future Eden*, 75; "Entre mes mains les malades peuvent perdre la vie;—jamais l'espoir!" *L'Ève future*, 101.

165. "À mon Docteur Dieu, à l'Être que j'adore et admire et auquel je donnerais avec joie ma vie. Sarah Bernhardt." Bernhardt to Pozzi, 1898, quoted in Vanderpooten, *Samuel Pozzi*, 227.

166. "[D]ans la chambre à coucher du professeur—son cabinet étant en réparation." "Le professeur Pozzi tombe victime d'un dément," 2.

167. Joseph, "Marcel Proust et 'Docteur Dieu,'" 21. The motive for the crime was to "teach doctors who don't want to obey their clients" a lesson, and after shooting Pozzi the deranged man (named Maurice Machu) used two more bullets to kill himself. "[P]our apprendre aux médecins qui ne veulent pas obéir à leurs clients." "Le professeur Pozzi tombe victime d'un dément," 2.

168. "Il s'était confie aux soins du docteur de Martel, son ami, mais en fait, c'est lui-même qui dirigea l'opération, avec un grand sang-froid"; "Il fallut pratiquer douze fois l'opération"; "au moment où l'on croyait l'opération terminée, et où le docteur paraissent reposer avec calme, on le vit tout à coup trésaillir. Un soubresaut . . . un caillot de sang sortit de sa buoche . . . il était mort. On suppose qu'une treizième perforation a dû passer inaperçue"; "ainsi le docteur Pozzi devait mourir de la blessure qu'il avait lui-même tant de fois guérie." "Le professeur Pozzi tombe victime d'un dément," 2, ellipses in original.

169. "[R]econnaître." Vanderpooten, *Samuel Pozzi*, 87.

170. Lucia Fairchild, diary entry for Paris 1891, quoted in Miller, "Sargent in the Diaries of Lucia Fairchild," 12.

171. Christian Brinton, "Sargent and His Art," *Munsey's Magazine*, December 1906, 284.

172. James, "John S. Sargent," 684. In Merleau-Pontian terms, his look "envelops, palpates, espouses the visible things." "Intertwining," 133.

173. Royal Cortissoz, "The Field of Art," *Scribner's Magazine*, March 1924, 348. For a contextualization of these tropes, see Sarah Burns, "The 'Earnest, Untiring Worker' and the Magician of the Brush: Gender Politics in the Criticism of Cecilia Beaux and John Singer Sargent," *Oxford Art Journal* 15, no. 1 (1992): 48.

174. Edwin Blashfield, "John Sargent," in *Commemorative Tributes to Cable by Robert Underwood Johnson, Sargent by Edwin Howland Blashfield, and Pennell by John Charles van Dyke* (New York: American Academy of Arts and Letters, 1927), 11.

175. Royal Cortissoz, "John S. Sargent," *Scribner's Magazine*, November 1903, 524.

176. Recorded in a manuscript by Sargent's assistant, Thomas A. Fox. "Painted Portraits are 'Paughtraits,'" box 2, folder 26, John Singer Sargent–Thomas A. Fox Papers, Boston Athenaeum.

177. Merleau-Ponty, "Working Notes," 209.

178. Émile Zola, *The Masterpiece,* trans. Thomas Walton and Roger Pearson (Oxford: Oxford University Press, 1999), 407. "Nos mains n'ont-elles donc pas la puissance de créer des êtres?" Zola, *L'oeuvre,* 380.

179. "[C]haque touche d'un John Sargent est une contribution à la vie." Camille Mauclair, "John Sargent," *L'Art et les Artistes* 4 (October 1906–March 1907): 374.

180. Robert Ross, "The Wertheimer Sargents," *Art Journal* 73 (January 1911): 10.

181. Albert Boime, "Sargent in Paris and London: A Portrait of the Artist as Dorian Gray," in Hills, *John Singer Sargent,* 88.

182. Ibid., 88. "Lavender or chlorate-of-potash-lozenge" (vegetable ash) skin, Sargent told Vernon Lee in a letter of 10 February 1883, Ormond Family Archive, quoted in CP 1:113.

183. Hamish Bowles, "The Madame X Files," *Vogue* magazine, January 1999, 179.

184. "[Q]uasi extrahumaine." Louis Fourcaud, "Le Salon de 1884," *Gazette des Beaux-Arts* 29 (May 1884): 483.

185. "[C]e teint où l'héliotrope est pétri avec la rose." Judith Gautier, "Le Salon (premier article)," *Le Rappel,* 1 May 1884, 1. Various writers expounded upon the floral charms of unusual skin. Proust's rare shrub of a page boy attracts "the eye no less by the unusual and harmonious colouring of his hair than by his plant-like epidermis." RTPE 1:758–59; il "ne frappait pas moins les yeux par l'harmonie singulière de ses cheveux colorés que par son épiderme de plante," RTPF 1:706. And Delarue-Mardrus's Geraldine, in *Hortensia dégénéré,* with "la couleur inouïe de son teint" (the unheard of color of her complexion), resembles a "corolle immaculée au pollen vert pâle" (immaculate corolla with pale green pollen). Lucie Delarue-Mardrus, *Hortensia dégénéré* (Paris: J. Ferenczi et Fils, 1929), 126.

186. RTPE 2:624; "de même la fleur-femme qui était ici, si l'insecte venait, arquerait coquettement ses 'Styles,' et pour être mieux pénétrée par lui ferait imperceptiblement, comme une jouvencelle hypocrite mais ardente, la moitié du chemin," RTPF 2:603–4.

187. See Trevor Fairbrother, "The Shock of John Singer Sargent's 'Madame Gautreau,'" *Arts Magazine* (January 1981): 90–97.

188. Olson, *John Singer Sargent,* 101.

189. Boime describes the portrait as the result of a collaborative effort by Sargent and Gautreau, who "shared a similar concept about the final work." "The startling innovation of [her] pose had to have been arranged between them." "Sargent in Paris and London," 91.

190. "[I]ndécent," "cadavérique." Joséphin Péladan, "Salon de 1884: Peinture," *L'Artiste* 54 (June 1884): 441.

191. "Eccentricities of French Art," *Art Amateur* 11 (August 1884): 52.

192. Claude Phillips, "The Salon II," *Academy* 25 (14 June 1884): 427. For more on the reception of *Madame X,* see Fairbrother, "Shock of Sargent's 'Madame Gautreau'"; Boime, "Sargent in Paris and London"; Simpson, Ormond, and Weinberg, *Uncanny Spectacle,* 140–41; and Susan Sidlauskas, "Painting Skin: John Singer Sargent's *Madame X,*" *American Art* 15 (Fall 2001): 8–33.

193. Sidlauskas, "Painting Skin," 23.

194. William Sharp, "The Paris Salon," *Art Journal* 46 (June 1884): 180.

195. "Un certain goût moins sensible aux franches floraisons saines qu'aux plantes artificielles de boudoir." André Michel, "Le Salon de 1884," *L'Art* 37 (1884): 14.

196. Sharp, "Paris Salon," 180.

197. "[T]out ce délicieux je ne sais quoi de l'épiderme de la femme, qu'on dirait fait avec le dessous de l'aile des colombes, l'intérieur des roses blanches." Quoted in Jean-Pierre Richard, "Deux écrivains épidermiques: Edmond et Jules de Goncourt," in Richard, *Littérature et sensation* (Paris: Éditions de Seuil, 1954), 268.

198. "[L]e dedans tel qu'il peut se laisser apercevoir et appréhender du dehors." Ibid., 268.

199. Sally Promey discusses the painter's depiction of costumes, tents, umbrellas, mosquito nets, and veils as "membrane[s] between the individual and the world"; Sargent marks "in an obvious fashion the sometimes all too insubstantial boundary between interior and exterior." Promey, *Painting Religion in Public,* 257–58.

200. Sidlauskas, "Painting Skin," 25.

201. Quoted in CP 1:113.

202. Sidlauskas, "Painting Skin," 28–29.

203. He also saw the right arm as dislocated: "le bras droit désarticulé, la main désossée." Henry Houssaye, "Le Salon de 1884," *Revue de Deux Mondes* 54 (June 1884): 589.

204. The exact profile of Madame Gautreau required a series of adjustments to the head and arms. Studies for the painting suggest Sargent's "obsession with the contours." Dorothy Mahon and Silvia Centeno, "A Technical Study of John Singer Sargent's Portrait of Madame Pierre Gautreau," *Metropolitan Museum Journal* 40 (2005): 123. Line suggests clinical precision here, but color could be as surgical: one critic remarked that *Olympia*'s "coloris au verjus, aigre et acide, pénètre dans l'oeil comme la scie d'un chirurgien dans les chairs" (color of verjuice, bitter and acid, penetrates the eye like the saw of a surgeon penetrating flesh). Geronte [Louis Leroy?], "Le Salon en 1865," *Gazette de France,* 30 June 1865, quoted in Michael Fried, *Manet's Modernism, or The Face of Painting in the 1860s* (Chicago: University of Chicago Press, 1996), 587n186.

205. Edward Simmons, *From Seven to Seventy: Memories of a Painter and a Yankee* (New York: Harper and Brothers, 1922), 127.

206. Ovid, *Metamorphoses,* trans. A. D. Melville (Oxford: Oxford University Press, 1987), 232.

207. Lacan, *Four Fundamental Concepts of Psycho-Analysis,* 82.

208. Sidlauskas, "Painting Skin," 29. Stanley Olson tells a different version: "when she went bathing, she climbed into the bathing machine, wheeled out a good distance, to protect her from spectators, but the fleeting glimpse of her shoulders showed the white colour intact." He attributes this to her possible use of arsenic as a whitener. *John Singer Sargent,* 102.

Chapter 4

1. Edmund Gosse to Evan Charteris, c. 1925, quoted in Charteris, *John Sargent*, 74.

2. "M. Paul Desjardins, in an amusing and illuminating phrase, speaks of 'la cinématographie d'une abeille dans le mystère de la mellification.' This, I confess, is what I like best in a literary biography, and it is what I have attempted to produce." Edmund Gosse, preface to *Portraits and Sketches* (1912) (Freeport, N.Y.: Books for Libraries Press, 1969), viii. The aptly named Desjardins had been Proust's philosophy lecturer when studying for his *licence en droit*. Painter, *Marcel Proust*, 1:223.

3. Sir George Henschel recalled of Sargent's back-and-forth movements that he would "recede about a dozen steps from the easel, look at me steadfastly, stop for a moment and suddenly, the brush lifted ready for action and without ever taking his eyes off me, made [*sic*] a dash for the canvas on which he then recorded his impression, generally accompanying the act by contentedly humming a little tune." George Henschel, *Musings and Memories of a Musician* (London: Macmillan, 1918), 332–33, quoted in CP 1:185. See also, for example, the letter of 13 September 1933 from Frederick Olmsted (Jr.) to Thomas A. Fox, which describes the "very springy motion" of the painter's swift dashes between canvas, model, and a spot twenty feet away. Quoted in Laura Wood Roper, *FLO: A Biography of Frederick Law Olmsted* (Baltimore: Johns Hopkins University Press, 1983), 467–68.

4. Charteris, *John Sargent*, 28.

5. Simpson, Ormond, and Weinberg, *Uncanny Spectacle*, 131. That his art required full mobility is confirmed by the fact that when Sargent once injured his foot, he couldn't paint. Charles Merrill Mount, *John Singer Sargent: A Biography* (London: Cresset Press, 1957), 272.

6. B. Purvis, "Sargent and His Sitters," letter to the editor, *Times* (London), 21 February 1926, 7, quoted in Fairbrother, *John Singer Sargent: The Sensualist*, 43.

7. Elaine Scarry, *Dreaming by the Book* (New York: Farrar, Straus and Giroux, 1999), 162.

8. There are also, of course, mineral pigments, but the correspondence of flowers and paints was more prominent in nineteenth-century writings on art (some key sources are quoted below) and certainly more important to the artists I examine. Baudelaire described plant life *as* color: "La sève monte et, mélange de principes, elle s'épanouit en tons mélangés" (the sap rises and, mixing principles, it blossoms in mixed tones). Charles Baudelaire, "Salon de 1846," in *Curiosités esthétiques*, ed. H. Lemaître (Paris: Garnier, 1962), 105.

9. The story is based on Carlo Ridolfi's "Life of Titian," in his 1648 *Le meraviglie dell'arte*. William Dyce's painting attests to the tale's nineteenth-century currency. Marcia Pointon, *William Dyce, 1806–1864: A Critical Biography* (Oxford: Clarendon Press, 1979), 145.

10. Quoted and translated by Carol Armstrong in "Facturing Femininity: Manet's *Before the Mirror*," *October* 74 (Fall 1995): 90. "Elle broie sur sa palette des pétales de fleurs, pour les étaler ensuite sur la toile en touches spirituelles, soufflées, jetées un peu au hasard." Charles Ephrussi, "Exposition des artistes indépendants," *Gazette des Beaux-Arts* 21 (May 1880): 487.

11. Rubin, *Manet's Silence and the Poetics of Bouquets*, 158.

12. Even the name of the company—Maule's—is painterly: a maulstick is a tool used by the painter to steady his or her hand.

13. Charles Baudelaire, "The Life and Work of Eugène Delacroix," in *The Painter of Modern Life and Other Essays*, ed. and trans. Jonathan Mayne (London: Phaidon, 1995), 48. "Je n'ai jamais vu de palette aussi minutieusement et aussi délicatement préparée que celle de Delacroix. Cela ressemblait à un bouquet de fleurs, savamment assorties." Baudelaire, "L'oeuvre et la vie d'Eugène Delacroix" (1863), *Oeuvres complètes*, ed. Claude Pichois, 2 vols. (Paris: Gallimard, 1975–76), 2:748.

14. Sargent's "encounters with pigment" were both "passionate and eccentric." Charles Merrill Mount, "A Phoenix at Richmond," *Arts in Virginia* 18 (Winter 1978): 7.

15. Olson, *John Singer Sargent*, 131.

16. Cecil Chard, "John S. Sargent, R.A.: The Work of a Great Portrait-Painter," *Pall Mall Magazine*, June 1907, 651.

17. I have slightly modified Rubin's translation in *Manet's Silence and the Poetics of Bouquets*, 165. "Rien qu'un égrenure de blanc et de bleu"; "les fleurs se lèvent de la toile; le dessin feuillu de l'oeillet, le coeur de la fleur, son ombre tendre, son chiffonnage, son déchiquetage, tout s'assemble et s'épanouit." Edmond de Goncourt and Jules de Goncourt, "Chardin" (1864), in *L'art du dix-huitième siècle*, 3rd ed., 2 vols. (Paris: A. Quantin, 1880–82), 1:72.

18. Hale, "The Sargent I Knew," 566.

19. Statement of Julie Heyneman quoted in Charteris, *John Sargent*, 182.

20. John Singer Sargent, preface to Goupil Gallery exhibition catalogue of Brabazon's works (December 1892), reprinted in C. Lewis Hind, *Hercules Brabazon Brabazon, 1821–1906: His Art and Life* (London: George Allen, 1912), 86.

21. They also exhibited together in group shows at the Cercle des arts libéraux in 1881 and in 1885 at the Galerie Georges Petit. CP 1:155.

22. Monet's palette was of great interest to Sargent. In a letter of 20 October 1885 to Alice Hoschedé, Monet complains of the time required to respond to one of Sargent's "extraordinary enquir[ies] about the use of yellow and green." "[U]n renseignement extraordinaire sur l'emploi du jaune et du vert," reprinted in Wildenstein, *Claude Monet*, 2:262, letter 592.

23. I thank Sally Promey for her observation that the figure on the right, one of Monet's stepdaughters (possibly Suzanne Hoschedé, CP 1:157), appears as another example of his floral creation: a giant lily in a classic Monet pose; compare her figure with the lilies in *Carnation, Lily, Lily, Rose* and *Garden Study of the Vickers Children* (fig. 151).

24. According to Violet Ormond, Sargent met "the 'botanizing priest'" in the Alps. "I think he had more or less been banished to this obscure little parish, his theological ideas not being strict enough." Ormond to

David McKibbin, 1951, quoted in Doreen Bolger Burke, *American Paintings in the Metropolitan Museum of Art,* vol. 3, ed. Kathleen Luhrs (New York: Metropolitan Museum of Art, 1980), 261.

25. "He broke with the R[oman] C[atholic] Church, & went to America. [I] know my brother was very kind & helpful. I don't know what became of him." Ibid.

26. See Simpson's "Cast of Characters" in RGA, 35–227. Regarding wildflowers, see Kate Millet to her family, 26 April 1886, Archives of American Art, Smithsonian Institution, quoted in RGA, 347.

27. Lucia Millet catalogued for her family what was in bloom; Alfred Parsons, a landscape painter, also made botanical prints and later became a landscape designer, and so on. RGA, 345–46, 221.

28. Sargent to Mrs. Charles Fairchild, 6 March 1892, John Singer Sargent–Thomas A. Fox Papers, Boston Athenaeum.

29. Jacqueline Ridge and Joyce Townsend note that he virtually never drew onto the canvas before painting. Ridge and Townsend, "John Singer Sargent's Later Portraits: The Artist's Technique and Materials," *Apollo* 148 (September 1998): 24. For more on Sargent's technique, see Patricia Hills, "The Formation of a Style and Sensibility," in Hills, *John Singer Sargent,* 27–47.

30. As Ormond remarks, "in spite of the shortness of the effect, and the long period over which the picture was painted, immediacy of touch is everywhere apparent." Richard Ormond, *John Singer Sargent: Paintings, Drawings, Watercolors* (New York: Harper and Row, 1970), 39.

31. Sargent to Emily Sargent, late September or early October 1885, reproduced in Charteris, *John Sargent,* between 76 and 77.

32. Or that if it doesn't, it should. The setting, according to one of Sargent's friends and fellow artists, leads one to expect "fantastic lovers." T. Martin Wood, *Sargent* (London: T. C. and E. C. Jack, 1909), 29–30.

33. Martha Kingsbury, "John Singer Sargent: Aspects of His Work" (PhD diss., Harvard University, 1969), 72.

34. "[L]anterne," de Landes, *Glossaire érotique,* 227.

35. Richard Ormond, catalogue entry in Elaine Kilmurray and Richard Ormond, eds., *John Singer Sargent,* exh. cat. (Princeton: Princeton University Press, 1998), 78; Ormond, "Modern Life Subjects," in Robertson, *Sargent and Italy,* 72.

36. Blashfield, "John Singer Sargent—Recollections," 644.

37. Lucia Fairchild recorded in her diary that Sargent especially loved the Flower Maidens music; he went to Bayreuth to see *Parsifal* in 1886. Miller, "Sargent in the Diaries of Lucia Fairchild," 9; see also Lucia Millet to her family, 25 July 1886, quoted in RGA, 368.

38. Theodore Child, "American Artists at the Paris Exhibition," *Harper's New Monthly Magazine,* September 1889, 506. If the girls mothify, the lilies are in turn humanized, to both critical acclaim and consternation. John van Dyke wrote approvingly, "the painting of the lilies shows the same exactness of observation that characterizes the faces. They are portraits of lilies." Another critic resented the flowers' "abnormal size and staring hues." John C. van Dyke, *American Painting and Its Tradition* (New York: Charles Scribner's Sons, 1919), 261; *Daily Telegraph,* 5 May 1887, 8, quoted in Anne Helmreich, "John Singer Sargent, *Carnation, Lily, Lily, Rose,* and the Condition of Modernism in England, 1887," *Victorian Studies* 45 (Spring 2003): 441.

39. By the popular essayist and flower-poet Leigh Hunt, for example: "the nice-leaved lesser Lilies, / Shading, like detected light, / Their little green-tipt lamps of white." Quoted in Lucy Hooper, ed., *The Lady's Book of Flowers and Poetry* (New York: Derby and Jackson, 1860), 94, lines 2–4. For other luminous flowers, see the chapter "Poetic Images of the Flame in Plant Life," in Gaston Bachelard, *The Flame of a Candle,* trans. Joni Caldwell (Dallas: Dallas Institute of Humanities and Culture, 1988). Walter Crane included tulip torches and primrose lamps in his book *Flora's Feast: A Masque of Flowers* (London: Cassell, 1889), 14.

40. Ormond, *John Singer Sargent,* 39. While the lantern metaphor suggests instantaneity (art flaring into being like a match's flame), raking light photographs reveal that the lanterns are the most heavily worked areas on the canvas. Given their centrality and metapictorial charge, it is not surprising that Sargent should have labored to perfect them. Tate conservation file 1615, Tate Britain, London.

41. Stevenson Papers, Beinecke Rare Book and Manuscript Library, Yale University Library, quoted in Richard Ormond, "*Carnation, Lily, Lily, Rose,*" in *Sargent at Broadway: The Impressionist Years,* ed. Stanley Olson, Warren Adelson, and Richard Ormond (New York: Universe and Coe Kerr Gallery, 1986), 67.

42. Blashfield, "John Sargent," 18. Jacques-Émile Blanche compared Sargent's travails to those of Degas, who with every work experienced "les douleurs maternelles de la parturition" (the maternal sufferings of parturition). Blanche, "Un grand Américain: John Singer Sargent," in *Propos de peintre: De Gauguin à la Revue nègre,* 3rd ed. (Paris: Éditions Émile-Paul Frères, 1928), 68.

43. Abbey to Charles Parsons, 28 September 1885, quoted in E. V. Lucas, *Edwin Austin Abbey, Royal Academician: The Record of His Life and Work,* 2 vols. (London: Methuen; New York: Charles Scribner's Sons, 1921), 1:151.

44. Zeus swallowed Metis ("thought"), got a headache, and had Hephaistos take a hammer and spike to his skull; Athena sprang out. Susan Stanford Friedman discusses the strangeness of the "head-birth" model in "Creativity and the Childbirth Metaphor," 78.

45. Artists "[n]'ont qu'un espoir . . . / C'est que la Mort, planant comme un soleil nouveau, / Fera s'épanouir les fleurs de leur cerveau!" Charles Baudelaire, "La Mort des Artistes," in *Les fleurs du mal,* 175, lines 12–14.

46. Interestingly, however, these and the other dark red flowers are the most fragile spots on the canvas, owing to cleavage between the oil layers, which were applied wet on wet (as blood and water mingled). Tate conservation file 1615, Tate Britain, London.

47. De Lorris and de Meun, *Romance of the Rose*, 53; "Se chapel en peüse faire, / je n'amasse tant nul afaire," *Roman de la Rose*, 1:51, lines 1651–52.

48. Burne-Jones finished *The Pilgrim at the Gate of Idleness* that year and was working on *The Heart of the Rose*, which he would complete in 1889. *The Heart of the Rose* was based on a drawing from 1881 that Sargent may also have seen. Quests involving vegetable penetration inspired much of Burne-Jones's oeuvre: he depicted scenes from various fairy tales and romances that stage them, including *Briar Rose* and *Sleeping Beauty*. On an earlier visit to his studio, Sargent and Vernon Lee saw some of these works; Lee wrote to her mother that "a set of unfinished things of the Sleeping Beauty, I admired extremely, & so did John." Lee to her mother, 20 June 1881, *Vernon Lee's Letters*, 63. Burne-Jones's studio, intriguingly, was the site of pollinating patronage as well as flowery paintings: *The Heart of the Rose* and *The Pilgrim at the Gate of Idleness* were both commissioned by the bee-identified William Connal (fig. 6). Stephen Wildman and John Christian, *Edward Burne-Jones: Victorian Artist-Dreamer*, exh. cat. (New York: Harry N. Abrams and the Metropolitan Museum of Art, 1998), 184.

49. For example, at Tessa Gosse's birthday party in September 1885, where all the children were so bedecked, and Lily Millet's birthday part in September 1886. Sargent painted Tessa in her wreath of autumn colchicum (meadow saffron). CP 1:166–67. Lucia Millet's letters to her parents catalogue the different wreaths; at the latter party, "Frank wore one made of large double red dahlias, Lily's was tea roses, Mrs. Barnard's was made of sweet peas, her husband's of single dahlias. Mr. Gosse wore pink roses & Mrs. Gosse nasturtiums Mr. Blashfield's was like Frank's. Mr. Carr wore yellow marigolds Mrs. Carr the passion flower Mr. Parsons wreath was a most successful one made of dark red roses Mrs. Williams wore an ivy wreath and mine was made of single yellow dahlias. I see I have left out Mrs. Blashfield, she was most lovely in a white dress with a wreath of huge white roses." Although she doesn't mention the flowers of Sargent's wreath, he was in attendance. He drew most of the caricatural place cards for the dinner, Lucia Millet recorded his gift, and Alice Comyns Carr recalled that he proposed a game of throwing silhouettes after the dinner. Lucia Millet to her parents, 15 September 1885 and 19 September 1886; Alice Carr, *J. Comyns Carr: Stray Memories by His Wife* (London: Macmillan, 1920), 175–77; all quoted in RGA, 248–49, 382–83n82.

50. "A wreath around her head she wore, Carnation, Lily, Lily, Rose." Joseph Mazzinghi, *The Wreath: A Pastoral Glee for Three Voices* (London: Goulding, Phipps and D'Almaine, n.d.), unpaginated.

51. I was unable to obtain permission to reproduce this painting. It will appear in a forthcoming volume of the catalogue raisonné; for a black-and-white illustration, see Olson, Adelson, and Ormond, *Sargent at Broadway*, 68, fig. 43.

52. In Sargent's *Garden Study with Lucia and Kate Millet* from 1885, a floating wreath of red roses hovers above the figure of Lucia, whose pose evokes that of the bather figure in Manet's *Déjeuner sur l'herbe* and thus functions, like the lantern, as a reassuringly mediated reference to water, glimpsed through an art historical screen. For an illustration, see Gallati, Hirshler, and Ormond, *Great Expectations*, 94, plate 25.

53. This caricature decorated Sargent's place card for Lily Millet's birthday party. On the reverse, a couplet by J. Comyns Carr reads: "And all mindful of the evening dew / Still to his faith in nature he was true / Till once reclining upon nature's breast / He woke and found a cold upon his chest." Francis David Millet and Millet Family Papers, Archives of American Art, Smithsonian Institution.

54. Henry James, who oscillated between the vegetal and human procreation models (in his letters he often equated artistic creation and pregnancy), was also aware of the autoerotic quality of artistic labor: his stories phallically "strain[ed] to shape under [his] hand" until his "thing had 'come.'" James, preface, *Spoils of Poynton*, xxxix, xliv.

55. I was unable to obtain permission to reproduce this painting. See Olson, Adelson, and Ormond, *Sargent at Broadway*, 66, fig. 40.

56. Perhaps the association of white roses with sterility and syphilis dictated the painter's reversion to red. On "fleurs blanches" and gynecological disorders, see the entry for "fleurs" in Larousse, *Grand dictionnaire universel*.

57. Art in embryo, as it were. It is also possible to read the lanterns as cocoons. I thank Therese O'Malley for suggesting this interpretation.

58. Charles Caffin, *American Masters of Painting* (New York: Doubleday, Page, 1902), 67.

59. To cite one example, A. D. Patterson reported that "Sargent hated his own appearance and others' portraits of him, as evidenced by a letter from the artist near the end of his life." A. D. Patterson, "Sargent: A Memory," *Canadian Magazine* 65 (March 1926): 41, summarized in Jordan, "John Singer Sargent's Images of Artists," 77n187.

60. On this prohibition, see Jordan, "John Singer Sargent's Images of Artists," 76–78. There are, nevertheless, a few sketches of Sargent by others. Albert de Belleroche, whom Sargent drew and painted many times, caught Sargent's likeness while he was asleep on a train. See Dorothy Moss, "John Singer Sargent, 'Madame X,' and 'Baby Millbank,'" *Burlington Magazine* 143 (May 2001): 273, fig. 18. James Carroll Beckwith also sketched Sargent several times when they were students together in Paris. On these drawings, now in the New-York Historical Society, see Roberta Olson, "John Singer Sargent and James Carroll Beckwith, Two Americans in Paris: A Trove of Their Unpublished Drawings," *Master Drawings* 43 (Winter 2005): 415–39.

61. See chapter 2. Sargent also posed for a figure in Besnard's École de pharmacie ceilings. Ormond suggests that it may have been one in *Le promenade botanique*, but (alas!) it cannot be identified; Vernon Lee imagines he is a geologist on the glacier in the mural. O'Hare, "Fashioning Manhood," 32; CP 1:158; Lee, "J. S. S.: In Memoriam," 236.

62. "I have long been sick and tired of portrait painting, and while I was painting my own 'mug' I firmly resolved to devote myself to other branches of art as soon as

possible." Quoted in Frederick W. Coburn, "The Sargent Decorations in the Boston Public Library," *American Magazine of Art* 8 (February 1917): 136.

63. Kingsbury, "John Singer Sargent," 22; and Jordan, "John Singer Sargent's Images of Artists," 219–20.

64. Leigh Culver, "Performing Identities in the Art of John Singer Sargent" (PhD diss., University of Pennsylvania, 1999), 43, 46. Charteris claims that Sargent "painted the [Uffizi] portrait of himself by fixing a mirror to the trunk of a tree" (*John Sargent*, 170). If this is true (which seems unlikely, though Kilmurray repeats the story in Kilmurray and Ormond, *John Singer Sargent*, 169), Sargent has certainly and atypically erased all evidence of the circumstances of the painting's making. I suspect that the 1892 self-portrait for the National Academy of Design in New York is the picture to which Charteris refers: the artist is wearing some sort of cap, and the background could more conceivably be outdoors. The story of the artist literally projecting his image onto a tree to paint a self-portrait fits well with the arguments of this book, though the painting is disappointingly dry.

65. Jordan, "John Singer Sargent's Images of Artists," 204; and Ormond, "In the Alps," 83. Alice Meynell includes *Padro Sebastiano* in her list of Sargent's "romantic" portraits. Alice Meynell and William Manson, *The Work of John Sargent, R.A.* (1903) (London: William Heinemann, 1927), unpaginated.

66. Caffin, *American Masters of Painting*, 64. Sargent did use Kate Millet, his first model for *Carnation, Lily, Lily, Rose*, more or less as a placeholder. He had her pose in a blond wig until models with better coloring—Dolly and Polly Barnard—came along.

67. Sidlauskas discusses Sargent's identification with the girls pictured in the 1882 *Daughters of Edward Darley Boit* in *Body, Place, and Self in Nineteenth-Century Painting* (Cambridge: Cambridge University Press, 2000). Sally Promey analyzes the function of the child as "a figure of subjectivity" in the Boston Public Library murals; see her *Painting Religion in Public*, 140–41. Interestingly, Caffin identifies *Carnation, Lily, Lily, Rose* and the murals as the most striking instances of self-portraiture in Sargent's oeuvre. Caffin, *American Masters of Painting*, 67. On the child as a figure for interiority in the nineteenth century in general, see Carolyn Steedman, *Strange Dislocations: Childhood and the Idea of Human Interiority, 1780–1930* (London: Virago, 1995).

68. M. G. van Rensselaer, "John S. Sargent," *Century Magazine*, March 1892, 798.

69. Sargent was angered by the cruel comments Sir George Sitwell made about the aesthetic defects of his daughter Edith during their sittings, for example. Her brother Osbert recalled that Sargent was "obviously a very kind and considerate man" who "showed plainly that he regarded this as no way in which to speak of her personal aspect in front of a very shy and supersensitive child of eleven." Sir George "pointed out to the painter that my sister's nose deviated slightly from the perpendicular, and hoped that he would emphasise this flaw"; Sargent instead "made her nose straight in his canvas and my father's nose crooked, and absolutely refused to alter either of them." Osbert Sitwell, *Left Hand, Right Hand!* (London: Macmillan, 1945), 222–23.

70. Kingsbury, "John Singer Sargent," 50–60. James described the artist as both innocent and precocious: "his character is disarmingly naïf, but not his talent." Henry James to Elizabeth Boott, 2 June 1884 [Whitsunday], bMS Am 1094, ser. 2, no. 565, Houghton Library, Harvard University.

71. Sargent's juvenile subjects do not fall into the category of the "innocent child," which T. J. Jackson Lears has demonstrated offered to some *fin-de-siècle* artists "a vision of psychic wholeness, a 'simple, genuine self' in a world where selfhood had become problematic and sincerity seemed obsolete." While Lears recognizes "the child's capacity for unrepressed emotional and imaginative experience" and "a link between childhood and the unconscious," he does not discuss the potentially desublimating quality of these connections. T. J. Jackson Lears, *No Place of Grace: Antimodernism and the Transformation of American Culture, 1880–1920* (Chicago: University of Chicago Press, 1994), 146. Charteris certifies that Sargent "was no indiscriminating worshipper of childhood, and sentimental generalizations on the subject found him an unsympathetic hearer"; he also occasionally longed "for the good old days of Herod." *John Sargent*, 75.

72. They work, therefore, to license alternative adult sexual identities. Repressing childhood sexuality does the opposite. Jacqueline Rose explains the phenomenon: "The child is sexual, but its sexuality (bisexual, polymorphous, perverse) threatens our own at its very roots. Setting up the child as innocent is not, therefore, repressing its sexuality—it is above all holding off any possible challenge to our own." Rose, *The Case of Peter Pan, or The Impossibility of Children's Fiction* (London: Macmillan, 1985), 4.

73. See Sally Shuttleworth, "The Psychology of Childhood in Victorian Literature and Medicine," in *Literature, Science, Psychoanalysis, 1830–1970: Essays in Honour of Gillian Beer*, ed. Helen Small and Trudi Tate (Oxford: Oxford University Press, 2003), 96–98.

74. Sidlauskas attributes the common contemporary equation of childhood and subjectivity to "the sheer number of biological, psychological, and cognitive discoveries about human beings [that] multiplied during these years. People thus became ever more preoccupied with the profound significance of growth being studied in the science of physiology, and with the human figure who best exemplified the trajectory of growth—the child." *Body, Place, and Self*, 64.

75. "Le rire des enfants est comme un épanouissement de fleur. C'est la joie de recevoir, la joie de respirer, la joie de s'ourvrir, la joie de contempler, de vivre, de grandir. C'est une joie de plante" (The laughter of children is like the blossoming of a flower. It is the joy of receiving, the joy of breathing, the joy of opening, the joy of contemplating, of living, of growing. It is a plant's joy). Baudelaire, "De l'essence du rire," *Oeuvres complètes*, 2:534. For the perceived connections between childhood and plant life earlier in the nineteenth century, see "Figures and Physiology,"

chapter 3 of Steedman's *Strange Dislocations*. As common symbols of metamorphic potential, plants were figures of both growth and creativity.

76. FitzWilliam Sargent to Emily Sargent, 16 September 1861, roll D317, Archives of American Art, Smithsonian Institution, quoted in Kathleen Lois Butler, "Tradition and Discovery: The Watercolors of John Singer Sargent" (PhD diss., University of California, Berkeley, 1994), 71n25. Sargent had a taste for nominal puns. At Broadway he gave Edmund Gosse a hat called a "'Gossamer' or 'goss'" for his birthday. Olson, "Sargent at Broadway," in Olson, Adelson, and Ormond, *Sargent at Broadway*, 18. On the Boit/boîte pun and others in *The Daughters of Edward Darley Boit*, see David Lubin, "The Boit Children," in his *Act of Portrayal: Eakins, Sargent, James* (New Haven: Yale University Press, 1985), 83–121.

77. See, for example, FitzWilliam Sargent to Winthrop Sargent, 15 September 1869, Archives of American Art, Smithsonian Institution, quoted in Stephen D. Rubin, *John Singer Sargent's Alpine Sketchbooks: A Young Artist's Perspective* (New York: Metropolitan Museum of Art, 1991), 9.

78. Baudelaire, *Painter of Modern Life*, 8; "Le génie n'est que l'*enfance retrouvée* à volonté," "Le peintre de la vie moderne," in *Oeuvres complètes*, 2:690.

79. Helmreich, "John Singer Sargent, *Carnation, Lily, Lily, Rose*," 437.

80. Similarly, J. M. Barrie's 1911 *Peter and Wendy* "acts out the question of whether or not to take up a position in language, whether or not to settle down and find a place (whether or not to grow up)." Rose, *Case of Peter Pan*, 74.

81. Earlier artists had experimented with re-creating the child's vision. Otto Runge, for example, was one of the first to "voyage into the mysterious world of a child's scale, psychology, and physiological closeness to the vital energies of nature"; in his painting *The Artist's Parents*, the children are presented as akin to blossoming flowers that are not part of the adult world. Robert Rosenblum, *The Romantic Child from Runge to Sendak* (London: Thames and Hudson, 1988), 33, 39. Significantly, Runge's "new symbolic-pictorial language" was rooted in "his invention of the child figure, the flower, and the child figure-flower composite symbols." Rudolf Bisanz, *German Romanticism and Philipp Otto Runge: A Study in Nineteenth-Century Art Theory and Iconography* (DeKalb: Northern Illinois University Press, 1970), 99.

82. Edwin Blashfield, "John Sargent," 19; William H. Gerdts, "The Square Format and Proto-Modernism in American Painting," *Arts Magazine* (June 1976): 72.

83. Rose, *Case of Peter Pan*, 12.

84. The number of pots Sargent painted, especially in his garden scenes, is extraordinary. One need only turn to *The Daughters of Edward Darley Boit* for another painted articulation of the necessity of pots to a nomadic childhood: the gigantic vases pictured crossed the ocean with the girls sixteen times. Sidlauskas, *Body, Place, and Self*, 72.

85. To water a garden or flower was slang for the sexual act (see chapter 1).

86. RTPE 2:653; "qui ne peuvent être fécondés par eux-mêmes, mais peuvent l'être par d'autres hermaphrodites," RTPF 2:629.

87. The canvas is inscribed "V and Katie Vickers." "V" could refer to either Violet Paget/Vernon Lee or Violet Sargent—probably the latter, but either way a flower. See also *The Misses Vickers* (fig. 57) for another pairing of black- and white-clad creatures—the two siblings entwined in a wicker garden chair.

88. See, for example, *Venetian Onion Seller* (fig. 58).

89. Richard Ormond, "Sargent's Childhood," in Gallati, Hirshler, and Ormond, *Great Expectations*, 46.

90. Olson, *John Singer Sargent*, 198.

91. Steedman argues compellingly that "it is not at all clear that nineteenth-century children were distributed across the sex-gender system that modern scholarship has outlined," and in her excellent study Catherine Robson points out that until the age of six or seven boys and girls alike wore dresses, often had long hair, stayed in the nursery, studied at home, and were not much distinguished. Alan J. L. Busst contends that as the nineteenth century progressed, "the hermaphrodite and the virgin [became] more and more confused in a single image"; sex and orientation were only "resolved" with the first sexual encounter. Steedman, *Strange Dislocations*, 8; Robson, *Men in Wonderland: The Lost Girlhood of Victorian Gentlemen* (Princeton: Princeton University Press, 2001), 4; Busst, "The Image of the Androgyne in the Nineteenth Century," in *Romantic Mythologies*, ed. Ian Fletcher (London: Routledge and Kegan Paul, 1967), 43.

92. See Introduction and chapter 3.

93. Busst, "Image of the Androgyne."

94. In the medical world, "hermaphroditism" was usually employed to refer to anomalous biological traits or sexual deviance (as in Ellis's term "psychosexual hermaphroditism"). This was not universally the case, however. Dr. Pierre Garnier considered hermaphroditism "synonymous with androgyny." Garnier's *Human Sterility and Hermaphroditism*, first published in 1883, went through at least six editions over the next twenty years. Pierre Garnier, *La stérilité humaine et l'hermaphrodisme* (Paris: Garnier Frères, 1883), 505, quoted in Albert, "From Myth to Pathology," 120.

95. Weil, *Androgyny and the Denial of Difference*, 63.

96. CP 4:63. The flowers are too large to be cherry blossoms and too small to be magnolias.

97. Ibid. The stamp, like the fruit, is named after the Satsuma Province of Japan.

98. The letters of the painted signature are composed of straight lines, the tapering sides of which suggest that Sargent was trying to imitate a process of excision. I thank Dean Zimmerman, chief curator at the Western Reserve Historical Society, for sending me a high-resolution jpeg image of the signature. The clumsy "oriental" signature is not repeated in Sargent's oeuvre.

99. Plato, "The Symposium," *Collected Dialogues of Plato*, 543.

100. Carpenter uses an apian example to naturalize this prospect: "At some past

period of evolution the worker-bee was without doubt differentiated from the two ordinary bee-sexes." Carpenter, *Intermediate Sex,* 11.

101. As suggested above, Weil's distinction between the hermaphrodite and the androgyne is useful but too absolute; in practice, many artists imagined intermediary figures.

102. Eve Kosofsky Sedgwick, *Epistemology of the Closet* (Berkeley and Los Angeles: University of California Press, 1990), 157.

103. James similarly described his reception of story-pollen as "the happiest chance for the indestructible," for continuance or perpetuation. Preface, *Spoils of Poynton,* xl.

104. Gagnier, *Idylls of the Marketplace,* 159. Martha Vicinus has also examined the elevation of art over sexual reproduction in the late nineteenth century. "Men and women whose sexual lives were in opposition to biological reproduction did not defy its hegemony, but rather insisted on a superior option—art. Physical reproduction is replaced with meta-physical and artistic generativity." Martha Vicinus, "The Adolescent Boy: Fin-de-Siècle Femme Fatale?" in *Victorian Sexual Dissidence,* ed. Richard Dellamora (Chicago: University of Chicago Press, 1999), 86.

105. Regenia Gagnier, "Production, Reproduction, and Pleasure," in Dellamora, *Victorian Sexual Dissidence,* 138.

106. Remy de Gourmont, *Physiology of Love* (New York: Rarity Press, 1932), 136, 109, 4, 6, translation modified. "Il n'y a d'êtres vivants que ceux qui peuvent perpétuer la vie"; "On cherche l'origine de l'art: la voilà, dans ce jeu sexuel d'un oiseau"; "le but est le même et le résultat le même: la copulation, la fécondation, la ponte"; "la vie est caractérisée sur la terre par l'existence d'individus groupés en espèces, c'est-à-dire ayant le pouvoir, un mâle s'étant uni à une femelle, de reproduire leur semblable." Gourmont, *Physique de l'amour* (Paris: Mercure de France, 1903), 214, 173, 10, 14–15.

107. Harrison Morris, "American Portraiture of Children," *Scribner's Magazine,* December 1901, 641.

108. In a letter of introduction he wrote for de Forest Brush, Sargent refers to the painter as "my very old friend and fellow artist." Letter of 18 January 1914, box 1, folder 18, John Singer Sargent–Thomas A. Fox Papers, Boston Athenaeum.

109. Morris, "American Portraiture of Children," 645.

110. Quoted in Mary Ellmann, *Thinking About Women* (New York: Harcourt, Brace and World, 1968), 15.

111. Sarah Burns, "Under the Skin: Reconsidering Cecilia Beaux and John Singer Sargent," *Pennsylvania Magazine of History and Biography* 124 (July 2000): 347.

112. Or the portable pot its own potential flight: it also bee-stripes itself tawny yellow and black and protracts luteofulvous lines of antennae.

113. According to the object file (55.1116) in the Museum of Fine Arts, Boston, Sargent changed the position of the hands, sleeves, and central flowers in *Helen Sears* at least once; their particular conjunction was of concern to him.

114. Mount notes finger painting in *The Sketchers* and the portraits of Peter Chardon Brooks and Teddy Roosevelt, among others. Mount, *John Singer Sargent,* 207, 236; "Phoenix at Richmond," 19. Art as child's play remained an alluring idea for Sargent. When working on reliefs for the mural projects, he enjoyed making what he called his "mud pies." Sargent to Isabella Stewart Gardner, n.d., Isabella Stewart Gardner Museum Archives.

115. The flower was named for Hortense Lepaute, a sterile eighteenth-century Frenchwoman, by the botanist Commerson; Lepaute was the wife of a friend of his and an astronomer (she features in Judy Chicago's *Dinner Party*). In the language of flowers, hortensias mean "you are cold." Seaton, *Language of Flowers,* 180.

116. Helen's reflection in the pot disappears into a leaf above, becomes the plant; in response, the leaves that reach over Helen's dress become transparent, part of her veils. I thank Sally Promey for calling my attention to these details.

117. Shuttleworth, "Psychology of Childhood in Victorian Literature," 100.

118. Kingsbury, "John Singer Sargent," 62, 97.

119. "A dreamy mental state is memory diving to a lower than its accustomed level, and bringing to the surface deeply-submerged personal annals, or even still more unfathomable ancestral traits." James Crichton-Browne, *The Cavendish Lecture: On Dreamy Mental States* (London: Baillière, Tindall and Cox, 1895), 5, 6, 7–8, 9. In the text Crichton-Browne identifies numerous artists who suffered from dreamy mental states, but John Addington Symonds is his primary example.

120. "Becky & Fowler are vegetating in N. Y. Cox & Robinson are I believe pretty flourishing." Sargent to E. Russell, 10 September 1885, Tate Gallery Archives, London, quoted in RGA, 245. Fowler is an anagram of flower; Cox and Robinson may flourish because they are bird- as well as flower-identified.

121. Sargent read *Punch* and clipped out cartoons for his scrapbooks from the 1870s on. See the scrapbook in the Metropolitan Museum of Art, 150.130.154.

122. OED, s.v. "namby-pamby."

123. A 1904 caricature by Edward Tennyson Reed also caricatures Burnand as a bee; the insect-editor is labeled "Punchiboss or EphseeBee."

124. The text below the cartoon, describing the exhibition at the Grosvenor, refers to the artists as "patients."

125. Ten years later a *Punch* diatribe called "Wilde Flowers" again mocked the writer's creative and compositional energies. The typography of *The Picture of Dorian Gray*'s preface is described as "a literary flower-bed, where rows of bright asterisks are planted between lines of brilliant aphorisms. The rule of the arrangement seems to be,—'when in doubt, plant asterisks.'" "Wilde Flowers," *Punch, or the London Charivari,* 14 March 1891, 125.

126. "Littré, la fleur de la science, / D'un Dictionnaire accoucha; / À ce travail de patience, / La pauvre fleur se dessécha." Le Petit, *Fleurs, fruits et légumes du jour,* 11.

127. Diary entry of Lucia Fairchild for 2 October 1890, Boston, quoted in Miller, "Sargent in the Diaries of Lucia Fairchild," 8.

128. James, preface, *Spoils of Poynton,* xli, xliv, xlv. He refers to the germ as a virus on xxxix and xli.

129. Edmund Burke best expresses the yearning to be perennial in a letter to the duke of Richmond: "You people of great families and hereditary trusts and fortunes are not like such as I am, who whatever we may be by the Rapidity of our growth and of the fruit we bear, flatter ourselves that while we creep on the Ground we belly into melons that are exquisite for size and flavour, yet still we are but annual plants that perish with our season and leave no sort of traces behind us. You, if you are what you ought to be, are the great Oaks that shade a country and perpetuate your benefits from Generation to Generation." Burke to the duke of Richmond, 15 November 1772, quoted in Alexander Nemerov, *The Body of Raphaelle Peale: Still Life and Selfhood, 1812–1824* (Berkeley and Los Angeles: University of California Press, 2001), 136.

130. Eighteen ninety-five was also the year that Max Nordau's *Degeneration* was translated into English.

131. According to Edgar Munhall, this was when they first hit the Paris market. Munhall, *Whistler and Montesquiou: The Butterfly and the Bat* (New York: Frick Collection; Paris: Flammarion, 1995), 36. Michelet recalls their introduction to France when he was forty (1838) in *Love,* 332.

132. For one of Montesquiou's drawings of blue hortensias, see Munhall, *Whistler and Montesquiou,* 35. Gallé and Montesquiou co-designed a hydrangea dresser and exhibited it at the 1892 Salon. Silverman, *Art Nouveau in Fin-de-Siècle France,* 233–34. Montesquiou also had himself photographed numerous times with hortensias; one print from 1885, *Les mains de Robert de Montesquiou,* is a close-up of his ringed hands and one umbel. See Philippe Thiébaut, ed., *Robert de Montesquiou ou l'art de paraître,* exh. cat. (Paris: Éditions de la Réunion des Musées Nationaux, 1999), 54, 56.

133. Robert de Montesquiou, *Paul Helleu, peintre et graveur* (Paris: H. Floury, 1913), 46, quoted in Munhall, *Whistler and Montesquiou,* 130.

134. The volume was illustrated with etchings of the flowers by Helleu. Painter, *Marcel Proust,* 2:92. On 21 May 1897, to celebrate the publication of these verses, Montesquiou held a party that Dr. Pozzi attended: "La salle est entièrement décorée sur le thème des Hortensias bleus, nappe, argenterie, assiettes s'ornent de la fleur choisie par le Maître de céans. . . . Les meubles de marqueterie montrent des incrustations d'hortensias. . . . Les vases, les rideaux sont à motifs d'hortensias" (The room was entirely decorated on the theme of blue hortensias, tablecloth, silverware, plates ornamented with the chosen flower of the master of the house. . . . Inlaid furniture showed incrustations of hortensias. . . . Vases, curtains had hortensia motifs). Ellipses in original. Vanderpooten, *Samuel Pozzi,* 206.

135. Flament, *Le bal du Pré-Catelan,* 228.

136. "Je sentis que ce végétal réfractaire associé à cet oiseau révolté dominerait ma vie, parce que tous deux devaient user de moi, pour extraire, l'un, de son azur anormal, l'autre, de son inquiétude incolore, les mille et une, peut-être même les mille et trois raisons qu'ils avaient d'être désignés, entre tous, et à tout jamais, pour le double signe de ralliement de la *Dissemblance* et de la *Mélancolie.* . . . J'approche de mon crépuscule, une heure à laquelle, mieux que jamais, convient la compagnie du volatile de ténèbres, et de la fleur où le clair de lune se prolonge. . . . Et tout autour de ces corymbes troublants, se perpétuerait le vol tournoyant de mes 'soeurs grises,' jusqu'au moment venu, pour les uns, de s'effeuiller, pour les autres, de se disperser, autour du dernier théâtre d'action de leur jardinier et de leur pâtre" (I felt that this refractory plant associated with this rebellious bird [the bat] would dominate my life, because the two together would make use of me, to extract, for the one, in its abnormal azure blue, and the other, in its colourless anxiety, the thousand and one, perhaps even the thousand and three reasons why they had been designated, among all others, and for ever and ever, to represent the double sign of the joining of *Dissimilarity* and *Melancholy.* . . . I am approaching my own twilight, an hour when, more than ever, the company of the winged creature of dusk is appropriate, as well as that of the flower on which the moonlight lingers. . . . And all about these troubling clusters would continue the wheeling flight of my 'grey sisters' until the moment has come, for the one, to shed its petals, and for the others, to scatter about the last scene of action of their gardener and of their shepherd." Montesquiou, *Les pas effaces,* 2:103–5, quoted and translated in Munhall, *Whistler and Montesquiou,* 36–38.

137. "Tous ces hommes imberbes à la figure de femme, couronnés d'hortensias." Marcel Proust, "Gustave Moreau," in *Contre Sainte-Beuve,* 389.

138. "Tes grappes mauves sont des corolles avortées. Ta tige est trop grosse et tes feuilles trop lourdes pour ta frêlerie. Tu n'as jamais eu de parfum. Mais la qualité de ta blancheur malade, l'élégance de ta courbe, la forme inconnue que t'a conféré l'abandon, c'est là ce qui me charme" (Your mauve clusters are aborted corollas. Your stem is too big and your leaves too heavy for your frailty. You have never had a perfume. But the quality of your sickly paleness, the elegance of your curve, the unknown form that has conferred abandon on you, it's those things that charm me). Delarue-Mardrus, *Hortensia dégénéré,* 110. Anna Livia has argued that although Delarue-Mardrus wrote in the *années folles,* her work remained "characteristic of the Belle Époque, which represented the tastes and passions of her youth." Introduction to Lucie Delarue-Mardrus, *The Angel and the Perverts,* trans. Anna Livia (New York: New York University Press, 1995), 28.

139. The image I reproduce is held in the Bibliothèque nationale. A copy also appears in a four-volume collection of photographs Montesquiou compiled of himself entitled "Ego Imago." In the nearly two hundred pictures taken over forty years, Montesquiou appears in a fantastic array of costumes and characters, from John the Baptist's head on a platter to Coppée's Zanetto. The series implies that identity is pure performance or masquerade, and Thiébaut compares Montesquiou to the comtesse de Castiglione, whom the count adored and emulated. Thiébaut, *Robert de Montesquiou et l'art de paraître,* 6.

Montesquiou's celebration of identity as a floating signifier is at odds with Sargent's desire for roots. Unlike the nomadic artist, however, the count had the security that comes with an aristocratic name, heritage, and land—in other words, autochthony.

140. In the overexposed bottom-left-hand corner of the photograph in "Ego Imago"—not the one reproduced here—Montesquiou has written, "'Au clair de la Lune mon ami Pierrot.' RM 1885." The swirls of Montesquiou's umbelliflorous monogram echo the form of the flowers. The addressee of this inscription may be read as "mon ami" or "mon amie" because of Montesquiou's flourishes; the count thus underlines the long-standing association of the figure of Pierrot with that of the hermaphrodite. The script in quotation marks is the first line of a folk song (a poem by Philippe Quinault set to music in the seventeenth century by the composer J. B. Lully). In the song someone knocks at Pierrot's door late at night asking for a candle and a pen. Pierrot refuses him the materials (with which to light lanterns, to make art) and sends him next door, where he ends up having sex with the neighbor. Montesquiou, however, has himself portrayed with neither bed nor flame. Pierrot was also associated with the figure of the gardener and, occasionally, the plant: Jules Laforgue thought that Pierrot's anaemic appearance, like Darwin's monstrous plants, gave him "the air of a hydrocephalic asparagus." See A. G. Lehmann, "Pierrot and Fin de Siècle," in Fletcher, *Romantic Mythologies,* 212. "Un air d'hydrocéphale asperge," Jules Laforgue, "Pierrots" (1886), in *The Penguin Book of French Poetry, 1820–1950,* ed. and trans. William Rees (Harmondsworth: Penguin, 1990), 339, line 4.

141. Maureen C. O'Brien, "John Singer Sargent: Portrait of Ernest-Ange Duez," *The American Painting Collection of the Montclair Art Museum, Research Supplement 1* (Montclair, N.J.: Montclair Art Museum, 1979), 13. Richard Ormond first noted that the background of Sargent's portrait was one of Duez's paintings. *John Singer Sargent,* 21.

142. O'Brien, "John Singer Sargent: Portrait of Ernest-Ange Duez," 13. On Sargent's painterly exchanges, see Jordan, "John Singer Sargent's Images of Artists."

143. Susan Strickler, "John Singer Sargent and Worcester," *Worcester Art Museum Journal* 6 (1982–83): 25.

144. "Verwaschnes wie an einer Kinderschürze." Rainer Maria Rilke, "Blaue Hortensie" (written 1906), quoted and translated in Judith Ryan, *Rilke, Modernism, and Poetic Tradition* (Cambridge: Cambridge University Press, 1999), 61. Rilke's poem continues to resonate with Sargent's painting: the blue of the foreground flowers is "faint as if it were reflected from far off," and the hortensias somehow convey how deeply one feels the brevity of a little life ("die ein Blau nicht auf sich tragen, nur von ferne spiegeln," "wie fühlt man eines kleinen Lebens Kürze"). Rilke's blue hydrangeas were in part a response to Montesquiou's; Ryan argues that in the poem Rilke rejects "the aestheticist conception of poetry as an exotic hothouse flower" but "suggests that the material of eighteen-nineties poetry should not so much be discarded as placed in the service of a new aesthetic" (62, 65).

145. Grandville's *Dried Perwinkle* already alludes to the desiccation of the rhetoric of the innocent *femme-fleur.* Beverly Seaton argues that *Les fleurs animées* is an unrecognized parody of books on the language of flowers; its message is that "the sentimental flower book is dead." Seaton, *Language of Flowers,* 153.

146. Erica E. Hirshler, "The Fine Art of Sarah Choate Sears," *Magazine Antiques* (September 2001): 325–26. See also the exhibition catalogue by Hirshler, *A Studio of Her Own: Women Artists in Boston, 1870–1940* (Boston: Museum of Fine Arts, 2001).

147. Sargent to Sarah Choate Sears, 7 August 1895, private collection, quoted in Hirshler, "Fine Art of Sarah Choate Sears," 323. This threat was a recurring one. In 1881 Sargent wrote to Vernon Lee, "I am convinced that the likeness of Miss Robinson is horrid. . . . It is really hideous. Let Heath Wilson retouch it with a hot poker, and put an umbrella into it and open the umbrella." Ormond family archive, quoted in Charteris, *John Sargent,* 55.

148. An 1864 lithograph by Daumier seems relevant here, as it mocks the aspirations of would-be fruitful artists. Spying a landscape painter, two peasants comment, "En v'la un qu'est un drôle de corps . . . y tire l'portrait d'un arbre qui n'produit seulement point des pommes . . . il est fou bé sûr!" (What a queer fellow . . . he's painting an apple tree which never bears fruit . . . he must be mad!). *Le Charivari,* 9 December 1864.

149. Gauguin's *Still Life with Japanese Print* (fig. 145) may also attest to the rivalry the artist felt with a mother. Various historians have pointed out that the silhouette of the vase on the left resembles the curve of Mme. Shuffenecker's breasts in Gauguin's portrait-vase from the same year. In the painting her vase overflows with luxuriant flowers; Gauguin's look a bit ratty and meager in comparison. Merete Bodelsen, *Gauguin's Ceramics: A Study in the Development of His Art* (London: Faber and Faber; Copenhagen: Mordisk Sprog-Og Kulturforlag, 1964), 119, 217n64.

150. I agree with Barbara Dayer Gallati's suggestion that the photograph was taken after Sargent had finished his painting, given the overt references to both the portrait and to *Carnation, Lily, Lily, Rose.* Gallati, "Posing Problems: Sargent's Model Children," in Gallati, Hirshler, and Ormond, *Great Expectations,* 203.

151. Burns, "The 'Earnest, Untiring Worker' and the Magician of the Brush," 48. Sargent's response to Sears's photographs of Helen evokes Hilda's experience of art after losing her innocence in *The Marble Faun,* dear to the painter from childhood: "She grew sadly critical, and condemned almost everything that she was wont to admire. Heretofore, her sympathy went deep into a picture, yet seemed to leave a depth which it was inadequate to sound; now, on the contrary, her perceptive faculty penetrated the canvas like a steel probe, and found but a crust of paint over an emptiness." Hawthorne, *Marble Faun,* 341.

152. "The Royal Academy (First Notice)," *Athenaeum* 3105 (30 April 1887): 580; "Art Chronicle," *Portfolio* 38 (June 1887): 125.

153. John to Emily Sargent, late September or early October 1885, reproduced in Charteris, between 76 and 77; Kingsbury,

"John Singer Sargent," 43. Years later, Sargent wrote to Isabella Stewart Gardner from London that his dogwood trees were leafy but that people "say it would be a miracle if they flowered in England. They are doing their best to. There is always the resource of paper flowers, as you & I know." Sargent to Gardner, 28 June 1904, Isabella Stewart Gardner Museum Archives.

154. Fifty "to plant, twenty to be forced in pots." Olson, "Sargent at Broadway," 19.

155. Ormond, "*Carnation, Lily, Lily, Rose*," 64.

156. In 1885 Lady Archie Campbell staged Fletcher's *Faithful Shepherdess* at her estate at Coombe. Vernon Lee described it to her mother in a letter of 19 July: "I have never seen anything more beautiful or charming than this play, Fletcher's *Faithful Shepherdess*, performed in this way by a number of exquisitely dressed amateurs in a real park, walking on hay, among the rustle of leaves, with little groups lounging quite far in the distance, beautiful spots of colour among the green, and these people acting in a graceful, childish, immature way, as if there to amuse themselves rather than others." Lady Campbell "acts the chief male part, looking perfectly charming & modest, in a most beautiful green dress; a very tall . . . strong but extremely supple & graceful creature, with mobile nervous face & hair cropped short, fantastic, chivalrous looking." In a letter of 25 June she wrote, "John has come over for the Coombe play," and two days later that he "is full of the Coombe play & Lady A. Campbell." *Vernon Lee's Letters*, 180, 183, 184. The previous year Whistler had painted Lady Campbell as Orlando; the oil on wood panel, called *Note in Green and Brown: Orlando at Coombe*, c. 1884, is in the Hunterian Museum and Art Gallery. The freedom represented by the cross-dressing Lady Campbell and the casual, even "childish" air of the activities at Coombe were key to the pastoral experience.

157. Jenkins, "Pre-Raphaelite and Impressionist," 12.

158. The title of Simpson's dissertation derives from a letter of 1 October 1886 from Henry James to Lily Millet, in which the author avows that the Americans had "reconstructed the golden age" at Broadway. Simpson's work is the "chronicle of that pastoral achievement." Letter in the Millet family collection, quoted in RGA, 1, 2.

159. Edmund Gosse, *More Books on the Table* (London: William Heinemann, 1923), 210–11, quoted in RGA, 36.

160. Caffin, *American Masters of Painting*, 85.

161. Edwin Austen Abbey to Charles Parsons, 28 September 1885, quoted in Lucas, *Edwin Austin Abbey*, 150.

162. "F. G.," of the Houghton Mifflin publishing firm, to John A. P. Millet, n.d., quoting Sargent, RGA, 92n122. Two years later, as a result of these ambrosial nights among his fellow men, Sargent would even be able to perpetuate his name through a real child in addition to his picture-daughter. Also a hybrid, John Alfred Parsons Millet was named after the artist friends of his artist father (Frank Millet): John Sargent and Alfred Parsons. CP 2:60.

163. Helmreich, "John Singer Sargent, *Carnation, Lily, Lily, Rose*," 435. For discussions of Sargent's nationality, see Trevor Fairbrother, *John Singer Sargent and America* (New York: Garland, 1986), 364–78; and Culver, "Performing Identities," 25–28.

164. "Art Notes and Reviews," *Art Journal* 48 (January 1886): 29.

165. *Illustrated London News*, 5 May 1888, 473, quoted in RGA, 549–50.

166. Roger Fry, "J. S. Sargent as Seen at the Royal Academy Exhibition of His Works, 1926, and in the National Gallery," in Fry, *Transformations: Critical and Speculative Essays on Art* (London: Chatto and Windus, 1926), 125.

167. See Jenkins, "Pre-Raphaelite and Impressionist"; and Ormond, "*Carnation, Lily, Lily, Rose*." Sargent's interest in the Pre-Raphaelites caused some of his friends distress. Daniel Curtis wrote to his parents in May 1884, before Sargent's visit to England, "He goes to Eng. in 3 weeks. I fear là bas he will fall into Pre-R. influence wh. has got a strange hold of him he says *since Siena*." Quoted in Charteris, *John Sargent*, 62.

168. Many European critics considered America, like Michelet, "a thing *relative;* it yearns after, it urgently needs the globe, and stretches beyond its own individuality." Michelet, *Mountain*, 131; "L'Amérique est un relatif; elle aspire, elle a besoin du globe, et tend hors de soi," *La montagne*, 145–46.

169. James S. Harding, "The Royal Academy Exhibition," *Art Amateur* 17 (July 1887): 30, quoted in RGA, 452. Unable to see "why, in the title of the picture the word 'lily' should be duplicated," one confused writer judged it confirmation that "what the picture means and why it was painted are simply, to us, inscrutable mysteries." *Daily Telegraph*, 5 May 1887, 8, quoted in Helmreich, "John Singer Sargent, *Carnation, Lily, Lily, Rose*," 441. No critics recognized the source of the title in Mazzinghi's rather obscure art song. "The Wreath" had been popular earlier in the century; Reverend Crisparkle sings it in Dickens's *The Mystery of Edwin Drood*. Charles Dickens, *The Mystery of Edwin Drood* (1870), ed. Arthur J. Cox (Harmondsworth: Penguin, 1985), 43.

170. "Homer Watercolors to Be on View To-day," *New York Sun*, 16 October 1915, quoted in Linda Ferber, "Watercolors by Winslow Homer at the Brooklyn Museum of Art," in *Masters of Color and Light: Homer, Sargent, and the American Watercolor Movement*, exh. cat. (Brooklyn, N.Y.: Brooklyn Museum of Art; Washington, D.C.: Smithsonian Institution Press, 1998), 112.

171. Vernon Lee to her mother, 16 June 1881, *Vernon Lee's Letters*, 6. Irene Cooper Willis recalls Lee likewise "often describing herself as an alien, having no ties, of nation, blood, class or profession. She was just an individual, she said, and to be an individual was a weak thing." Ibid., xii.

172. "Ce fut, je pense, excellent pour l'artiste en herbe." Pailleron, *Paradis perdu*, 151.

173. FitzWilliam Sargent to his parents, 10 October 1870, Archives of American Art, Smithsonian Institution, quoted in Mount, *John Singer Sargent*, 22.

174. Olson, *John Singer Sargent*, 84.

175. This bred "slightly distorted notions of normality"; it "used a sort of compost which also made him perfectly self-contained. He had flourished in a rarified but highly intimate atmosphere." Stanley Olson, "On the Question of Sargent's Nationality," in Hills, *John Singer Sargent,* 19.

176. Olson, *John Singer Sargent,* 84.

177. Genet, *Our Lady of the Flowers,* 198; "Elle se cherchait des racines. Elle voulait sentir, la traînant à ses pieds, la force dynastique, qui fût brutale, musculaire, fécondante," *Notre-Dame-des-Fleurs,* 220.

178. Wood, *Sargent,* 30.

179. Child, "American Artists at the Paris Exhibition," 504.

180. Fairy lights are lamps containing a candle, often hung among trees, and especially a kind of nightlight. OED, s.v. "fairy," c.2.

181. Camille Mauclair called him one of the "filleuls des fées." Mauclair, "John Sargent," 369.

182. OED, s.v. "fairy," 5.c.

183. After many unrequited infatuations, Andersen's romantic friendship with the grand duke of Saxe-Weimar-Eisenach prompted this remark. His diary entry for 20 August 1846 reads, "The Hereditary Grand Duke crossed the courtyard arm in arm with me to my room in the evening, kissed me affectionately, begged me always to love him although he was only an ordinary person, asked me to stay with him this winter. Unhappy me, *I* have the feeling in all this that this great love can't last. Went to sleep with wistful happy feelings that I was the guest of this strange prince at his castle and he loved me; the footmen call me: *gnädiger Herr!* (Honoured Sir.) It is like a fairy tale [*Eventyr*]." Quoted in Jackie Wullschlager, *Hans Christian Andersen: The Life of a Storyteller* (London: Allen Lane, 2000), 278. Andersen's first physical relationship was with the ballet dancer Harald Scharff of the Copenhagen Royal Theater Company. The two met in 1860, and Andersen records in his diary that the relationship "blossomed into a love-affair" in 1862. The floral nature of the pair's amorous exchanges is suggested by the gift Scharff gave Andersen for his sixty-second birthday: "an incomparably beautiful" white azalea. Alison Prince, *Hans Christian Andersen: The Fan Dancer* (London: Allison and Busby, 1998), 303, 314, 333.

184. Oscar Wilde's fairy tales, for example, "not only teach and embody Paterian aestheticism, but also the homosexual counter-discourse it inspired." Naomi Wood, "Creating the Sensual Child: Paterian Aesthetics, Pederasty, and Oscar Wilde's Fairy Tales," *Marvels and Tales: Journal of Fairy-Tale Studies* 16, no. 2 (2002): 167. On fairy tales and homosexuality, see also chapter 8, "Fairy Tales," in Robb, *Strangers.* In his 1893 Slade lecture, entitled "Fairyland," Ruskin articulated what must be a partial basis for this phenomenon: "A man can't always *do* as he likes, but he can always *fancy* what he likes." Quoted in Stella Beddoe, "Fairy Writing and Writers," in Maas et al., *Victorian Fairy Painting,* 31.

185. Carole G. Silver, *Strange and Secret Peoples: Fairies and Victorian Consciousness* (Oxford: Oxford University Press, 1999), 167.

186. Henry James to Hendrik Andersen, 12 January 1901, Virginia MSS 6251 (13), quoted in *Dearly Beloved Friends: Henry James's Letters to Younger Men,* ed. Susan E. Gunter and Steven H. Jobe (Ann Arbor: University of Michigan Press, 2004), 34.

187. "Royal Academy (First Notice)," 580.

188. Morris, "American Portraiture of Children," 651.

189. She describes "the working of the fairy 'life' in the primrose or the bee." Arabella Buckley, *The Fairy-Land of Science* (New York: D. Appleton, 1883), 237.

190. Maeterlinck, *Intelligence of the Flowers,* 100; "féerique biographie," *L'intelligence des fleurs,* 59.

191. "To Charles Darwin from the Author and Artist, Ó Kόσμος αλλοίωσis, Ó βίos ὺπόληφi" (the universe is change; life is what our thoughts make it). John Richard de Capel Wise, with illustrations by Walter Crane, *The First of May: A Fairy Masque* (Boston: James R. Osgood, 1881), 2. Rabun Taylor kindly translated this (and pointed out that the last phrase could be rendered more literally as "life is assumption") and identified its source as Marcus Aurelius's *Meditations.*

192. Reported in Lee to her mother, 12 June 1894, *Vernon Lee's Letters,* 372.

193. Lee to her mother, 3 July 1894, ibid., 375.

194. Silver, *Strange and Secret Peoples,* 51. Fairies were linked to evolutionary scenarios in Pierre Loti's mind. As a child, Loti, who professed himself "haunted" by the "antediluvian world," spent years making elaborate sets and costumes for a production of Charles Perrault's fairy tale "Donkey's Skin" in his model theatre. He set his "fairy spectacle" in a fantastic primeval landscape "representing the liassic period," consisting of "a dismal swamp overshadowed by lowering clouds, where, in the shave-grass and the gigantic ferns, strange extinct beasts wandered slowly." "Le monde antédiluvien, qui déjà hantait mon imagination, devint un de mes plus habituels sujets de rêve. . . . Bientôt aussi un nouveau décor de Peau-d'âne s'ébaucha, qui représentait un site de la période du lias: c'était, dans une demi-obscurité, sous d'accablantes nuées, un morne marécage où, parmi des prêles et des fougères, remuaient lentement des bêtes disparues." Pierre Loti, *The Story of a Child,* trans. Caroline Smith (Boston: C. C. Birchard, 1901), 229–30; Loti, *Le roman d'un enfant* (1890) (Paris: Calmann-Lévy, 1891), 234.

195. See, for example, J. A. Fitzgerald's *Fairyland* in the *Illustrated London News,* 21 December 1867, 688, and many of the paintings and illustrations represented in Maas et al., *Victorian Fairy Painting.*

196. Silver, *Strange and Secret Peoples,* 52–53.

197. Quoted in Arthur Conan Doyle, *The Coming of the Fairies* (1922) (London: Pavilion, 1997), 100, 9.

198. "Pariser Passagen: Eine dialektische Feerie." Translators' foreword to Walter Benjamin, *The Arcades Project,* trans. Howard Eiland and Kevin McLaughlin (Cambridge: Belknap Press of Harvard University Press, 1999), ix.

199. Susan Buck-Morss, *The Dialectics of Seeing: Walter Benjamin and the Arcades Project* (Cambridge: MIT Press, 1991), 271.

Vernon Lee, on the other hand, might have. In a letter to her mother of 28 August 1894 she noted "a grotesque fairy story element in my arrival at the station here, yesterday. Miss Smyth has no vehicle, so she made me tramp to the workman's cottage she has bought (her father has recently died), fastened my small bag on her bicycle & said to the porter, pointing to my portmanteau, 'The Empress will call for that about five' & accordingly the cart with the eagle harness came round with my effects." *Vernon Lee's Letters,* 383. The empress was Eugénie ("La rose impériale," fig. 43), who lived at Farnborough after the fall of the Second Empire and the death of Napoleon III in 1873. Benjamin described the fairy tale "as a mode of cultural inheritance that, far from participating in the ideology of class domination, keeps alive the promise of liberation by showing that nature—animals and animated forces—prefers not to be 'subservient to myth,' but rather, 'to be aligned with human beings' against myth." Buck-Morss, *Dialectics of Seeing,* 49, 275.

200. In this respect *Carnation, Lily, Lily, Rose* has been said to recall Whistler's "fairy-land" on the Thames. Elizabeth Prettejohn, *Interpreting Sargent* (New York: Stewart, Tabori and Chang, 1998), 50. In Whistler's 1885 (i.e., exactly contemporary) "Ten O'Clock Lecture," he conjured up the moment "when the evening mist clothes the riverside with poetry . . . and fairy-land is before us." James Abbott McNeill Whistler, *The Gentle Art of Making Enemies* (1st ed. 1890, 2nd ed. 1892) (New York: Dover, 1967), 144.

201. The gardens contained only plants native to the area and were dominated by rhododendrons, mountain laurel (Kalmia), and dogwood. CP 2:101.

202. Caffin, *American Masters of Painting,* 65.

203. Roper, *FLO,* 467.

204. CP 2:101.

205. Jordan, "John Singer Sargent's Images of Artists," 183.

206. When the Metropolitan Museum of Art bought the painting (*Il Solitario*), it was given an English title, *The Hermit.* On this matter Sargent wrote to Edward Robinson, director of the museum, "'Hermit' is alright . . . I wish there were another simple word that did not bring with it any Christian association, and that rather suggested quietness or pantheism; but it might seem affected, and at any rate I can't think of it." Sargent to Robinson, 16 March 1911, Archives of the Metropolitan Museum of Art, quoted in Promey, *Painting Religion in Public,* 258. According to his manservant, Sargent was "a freethinker" who "did not 'go in' for religion." Nicola D'Inverno, "The Real John Singer Sargent, as His Valet Saw Him," *Boston Sunday Advertiser,* 7 February 1926, 3. Vernon Lee also averred that John was "quite emancipated from all religious ideas." Lee to her mother, 20 June 1881, *Vernon Lee's Letters,* 63.

207. Max Beerbohm, "A Play and an Actress," in *Last Theatres: 1904–1910* (London: Hart-Davis, 1970), 471, quoted in CP 2:83n12. P. G. Konody also saw the painting as evidence of Sargent's desire for "the simple life." Konody, "Art Notes: The New English Art Club," *Observer* (London), 30 May 1909, 4, quoted in Jordan, "John Singer Sargent's Images of Artists," 192.

208. Langtry, *Days I Knew,* 210.

209. I thank Elizabeth Legge for this suggestion.

210. James George Frazer, *Lectures on the Early History of the Kingship* (London: Macmillan, 1905).

211. James George Frazer, *The Golden Bough: A Study in Magic and Religion,* abridged ed. 1922 (New York: Touchstone, 1996), 5, 9.

212. Actually one stuffed gazelle, "bought at Rowland Ward's," reinvented as a deer, repositioned, and painted twice. Charteris, *John Sargent,* 170.

Chapter 5

1. The poem from which the epigraph is taken—Edmund Gosse, "Discipline," in *Collected Poems,* 239, lines 1–4—was first published in Gosse's anthology *In Russet and Silver* (London: William Heinemann, 1894), 8.

2. OED, s.v. "wagtail," 3.b. Other pollinator personae had lascivious connotations: as previously mentioned, a butterfly was a libertine, a ladybird a "wanton or a lewd girl," and a columbine a prostitute; a bat was "a prostitute who only walks the streets at night," and a "moth" was also a prostitute. Barrère and Leland, *Dictionary of Slang,* 2:398, 2:1, 1:265, 1:86. See also Farmer, *Slang and Its Analogues,* 1:140, 4:141, 4:360, and 7:283.

3. "H. Smith" [William Dugdale], "A few words about margeries—the way to know the beasts—their haunts, etc.," from *Yokel's Preceptor, or More Sprees in London!* c. 1850, quoted in White, *Nineteenth-Century Writings on Homosexuality,* 14.

4. RTPE 3:843; "enchaîné sur un lit . . . recevant les coups d'un martinet en effet planté de clous que lui infligeait Maurice . . . tout en sang, et couvert d'ecchymoses qui prouvaient que le supplice n'avait pas lieu pour la première fois . . . M. de Charlus," RTPF 3:815.

5. Robert Musil, *The Confusions of Young Törless,* trans. Shaun Whiteside (London: Penguin, 2001), 121. "Ein Jagen und Hasten, das sich tausendfältig, wie mit samtenen Fühlfäden von Schmetterlingen an seinem Körper stieß"; "der Anblick Basinis, seiner nackten, leuchtenden Haut, duftete wie ein Fliederstrauch in das Dämmern der Empfindungen, das dem Schlafe vorausging. Sogar aller moralische Abscheu verlor sich." Musil, *Die Verwirrungen des Zöglings Törleß* (1906) (Hamburg: Rowohlt, 2007), 122, 151.

6. RTPE 2:649; "aucune suspicion n'effleure, mais qui au carrefour se voit renversé sur l'herbe, sans une parole, par l'alpiniste bientôt père," RTPF 2:626.

7. "[L]e rendez-vous de tous les vices errants," Jean Lorrain, *L'art d'aimer* (Paris: Baudinière, 1929), 193; "de fortes senteurs d'ail et d'humanité rance s'exhalent," "dans les herbes mouillées," Jean Lorrain, "Le réveil des faunes," *L'Écho de Paris,* 10 July 1891, unpaginated, both quoted in José Santos, "Réalité et imaginaire des parcs et des jardins dans la deuxième moitié du XIXe siècle," *Nineteenth-Century French Studies* 31 (Spring–Summer 2003): 289.

8. Haldane Macfall, *Whistler: Butterfly, Wasp, Wit, Master of the Arts, Enigma* (Boston: John W. Luce, 1906), 11. Whistler was a whipster in every sense: a "sly,

cunning fellow" as well as a devotee of the rod and cane. Barrère and Leland, *Dictionary of Slang*, 2:409.

9. Julian Weir to his parents, 1878, in Dorothy Weir Young, *The Life and Letters of J. Alden Weir* (New Haven: Yale University Press, 1960), 137, quoted in Laura Beatty, *Lillie Langtry: Manners, Masks, and Morals* (London: Vintage, 2000), 125.

10. Quoted in E. R. Pennell and J. Pennell, eds., *The Whistler Journal* (Philadelphia: J. B. Lippincott, 1921), 168.

11. Schneider, "Butterfly or Scorpion," 27.

12. Philippe Jullian, *Prince of Aesthetes: Count Robert de Montesquiou, 1855–1921*, trans. John Haylock and Francis King (New York: Viking, 1967), 93.

13. Munhall, *Whistler and Montesquiou*, 80.

14. Jullian, *Prince of Aesthetes*, 29.

15. Ibid., 94.

16. Munhall, *Whistler and Montesquiou*, 36.

17. Jullian, *Prince of Aesthetes*, 54. See also Pierre Loti's extended description of the actress's bedroom furnishings, quoted in Arthur Gold and Robert Fizdale, *The Divine Sarah: A Life of Sarah Bernhardt* (New York: Knopf, 1991), 217–18.

18. Bram Stoker, *Dracula* (1897), ed. Glennis Byron (Peterborough, Ont.: Broadview Press, 1998), 276.

19. Bjerregaard, "Philosopher and a Butterfly," 382.

20. Marquis de Sade, *Justine, Philosophy in the Bedroom, and Other Writings*, 637; "l'image d'un beau lis où l'abeille a fait quelques taches," *Justine*, 265.

21. Marquis de Sade, *Justine, Philosophy in the Bedroom, and Other Writings*, 708; "La séance fut longue; plus la fleur était fraîche, plus ce frelon impur aimait à la pomper. À ces suçons multipliés succéda l'examen du cou; et je sentis qu'en le palpant, le membre que j'excitais prenait encore plus d'énergie," *Justine*, 369. The association of vampires and sexual impropriety—never mind *Dracula*'s "incestuous, necrophilous, oral-anal-sadistic all-in wrestling match," in Richardson's famous description—is confirmed by late nineteenth-century slang, in which a "vampire" is someone who blackmails others for their sexual indiscretions. Maurice Richardson, "The Psychoanalysis of Ghost Stories," *Twentieth Century* 166 (December 1959): 427; Barrère and Leland, *Dictionary of Slang*, 2:394.

22. Arthur Jerome Eddy, *Recollections and Impressions of James A. McNeill Whistler* (Philadelphia: Lippincott, 1903), 213–14, quoted in Munhall, *Whistler and Montesquiou*, 78. Caffin also attests that Whistler kept "his studio dimly lighted." Charles Caffin, "How to Study Pictures," *St. Nicholas* 32 (October 1905): 1097.

23. Munhall, *Whistler and Montesquiou*, 156.

24. Jullian, *Prince of Aesthetes*, 97. The Goncourts' journal entry of 7 July 1891, which Montesquiou quotes in his memoirs, reads: "des séances, où il semblait à Montesquiou, que Whistler, avec la fixité de son attention, lui prenait sa vie . . . et, à la fin, il se sentait tellement aspiré, qu'il éprouvait comme une contracture de tout son être, et qu'heureusement il avait découvert un certain vin de *coca* qui le remettait de ces terribles séances." Montesquiou, *Les pas effacés*, 2:217. Others found their posing for Whistler both physically and psychologically exhausting: Logan Pearsall Smith recalled that "when I had stood until I felt I should die if I stood there longer, and would beg for a little rest or some change of position, 'In a moment, in just a moment,' Whistler would cheerfully answer, and then would go on painting. His method, as I observed it, was first of all to arrange his subject with incredible pains and care, so that every detail was to his liking, and to paint it with infinite touches and retouches; and then, when it seemed finished and perfect in execution, to stand back, gaze at it, and cry 'Ha!' and rush at it in a kind of fury and paint the whole thing out." Logan Pearsall Smith, *Unforgotten Years* (Boston: Little, Brown, 1939), 207–8, quoted in Munhall, *Whistler and Montesquiou*, 76.

25. "[P]ortraits-fantômes." Quoted in Symons, "Whistler," in *Studies in Seven Arts*, 140.

26. Christian Brinton, *Modern Artists* (New York: Baker and Taylor, 1908), 102.

27. "C'est sensation très simple d'un être aspiré tout vivant, Edgar Poë l'a d'ailleurs tout au long définie dans un conte intitulé *le Portrait*." Raitif de la Bretonne, "Pall-Mall Semaine," *L'Écho de Paris*, 16 May 1895, 2, translated in Munhall, *Whistler and Montesquiou*. Poe was often invoked in descriptions of Whistler's canvases (as was the case with Sargent's): Mallarmé avowed that Poe had "appeared to [him] through Whistler," the two artists being similar in person as well as in their works' "tragic flirtation in black, disturbed and discreet." Mallarmé also linked the two Americans to Villiers de l'Isle-Adam (see chapter 2), whose *L'Ève future* is the story of a soul being abstracted, then re-embodied into Edison's work of art. Stéphane Mallarmé, "Edgar Poe," in *Mallarmé in Prose*, ed. Mary Ann Caws (New York: New Directions, 2001), 74.

28. Edgar Allan Poe, "The Oval Portrait" (1850), *Complete Tales and Poems*, 291–92.

29. For more literary examples of the artist as vampire, see James B. Twitchell, *The Living Dead: A Study of the Vampire in Romantic Literature* (Durham: Duke University Press, 1981), chapter 5.

30. "[L]es innombrables et interminables séances (j'ai posé plus de cent fois)"; "semble subir les lois mêmes de la croissance humaine." Montesquiou, *Les pas effacés*, 2:260.

31. John Singer Sargent to Isabella Stewart Gardner, 28 October 1917, Isabella Stewart Gardner Museum Archives.

32. Quoted and translated in Munhall, *Whistler and Montesquiou*, 69, translation modified. "[I]l se recule, on dirait pour puiser une touche dans l'essence même du modèle, et la reporter sur la toile sans lui laisser le temps de s'évaporer ou de s'évanouir." Montesquiou, *Les pas effacés*, 2:260.

33. Joy Newton, ed., *La chauve-souris et le papillon, correspondance Montesquiou-Whistler* (Glasgow: University of Glasgow, French and German Publications, 1990), 75, quoted in Munhall, *Whistler and Montesquiou*, 73.

34. *Le Petit Parisien,* 24 April 1894, and *Le XIXe siècle,* 25 April 1894, quoted in Munhall, *Whistler and Montesquiou,* 159. Another critic wrote of the black paintings that "all these pictures . . . seem like half-materialised ghosts at a spiritualistic *séance.* I cannot help wondering when they will gain substance and appear more clearly out of their environing fog, or when they will melt altogether from my attentive gaze." From *The Echo,* quoted in Whistler, *Gentle Art of Making Enemies,* 323. Munhall argues that the "mediumistic interpretation of the black portraits by so many writers reflects the popularity of experiments with the occult at the time." Whistler and Montesquiou both dabbled in spiritualism, attending séances and table-turnings with some regularity during the 1880s, and Montesquiou's grandfather found young Robert a soul most sensitive to vibrations from the other world. Munhall, *Whistler and Montesquiou,* 159. There was something that affected Whistler and Montesquiou profoundly in mesmerism, totemism, vampirism, and various of the other fads that popped up like the Roman mystery cults in response to *fin-de-siècle* anxiety, all seeking alternate access to and comprehension of both the internal and the eternal. These invert artists' pollinator personae, then, were predicated on more than their fertilizing functions: Whistler's butterfly was associated with the life of the soul beyond that of the earthly body (*psyche* means both soul and butterfly in Greek), and Montesquiou's totem bat was a creature capable of navigating the night and charged with the death-defying associations of vampirism. The count considered both himself and his pet bat metempsychoses of Ludwig II of Bavaria: "la délicatesse la rendait si précieuse, et dont les demi-ténèbres donnaient tant de mélancolique majesté à sa tête tiarée, que nous en étions venus à la tenir pour une métempsychose de Louis II de Bavière." Montesquiou, *Les pas effacés,* 2:303.

35. Eddy, *Recollections,* 235–36, quoted in Munhall, *Whistler and Montesquiou,* 70.

36. Quoted in Henri Ellenberger, *The Discovery of the Unconscious: The History and Evolution of Dynamic Psychiatry* (New York: Basic Books, 1970), 160.

37. Du Maurier, *Trilby;* Nicolas Daly, "The Woman in White: Whistler, Hiffernan, Courbet, Du Maurier," *Modernism/Modernity* 12, no. 1 (2005): 15–17.

38. "[Q]ui semblait toujours dire '*surge et veni!*' à la ressemblance en formation." Montesquiou, *Les pas effacés,* 2:261–62.

39. *The Thousand Nights and One Night* stood for Eastern liberties and proclivities, and unsurprisingly Montesquiou had himself photographed as Prince Houssein in 1885. In the cyanotype the count holds up a giant pansy, nearly the size of his own head, as though to lure the painter-butterfly, whose services he had plotted for some time to secure. See Thiébaut, *Robert de Montesquiou ou l'art de paraître,* 55.

40. Montesquiou habitually carried long, thin canes. When he duelled with Henri de Régnier in 1897, a witness described his accessory as "a shadow, a puff, a nothing, a gossamer thread, the phantom of a rod, a ghost of a stick! Something so light, so thin, and so delicate, so attenuated! ah! . . . of a wood so tenderly etiolated and so sveltely flexible that a stem of a poppy would have mastered it rather than have its head cut off." L. Marsoleau, "En Passant," *Le Rappel,* 12 June 1897, 52, quoted in Munhall, *Whistler and Montesquiou,* 150.

41. L. K., "The Champ de Mars Salon," *New York Times,* 20 May 1894, 20; *Le XIXe Siècle,* 25 April 1894, quoted in Munhall, *Whistler and Montesquiou,* 159.

42. L. K., "Champ de Mars Salon," 20.

43. Eddy, *Recollections,* 236, quoted in Munhall, *Whistler and Montesquiou,* 70.

44. Vernon Lee, "The Blame of Portraits," in Lee, *Hortus vitae: Essays on the Gardening of Life* (London: John Lane, the Bodley Head, 1904), 143, 140 (emphasis added).

45. Newton, *La chauve-souris et le papillon,* 76, quoted in Munhall, *Whistler and Montesquiou,* 74.

46. "Tout ça bourdonne là-dessous, la toile n'est pas encore contente. . . ." Montesquiou, *Les pas effacés,* 2:259 (ellipsis in original).

47. Jullian, *Prince of Aesthetes,* 93.

48. "[A]lors, paraissant réaliser en elle la rencontre du Wilson de Poë, avec son double, du Zoroastre de Shelley, avec soi-même, ou de Musset, avec le 'jeune homme vêtu de noir, qui lui ressemblait comme un frère,' une telle apparition faisait clamer à l'artiste qui me l'a lancé, dans un cri . . . '*Regardez-moi encore, un instant,* ET VOUS REGARDEREZ POUR TOUJOURS!'" Montesquiou, *Les pas effacés,* 2:262.

49. Wilde, *Picture of Dorian Gray,* 28.

50. R. A. M. Stevenson, "J. S. Sargent," *Art Journal* 40 (March 1888): 69. Torture comes from the latin *torquere,* to twist; the painting literally portrays the act of twisting.

51. Stoker, *Dracula,* 328.

52. "[M]arotte"; "J. S. Sargent les guettait sans lassitude derrière son volet, comme un assassin, et quand il pouvait s'en saisir, asphyxiait sa victime en lui soufflant sous un verre la fumée de son cigare." Sargent, who was painting the Pailleron family at the time, showed Marie-Louise his collection, kept in crystal boxes in his Paris studio, and "taught her their strange names." "[I]l les exposait dans de grandes boîtes de cristal"; "Sargent me les avait montrés dans son atelier et m'avait appris leur noms étranges." Pailleron, *Paradis perdu,* 157–58.

53. Swettenham's books include *Malay Sketches* (1895), *The Real Malay* (1899), *British Malaya* (1906), *Also and Perhaps* (1912), *Arabella in Africa* (1925), and *Footprints in Malaya* (1942). On his painting, see Lim Chong Keat and Henry Barlow, *Frank Swettenham and George Giles: Watercolours and Sketches of Malaya, 1880–1894* (Kuala Lumpur: Malaysian-British Society, 1988).

54. Letter of 1898, reprinted in Sir Frank Swettenham, *Unaddressed Letters,* 2nd ed. (London: John Lane, the Bodley Head, 1900), 205.

55. Ibid., 207–8.

56. Maurice Maeterlinck, *The Life of the Bee,* trans. Alfred Sutro (New York: Dodd, Mead, 1919), 153; "fantastique comme une arme chinoise," Maeterlinck, *La vie des abeilles* (1901) (Paris: Fasquelle, 1914), 109.

57. "The Hymenopteron, paralysing its victim, acts like the surgeon, who directs his scalpel with a skilled hand; the wasp

when slaying acts like a common assassin stabbing blindly in a struggle"; the wasp's mandibles are like "a pair of scissors . . . cutting, disembowelling, and dividing." Jean-Henri Fabre, *Insect Life: Souvenirs of a Naturalist,* trans. Margaret Roberts, ed. F. Merrifield (London: Macmillan, 1913), 129. "L'hyménoptère paralysant sa victime agit en chirurgien, dont une main habile dirige le scalpel; la guêpe tuant sa proie agit en vulgaire assassin, qui, dans la lutte, poignarde au hasard"; "Ces ciseaux sont les mandibules de la guêpe, taillant, éventrant, dépéçant." "Les trois coups de poignard," "Un savant tueur," J.-H. Fabre, *Souvenirs entomologiques: Études sur l'instinct et les moeurs des insectes,* vol. 1 (Paris: Librairie Ch. Delagrave, 1879), 129–30, 93, 67.

58. As described by Fairbrother, *John Singer Sargent,* 117.

59. John Shute Duncan, *Hints to the Bearers of Walking-Sticks and Umbrellas* (London: C. Squire for J. Murray, 1809), 25–26.

60. CP 3:78.

61. Fairbrother, *John Singer Sargent,* 121.

62. CP 3:76.

63. D. S. MacColl, "The New Gallery and Two Others," *Saturday Review* 89 (5 May 1900): 555. Victoria would later marry a flower: Neil Primrose. CP 2:160. Neil was the son of Archibald Primrose, fifth earl of Rosebery and prime minister from 1894 to 1895, whom Queensberry characterized as a "Snob Queer." Richard Ellmann, *Oscar Wilde* (New York: Vintage, 1988), 426.

64. Lucia Fairchild, diary entry for 1891 (Paris), quoted in Miller, "Sargent in the Diaries of Lucia Fairchild," 11.

65. Wood, *Sargent,* 19.

66. Chard, "John Singer Sargent, R.A.," 644.

67. D. S. MacColl, "The Academy II: The Rape of Painting," *Saturday Review* 91 (18 May 1901): 632–33; MacColl, "New Gallery and Two Others," 555.

68. Vernon Lee to her mother, describing the "insolent kind of magnificence" of Dr. Pozzi's portrait, 16 June 1882, *Vernon Lee's Letters,* 87.

69. Symons, "Watts," in *Studies in Seven Arts,* 94. Marion Hepworth Dixon similarly argued that for Sargent "the sitter is at once a problem and a 'subject,' a subject out of whom [he] must needs wring the most vital confessions." Dixon, "Mr. John S. Sargent as a Portrait Painter," *Magazine of Art* (January 1899): 112.

70. See chapter 4 for a discussion of Sargent's self-representation as little girls.

71. MacColl, "New Gallery and Two Others," 555.

72. In this way Sargent was like other nineteenth-century writers and painters who responded to the "New Woman" threat by emphasizing the passivity of *femmes-fleurs.*

73. James Lomax and Richard Ormond, *John Singer Sargent and the Edwardian Age,* exh. cat. (Leeds: Leeds Art Galleries, 1979), 44. Apparently, "in later life" the sitter became "noted for her moral rectitude." CP 1:144.

74. Henry James to Elizabeth Boott, 7 January 1886, in *Henry James Letters,* vol. 3 (1883–95), ed. Leon Edel (Cambridge: Belknap Press of Harvard University Press, 1980), 107, quoted in CP 1:144.

75. *Evesham Journal and Four Shires Advertiser,* 2 May 1886, quoted in CP 1:144.

76. Huysmans, *Against Nature,* 75, 81; "des Nidularium, ouvrant, dans des lames de sabres, des fondements écorchés et béants"; "la hideuse blessure"; "en saignant," *À rebours,* 121, 131.

77. "[S]'escrimant avec son crayon, sa plume, son pinceau." Baudelaire, "Le peintre de la vie moderne," *Oeuvres complètes,* 2:693.

78. Journal entry of 27 November 1890, quoted in CP 1:191.

79. Elaine Kilmurray, "A Madonna Without a Child," *ARTNews* 98 (April 1999): 76.

80. *Times* (London), 8 May 1891, 13, quoted in CP 1:191.

81. R. H. Ives Gammell, "The Enigma of John Sargent's Art," *Classical America* 4 (1977): 154.

82. Mount, "Carolus-Duran and the Development of Sargent," 397. Priestley studied at the Académie Jullian. CP 1:226.

83. "The Society of American Artists Exhibition," *Art Amateur* 23 (June 1890): 3, quoted in Culver, "Performing Identities," 25.

84. For the Wilde quotation, see chapter 2. With his rendering of Flora's hands, Sargent seems to partake of something akin to "the Chinese custom of mutilating the female foot and then revering it like a fetish after it has been mutilated." Freud's gloss is that this reverence reads "as though the Chinese male wants to thank the woman for having submitted to being castrated." Sigmund Freud, "Fetishism" (1927), in *On Sexuality,* 357.

85. Wilde, *Picture of Dorian Gray,* 80. Wilde wrote to a friend, "If I do live again I would like it to be as a flower—no soul but perfectly beautiful. Perhaps for my sins I shall be made a red geranium!!" Wilde to Harry Marillier, 8 November 1885, in *Complete Letters of Wilde,* 267.

86. Letter quoted in Mount, "Carolus-Duran and the Development of Sargent," 405.

87. "[U]n être qui s'enferme dans une cuirasse, ne se déshabille jamais devant vous, comme si sous sa chemise il portait un pansement." Blanche, *Propos de peintre,* 69.

88. The phantasmatic infliction of a wound on unruly peers has parallels in the creative acts of other frustrated pollinators. In his poem "The Bee," the syphilitic Valéry urges the insect to "Pique du sein la gourde belle, / Sur qui l'Amour meurt ou sommeille, / Qu'un peu de moi-même vermeille / Vienne à la chair ronde et rebelle!" (prick the lovely gourd of the breast, where Love lies dead or sleeping, so that a little of myself may rise rosy-red in the round, rebellious flesh!). Paul Valéry, "L'abeille," in Rees, *Penguin Book of French Poetry,* 486, lines 5–8.

89. Quoted in Painter, *Marcel Proust,* 1:178–79.

90. Ibid., 1:179.

91. Flora's plaquey skin also connotes syphilis. Vampirism, like inversion, was associated with syphilis in the late nineteenth-century imagination; both conditions were spread through an

intimate, sexual or sexualized exchange of body fluids. As Alexandra Warwick points out, Bram Stoker's "probable death from tertiary syphilis is recorded by his biographer, but even this putative personal connection is unnecessary in the analysis of the appearance of the disease in these texts; syphilophobia was common in the period." Warwick, "Vampires and the Empire: Fears and Fictions of the 1890s," in *Cultural Politics at the Fin de Siècle,* ed. Sally Ledger and Scott McCracken (Cambridge: Cambridge University Press, 1995), 209. Pollinators could be at once vampiric and disease carrying. Sargent's own collection of butterflies, for example, included infectious lepidoptera: the painter told Marie-Louise Pailleron that "the large azure wings" of the fairylike *Morpho-Menelaus* conceal "the plague" under their "celestial blue robe hemmed with gold." ("Je n'avais jamais, alors, rien vu de si féérique que ces larges ailes azurés aux reflets changeant, telles que celles de ce *Morpho-Menelaus* qui cache, dit-on, la peste sous sa robe bleu céleste ourlée d'or." Pailleron, *Paradis perdu,* 157). If Flora and John, Marcel and Albertine were stigmatized by an inherent vice, though, it was nonetheless their very infection that opened up the possibility of immortalization through art. In the words of Hawthorne, from Sargent's childhood reading, "romance and poetry, like ivy, lichens, and wall-flowers, need Ruin to make them grow." Hawthorne, *Marble Faun,* 3.

92. "American Figure Paintings," *New York Daily Tribune,* 14 February 1890, 7, quoted in RGA, 606.

93. Quoted in Fairbrother, *John Singer Sargent,* 84.

94. Meynell and Manson, *Work of John Sargent,* unpaginated.

95. Sargent to Henry James, 29 June 1885, bMS Am 1094 (396), Houghton Library, Harvard University; Stevenson to W. E. Henley, 17 December 1884, *Selected Letters,* 273.

96. Meynell and Manson, *Work of John Sargent.*

97. Sidney Colvin, *Memories and Notes of Persons and Places, 1852–1912* (New York: C. Scribner's Sons, 1921), 100, quoted in R. C. Terry, ed., *Robert Louis Stevenson: Interviews and Recollections* (Houndmills, Basingstoke: Macmillan, 1996), 62.

98. Ovid, *Metamorphoses,* 17.

99. Sargent to Stevenson, November 1885, Stevenson Papers, Beinecke Rare Book and Manuscript Library, Yale University, MS B 5427, quoted in CP 1:168.

100. Fanny Stevenson to Sidney Colvin, spring 1885, in Sidney Colvin, ed. "Robert Louis Stevenson and Henry James," *Scribner's Magazine,* March 1924, 322.

101. Letter of 1884 regarding Symonds's *Writings on Italy,* in *Henry James's Letters,* 3:30, quoted in Wendy Graham, *Henry James's Thwarted Love* (Stanford: Stanford University Press, 1999), 31.

102. See Robert Fraser on the case of Jupien and Charlus in particular. Fraser, *Proust and the Victorians,* 213.

103. RTPE 3:462; "corolle du visage," RTPF 3:453.

104. Sonnet to the comtesse Greffulhe, quoted in Painter, *Marcel Proust,* 1:149.

105. Proust, *Jean Santeuil,* trans. Gerard Hopkins, 529; "sur les étamines déliées de son regard recueilli le pollen ardent et léger de sa sensibilité," *Jean Santeuil,* 3:57. As with hands, watery depths lie below such floral visions of eyes. In *Les faux-monnayeurs,* when Édouard catches his precocious nephew Georges Molinier stealing a book from a stall, the older man looks into his "charming eyes" and sees "every kind of feeling wavering in their depths like water weeds at the bottom of a stream." The two have a private tête-à-tête immediately afterward, in which the youngster, fully aware of his uncle's proclivities, asks him if he often goes "about picking up schoolboys." André Gide, *The Counterfeiters,* trans. Dorothy Bussy (New York: Vintage, 1973), 86, 88. "[Q]uel joli regard il avait! J'y voyais toute sorte de sentiments s'agiter comme des herbes au fond d'un ruisseau." "Ça vous arrive souvent de reluquer les lycéens?" Gide, *Les faux-monnayeurs* (1925), in *Romans,* ed. Yvonne Davet and Jean-Jacques Thierry (Paris: Gallimard, 1961), 999, 1001.

106. A contemporary described Stevenson's mouth as a "mobile" one, with a "lurking suggestion of the great god Pan at times." Will H. Low, *A Chronicle of Friendships, 1873–1900* (New York: Charles Scribner's Sons, 1908), 53.

107. Stevenson went by the soubriquet "Velvet Coat." Ian Bell, *Dreams of Exile: Robert Louis Stevenson, a Biography* (Edinburgh: Mainstream, 1992), 77.

108. Henry James, "Robert Louis Stevenson," *Century Magazine,* April 1888, 869.

109. In the portrait Fanny Stevenson is barefoot but not pregnant. Others besides Sargent engaged in graphically vegetable flirtations with the writer, some offering to bear his vegetable offspring. Gosse, ever the coquette, wrote to RLS that he was "training infancy in the shape of sweet peas to trail along my leads, and I am watching pots of mignonette and eschscholtzia as though they were Scottish families from which fantastic poets might be expected to burst forth." Edmund Gosse to Robert Louis Stevenson, 7 May 1884, quoted in Evan Charteris, *The Life and Letters of Sir Edmund Gosse* (London: William Heinemann, 1931), 162.

110. James, "Robert Louis Stevenson," 870.

111. Stevenson, *Dr. Jekyll and Mr. Hyde,* 107, 109.

112. Ibid., 39, 20, 42.

113. Bell, *Dreams of Exile,* 87.

114. Stevenson, *Dr. Jekyll and Mr. Hyde,* 37, 84, 138.

115. Stevenson to Will Low, 6 August 1887, quoted in Low, *Chronicle of Friendships,* 372. In accordance with the wayward surfacing of the death drive, Sidney Colvin noted that Stevenson "comprised within himself, and would flash on you in the course of a single afternoon, all the different ages and half the different characters of man." Colvin, *Memories and Notes of Persons and Places,* 101, quoted in Terry, *Stevenson: Interviews and Recollections,* 63.

116. Stevenson, *Dr. Jekyll and Mr. Hyde,* 99.

117. James, preface, *Spoils of Poynton,* xlv. The floral reading of Stevenson and his creation is not inapt: Stevenson took the

name "Jekyll" from his friend Walter Jekyll, brother of Gertrude the gardener. Mr. Hyde might be seen as a sort of anemic hyde-rangea.

118. The fine picture hanging in Hyde's Soho digs in particular caused much consternation among readers and critics. Frederick William Henry Myers thought this made him too much of a "jaded voluptuar[y]"; he wanted Hyde to remain in the world of "primary stimuli." "I cannot fancy Hyde looking in at picture shops," he added, though Myers might not have realized that picture shops, as we saw above, were realms of "primary stimuli" for the invert. Myers's notes of 17 March 1886, reprinted in Paul Maixner, ed., *Robert Louis Stevenson: The Critical Heritage* (London: Routledge and Kegan Paul, 1981), 220.

119. Charles Guthrie, *Robert Louis Stevenson—Some Personal Recollections by the Late Lord Guthrie* (Edinburgh: W. Green and Son, 1920), quoted in Bell, *Dreams of Exile*, 78.

120. "Il rappelait étrangement ce personnage double de Stevenson, tantôt excellent et bienfaisant sous les traits du Dr Jekyll, tantôt implacable et bestial sous le masque de master Hyde. C'était, en somme, un hérédo type, chargé d'un poids ancestral trop lourd pour un moignon de volonté." Léon Daudet, *L'entre-deux-guerres*, part 3 of *Souvenirs des milieux littéraires, politiques, artistiques et médicaux*, 1:341. "Personnage" can mean character, persona, or literary character.

121. Pater, *Renaissance*, 123.

122. Eric Savoy, "Theory *a Tergo* in *The Turn of the Screw*," in *Curiouser: On the Queerness of Children*, ed. Steven Bruhm and Natasha Hurley (Minneapolis: University of Minnesota Press, 2004), 260.

123. RTPE 1:789; "une grimace de plaisir s'emparait irrésistiblement de son visage; la peau trop fine de ses joues laissait transparaître une vive rougeur, ses yeux reflétaient la confusion et la joie. . . . Mais j'ai connu un autre être, et il y en a beaucoup, chez lequel la sincérité physiologique de cet incarnat passager n'excluait nullement la duplicité morale," RTPF 1:735. The connection Sargent felt with Stevenson seems to have been shared by Proust. Swann considers the Scotsman "a very great" writer, "equal to the greatest," and Robert Louis Balfour Stevenson (RLBS) may have been one of the models for Proust's Robert de Saint-Loup-en-Bray (RSLB), both of whom had wives and mistresses but were also inverts, and are of matching description in terms of physical delicacy, modeling, and grace; habits of dress (both wore clothes "of a supple elegance," with "no stiffness or starch"); and psychology. RTPE 3:736, 1:789, translation slightly modified; "un grand écrivain . . . un très grand, l'égal des plus grands," "élégance souple," "sans raideur et sans empois," RTPF 3:716, 1:735.

124. Kingsbury, "John Singer Sargent," 253n54, 247n11.

125. Thomas Fox, "Collections and Recollections" of Sargent, box 2, folder 28, John Singer Sargent–Thomas A. Fox Papers, Boston Athenaeum.

126. This exchange occurred well before Wilde's trial and imprisonment. Lucia Fairchild, diary entry for 2 October 1890 (Boston), quoted in Miller, "Sargent in the Diaries of Lucia Fairchild," 8.

127. For example, in Paris in June 1884, as noted in a letter of 3 June 1884 from Constance Wilde to Otho Holland Lloyd. *Complete Letters of Wilde*, 229.

128. "To my friend John S. Sargent, in deep admiration of his work." On the title page he added, "Rien n'est vrai, que le beau" (nothing is true but the beautiful). Albert Boime, "Sargent in Paris and London: A Portrait of the Artist as Dorian Gray," in Hills, *John Singer Sargent*, 76.

129. Fairbrother, *John Singer Sargent*, 83.

130. RTPE 2:655; "ils forment dans tous les pays une colonie orientale, cultivée, musicienne, médisante," RTPF 2:632.

131. W. Graham Robertson, *Life Was Worth Living* (New York: Harper and Brothers, 1931), 40, 235.

132. Kerrison Preston, Introduction, *Letters from Graham Robertson* (London: Hamish Hamilton, 1953), xvi.

133. Robertson, *Life Was Worth Living*, 235.

134. RTPE 2:626; "donnait à sa taille un port avantageux, posait avec une impertinence grotesque son poing sur la hanche," RTPF 2:604. With his handle out (hand on hip), Robertson assumes the teapot pose discussed in chapter 2.

135. Dorian Gray becomes a kind of queer vegetable vampire, sucking to maintain his bloom, after being infected by Henry Wotton. Wotton lures Dorian into vice by taunting him with his mortality: he tells Dorian that although flowers bloom year after year, human beauty passes away forever once "rose-red youth" and "rose-white boyhood" are gone, that time "wars against" his "lilies" and "roses." Following Wotton's example, Dorian learns to steal his victims' innocence and goodness to maintain his own fresh bloom, leaving a trail of bodies—suicides and murders—in his wake. Sibyl Vane, Alan Campbell, Adrian Singleton, and a "wretched boy in the Guards" kill themselves after Dorian has dashed their hopes or ruined their reputations; Dorian murders the naïve and idealistic painter Basil by emptying his jugular ("the great vein that is behind the ear"). Although the first of these victims is female (a cross-dressing one), the rest are all male; his "friendship" is particularly "fatal to young men." *Picture of Dorian Gray*, 42, 46, 183, 192, 183. On the idea of Dorian as a vampire, see Twitchell, *Living Dead*, 172–78; and Judith Halberstam, *Skin Shows: Gothic Horror and the Technology of Monsters* (Durham: Duke University Press, 1995), 76–77.

136. Dorian's picture is a "withered, wrinkled," buried thing, like a coffin (264). It is the root of his power, the font of his youth. While Dorian protects it, the picture alone ages, as he knew it would: "Its gold would wither into grey. Its red and white roses would die" (121). Once he stabs it, though—staking the picture for his life, in a way—he ages apocalyptically quickly; the lilies and roses of his golden youth give way to age, as his prophetic name foretold, moving from gold—*d'or*—to grey. Robert Fraser argues that the message Wotton brought Dorian "was not pollen at all, but a sting." Fraser, *Proust and the Victorians*, 207. (Richard Le Gallienne made a similar comment about Wilde in his 1892 book *English Poems;* in the prefatory poem "To the Reader," he denounced "these strange

green flowers that spring / From daisy roots and seem to bear a sting." Quoted in Matthew Sturgis, *Passionate Attitudes: The English Decadence of the 1890s* [London: Macmillan, 1995], 97, lines 14–15.) The image of gold fading to grey appears in other texts by Wilde, including "Athanasia" (Immortality), a poem about a seemingly immortal flower grown from a seed found in the hand of an Egyptian mummy. "To this bright flower a thousand years / Seemed but the lingering of a summer's day, / It never knew the tide of cankering fears / Which turn a boy's gold hair to withered grey, / The dread desire of death it never knew, / Or how all folk that they were born must rue." Wilde, "Athanasia," in *Complete Poetry*, ed. Isobel Murray (Oxford: Oxford University Press, 1997), 89, lines 43–48.

137. Will Low saw "an indication of a portfolio on its stand," but this could also be read as two framed works turned toward the wall. *Chronicle of Friendships*, 464.

138. In this light, the earlier recounted story of Robertson's "squeezing" by a rather sadistic Sargent brings to mind Fabre's description of the philanthus, a type of predatory wasp, wringing the honey out of a bee: "If the Philanthus feels that the bee is full of honey, it does not fail to squeeze it before storing it, . . . to make it disgorge the delicious liquid; this it drinks by licking the tongue of the unfortunate bee, which, dying, extends it at full length. This profanation of a dying creature, squeezed by its murderer to empty its body and enjoy the contents, has something so hideous that I should call it a crime if a Philanthus could be in the wrong." Although he would like to consider the tongue-licking, honey-sucking activity of the insect a hideous crime, he cannot, as it is a perfectly natural phenomenon. Fabre, *Insect Life*, 166, translation modified. "Si le Philanthe qui vient de faire capture sent son abeille gonfle de miel, il ne manque guère avant de l'emmagasiner, de lui presser le jabot, . . . pour lui faire dégorger la délicieuse purée, dont il s'abreuve en léchant la langue de la malheureuse, qui, agonisante, l'étale dans toute sa longueur hors de la bouche. Cette profanation d'un mourant, dont le meurtrier presse le ventre pour le vider et faire régal du contenu, a quelque chose de hideux dont je ferais un crime au Philanthe si la bête pouvait avoir tort." Fabre, *Souvenirs entomologiques*, 167.

139. "[L]e miel artiste," "gris de perle," "miels fins et gris." Comte Robert de Montesquiou, "Pecus," in *Les chauves-souris* (Paris: G. Richard, 1893), 96–97, lines 19, 5, 4; "miel solitaire et féroce," Montesquiou, "Avis," in *Les hortensias bleus* (1896) (Paris: Éditions des Autres, 1979), 186, line 12.

140. "Faisant honte à l'abeille et peur aux passereaux, / Et n'attirant . . . / Que des chauves-souris." Montesquiou, "Avis," 186, lines 6–8.

141. Alfred, Lord Tennyson, "Epilogue—to the Queen" (1873), *Idylls of the King*, in *The Works of Tennyson: The Eversley Edition*, ed. Hallam Tennyson, 9 vols. (London: Macmillan, 1907–8), 5:421, line 56.

142. Wilde, *Picture of Dorian Gray*, 155.

143. James Huneker, review of George Moore's work, in *Musical Courier* 36 (22 June 1898): 18–20, quoted in Arnold T. Schwab, "Irish Author and American Critic: George Moore and James Huneker, Part One," *Nineteenth-Century Fiction* 8 (March 1954): 261.

144. Edmund Gosse, "A First Sight of Verlaine," *Savoy* 2 (April 1896): 113, quoted in Phil Baker, *The Dedalus Book of Absinthe* (Sawtry, Cambridgeshire: Dedalus, 2001), 70.

145. "Abeille de silence." Montesquiou, "Avis," 186, line 11; "L'abeille de la brune . . . / Est bien votre symbole, Êtres désorbités," "Essence," *Les chauves-souris*, 56, lines 29–30.

146. Charles Maturin, *Melmoth the Wanderer* (1820) (Harmondsworth: Penguin, 2000), 393.

147. Walter Sickert, "Impressionism," *New Age* (30 June 1910), reprinted in *Walter Sickert: The Complete Writings on Art*, ed. Anna Gruetzner Robins (Oxford: Oxford University Press, 2000), 253.

148. RTPE 1:874; "abeille engourdie par la fumée du tabac, qui n'a plus le souci de préserver la provision de ses efforts accumulés et l'espoir de sa ruche," RTPF 1:816.

149. "Entrée des artistes," Delvau, *Dictionnaire érotique moderne*, 153; "porte de derrière," "fenêtre de derrière," de Landes, *Glossaire érotique*, 390, 380; "back-door," Farmer, *Slang and Its Analogues*, 1:87.

150. *Boston Transcript*, 12 June 1903, quoted in CP 3:109.

151. Anton Joseph Kerner von Marilaun, *The Natural History of Plants* (1894), trans. F. W. Oliver et al., 2 vols. (London: Blackie and Son, 1895), 2:238.

152. Darwin, *Effects of Cross and Self Fertilisation*, 430.

153. Darwin, *Various Contrivances*, 2nd ed., 275.

154. Kerner von Marilaun, *Natural History of Plants*, 2:239.

155. Henry James to Hendrik Andersen, 1 July 1901, Virginia MSS 6251 (15), quoted in *Dearly Beloved Friends*, 35.

156. Wilde et al., *Teleny*, 117.

157. Stevenson, *Dr. Jekyll and Mr. Hyde*, 1, 9, 11, 19. Robert Fraser notes that Proust's taste in Stevenson "seems eccentric" (he recommended *The Dynamiter*, *The Suicide Club*, and *Dr. Jekyll and Mr. Hyde* to the princesse de Caraman-Chimay) until "one realises how precisely they are all novels" exploring "inner spaces that are initially sealed." Fraser, *Proust and the Victorians*, 179–80.

158. Stevenson, *Dr. Jekyll and Mr. Hyde*, 21, 31, 20.

159. He may even have been blackmailed for them. On one occasion Alfred's father discovered that his son had taken a substantial sum of money from him; when questioned, Alfred "very hotly . . . claimed that he had lent it to someone and would say no more." Sargent lunched with Alfred, was told a story that "took [his] breath away," and offered to pretend to be the recipient of the funds to smooth things over with Asher. We don't know if Alfred took up Sargent's offer, but he was sent to South Africa shortly after the portrait was painted and died there the same year. Mount, *John Singer Sargent*, 195.

160. Jekyll's project fails. Stevenson similarly queries the ability of painting/pollination to create or extend life. His story

"The Yellow Paint" is about a physician who sells pollen-colored paint "of so singular a virtue that whoso was bedaubed with it from head to heel was set free from the dangers of life, and the bondage of sin, and the fear of death for ever," though this turns out to be a hoax. Robert Louis Stevenson, "The Yellow Paint," in *The Complete Short Stories,* centenary ed., vol. 2, ed. Ian Bell (Edinburgh: Mainstream, 1993), 412. I thank Elaine Scarry for this reference.

161. To ensure that we (the readers) can be witness to what strange events occur within it, "the whole southern face of the pentagon" is occupied by the room's "sole window," an "immense sheet of unbroken glass from Venice." The tower functions as a kind of observation hive. The wall coverings are strangely wrought with patterns of the sort that appear "in the guilty slumbers of the monk," foregrounding the theme of perversion. Poe, "Ligeia," 660–61.

162. Sargent to Ralph Curtis, 18 November 1890, box 1, folder 3, John Singer Sargent–Thomas A. Fox Papers, Boston Athenaeum.

163. Quoted in Fairbrother, "Warhol Meets Sargent at Whitney," 68. Similarly, Comte Robert de Montesquiou described the style, if not the plot, of *À la recherche du temps perdu* as "a mixture of litanies and sperm," a phrase that also nicely encapsulates Sargent's *Hell* lunette. Quoted in Edmund White, *Marcel Proust* (New York: Viking, 1999), 3.

164. "[L]a phosphorescence de la pourriture" (the phosphorescence of decay). Baudelaire, "Edgar Poe, sa vie et ses oeuvres" (1856), *Oeuvres complètes,* 2:317–18.

165. "Il se plaint en riant que la charcuterie de Pozzi dans son corps y a apporté une phosphorescence, qu'il a un besoin enragé de coït." Edmond and Jules de Goncourt, *Journal,* entry for 29 April 1896, 3:1273.

166. Trevor Fairbrother, "*Man Screaming* by John Singer Sargent," *American Art* 15 (Spring 2001): 84.

167. Most of this chapter's examples have been of vampiric usurpations of life by pollinators, plants, and portraits. There is also, however, a Frankensteinian model of queer life giving. Lacking the space to consider it in depth here, I leave you to brood on the image of William James urging Henry to go to an "electro-therapeutist" to have "one pole" of an electrical battery "put inside the rectum" and "a strong *galvanic* current"—not a "piddling" one—applied, to cure his intestinal problems and give him a second lease on life. William James to Henry James, 25 October 1869, in *The Correspondence of William James,* vol. 1, ed. Ignas Skrupskelis and Elizabeth Berkeley (Charlottesville: University Press of Virginia, 1992), 113, quoted in Graham, *Henry James's Thwarted Love,* 48.

168. RTPE 3:64; "en dormant, elle était devenue une plante," "une longue tige en fleur," "quand elle dormait, je n'avais plus à parler, je savais que je n'étais plus regardé par elle, je n'avais plus besoin de vivre à la surface de moi-même," "le pouvoir de rêver que je n'avais qu'en son absence, je le retrouvais à ces instants," RTPF 3:69–70.

169. Ben Jonson, *The New Inn* (1631), ed. Michael Hattaway (Manchester: Manchester University Press, 1984), 2.4.37, p. 115.

170. Thus Sargent dubbed his female models, who were nieces and friends: Rose-Marie, Polly, and Dolly, with an extra Violet or Flora added to the bunch here and there. Fairbrother, "Sargent's Genre Paintings," 41.

171. RTPE 2:89; "je me sentais attaché à un sol invisible et profond par les articulations, que la fatigue me rendait sensibles, de radicelles musculeuses et nourricières," RTPF 2:91.

Conclusion

1. The epigraph is from Andy Warhol, *The Philosophy of Andy Warhol (from A to B and Back Again)* (San Diego: Harcourt, Brace, 1975), 48.

2. Kupka's *Amorpha: Fugue for Two Colors* and *Amorpha: Warm Chromatics* were exhibited at the Salon d'Automne.

3. Virginia Spate has argued that "the connection between the evolution of matter and the sexual generation of man" is made "explicit in the association between writhing, embracing human forms, fetal shapes, and unspecific organic structures." Spate, *Orphism: The Evolution of Non-Figurative Painting in Paris, 1910–1914* (Oxford: Clarendon Press, 1979), 119.

4. Anne Middleton Wagner, *Three Artists (Three Women): Modernism and the Art of Hesse, Krasner, and O'Keeffe* (Berkeley and Los Angeles: University of California Press, 1996), 52.

5. Ibid., 72.

6. Deni Brown, *Aroids: Plants of the Arum Family* (Portland, Ore.: Timber Press, 2000), 82–84.

7. Paul Rosenfeld, "The Paintings of Georgia O'Keeffe: The Work of the Young Artist Whose Canvases Are to Be Exhibited in Bulk for the First Time This Winter," *Vanity Fair,* October 1922, quoted in Marcia Brennan, *Painting Gender, Constructing Theory: The Alfred Stieglitz Circle and American Formalist Aesthetics* (Cambridge: MIT Press, 2001), 3.

8. Wagner, *Three Artists,* 64.

9. William A. Ewing, *Flora Photographica: Masterpieces of Flower Photography, 1835 to the Present* (New York: Simon and Schuster, 1991), 19.

10. Lawrence Alma-Tadema, for example, a painter and friend of Sargent's, used von Gloeden's photographs as a source for some of his neoclassical subjects (Pearsall, *Worm in the Bud,* 452). At Robert Browning's funeral in Westminster Abbey, Edmund Gosse couldn't stop staring at one of von Gloeden's photographs he had just been sent by John Addington Symonds: "As I sat in the Choir, with George Meredith at my side, I peeped at it again and again, and at last, while waiting in the deep silence for the ceremonial to begin, with many thoughts of love and life and genius and decay, moving in my mind, this sextain formed itself—I hardly know how—and I send it to you as the onlie begetter of all that sehnsucht [yearning]: 'Dark-stamen'd flower, across thy beauty / Sighing, I cast the veil. / In Youth's high spring-tide Love's a duty / And rose-crowned hopes prevail; / But autumn comes, and brings I see / No rose, but rosemary for me.'" Gosse to Symonds, 31 December 1889, quoted in Anne Thwaite, *Edmund Gosse: A Literary*

Landscape, 1849–1928 (London: Secker and Warburg, 1984), 324.

11. The label "sculptors" at the bottom right of the postcard reinforces the suggestion of Bernini's *Apollo and Daphne*.

12. Forbes Watson, "Sargent, Boston, and Art," *Arts and Decoration* (February 1917): 194.

BIBLIOGRAPHY

Archival Sources

Edward Lear Collection. Houghton Library, Harvard University.

Francis Davis Millet and Millet Family Papers. Archives of American Art. Smithsonian Institution, Washington, D.C.

Isabella Stewart Gardner–John Singer Sargent Correspondence. Isabella Stewart Gardner Museum Archives, Boston.

James, Henry. Letters. Houghton Library, Harvard University.

John Singer Sargent Catalogue and Conservation Files. Tate Britain, London.

John Singer Sargent Letters, 1887–1922. Archives of American Art. Smithsonian Institution, Washington, D.C.

John Singer Sargent Object Files. Museum of Fine Arts, Boston.

John Singer Sargent–Thomas A. Fox Papers. Boston Athenaeum.

Lee, Vernon. Papers. Special Collections. Miller Library, Colby College, Waterville, Maine.

Other Sources

Abrams, Meyer Howard. *The Mirror and the Lamp: Romantic Theory and the Critical Tradition*. New York: W. W. Norton, 1958.

Adelson, Warren, et al. *Sargent Abroad: Figures and Landscapes*. New York: Abbeville Press, 1997.

Albaret, Celeste. *Monsieur Proust*. Ed. Georges Belmont. Trans. Barbara Bray. New York: McGraw-Hill, 1976.

Albert, Nicole. "From Myth to Pathology: Perversions of Gender-Types in Late Nineteenth-Century Literature and Clinical Medicine." *Diogenes* 52 (November 2005): 114–26.

Alcott, Louis May. *Rose in Bloom*. 1876. N.p.: Dodo Press, 2005.

Alibert, Jean-Louis. *Clinique de l'hôpital Saint-Louis ou traité complet des maladies de la peau*. Paris: Cormon et Blanc, 1833.

Allen, Grant. *Physiological Aesthetics*. 1877. London: Elibron Classics, 2005.

"American Figure Paintings." *New York Daily Tribune*, 14 February 1890, 7.

Andersen, Hans Christian. *The Complete Fairy Tales and Stories*. Trans. Erik Christian Haugaard. New York: Anchor, 1983.

———. *Eventyr*. Copenhagen: C. A. Reitzel, 1850.

Appelbaum, Stanley, ed. *Bizarreries and Fantasies of Grandville*. New York: Dover, 1974.

Apter, Emily. "The Garden of Scopic Perversion from Monet to Mirbeau." *October* 47 (Winter 1988): 91–115.

Aquarone, Stanislas. *The Life and Works of Émile Littré, 1801–1881*. Leyden: A. W. Sythoff, 1958.

Argencourt, Louise d'. *European Paintings of the Nineteenth Century*. Vol. 1, *Aligny–Gros*. Cleveland: Cleveland Museum of Art, 1974.

Armstrong, Carol. "Facturing Femininity: Manet's *Before the Mirror*." *October* 74 (Fall 1995): 75–104.

Armstrong, Mrs. M. F. *On Habits and Manners*. Rev. ed. Hampton, Va.: Normal School Press, 1888.

"Art Chronicle." *Portfolio* 38 (May 1887): 104–5.

"Art Chronicle." *Portfolio* 38 (June 1887): 125.

The Artistic Language of Flowers. London: George Routledge and Sons, [1888?].

"Art Notes and Reviews." *Art Journal* 48 (January 1886): 29–32.

Asendorf, Christoph. *Batteries of Life: On the History of Things and Their Perception in Modernity*. Trans. Don Reneau. Berkeley and Los Angeles: University of California Press, 1993.

Asleson, Robyn. *Albert Moore*. London: Phaidon, 2000.

Auerbach, Nina. *Our Vampires, Ourselves*. Chicago: University of Chicago Press, 1995.

Bachelard, Gaston. *The Flame of a Candle*. Trans. Joni Caldwell. Dallas: Dallas Institute of Humanities and Culture, 1988.

———. *La poétique de l'espace*. Paris: Presses Universitaires de France, 1961.

Bain, Francis William. *A Digit of the Moon: A Hindoo Love Story*. 1899. 5th ed. London: James Parker, 1906.

Baker, Phil. *The Dedalus Book of Absinthe*. Sawtry, Cambridgeshire: Dedalus, 2001.

Baldry, Alfred Lys. "The Art of J. S. Sargent, R.A.: Part I." *Studio* 19 (15 February 1900): 3–21.

———. "The Art of J. S. Sargent, R.A.: Part II." *Studio* 19 (March 1900): 107–19.

Balzac, Honoré de. *A Harlot High and Low (Splendeurs et misères des courtisanes)*. Trans. Rayner Heppenstall. Harmondsworth: Penguin, 1970.

———. *Illusions perdues*. 1837–43. Paris: Garnier, 1961.

———. *The Lily of the Valley*. Trans. Lucienne Hill. New York: Carroll and Graf, 1989.

———. *Lost Illusions*. Harmondsworth: Penguin, 1971.

———. *Le lys dans la vallée*. 1836. Ed. Moise Le Yaouanc. Paris: Garnier, 1966.

———. *Splendeurs et misères des courtisanes*. 1847. Ed. A. Adam. Paris: Garnier, 1958.

Banner, Delmar Harmood. "John Singer Sargent." *Nineteenth Century and After*, February 1926, 298–312.

Baring-Gould, Sabine. *A Book of Ghosts*. 1904. Penyffordd, Chester: Ash-Tree Press, 1996.

Barnes, Robert. *Traité clinique des maladies des femmes*. Trans. A. Cordes. Paris: G. Masson, 1876.

Barrère, Albert, and Charles Leland, eds. *A Dictionary of Slang, Jargon, and Cant*. 2 vols. London: Ballantyne Press, 1889–90.

Barrie, James Matthew. *Peter Pan*. 1911. New York: Bantam Books, 1985.

Bartlett, Neil. *Who Was That Man? A Present for Mr. Oscar Wilde*. London: Serpent's Tail, 1988.

Bataille, Georges. "Le langage des fleurs." 1929. In *Oeuvres complètes*, 12 vols., 1:173–78. Paris: Gallimard, 1970–88.

———. "The Language of Flowers." Trans. Roger Banner and Allan Stoekl. In *Karl Blossfeldt: Art Forms in Nature, the Complete Edition*, 7–11. Munich: Schirmer Art Books, 1999.

Bätschmann, Oskar. *The Artist in the Modern World: The Conflict Between Market and Self-Expression*. Cologne: Dumont, 1997.

Baudelaire, Charles. *Les fleurs du mal*. 1861. Ed. Claude Pichois. Paris: Gallimard, 1996.

———. *Oeuvres complètes*. Ed. Claude Pichois. 2 vols. Paris: Gallimard, 1975–76.

———. *The Painter of Modern Life and Other Essays*. Ed. and trans. Jonathan Mayne. London: Phaidon, 1995.

———. "Salon de 1846." In *Curiosités esthétiques, l'art romantique, et autres oeuvres critiques*, ed. H. Lemaître, 97–200. Paris: Garnier, 1962.

Bauer, George. "Gay Incipit: Botanical Connections, Nosegays, and Bouquets." In *Articulations of Difference: Gender Studies and Writing in French*, ed. Dominique Fisher and Lawrence Schehr, 64–82. Stanford: Stanford University Press, 1997.

Baugé, Isabelle. *Pantomimes*. Quercy: Cicero, 1995.

Bazin, Nancy Topping, and Alma Freeman. "The Androgynous Vision." *Women's Studies* 2 (1974): 185–215.

Beatty, Laura. *Lillie Langtry: Manners, Masks, and Morals*. London: Vintage, 2000.

Beaumont, Cyril. *Ballet Design, Past and Present*. London: The Studio, 1946.

———. *Complete Book of Ballets: A Guide to the Principal Ballets of the Nineteenth and Twentieth Centuries*. Garden City, N.Y.: Garden City Publishing, 1941.

Beaver, Harold. "Homosexual Signs (in Memory of Roland Barthes)." *Critical Inquiry* 8 (Autumn 1981): 99–119.

Beer, Gillian. *Darwin's Plots: Evolutionary Narrative in Darwin, George Eliot, and Nineteenth-Century Fiction*. 1983. Cambridge: Cambridge University Press, 2000.

Beerbohm, Max. *A Book of Caricatures*. London: Methuen, 1907.

Bell, Charles. *The Hand: Its Mechanism and Vital Endowments as Evincing Design*. London: W. Pickering, 1833.

Bell, Clive. "Sargent." *New Republic*, 20 May 1925, 340.

Bell, Ian. *Dreams of Exile: Robert Louis Stevenson, a Biography*. Edinburgh: Mainstream, 1992.

Bell, Quinton. "John Singer Sargent and Roger Fry." *Burlington Magazine* 99 (November 1957): 380–82.

Belleroche, Albert de. "The Lithographs of Sargent." *Print Collector's Quarterly* 13 (February 1926): 31–45.

Bellow, Juliet. "Engendering the Masculine Interior: John Singer Sargent's *Dr. Pozzi at Home*." Master's thesis (abridged), University of Pennsylvania, 2000.

Benjamin, Walter. *The Arcades Project*. Trans. Howard Eiland and Kevin McLaughlin. Cambridge: Belknap Press of Harvard University Press, 1999.

———. *Illuminationen*. 1955. Frankfurt am Main: Suhrkamp Verlag, 1974.

———. *Illuminations*. Ed. Hannah Arendt. Trans. Harry Zohn. New York: Schocken Books, 1968.

———. *Das Kunstwerk im Zeitalter seiner technischen Reproduzierbarkeit*. 1936. Frankfurt am Main: Suhrkamp Verlag, 1977.

Bennett, Jennifer. *Lilies of the Hearth: The Historical Relationship Between Women and Plants*. Camden East, Ont.: Camden House, 1991.

Bennett, Paula. "Critical Clitoridectomy: Female Sexual Imagery and Feminist Psychoanalytic Theory." *Signs* (Winter 1993): 235–59.

———. "'Pomegranate-Flowers': The Phantasmic Productions of Late-Nineteenth-Century Anglo-American Women Poets." In *Solitary Pleasures: The Historical, Literary, and Artistic Discourses of Autoeroticism*, ed. Paula Bennett and Vernon Rosario, 189–213. New York: Routledge, 1995.

Bennett, Paula, and Vernon Rosario, eds. *Solitary Pleasures: The Historical, Literary, and Artistic Discourses of Autoeroticism*. New York: Routledge, 1995.

Berenson, Bernard. *Sunset and Twilight, from the Diaries of 1947–1958*. Ed. Nicky Mariano. New York: Harcourt, Brace and World, 1963.

Berliocchi, Luigi. *The Orchid in Lore and Legend*. Trans. Lenore Rosenberg and Anita Weston. Ed. Mark Griffiths. Portland, Ore.: Timber Press, 2000.

Bernheimer, Charles. "Manet's *Olympia:* The Figuration of Scandal." *Poetics Today* 10 (Summer 1989): 255–77.

Berry, Rose V. "John Singer Sargent: Some of His American Work." *Art and Archaeology* 18 (September 1924): 83–112.

Bertucci, Paola, and Giuliano Pancaldi, eds. *Electric Bodies: Episodes in the History of Medical Electricity*. Bologna: Università di Bologna, 2001.

Birds, Bees, and Gardens: Romantic Visions of Victorian England. Exh. cat. New York: Hirschl and Adler Galleries, 1983.

Birken, Lawrence. *Consuming Desire: Sexual Science and the Emergence of a Culture of Abundance, 1871–1914*. Ithaca: Cornell University Press, 1988.

Birnbaum, Martin. *John Singer Sargent: A Conversation Piece*. New York: William E. Rudge's Sons, 1941.

Bisanz, Rudolf. *German Romanticism and Philipp Otto Runge: A Study in Nineteenth-Century Art Theory and Iconography*. DeKalb: Northern Illinois University Press, 1970.

Bjerregaard, Carl Hendrik Andreas. "The Elementals, the Elementary Spirits, and the Relationship Between Them and Human Beings." *Path* 1 (January 1887): 289–331.

———. "A Philosopher and a Butterfly." *Monthly Illustrator and Home and Country* 3, no. 11 (1895): 380–84.

Black, Joel. "The Aesthetics of Gender: Zeuxis' Maidens and the Hermaphroditic Ideal." *New York Literary Forum* 8–9 (1981): 189–208.

Blanc, Charles. *Grammaire des arts du dessin*. 1867. Paris: Librairie Renouard, 1880.

Blanc, Olivier. "Visibilité du libertinage féminin sous Louis XVI." In *Femmes et libertinage au xviiie siècle ou Les caprices de Cythère*, ed. Anne Richardot, 45–54. Rennes: Presses Universitaires de Rennes, 2003.

Blanchard, Mary Warner. *Oscar Wilde's America: Counterculture in the Gilded Age*. New Haven: Yale University Press, 1998.

Blanche, Jacques-Émile. *Mes modèles*. 1928. Paris: Stock, 1984.

———. *La pêche aux souvenirs*. Paris: Flammarion, 1949.

———. *Portraits of a Lifetime: The Late Victorian Era, the Edwardian Pageant, 1870–1914*. Trans. and ed. Walter Clement. New York: Coward-McCann, 1938.

———. *Propos de peintre: De Gauguin à la Revue nègre*. 3rd ed. Paris: Éditions Émile-Paul Frères, 1928.

Bland, Lucy, and Laura Doan, eds. *Sexology in Culture: Labelling Bodies and Desires*. Chicago: University of Chicago Press, 1998.

———. *Sexology Uncensored: The Documents of Sexual Science*. Chicago: University of Chicago Press, 1998.

Bland-Sutton, J. *Surgical Diseases of the Ovaries and Fallopian Tubes, Including Tubal Pregnancy*. London: Cassell, 1896.

Blashfield, Edwin. "John Sargent." In *Commemorative Tributes to Cable by Robert Underwood Johnson, Sargent by Edwin Howland Blashfield, and Pennell by John Charles van Dyke*, 9–44. New York: American Academy of Arts and Letters, 1927.

———. "John Singer Sargent—Recollections." *North American Review* (June–August 1925): 641–53.

Bodelsen, Marete. *Gauguin's Ceramics: A Study in the Development of His Art*. London: Faber and Faber; Copenhagen: Nordisk Sprog-Og Kulturforlag, 1964.

Boime, Albert. "Sargent in Paris and London: A Portrait of the Artist as Dorian Gray." In *John Singer Sargent*, exh. cat., ed. Patricia Hills, 75–109. New York: Whitney Museum of American Art and Harry N. Abrams, 1986.

Booth, Michael. *Victorian Spectacular Theatre, 1850–1910*. Boston: Routledge and Kegan Paul, 1981.

Borsi, Franco, and Ezio Godoli. *Paris 1900*. New York: Rizzoli, 1977.

Bourget, Paul. *Outre-Mer: Impressions of America*. London: T. Fisher Unwin, 1895.

Bowles, Hamish. "The Madame X Files." *Vogue* magazine, January 1999, 174–81, 205.

Bradstreet, Christina. "'Wicked with Roses': Floral Femininity and the Erotics of Scent." *Nineteenth-Century Art Worldwide* 6 (Spring 2007). http://www.19thc-artworldwide.org/spring_07/articles/brad.shtml.

Brennan, Marcia. *Painting Gender, Constructing Theory: The Alfred Stieglitz Circle and American Formalist Aesthetics*. Cambridge: MIT Press, 2001.

Bretonne, Raitif de la. "Pall-Mall Semaine." *L'Écho de Paris*, 16 May 1895, 2.

Breuer, Josef, and Sigmund Freud. *Studies on Hysteria*. Trans. James Strachey and Anna Freud. New York: Basic Books, 1957.

Brinton, Christian. *Modern Artists*. New York: Baker and Taylor, 1908.

———. "Sargent and His Art." *Munsey's Magazine*, December 1906, 265–84.

Brown, Deni. *Aroids: Plants of the Arum Family*. Portland, Ore.: Timber Press, 2000.

Brugmans, Linette F., ed. and trans. *The Correspondence of André Gide and Edmund Gosse, 1904–1928*. New York University Studies in Romance Languages and Literature 2. New York: New York University Press, 1959.

Bruhm, Steven, and Natasha Hurley, eds. *Curiouser: On the Queerness of Children*. Minneapolis: University of Minnesota Press, 2004.

Buchmann, Stephen L., and Gary Paul Nabhan. *The Forgotten Pollinators*. Washington D.C.: Island Press; Covelo, Calif.: Shearwater Books, 1996.

Buckley, Arabella. *The Fairy-Land of Science*. New York: D. Appleton, 1883.

Buck-Morss, Susan. *The Dialectics of Seeing: Walter Benjamin and the Arcades Project*. Cambridge: MIT Press, 1991.

Burke, Doreen Bolger. *American Paintings in the Metropolitan Museum of Art*. Vol. 3. Ed. Kathleen Luhrs. New York: Metropolitan Museum of Art, 1980.

Burns, Sarah. "The 'Earnest, Untiring Worker' and the Magician of the Brush: Gender Politics in the Criticism of Cecilia Beaux and John Singer Sargent." *Oxford Art Journal* 15, no. 1 (1992): 36–53.

———. *Inventing the Modern Artist: Art and Culture in Gilded Age America*. New Haven: Yale University Press, 1996.

———. "Under the Skin: Reconsidering Cecilia Beaux and John Singer Sargent." *Pennsylvania Magazine of History and Biography* 124 (July 2000): 317–47.

Burton, Richard, and L. C. Smithers. *Priapeia, sive, Diversorum poetarum in Priapum lusus, or Sportive epigrams on Priapus*. 1890. Ware, Hertfordshire: Wordsworth Editions, 1995.

Busst, Alan J. L. "The Image of the Androgyne in the Nineteenth Century." In *Romantic Mythologies*, ed. Ian Fletcher, 1–95. London: Routledge and Kegan Paul, 1967.

Butler, Kathleen Lois. "Tradition and Discovery: The Watercolors of John Singer Sargent." PhD diss., University of California, Berkeley, 1994.

Caffin, Charles. *American Masters of Painting*. New York: Doubleday, Page, 1902.

———. "How to Study Pictures." *St. Nicholas* 32 (October 1905): 1094–98.

———. "John S. Sargent: The Greatest Contemporary Portrait Painter." *World's Work* 7 (November 1903): 4099–116.

Calabrella, E. C., Baroness de. *The Ladies' Science of Etiquette, by an English lady of rank: To which is added, The Ladies' Handbook of the Toilet, a Manual of Elegance and Fashion.* Philadelphia: T. B. Peterson, 1853.

Carlier, F. *Études de pathologie sociale: Les deux prostitutions.* Paris: E. Dentu, 1887.

Carpenter, Edward. *The Intermediate Sex.* 1908. London: Allen and Unwin, 1918.

Carr, Alice Comyns. *Mrs. J. Comyns Carr's Reminiscences.* Ed. Eve Adam. London: Hutchinson, 1926.

Carroll, Lewis. *Alice's Adventures in Wonderland and Through the Looking Glass.* Oxford: Oxford University Press, 1982.

Carter, Tom. *The Victorian Garden.* Salem, N.H.: Salem House, 1985.

Castle, Terry. "Lab'ring Bards: Birth Topoi and English Poetics, 1660–1820." *Journal of English and Germanic Philology* 78 (April 1979): 193–208.

Catalogue des objets d'art et d'ameublement: Composant la première partie de la collection de feu le Dr. Prof. S. Pozzi. Paris: Galerie Georges Petit, 23–24 June 1919.

Catalogue of the Library of the Late John Singer Sargent, R.A., D.C.L., L.L.D. London: Messrs. Christie, Manson and Wood, 29 July 1925.

Catalogue of Ten Sargents from Sargent's Own Collection. Exh. cat. New York: Scott and Fowles, 1948.

Cater, Harold, ed. *Henry Adams and His Friends: A Collection of His Unpublished Letters.* Boston: Houghton Mifflin, 1947.

Champeau, Stéphanie. "Le végétal dans le *Journal* des Goncourt." In *Le végétal,* ed. Jean-Pierre Cléro and Alain Niderst, 181–249. Rouen: Publications de l'Université de Rouen, 1999.

Chard, Cecil. "John S. Sargent, R.A.: The Work of a Great Portrait-Painter." *Pall Mall Magazine,* June 1907, 643–51.

Charteris, Evan. *John Sargent.* New York: Charles Scribner's Sons, 1927.

———. *The Life and Letters of Sir Edmund Gosse.* London: William Heinemann, 1931.

Chauncey, George. *Gay New York: Gender, Urban Culture, and the Making of the Gay Male World, 1890–1940.* New York: Basic Books, 1994.

Child, Theodore. "American Artists at the Paris Exhibition." *Harper's New Monthly Magazine,* September 1889, 489–521.

Clark, T. J. *The Absolute Bourgeois: Artists and Politics in France, 1848–1851.* Berkeley and Los Angeles: University of California Press, 1999.

Clay, Jean. "Ointments, Makeup, Pollen." *October* 27 (Winter 1983): 3–44.

Cleland, John. *The Memoirs of a Woman of Pleasure, Fanny Hill.* 1748–49. Ware, Hertfordshire: Wordsworth Editions, 1996.

Clitorides, Philogynes. *The Natural History of the Frutex Vulvaria, or Flowering Shrub.* London: W. James, 1732.

Coburn, Frederick W. "The Sargent Decorations in the Boston Public Library." *American Magazine of Art* 8 (February 1917): 129–36.

Cohen, William. "Manual Conduct in *Great Expectations.*" *ELH* 60 (Spring 1993): 217–59.

Collection S. Pozzi (deuxième partie): Art antique. Paris: Galerie Georges Petit, 25–27 June 1919.

Collins, J. P. "Sargent in London: His Haunts and Habits." *Boston Evening Transcript,* magazine section, 2 May 1925, 1, 8.

Colvin, David. *Aubrey Beardsley: A Slave to Beauty.* London: Orion, 1998.

Colvin, Sidney, ed. "Robert Louis Stevenson and Henry James." *Scribner's Magazine,* March 1924, 315–26.

Congrès international de dermatologie et de syphiligraphie: Comptes rendus. Paris: G. Masson, 1890.

Conrads, Margaret. *American Paintings and Sculpture at the Sterling and Francine Clark Art Institute.* New York: Hudson Hills Press, 1990.

Cooper, James Fenimore. *The Oak Openings, or The Bee-Hunter.* 1848. New York: Co-operative Publication Society, n.d.

Corbin, Alain. *The Foul and the Fragrant: Odor and the French Social Imagination.* Cambridge: Harvard University Press, 1986.

———. *Women for Hire: Prostitution and Sexuality in France After 1850.* Trans. Alan Sheridan. Cambridge: Harvard University Press, 1990.

Cortissoz, Royal. *Art and Common Sense.* New York: Charles Scribner's Sons, 1913.

———. "The Field of Art." *Scribner's Magazine,* March 1924, 345–52.

———. "John S. Sargent." *Scribner's Magazine,* November 1903, 515–32.

Courthion, Pierre, ed. *Courbet, raconté par lui-même et par ses amis.* 2 vols. Geneva: Pierre Cailler, 1948, 1950.

Crane, Walter. *Flora's Feast: A Masque of Flowers.* London: Cassell, 1890.

Crichton-Browne, Sir James. *The Cavendish Lecture: On Dreamy Mental States.* London: Baillière, Tindall and Cox, 1895.

Crompton, Louis. *Byron and Greek Love: Homophobia in Nineteenth-Century England.* Berkeley and Los Angeles: University of California Press, 1985.

Cullerier, Auguste. *Atlas of Venereal Diseases.* Trans. Freeman J. Bumstead. Philadelphia: Henry C. Lea, 1868.

Culver, Leigh. "Performing Identities in the Art of John Singer Sargent." PhD diss., University of Pennsylvania, 1999.

Current, Richard Nelson, and Marcia Ewing Current. *Loie Fuller: Goddess of Light.* Boston: Northeastern University Press, 1997.

Curry, David Park. "Portrait of Mrs. Albert Vickers." *American Art Review* 5 (Winter 1993): 109, 166–67.

Curry, Larry J. "*Madame Paul Poirson:* An Early Portrait by Sargent." *Bulletin of the Detroit Institute of Art* 51, no. 4 (1972): 96–104.

Dacier, Émile. *Gabriel de Saint-Aubin: Peintre, dessinateur et graveur.* 2 vols. Paris: Éditions G. van Oest, 1929, 1931.

Dalby, Richard, ed. *Dracula's Brood.* Wellingborough, UK: Equation, 1989.

Daly, Nicolas. "The Woman in White: Whistler, Hiffernan, Courbet, Du Maurier." *Modernism/Modernity* 12, no. 1 (2005): 1–25.

Darwin, Charles. *The Descent of Man, and Selection in Relation to Sex.* 2nd ed. 1879. Harmondsworth: Penguin, 2004.

———. *The Effects of Cross and Self Fertilisation in the Vegetable Kingdom.* London: John Murray, 1876.

———. *L'expression des émotions chez l'homme et les animaux.* Trans. Samuel Pozzi and René Bénoist. Paris: C. Reinwald, 1874.

———. *On the Origin of Species.* 1859. Cambridge: Harvard University Press, 1998.

———. *The Various Contrivances by which Orchids are Fertilised by Insects.* 2nd ed. London: John Murray, 1877.

Darwin, Erasmus. *The Loves of the Plants.* 1789. Oxford: Woodstock Books, 1991.

———. *The Temple of Nature, or The Origin of Society.* London: J. Johnson, 1803.

Daudet, Alphonse. *La doulou: Aux pays de la douleur.* 1929. Paris: Arléa, 1995.

———. *In the Land of Pain.* Ed. and trans. Julian Barnes. New York: Knopf, 2002.

Daudet, Léon. *Souvenirs des milieux littéraires, politiques, artistiques et médicaux.* 2 vols. Paris: Nouvelle Librairie Nationale, 1920.

Davis, Whitney. "Homoeroticism, Sexual Selection, and the Sense of Beauty." Keynote address, Anxious Flirtations: Homoeroticism, Art, and Aestheticism in Late-Victorian Britain conference, Institute of English Studies, University of London, 24–25 July 2001.

———. "The Image in the Middle: John Addington Symonds and Homoerotic Art Criticism." In *After the Pre-Raphaelites: Art and Aestheticism in Victorian England,* ed. Elizabeth Prettejohn, 188–216. New Brunswick: Rutgers University Press, 1999.

Day, Charles. *Etiquette, or A Guide to the Usages of Society.* Philadelphia: T. B. Peterson, 1843.

Delacroix, Eugène. *Correspondance générale.* Ed. André Joubin. 5 vols. Paris: Plon, 1935–38.

De Landes, Louis. *Glossaire érotique de la langue française.* 1861. Paris: Éditions de Paris, 2004.

Delarue-Mardrus, Lucie. *The Angel and the Perverts.* Trans. Anna Livia. New York: New York University Press, 1995.

———. *Hortensia dégénéré.* Paris: J. Ferenczi et Fils, 1929.

Delbourgo, James. *A Most Amazing Scene of Wonders: Electricity and Enlightenment in Early America.* Cambridge: Harvard University Press, 2005.

Deleuze, Gilles, and Félix Guattari. *A Thousand Plateaus: Capitalism and Schizophrenia.* Trans. Brian Massumi. Minneapolis: University of Minnesota Press, 1987.

De l'Impressionisme à l'art nouveau: Acquisitions du Musée d'Orsay 1990–1996. Exh. cat. Paris: Éditions de la Réunion des Musées Nationaux, 1996.

Dellamora, Richard, ed. *Victorian Sexual Dissidence.* Chicago: University of Chicago Press, 1999.

Delvau, Alfred. *Dictionnaire érotique moderne.* 1864. Geneva: Slatkine Reprints, 1968.

Desbuissons, Frédérique. "Voir le mal en face: 'L'origine du monde' de Gustave Courbet." *Ligeia* 19–20 (October 1996–June 1997): 14–22.

Dezarrois, André. "Une exposition d'art américain." *La Revue de l'Art Ancien et Moderne* 44 (June–December 1923): 143–55.

Dickens, Charles. *The Mystery of Edwin Drood.* Ed. Arthur J. Cox. Harmondsworth: Penguin, 1985.

Dickinson, Emily. *The Complete Poems of Emily Dickinson.* Ed. Thomas H. Johnson. Boston: Little, Brown, 1961.

Dini, Jane. "Public Bodies: Form and Identity in the Work of John Singer Sargent." PhD diss., University of California, Santa Barbara, 1998.

D'Inverno, Nicola. "The Real John Singer Sargent, as His Valet Saw Him." *Boston Sunday Advertiser,* 7 February 1926, 3.

Dixon, Marion Hepworth. "Mr. John S. Sargent as a Portrait Painter." *Magazine of Art* (January 1899): 112–20.

Donne, John. *The Complete English Poems.* Ed. C. A. Patrides. London: J. M. Dent, 1994.

Downes, William Howe. *John S. Sargent, His Life and Work.* Boston: Little, Brown, 1925.

Doyle, Arthur Conan. *The Adventures of Sherlock Holmes.* 1892. Ed. Richard Lancelyn Green. Oxford: Oxford University Press, 1994.

———. *The Coming of the Fairies.* 1922. London: Pavilion, 1997.

Dreger, Alice Domurat. *Hermaphrodites and the Medical Invention of Sex.* Cambridge: Harvard University Press, 1998.

Du Camp, Maxime. *Les convulsions de Paris.* 2nd ed. Vol. 2. Paris: Hachette, 1879.

Dumas, Alexandre (*père*). *The Black Tulip.* Ed. David Coward. Trans. Franz Demmler. Oxford: Oxford University Press, 2000.

———. *La tulipe noire.* 1850. Paris: Calmann-Lévy, 1938.

Dumas, Alexandre (*fils*). *La Dame aux Camélias.* Trans. David Coward. Oxford: Oxford University Press, 2000.

———. *La dame aux camélias.* 1848. Paris: Calmann-Lévy, 1931.

Du Maurier, George. *The Martian.* New York: Harper and Brothers, 1897.

———. *Trilby.* 1894. Oxford: Oxford University Press, 1998.

Duncan, Isadora. *My Life.* 1927. New York: Liveright, 1955.

Duncan, John Shute. *Hints to the Bearers of Walking-Sticks and Umbrellas.* London: C. Squire for J. Murray, 1809.

Dunham, O. M. *Manners: A Hand-Book of Social Customs.* Chicago: Westminster, 1888.

Eby, C. Stanley. "Stylistic Analysis of Selected Songs for Voice and Piano by Charles Martin Loeffler: *Quatres mélodies,* Poèsies de Gustave Kahn, Op. 10; and *Four Poems,* Op. 15." PhD diss., Boston University, 1997.

"Eccentricities of French Art." *Art Amateur* 11 (August 1884): 52.

Eells, Emily. *Proust's Cup of Tea: Homoeroticism and Victorian Culture.* Aldershot: Ashgate, 2002.

Elkins, James. "On Visual Desperation and the Bodies of Protozoa." *Representations* 40 (Autumn 1992): 33–56.

Ellenberger, Henri. *The Discovery of the Unconscious: The History and Evolution of Dynamic Psychiatry.* New York: Basic Books, 1970.

Ellis, Havelock, and John Addington Symonds. *Sexual Inversion.* 1897. New York: Arno Press, 1975.

Ellmann, Mary. *Thinking About Women.* New York: Harcourt, Brace and World, 1968.

Ellmann, Richard. *Oscar Wilde.* New York: Vintage, 1988.

Ephrussi, Charles. "Exposition des artistes indépendants." *Gazette des Beaux-Arts* 21 (May 1880): 485–88.

Epstein, Julia. *Altered Conditions: Disease, Medicine, and Storytelling.* New York: Routledge, 1995.

Esten, John. *John Singer Sargent: The Male Nudes.* New York: Universe, 1999.

———. *Sargent: Painting Out of Doors.* New York: Universe, 2000.

Ewing, William A. *Flora Photographica: Masterpieces of Flower Photography, 1835 to the Present.* New York: Simon and Schuster, 1991.

Exhibition of Drawings by John Singer Sargent. Exh. cat. New York: Grand Central Art, 1928.

Fabre, Jean-Henri. *Insect Life: Souvenirs of a Naturalist.* Trans. Margaret Roberts. Ed. F. Merrifield. London: Macmillan, 1913.

———. *Souvenirs entomologiques: Études sur l'instinct et les moeurs des insectes.* Vol. 1. Paris: Librairie Ch. Delagrave, 1879.

Fairbrother, Trevor. "The Complications of Being Sargent." In *John Singer Sargent Portraits of the Wertheimer Family,* exh. cat., ed. Norman L. Kleeblatt, 34–42. New York: Jewish Museum, 1999.

———. *John Singer Sargent.* New York: Harry N. Abrams, 1994.

———. *John Singer Sargent and America.* New York: Garland, 1986.

———. "John Singer Sargent's 'Gift' and His Early Critics." *Arts Magazine* (February 1987): 56–63.

———. *John Singer Sargent: The Sensualist.* New Haven: Yale University Press; Seattle: Seattle Art Museum, 2000.

———. "*Man Screaming* by John Singer Sargent." *American Art* 15 (Spring 2001): 84–89.

———. "Notes on John Singer Sargent in New York, 1888–1890." *Archives of American Art Journal* 22, no. 4 (1982): 27–32.

———. "Possessions and Props: The Collection of John Singer Sargent." *Antiques* (February 2001): 314–23.

———. "A Private Album: John Singer Sargent's Drawings of Nude Male Models." *Arts Magazine* (December 1981): 70–79.

———. "Sargent's Genre Paintings and the Issues of Suppression and Privacy." In *American Art Around 1900: Lectures in Memory of Daniel Fraad,* ed. Doreen Bolger and Nicolai Cikovsky Jr., 29–49. Washington, D.C.: National Gallery of Art, 1990.

———. "The Shock of John Singer Sargent's 'Madame Gautreau.'" *Arts Magazine* (January 1981): 90–97.

———. "Warhol Meets Sargent at Whitney." *Arts Magazine* (February 1987): 64–72.

Farmer, John S. *Slang and Its Analogues Past and Present.* 7 vols. London: John S. Farmer, 1890–1904.

Ferber, Linda. *Masters of Color and Light: Homer, Sargent, and the American Watercolor Movement.* Exh. cat. Brooklyn: Brooklyn Museum of Art; Washington, D.C.: Smithsonian Institution Press, 1998.

Field, Rachel. "John Sargent's Boyhood Sketches." *St. Nicholas* 53 (June 1926): 774–77.

Flament, Albert. *Le bal du Pré-Catelan.* Paris: Librairie Arthème Fayard, 1946.

Flaubert, Gustave. *Bouvard and Pécuchet, with The Dictionary of Received Ideas.* Trans. A. J. Krailsheimer. Harmondsworth: Penguin, 1976.

———. *Dictionnaire des idées reçues.* Paris: Jean Aubier, 1951.

———. *Madame Bovary.* Ed. and trans. Paul de Man. New York: W. W. Norton, 1965.

———. *The Temptation of Saint Anthony.* Trans. Kitty Mrosovky. Harmondsworth: Penguin, 1983.

———. *La tentation de Saint-Antoine—1874.* Vol. 4 of *Oeuvres complètes,* 18 vols. Paris: Conard, 1910–33.

Flexner, James Thomas. *American Painting: First Flowers of Our Wilderness.* Boston: Houghton Mifflin, 1947.

Forster, E. M. *Maurice.* Written 1913–14, first published 1971. New York: W. W. Norton, 1987.

Foshay, Ella M. *Art in Bloom.* New York: Phaidon, 1990.

———. "Charles Darwin and the Development of American Flower Imagery." *Winterthur Portfolio* 15 (Winter 1980): 301–14.

———. *Reflections of Nature: Flowers in American Art.* Exh. cat. New York: Knopf, and the Whitney Museum of American Art, 1984.

Foucault, Michel. *The History of Sexuality.* Vol. 1, *An Introduction.* Trans. Robert Hurley. New York: Vintage, 1990.

———. "Of Other Spaces." Trans. Jay Miskowiec. *Diacritics* 16 (Spring 1986): 22–27.

Fourcaud, Louis. "Le Salon de 1884." *Gazette des Beaux-Arts* 29 (May 1884): 465–92.

"The Four Doctors." *Johns Hopkins Alumni Magazine,* November 1913–June 1914, 22–26.

France, Anatole. *The Red Lily.* 1894. New York: Modern Library, 1917.

Fraser, Robert. *Proust and the Victorians: The Lamp of Memory.* New York: St. Martin's Press, 1994.

Frazer, James George. *The Golden Bough: A Study in Magic and Religion.* 1890. 2nd ed. 3 vols. New York: Macmillan, 1900.

———. *The Golden Bough: A Study in Magic and Religion.* Abridged ed. 1922. New York: Touchstone, 1996.

———. *Lectures on the Early History of the Kingship.* London: Macmillan, 1905.

Frémine, Charles. "Le Salon des Champs-Élysées." *Le Rappel,* 1 May 1894, 1.

Freud, Sigmund. "Beobachtungen über Gestaltung und feineren Bau der als Hoden beschriebenen Lappenorgane des Aals." *Sitzungsberichte der Kaiserlichen Akademie der Wissenschaften, Wien* (April 1877): 419–30.

———. *Beyond the Pleasure Principle.* 1920. Trans. James Strachey. New York: W. W. Norton, 1961.

———. "The Moses of Michelangelo." 1914. In *Art and Literature,* 253–82. Penguin Freud Library 14. Trans. James Strachey. Ed. Albert Dickson. Harmondsworth: Penguin, 1985.

———. *On Sexuality.* Penguin Freud Library 7. Trans. James Strachey. Ed. Angela Richards. Harmondsworth: Penguin, 1977.

Fried, Michael. *Courbet's Realism.* Chicago: University of Chicago Press, 1990.

———. *Manet's Modernism, or The Face of Painting in the 1860s.* Chicago: University of Chicago Press, 1996.

Friedman, Susan Stanford. "Creativity and the Childbirth Metaphor: Gender Difference in Literary Discourse." In *Speaking of Gender,* ed. Elaine Showalter, 73–100. New York: Routledge, 1989.

Fry, Roger. *Transformations: Critical and Speculative Essays on Art.* London: Chatto and Windus, 1926.

Gagnier, Reginia. *Idylls of the Marketplace: Oscar Wilde and the Victorian Public.* Stanford: Stanford University Press, 1986.

———. "Production, Reproduction, and Pleasure." In *Victorian Sexual Dissidence,* ed. Richard Dellamora, 127–45. Chicago: University of Chicago Press, 1999.

Gaillard, Françoise. "Allégorie d'un fantasme fin de siècle, Courbet: *L'origine du monde.*" In *Mimesis et semiosis: Littérature et représentation; Miscellanées offertes à Henri Mitterand,* ed. Philippe Hamon and Jean-Pierre Leduc-Adine, 427–33. Paris: Nathan, 1992.

Gallati, Barbara Dayer, Erica Hirshler, and Richard Ormond. *Great Expectations: John Singer Sargent Painting Children.* Exh. cat. New York: Brooklyn Museum and Bullfinch Press, 2004.

Gammell, R. H. Ives. "The Enigma of John Sargent's Art." *Classical America* 4 (1977): 153–59.

Garelick, Rhonda. "Electric Salomé: Loie Fuller at the Exposition Universelle of 1900." In *Imperialism and Theatre: Essays on World Theatre, Drama, and Performance,* ed. J. Ellen Gainor, 85–103. New York: Routledge, 1995.

———. *Electric Salomé: Loie Fuller's Performance of Modernism.* Princeton: Princeton University Press, 2007.

Gasking, Elizabeth. *Investigations into Generation, 1651–1828.* London: Hutchinson, 1967.

Gautier, Judith. "Le Salon (premier article)." *Le Rappel,* 1 May 1884, 1.

Gautier, Théophile. *Émaux et camées.* 1852. Paris: Charpentier, 1872.

———. *Lettre à la présidente: Voyage en Italie, 1850.* Naples: Imprimerie du Musée Secret du Roi de Naples, 1890.

———. *Mademoiselle de Maupin.* 1836. Fribourg: Éditions du Lac, 1973.

———. *Mademoiselle de Maupin.* [New York]: Pierre Louÿs Society, 1927.

Genet, Jean. *Notre-Dame-des-Fleurs.* 1943. Saint-Amand: Bussière, 1976.

———. *Our Lady of the Flowers.* Trans. Bernard Frechtman. New York: Grove Press, 1963.

Geoffroy Saint-Hilaire, Isidore. *Histoire générale et particulière des anomalies de l'organisation chez l'homme et les animaux.* 3 vols. Paris: J.-B. Baillière, 1832–37.

Gerdts, William H. *Down Garden Paths: The Floral Environment in American Art.* London: Associated University Presses, 1983.

———. "The Square Format and Proto-Modernism in American Painting." *Arts Magazine* 50 (June 1976): 70–75.

Getty, Clive. *Grandville: Dessins originaux.* Exh. cat. Nancy: Musée des Beaux-Arts, 1986.

Gibson, William Hamilton. *Blossom Hosts and Insect Guests: How the Heath Family, The Bluets, The Figworts, The Orchids, and Similar Wild Flowers Welcome The Bee, The Fly, The Wasp, The Moth, and Other Faithful Insects.* 1880. Ed. Eleanor E. Davie. New York: Newson, 1901.

Gide, André. *The Counterfeiters.* Trans. Dorothy Bussy. New York: Vintage, 1973.

———. *Les faux-monnayeurs.* 1925. In *Romans,* ed. Yvonne Davet and Jean-Jacques Thierry, 931–1248. Paris: Gallimard, 1961.

———. *Si le grain ne meurt.* 1924. Paris: Gallimard, 1928.

Gilbert, W. S., and Arthur Sullivan. *The Complete Annotated Gilbert and Sullivan.* Ed. Ian Bradley. Oxford: Oxford University Press, 1996.

Gilmore, Paul. "Romantic Electricity, or The Materiality of Aesthetics." *American Literature* 76 (September 2004): 467–94.

Ginzburg, Carlo. *Clues, Myths, and the Historical Method.* Trans. John Tedeschi and Anne C. Tedeschi. Baltimore: Johns Hopkins University Press, 1989.

Gladwell, Adèle Olivia, and James Havoc. *Blood and Roses: The Vampire in Nineteenth-Century Literature.* London: Creation Press, 1992.

Goethe, Johann Wolfgang von. *Erotic Poems.* Trans. David Luke. Oxford: Oxford University Press, 1997.

Gold, Arthur, and Robert Fizdale. *The Divine Sarah: A Life of Sarah Bernhardt.* New York: Knopf, 1991.

———. *Misia: The Life of Misia Sert.* New York: Vintage, 1992.

Goldfarb, Hilliard, Erica E. Hirshler, and T. J. Jackson Lears. *Sargent: The Late Landscapes*. Boston: Isabella Stewart Gardner Museum, 1999.

Goley, Mary Anne. "Sargent's 'Dr. Pozzi at Home' Exhibited in Brussels in 1884." *Bulletin des Musées Royaux des Beaux-Arts de Belgique* 30–33 (1981–84): 143–51.

Goncourt, Edmond de. *La Faustin*. Paris: Charpentier, 1882.

———. *La Faustin*. Trans. G. F. Monkshood and Ernest Tristan. New York: Howard Fertig, 1976.

Goncourt, Edmond de, and Jules de Goncourt. *L'art du dix-huitième siècle*. 3rd ed. 2 vols. Paris: A. Quantin, 1880–82.

———. *Idées et sensations*. Paris: Librairie Internationale, 1866.

———. *Journal: Mémoires de la vie littéraire*. Ed. Robert Ricatte. 3 vols. Paris: Robert Laffont, 1989.

Goody, Jack. *The Culture of Flowers*. Cambridge: Cambridge University Press, 1993.

Gopnik, Adam. "Sargent's Pearls: The Artist Who Makes the Perfect Case for Wasted Talent." *New Yorker*, 15 February 1999, 64–71.

Gosse, Edmund. *Aspects and Impressions*. 1922. Freeport, N.Y.: Books for Libraries Press, 1970.

———. *The Collected Poems of Edmund Gosse*. London: Heinemann, 1911.

———. *Father and Son: A Study of Two Temperaments*. 1907. Ed. Peter Abbs. Harmondsworth: Penguin, 1989.

———. *On Viol and Flute*. 1873. New York: Henry Holt, 1883.

———. *Portraits and Sketches*. 1912. Freeport, N.Y.: Books for Libraries Press, 1969.

———. *The Secret of Narcisse: A Romance*. New York: Tait, 1892.

Gosse, Philip Henry. *Actinologia Britannica: A History of the British Sea-Anemones and Corals*. London: Van Voorst, 1860.

———. *Evenings at the Microscope, or Researches Among the Minuter Organs and Forms of Animal Life*. 1859. New York: D. Appleton, 1889.

———. *A Naturalist's Rambles on the Devonshire Coast*. London: John van Voorst, 1853.

Gourmont, Remy de. *The Physiology of Love*. New York: Rarity Press, 1932.

———. *Physique de l'amour*. Paris: Mercure de France, 1903.

Graham, Wendy. *Henry James's Thwarted Love*. Stanford: Stanford University Press, 1999.

Grandville, J. J. *Un autre monde: Transformations, visions, incarnations, ascensions, locomotions, explorations, pérégrinations, excursions, stations . . .* Paris: H. Fournier, 1844.

———. *Les fleurs animées*. 1847. 2 vols. Text by Taxile Delord, Alphonse Karr, and Le Conte Foelix. Paris: Garnier Frères, 1867.

———. *Grandville: Das Gesamte Werk*. Zurich: Buchclub ex Libris, 1969.

Greene, Robert. *Mamillia, the Second Part of the Triumph of Pallas*. London: Th. C. for William Ponsonbie, 1593.

Greta. "Art in Boston." *Art Amateur* 18 (December 1887): 3–4.

Grew, Nehemiah. *The Anatomy of Plants*. London: W. Rawlins, 1682.

Griswold, Rufus, ed. *The Poetry of Flowers, Original and Selected*. New York: George A. Leavitt, [1848?].

Gubar, Susan. "The Birth of the Artist as Heroine: (Re)production, the *Künstlerroman* Tradition, and the Fiction of Katherine Mansfield." In *The Representation of Women in Fiction*, ed. Carolyn G. Heilbrun and Margaret R. Higgonet, 19–59. Baltimore: Johns Hopkins University Press, 1983.

Gubernatis, Angelo de. *La mythologie des plantes*. 1878–82. 2 vols. New York: Arno Press, 1978.

Guiraud, Pierre. *Dictionnaire historique, stylistique, rhétorique, étymologique, de la littérature érotique*. Paris: Payot, 1978.

The Habits of Good Society: A Handbook for Ladies and Gentlemen. New York: Carleton, 1869.

Haddad, Michèle. "Courbet, entre présence et absence, dans deux toiles limites: *Les baigneuses* et *L'origine du monde*." In *Courbet, l'amour . . .*, 127–35. Exh. cat. Ornans: Musée Gustave Courbet, 1996.

Halberstam, Judith. *Skin Shows: Gothic Horror and the Technologies of Monsters*. Durham: Duke University Press, 1995.

Hale, Mary. "The Sargent I Knew." *World Today*, November 1927, 565–70.

Hall, Donald, and Pat Corrington Wykes. *Anecdotes of Modern Art: From Rousseau to Warhol*. New York: Oxford University Press, 1990.

Hammer-Purgstall, Joseph. "Sur le langage des fleurs." *Fundgruben des Orients* 1 (1809): 32–42.

Hardie, Martin. *J. S. Sargent, R.A., R.W.S.* London: The Studio, 1930.

Harris, Frank. *Oscar Wilde, His Life and Confessions*. 2 vols. N.p.: privately printed, 1916.

Harrison, Charles, and Paul Wood, eds. *Art in Theory, 1900–2000: An Anthology of Changing Ideas*. Oxford: Blackwell, 2003.

Haweis, Mary (Mrs. H. R.). *The Art of Beauty*. London: Chatto and Windus, 1878.

Hawthorne, Melanie. *Rachilde and French Women's Authorship from Decadence to Modernism*. Lincoln: University of Nebraska Press, 2001.

Hawthorne, Nathaniel. *The Marble Faun, or The Romance of Monte Beni*. 1860. Harmondsworth: Penguin, 1990.

———. *The Scarlet Letter and Other Tales of the Puritans*. Ed. Harry Levin. Boston: Houghton Mifflin, 1961.

Hayden, Deborah. *Pox: Genius, Madness, and the Mysteries of Syphilis*. New York: Basic Books, 2003.

Hayes, Jarrod. "Proust in the Tearoom." *PMLA* 110 (October 1995): 992–1005.

Helmreich, Anne. "John Singer Sargent, *Carnation, Lily, Lily, Rose*, and the Condition of Modernism in England, 1887." *Victorian Studies* 45 (Spring 2003): 433–55.

Hennig, Jean-Luc. *Dictionnaire littéraire et érotique des fruits et légumes.* Paris: Albin Michel, 1994.

Hersey, George. *The Evolution of Allure: Sexual Selection from the Medici Venus to the Incredible Hulk.* Cambridge: MIT Press, 1996.

Hichens, Robert. *The Green Carnation.* 1894. New York: Dover, 1970.

Higginson, Thomas Wentworth. "Water-Lilies." *Atlantic Monthly,* September 1858, 465–73.

Higonnet, Anne. *Pictures of Innocence: The History and Crisis of Ideal Childhood.* London: Thames and Hudson, 1998.

Hills, Patricia, ed. *John Singer Sargent.* Exh. cat. New York: Whitney Museum of American Art and Harry N. Abrams, 1986.

———. "Thomas Eakins's *Agnew Clinic* and John S. Sargent's *Four Doctors:* Sublimity, Decorum, and Professionalism." *Prospects: A Journal of American Cultural Studies* 11 (1987): 217–30.

Hind, C. Lewis. *Hercules Brabazon Brabazon, 1821–1906: His Art and Life.* London: George Allen, 1912.

Hinton, Denys. "John Singer Sargent R.A.: A Re-evaluation?" *Connoisseur* 157 (December 1964): 249–51.

Hirshler, Erica. *Dennis Miller Bunker: American Impressionist.* Exh. cat. Boston: Museum of Fine Arts, 1994.

———. "The Fine Art of Sarah Choate Sears." *Magazine Antiques* (September 2001): 320–29.

———. *A Studio of Her Own: Women Artists in Boston, 1870–1940.* Exh. cat. Boston: Museum of Fine Arts, 2001.

Holmes, Oliver Wendell. *Elsie Venner: A Romance of Destiny.* 1861. Dublin: Nonsuch, 2007.

Hooper, Lucy, ed. *The Lady's Book of Flowers and Poetry; to which are added a Botanical Introduction, a Complete Floral Dictionary and a chapter on Plants in Rooms.* New York: Derby and Jackson, 1860.

Hoopes, Donelson. "John S. Sargent: The Worcestershire Interlude, 1885–89." *Brooklyn Museum Annual* 7 (1965–66): 74–89.

———. *The Private World of John Singer Sargent.* Washington, D.C.: Corcoran Gallery of Art, 1964.

———. *Sargent Watercolors.* New York: Watson-Guptill, 1976.

Hopkins, Gerard Manley. *Poems and Prose.* Ed. W. H. Gardner. Harmondsworth: Penguin, 1985.

Hoschedé, Jean-Pierre. *Claude Monet, ce mal connu: Intimité familiale d'un demi-siècle à Giverny de 1883 à 1926.* 2 vols. Geneva: Pierre Cailler, 1960.

Houssaye, Henri. "Le Salon de 1884." *Revue des Deux Mondes* 54 (June 1884): 560–95.

Hugo, Victor. *Les misérables.* Paris: Garnier, 1957.

Huneault, Kristina. *Difficult Subjects: Working Women and Visual Culture, Britain, 1880–1914.* Aldershot: Ashgate, 2002.

Hutcheon, Linda, and Michael Hutcheon. *Opera: Desire, Disease, Death.* Lincoln: University of Nebraska Press, 1996.

Huysmans, Joris-Karl. *Against Nature.* Trans. Margaret Mauldon. Oxford: Oxford University Press, 1998.

———. *À rebours.* 1884. Paris: Fasquelle, 1907.

———. *Becalmed.* Trans. Terry Hale. London: Atlas, 1992.

———. *En rade.* 1887. Paris: Plon-Nourrit, 1928.

Hyde, H. Montgomery, ed. *The Three Trials of Oscar Wilde.* New York: University Books, 1956.

The Illustrated Manners Book: A Manual of Good Behavior and Polite Accomplishments. New York: Leland, Clay, 1855.

"Imperfect Fertilization." *Vick's Magazine,* December 1883, 381.

Iota [Katherine Mannington]. *A Yellow Aster.* 1894. New York: Optimus Printing, n.d.

Jacobs, Frederika H. *The Living Image in Renaissance Art.* Cambridge: Cambridge University Press, 2005.

Jacomb-Hood, G. P. "John Sargent." *Cornhill Magazine,* September 1925, 280–90.

Jacques-Émile Blanche, peintre (1861–1942). Exh. cat. Rouen: Musée des Beaux-Arts, 1997.

James, Henry. *The Bostonians.* 1886. Oxford: Oxford University Press, 1984.

———. *Dearly Beloved Friends: Henry James's Letters to Younger Men.* Ed. Susan E. Gunter and Steven H. Jobe. Ann Arbor: University of Michigan Press, 2004.

———. *In the Cage and Other Tales.* Ed. Morton Dauwen Zabel. London: Rupert Hart-Davis, 1958.

———. "John S. Sargent." *Harper's New Monthly Magazine,* October 1887, 683–91.

———. *The Reverberator.* 1888. London: Rupert Hart-David, 1949.

———. "Robert Louis Stevenson." *Century Magazine,* April 1888, 869–79.

———. *Roderick Hudson.* 1875. Harmondsworth: Penguin, 1986.

———. *The Sacred Fount.* 1901. New York: New Directions, 1983.

———. *The Spoils of Poynton.* 1896. Oxford: Oxford University Press, 2000.

———. *The Tragic Muse.* 1889–90. Harmondsworth: Penguin, 1995.

———. *Watch and Ward.* 1871. New York: Grove Press, 1960.

James, William. *The Correspondence of William James.* Vol. 1. Ed. Ignas Skrupskelis and Elizabeth Berkeley. Charlottesville: University Press of Virginia, 1992.

Jekyll, Gertrude. *Colour Schemes for the Flower Garden.* 1914. Boston: Little, Brown, 1988.

Jenkins, David Fraser. "Pre-Raphaelite and Impressionist: Sargent's *Carnation, Lily, Lily, Rose.*" In Jenkins, *British Art in Focus: Patrons' Papers I.* London: Tate Gallery, 1998.

Jennings, Kate. *John Singer Sargent.* New York: Crescent Books, 1991.

John Singer Sargent, His Own Work. Exh. cat. New York: Coe Kerr Gallery and Wittenborn Art Books, 1980.

John Singer Sargent and the Edwardian Age: The Documents. Exh. cat. Detroit: Detroit Institute of Arts, 1979.

John Singer Sargent: Sargent in Japan. Exh. cat. Osaka: Yomiuri Shinbunsha, 1989.

Johnson, Barbara. *The Feminist Difference: Literature, Psychoanalysis, Race, and Gender*. Cambridge: Harvard University Press, 1998.

———. *Mother Tongues: Sexuality, Trials, Motherhood, Translation*. Cambridge: Harvard University Press, 2003.

Jonson, Ben. *The New Inn*. Ed. Michael Hattaway. Manchester: Manchester University Press, 1984.

Jordan, Lacey Taylor. "John Singer Sargent's Images of Artists in an International Context." PhD diss., Emory University, 1999.

Joseph, Lawrence. "Marcel Proust et 'Docteur Dieu': Lettres inédites à Samuel Pozzi." *Bulletin Marcel Proust* 51 (2001): 11–29.

Jullian, Philippe. *Jean Lorrain, ou Le Satiricon 1900*. Paris: Fayard, 1974.

———. *Prince of Aesthetes: Count Robert de Montesquiou, 1855–1921*. Trans. John Haylock and Francis King. New York: Viking, 1967.

K., L. "The Champ de Mars Salon." *New York Times*, 20 May 1894, 20.

Kasson, John. *Rudeness and Civility: Manners in Nineteenth-Century Urban America*. New York: Hill and Wang, 1990.

Keat, Lim Chong, and Henry Barlow. *Frank Swettenham and George Giles: Watercolours and Sketches of Malaya, 1880–1894*. Kuala Lumpur: Malaysian-British Society, 1988.

Kerner von Marilaun, Anton. *Flowers and Their Unbidden Guests*. Trans. W. Ogle. London: C. Kegan Paul, 1878.

———. *The Natural History of Plants*. Trans. F. W. Oliver et al. 2 vols. London: Blackie and Son, 1895.

Kilmurray, Elaine. "A Madonna Without a Child." *ARTnews* 98 (April 1999): 76.

Kilmurray, Elaine, and Richard Ormond, eds. *John Singer Sargent*. Exh. cat. Princeton: Princeton University Press, 1998.

———. "Sargent in Paris." *Apollo* 148 (September 1998): 13–22.

King, Amy. *Bloom: The Botanical Vernacular in the English Novel*. Oxford: Oxford University Press, 2003.

Kingsbury, Martha. "John Singer Sargent: Aspects of His Work." PhD diss., Harvard University, 1969.

Kingsley, Charles. *Glaucus, or The Wonders of the Shore*. 1855. London: Macmillan, 1890.

Kleeblatt, Norman, ed. *John Singer Sargent: Portraits of the Wertheimer Family*. Exh. cat. New York: Jewish Museum, 1999.

Knight, Philip. *Flower Poetics in Nineteenth-Century France*. Oxford: Clarendon Press, 1986.

Koerner, Lisbet. *Linnaeus: Nature and Nation*. Cambridge: Harvard University Press, 2001.

Krafft-Ebing, Richard von. *Psychopathia Sexualis*. 1886. Ed. Brian King. Trans. F. J. Rebman. Burbank, Calif.: Bloat, 1999.

Kramer, Hilton. "The Case of Madame's Shoulder Strap." *New York Times*, 1 February 1981, D23, D27.

Kris, Ernst, and Otto Kurtz. *Legend, Myth, and Magic in the Image of the Artist: A Historical Experiment*. 1934. Trans. Alastair Lang and Lottie Newman. New Haven: Yale University Press, 1979.

Kristeva, Julia. *Time and Sense: Proust and the Experience of Literature*. Trans. Ross Guberman. New York: Columbia University Press, 1996.

Lacan, Jacques. *Écrits, a Selection*. Trans. Alan Sheridan. New York: W. W. Norton, 1977.

———. *The Four Fundamental Concepts of Psycho-Analysis*. Ed. Jacques-Alain Miller. Trans. Alan Sheridan. New York: W. W. Norton, 1981.

Lang, Andrew, ed. *The Green Fairy Book*. London: Longmans, Green, 1892.

Langtry, Lillie. *The Days I Knew*. London: Hutchinson, 1925.

Laqueur, Thomas. *Making Sex: Body and Gender from the Greeks to Freud*. Cambridge: Harvard University Press, 1990.

Larousse, Pierre. *Grand dictionnaire universel du dix-neuvième siècle*. 15 vols. Paris: Administration du Grand Dictionnaire Universel, 1866–76.

Larson, Barbara. *The Dark Side of Nature: Science, Society, and the Fantastic in the Work of Odilon Redon*. University Park: Pennsylvania State University Press, 2005.

The Laws of Etiquette. Philadelphia: Carey, Lea, and Blanchard, 1836.

Lear, Edward. *The Complete Verse and Other Nonsense*. Ed. Vivien Noakes. Harmondsworth: Penguin, 2001.

Lears, T. J. Jackson. *No Place of Grace: Antimodernism and the Transformation of American Culture, 1880–1920*. Chicago: University of Chicago Press, 1994.

Ledger, Sally, and Scott McCracken, eds. *Cultural Politics at the Fin de Siècle*. Cambridge: Cambridge University Press, 1995.

Lee, Ruth Web. *A History of Valentines*. New York: Studio Publications, 1952.

Lee, Vernon [Violet Paget]. *Hauntings*. London: William Heinemann, 1890.

———. *Hortus vitae: Essays on the Gardening of Life*. London: John Lane, the Bodley Head, 1904.

———. "Imagination in Modern Art: Random Notes on Whistler, Sargent, and Besnard." *Fortnightly Review* 68 (1 October 1897): 513–21.

———. "J. S. S.: In Memoriam." In Evan Charteris, *John Sargent*, 235–55. New York: Charles Scribner's Sons, 1927.

———. *Vernon Lee's Letters*. Ed. Irene Cooper Willis. London: privately printed, 1937.

Lee, Vernon, and Clementina Anstruther-Thomson. *Beauty and Ugliness, and Other Studies in Psychological Aesthetics*. London: John Lane, 1911.

Le Fanu, Joseph Sheridan. *Best Ghost Stories of J. S. Le Fanu*. Ed. E. F. Bleiler. New York: Dover, 1964.

Lehmann, A. G. "Pierrot and Fin de Siècle." In *Romantic Mythologies*, ed. Ian Fletcher, 209–23. London: Routledge and Kegan Paul, 1967.

Le Petit, Alfred. *Fleurs, fruits et légumes du jour.* Paris: Au Bureau de l'Éclipse, 1871.

Leschey, Katrinka. "Pigment Analysis of John Singer Sargent's Palette, 1900–1920." Master's thesis, Strauss Center for Conservation, 1985.

Little, Carl. *The Watercolors of John Singer Sargent.* Berkeley and Los Angeles: University of California Press, 1998.

Lomax, James, and Richard Ormond. *John Singer Sargent and the Edwardian Age.* Exh. cat. Leeds: Leeds Art Galleries, 1979.

Looby, Christopher. "Flowers of Manhood: Race, Sex, and Floriculture from Thomas Wentworth Higginson to Robert Mapplethorpe." *Criticism* 37 (Winter 1995): 109–56.

———. "'The Roots of the Orchis, the Iuli of Chestnuts': The Odor of Male Solitude." In *Solitary Pleasures: The Historical, Literary, and Artistic Discourses of Autoeroticism,* ed. Paula Bennett and Vernon Rosario, 163–87. New York: Routledge, 1995.

Loring, William. "An American Art Student in London." *Archives of American Art Journal* 24, no. 3 (1984): 17–21.

Lorrain, Jean [Paul Duval]. *Buveurs d'âmes.* 1893. Paris: Charpentier, 1929.

———. *Correspondance: Lettres à Barbey d'Aurevilly, François Coppée, Oscar Méténier, Catulle Mendès, Edmond Deschaumes, Mecislas Golberg, etc.* Paris: Éditions Baudinière, 1929.

———. *Monsieur de Phocas.* 1901. Paris: Albin Michel, 1929.

———. *Monsieur de Phocas.* Trans. Francis Amery. Sawtry, Cambridgeshire: Dedalus, 1994.

———. *Nightmares of an Ether-Drinker.* Trans. Brian Stableford. Carlton-in-Coverdale: Tartarus, 2002.

———. *L'ombre ardente.* Paris: E. Fasquelle, 1897.

Lorris, Guillaume de, and Jean de Meun. *The Romance of the Rose.* Trans. Charles Dahlberg. Princeton: Princeton University Press, 1995.

———. *Le roman de la rose.* Written 1225–40, 1275. Ed. Félix Lecoy. 3 vols. Paris: Honoré Champion, 1965.

Loti, Pierre [Louis Viaud]. *An Iceland Fisherman.* 1886. Trans. Jules Cambon. New York: P. F. Collier and Son, 1902.

———. *Le roman d'un enfant.* 1890. Paris: Calmann-Lévy, 1891.

———. *The Story of a Child.* Trans. Caroline Smith. Boston: C. C. Birchard, 1901.

Louÿs, Pierre. *Aphrodite.* 1896. Trans. Willis Parker. New York: Three Sirens Press, 1932.

Lovejoy, Lucretia. *An Elegy on the Lamented Death of the Electrical Eel, or Gymnotus Electricus.* London: T. Hookham, 1779.

Lovell, Margaretta. *Venice: The American View, 1860–1920.* Exh. cat. San Francisco: Fine Arts Museums of San Francisco, 1984.

———. *A Visitable Past: Views of Venice by American Artists, 1860–1915.* Chicago: University of Chicago Press, 1989.

Low, Will H. *A Chronicle of Friendships, 1873–1900.* New York: Charles Scribner's Sons, 1908.

Loyrette, Henri. *Degas.* Paris: Fayard, 1991.

Lubin, David. *Act of Portrayal: Eakins, Sargent, James.* New Haven: Yale University Press, 1985.

Lucas, E. V. *Edwin Austen Abbey, Royal Academician: The Record of His Life and Work.* London: Methuen; New York: Charles Scribner's Sons, 1921.

Luft, Joanna. "Unfixing the Rose as a Fixture of the Female Sex in Guillaume de Lorris's and Jean de Meun's Thirteenth-Century *The Romance of the Rose.*" Paper presented at the Fixed Ideas conference, Harvard University, 1 March 2002.

Maas, Jeremy, et al. *Victorian Fairy Painting.* Exh. cat. London: Merrell Holberton, 1997.

MacColl, D. S. "The Academy II: The Rape of Painting." *Saturday Review* 91 (18 May 1901): 632–33.

———. "The New Gallery and Two Others." *Saturday Review* 89 (5 May 1900): 555–56.

MacDonald, Margaret F. *James McNeill Whistler Drawings, Pastels, and Watercolours: A Catalogue Raisonné.* New Haven: Yale University Press, 1995.

———. "Whistler's Designs for a Catalogue of Blue and White Nankin Porcelain." *Connoisseur* 198 (August 1978): 290–95.

Macfall, Haldane. *Whistler: Butterfly, Wasp, Wit, Master of the Arts, Enigma.* Boston: John W. Luce, 1906.

MacLeod, Catriona. *Embodying Ambiguity: Androgyny and Aesthetics from Winckelmann to Keller.* Detroit: Wayne State University Press, 1998.

Madden, Ed. "Say It with Flowers: The Poetry of Marc-André Raffalovich." *College Literature* 24 (February 1997): 11–27.

Maeterlinck, Maurice. *The Double Garden.* Trans. Alexander Teixeira de Mattos. New York: Dodd, Mead, 1904.

———. *Le double jardin.* Paris: Charpentier, 1904.

———. *Hothouses: Poems, 1889.* Trans. Richard Howard. Princeton: Princeton University Press, 2003.

———. *L'intelligence des fleurs.* Paris: Bibliothèque Charpentier, 1907.

———. *The Intelligence of the Flowers.* Trans. Alexander Teixeira de Mattos. Honolulu: University Press of the Pacific, 2001.

———. *The Life of the Bee.* Trans. Alfred Sutro. New York: Dodd, Mead, 1919.

———. *Pelleas and Mélisanda and The Sightless: Two Plays by Maurice Maeterlinck.* Trans. Laurence Alma-Tadema. London: Walter Scott Publishing, 1895.

———. *Pelléas et Mélisande.* 1892. Brussels: P. Lacomblez, 1908.

———. *La vie des abeilles.* 1901. Paris: Fasquelle, 1914.

Mahon, Dorothy, and Silvia Centeno. "A Technical Study of John Singer Sargent's Portrait of Madame Pierre Gautreau." *Metropolitan Museum Journal* 40 (2005): 121–29.

Maixner, Paul, ed. *Robert Louis Stevenson: The Critical Heritage.* London: Routledge and Kegan Paul, 1981.

Mallarmé, Stéphane. *Documents Stéphane Mallarmé.* Ed. Carl Paul Barbier. 7 vols. Paris: Nizet, 1968.

———. *Mallarmé in Prose*. Ed. Mary Ann Caws. New York: New Directions, 2001.

———. *Oeuvres complètes*. Ed. Henri Mondor and G. Jean-Aubry. Paris: Gallimard, 1945.

Maloney, Martin. "Girls, Girls, Girls." *Tate* 16 (Winter 1998): 28–29.

Malvasia, Carlo Cesare. *The Life of Guido Reni*. 1678. Trans. Catherine Enggass and Robert Enggass. University Park: Pennsylvania State University Press, 1980.

Mancoff, Debra N. *Flora Symbolica: Flowers in Pre-Raphaelite Art*. New York: Prestel, 2003.

Mansfield, Elizabeth C. *Too Beautiful to Picture: Zeuxis, Myth, and Mimesis*. Minneapolis: University of Minnesota Press, 2007.

Marmier, Xavier. *Souvenirs de voyages et traditions populaires*. Paris: P. Masgana, 1841.

Marvell, Andrew. *The Complete Poems*. Ed. Elizabeth Story Donno. Harmondsworth: Penguin, 1996.

Marx, Roger. "Loïe Fuller." *Les Arts de la Vie* 3 (May 1905): 265–73.

———. "Loïe Fuller." *Les Arts de la Vie* 3 (June 1905): 352–57.

Maryatt, Frederick. *A Diary in America, with Remarks on Its Institutions*. Ed. Sydney Jackman. New York: William Colyer, 1839.

Mason, Stuart [C. S. Millard]. *Oscar Wilde: Art and Morality; A Record of the Discussion Which Followed the Publication of "Dorian Gray."* London: Frank Palmer, 1912.

Maturin, Charles. *Melmoth the Wanderer*. 1820. Harmondsworth: Penguin, 2000.

Mauclair, Camille. "John Sargent." *L'Art et les Artistes* 4 (October 1906–March 1907): 369–79.

"A Maudle-In Ballad." *Punch, or the London Charivari*, 9 April 1881, 161.

Mauriès, Patrick. *Sur les papillonneries humaines*. Paris: Gallimard, 1996.

Mazzinghi, Joseph. *The Wreath: A Pastoral Glee for Three Voices*. London: Goulding, Phipps and D'Almaine, n.d.

McKibbin, David. *Sargent's Boston*. Exh. cat. Boston: Museum of Fine Arts, 1956.

McPherson, Heather. *Fin-de-Siècle Faces: Portraiture in the Age of Proust*. Exh. cat. Birmingham: University of Alabama at Birmingham, Visual Arts Gallery, 1998.

McSpadden, J. Walker. *Famous Painters of America*. New York: Dodd, Mead, 1916.

Mechlin, Leila. "John Singer Sargent." *American Magazine of Art* 16 (June 1925): 285–87.

———. "The Sargent Exhibition." *American Magazine of Art* 15 (April 1924): 168–90.

Meeuse, Bastiaan, and Sean Morris. *The Sex Life of Flowers*. New York: Facts on File, 1984.

Melville, Herman. *Billy Budd and Other Stories*. Harmondsworth: Penguin, 1986.

———. *Mardi, and a Voyage Thither*. 1849. Ed. Harrison Hayford et al. Evanston: Northwestern University Press, 1998.

———. *Moby-Dick*. 1851. Ed. Charles Child Walcutt. New York: Bantam Books, 1981.

———. *Pierre, or The Ambiguities*. 1852. Harmondsworth: Penguin, 1996.

Merleau-Ponty, Maurice. *The Visible and the Invisible*. Trans. Alphonso Lingis. Ed. Claude Lefort. Evanston: Northwestern University Press, 1968.

Merrill, Lynn. *The Romance of Victorian Natural History*. New York: Oxford University Press, 1989.

Méry, Joseph. *Monsieur Auguste*. Paris: Librairie Nouvelle, 1859.

Meynell, Alice, and William Manson. *The Work of John Sargent, R.A.* 1903. London: William Heinemann, 1927.

Michaelson, Alexander [Marc-André Raffalovich]. "Verglas and Pollen." *New Blackfriars* 9 (May 1928): 276–84.

Michel, André. "Le Salon de 1884." *L'Art* 37 (1884): 9–16.

Michelet, Jules. *L'amour*. 1858. Paris: Calmann-Lévy, 1882.

———. *The Insect*. London: T. Nelson and Sons, 1875.

———. *L'insecte*. Paris: Hachette, 1858.

———. *Love*. Trans. J. W. Palmer. New York: Rudd and Carleton, 1860.

———. *La mer*. 1861. Paris: Michel Lévy Frères, 1875.

———. *La montagne*. Paris: A. Lacroix, Verboeckhoven, 1868.

———. *The Mountain*. Trans. W. H. Davenport Adams. London: T. Nelson and Sons, 1875.

Millais, John Guille. *The Life and Letters of Sir John Everett Millais*. Vol. 1. Toronto: Morang, 1900.

Miller, Lucia. "John Singer Sargent in the Diaries of Lucia Fairchild, 1890 and 1891." *Archives of American Art Journal* 26, no. 4 (1986): 2–16.

Mills, Evan. "A Personal Sketch of Mr. Sargent." *World's Work* 7 (November 1903): 4116–18.

Minchin, Hamilton. *Some Early Recollections of Sargent*. Letchworth: Garden City Press, 1925.

Mirbeau, Octave. *Le jardin des supplices*. 1899. Paris: Fasquelle, 1957.

———. *Torture Garden*. Trans. Michael Richardson. Sawtry, Cambridgeshire: Dedalus, 1997.

Modern Etiquette in Public and Private. London: Frederick Warne, 1887.

"Le modernisme de Frans Hals." *L'Art Moderne* (Brussels) 3 (23 September 1883): 301–3.

Moll, Albert. *Die conträre Sexualempfindung*. Berlin: Fischer, 1891.

Montesquiou, Robert de. *Les chauves-souris*. Paris: G. Richard, 1893.

———. *Les hortensias bleus*. 1896. Paris: Éditions des Autres, 1979.

———. *Les pas effacés: Mémoires*. 3 vols. Paris: Émile-Paul Frères, 1923.

———. "Le pavé rouge: Quelques réflexions sur 'l'oeuvre' de M. Sargent." *Les Arts de la Vie* 3 (June 1905): 329–48.

Morand, Paul. *La fleur double*. Paris: Émile-Paul Frères, 1924.

Morelli, Giovanni [Ivan Lermolieff]. *Italian Painters: Critical Studies of Their Works*. Trans. Constance Jocelyn Ffoulkes. 2 vols. London: John Murray, 1892–93.

Moriarty, Christopher. *Eels: A Natural and Unnatural History.* New York: Universe, 1978.

Morris, Harrison. "American Portraiture of Children." *Scribner's Magazine,* December 1901, 641–56.

Moskowitz, Sam, ed. *Science Fiction by Gaslight: A History and Anthology of Science Fiction in the Popular Magazines, 1891–1911.* Cleveland: World Publishing, 1968.

Moss, Dorothy. "John Singer Sargent, 'Madame X,' and 'Baby Millbank.'" *Burlington Magazine* 143 (May 2001): 268–75.

Mount, Charles Merrill. "Carolus-Duran and the Development of Sargent." *Art Quarterly* 26 (Winter 1963): 384–417.

———. *John Singer Sargent: A Biography.* London: Cresset Press, 1957.

———. "John Singer Sargent and Judith Gautier." *Art Quarterly* 18 (Summer 1955): 136–45.

———. "New Discoveries Illumine Sargent's Paris Career." *Art Quarterly* 20 (Autumn 1957): 304–16.

———. "A Phoenix at Richmond." *Arts in Virginia* 18 (Winter 1978): 2–19.

———. "The Rabbit and the Boa Constrictor: John Singer Sargent at the White House." *Records of the Columbia Historical Society of Washington, D.C.* (1971–72): 618–56.

Munhall, Edgar. *Whistler and Montesquiou: The Butterfly and the Bat.* New York: Frick Collection; Paris: Flammarion, 1995.

Musil, Robert. *The Confusions of Young Törleß.* Trans. Shaun Whiteside. London: Penguin, 2001.

———. *Die Verwirrungen des Zöglings Törless.* 1906. Hamburg: Rowohlt, 2007.

Needham, John Turberville. "A Summary of Some Late Observations upon the Generation, Composition, and Decomposition of Animal and Vegetable Substances; Communicated in a Letter to Martin Folkes Esq; President of the Royal Society, by Mr. Turbervill Needham, Fellow of the same Society." *Philosophical Transactions of the Royal Society of London* 45 (1748): 615–66.

Nemerov, Alexander. *The Body of Raphaelle Peale: Still Life and Selfhood, 1812–1824.* Berkeley and Los Angeles: University of California Press, 2001.

Nesbit, Edith. *In the Dark.* Ashcroft, B.C.: Ash-Tree Press, 2000.

Nochlin, Linda. *Realism and Tradition in Art, 1848–1900.* Englewood Cliffs, N.J.: Prentice Hall, 1966.

Nordau, Max. *Degeneration.* 1st German ed. 1892; 1st English translation 1895. 2nd ed. Lincoln: University of Nebraska Press, 1993.

Normandy, Georges. *Jean Lorrain intime.* Paris: Albin Michel, 1928.

Novalis [Georg von Hardenberg]. "Blüthenstaub." *Athenäum* 1 (1798): 70–106.

———. "Miscellaneous Remarks (Original Version of Pollen)." Trans. Alexander Gelley. *New Literary History* 22 (Spring 1991): 383–406.

Oakley, Lucy. *The Exotic Flower: Constructions of Femininity in Late Nineteenth-Century French Art.* New Brunswick: Jane Voorhees/Zimmerli Museum, Rutgers University, 1998.

O'Brien, Maureen C. "John Singer Sargent: Portrait of Ernest-Ange Duez." In *The American Painting Collection of the Montclair Art Museum, Research Supplement 1,* 11–16. Montclair, N.J.: Montclair Art Museum, 1979.

O'Connor, Francis V. "Painting Women: Feinting with Eros." *Art Criticism* 7, no. 2 (1992): 76–78.

O'Hare, Mary-Kate. "Fashioning Manhood: John Singer Sargent and the Construction of Masculinity at the Turn of the Century." PhD diss., Rutgers University, 2004.

Olson, Roberta. "John Singer Sargent and James Carroll Beckwith, Two Americans in Paris: A Trove of Their Unpublished Drawings." *Master Drawings* 43 (Winter 2005): 415–39.

Olson, Stanley. *John Singer Sargent: His Portrait.* New York: St. Martin's Press, 1986.

———. "On the Question of Sargent's Nationality." In *John Singer Sargent,* exh. cat., ed. Patricia Hills, 13–25. New York: Whitney Museum of American Art and Harry N. Abrams, 1986.

Olson, Stanley, Warren Adelson, and Richard Ormond, eds. *Sargent at Broadway: The Impressionist Years.* New York: Universe and Coe Kerr Gallery, 1986.

Oppian. *Oppian's Halieuticks of the nature of fishes and fishing of the ancients in V. books. Translated from the Greek, with an account of Oppian's life and writings, and a catalogue of his fishes.* Trans. William Diaper. London: Rob. Shippen, 1722.

Oriel, J. D. *The Scars of Venus: A History of Venereology.* London: Springer-Verlag, 1994.

Ormond, Richard. "*Carnation, Lily, Lily, Rose.*" In *Sargent at Broadway: The Impressionist Years,* exh. cat., ed. Stanley Olson, 63–75. New York: Universe and Coe Kerr Gallery, 1986.

———. "In the Alps." In Warren Adelson et al., *Sargent Abroad: Figures and Landscapes,* 63–113. New York: Abbeville Press, 1997.

———. "John Singer Sargent and Vernon Lee." *Colby Library Quarterly* 9 (September 1970): 154–78.

———. *John Singer Sargent: Paintings, Drawings, Watercolors.* New York: Harper and Row, 1970.

———. "The Letters of Dr. FitzWilliam Sargent: The Youth of John Singer Sargent." *Archives of American Art Journal* 14, no. 1 (1974): 16–18.

———. "Modern Life Subjects." In *Sargent and Italy,* exh. cat., ed. Bruce Robertson, 51–79. Princeton: Princeton University Press and Los Angeles County Museum of Art, 2003.

———. "Sargent's Childhood." In Barbara Dayer Gallati, Erica Hirshler, and Richard Ormond, *Great Expectations: John Singer Sargent Painting Children,* exh. cat., 41–63. New York: Brooklyn Museum and Bullfinch Press, 2004.

Ormond, Richard, and Elaine Kilmurray, eds. *John Singer Sargent: The Early Portraits.* Vol. 1 of *Complete Paintings.* New Haven: Yale University Press, 1998.

———. *John Singer Sargent: Figures and Landscapes, 1874–1882.* Vol. 4 of *Complete Paintings.* New Haven: Yale University Press, 2006.

———. *John Singer Sargent: The Later Portraits.* Vol. 3 of *Complete Paintings.* New Haven: Yale University Press, 2003.

———. *John Singer Sargent: Portraits of the 1890s.* Vol. 2 of *Complete Paintings.* New Haven: Yale University Press, 2002.

Ormond, Richard, and Mary Pixley. "Sargent After Velázquez: The Prado Studies." *Burlington Magazine* 145 (September 2003): 632–40.

Ouida [Mary Louise Ramé]. *Moths.* 1880. Peterborough, Ont.: Broadview, 2005.

Ovid. *Metamorphoses.* Trans. A. D. Melville. Oxford: Oxford University Press, 1987.

Pailleron, Marie-Louise. *Le paradis perdu: Souvenirs d'enfance.* Paris: Éditions Albin Michel, 1947.

Painter, George. *Marcel Proust: A Biography.* 2 vols. New York: Vintage, 1978.

Papet, Edouard, ed. *À fleur de peau: Le moulage sur nature au XIXe siècle.* Paris: Réunion des Musées Nationaux, 2001.

Pater, Walter. *The Renaissance.* 1873. Oxford: Oxford University Press, 1998.

Peakman, Julie. *Mighty Lewd Books: The Development of Pornography in Eighteenth-Century England.* Houndmills, Basingstoke: Palgrave Macmillan, 2003.

Pearsall, Ronald. *The Worm in the Bud: The World of Victorian Sexuality.* London: Weidenfeld and Nicolson, 1969.

Péladan, Joséphin. "Salon de 1884: Peinture." *L'Artiste* 54 (June 1884): 414–54.

Peniston, William. *Pederasts and Others: Urban Culture and Sexual Identity in Nineteenth-Century Paris.* New York: Harrington Park Press, 2004.

Pennell, E. R., and J. Pennell, eds. *The Whistler Journal.* Philadelphia: J. B. Lippincott, 1921.

Pennisi, Elizabeth. "Much Ado About Eels." *Bioscience* 39 (October 1989): 594–98.

Petrarch, Francesco. *Petrarch, the first modern scholar and man of letters; a selection from his correspondence with Boccaccio and other friends, designed to illustrate the beginnings of the Renaissance.* Trans. and ed. James Harvey Robinson. New York: G. P. Putnam's Sons, 1898.

Phillips, Claude. "The Royal Academy II." *Academy* 993 (16 May 1891): 471–72.

———. "The Salon II." *Academy* 25 (14 June 1884): 427–28.

Piccolino, Marco. *The Taming of the Ray: Electric Fish Research in the Enlightenment from John Walsh to Alessandro Volta.* Florence: Leo S. Olschki, 2003.

Plato. *The Collected Dialogues of Plato, Including the Letters.* Ed. Edith Hamilton and Huntington Cairns. Princeton: Princeton University Press, 1989.

Poe, Edgar Allan. *The Complete Tales and Poems of Edgar Allan Poe.* New York: Vintage, 1975.

Pohlmann, Ulrich. *Wilhelm von Gloeden, Taormina.* Munich: Schirmer/Mosel, 1998.

Pointon, Marcia. *William Dyce, 1806–1864: A Critical Biography.* Oxford: Clarendon Press, 1979.

Ponsonailhe, Charles. "Salon de 1885: La peinture (Suite) III." *L'Artiste* 55 (June 1885): 424–55.

Porta, Giambattista della. *Phytognomonica.* Rothomagi: Berthelin, 1650.

Porten, Lillian. "The Mâle du Siècle: Creation and Procreation in French Romantic Literature." PhD diss., Harvard University, 1999.

Pozzi, Samuel. *Étude sur les fistules de l'espace pelvi-rectale supérieur ou fistules pelvi-rectales supérieures.* Paris: G. Masson, 1873.

———. *A System of Gynaecology.* Ed. Curtis M. Beebe. New York: J. B. Flint, 1894.

———. *Traité de gynécologie clinique et opératoire.* Paris: G. Masson, 1890; 2nd ed., Paris: G. Masson, 1892; 3rd ed., Paris: Masson et Cie, 1897.

Praz, Mario. *The Romantic Agony.* Trans. Angus Davidson. London: Oxford University Press, 1933.

Prettejohn, Elizabeth. *Interpreting Sargent.* New York: Stewart, Tabori and Chang, 1998.

———. "Sargent and the End of High Society." *Tate* 16 (Winter 1998): 23–27.

Prince, Alison. *Hans Christian Andersen: The Fan Dancer.* London: Allison and Busby, 1998.

Prince, Sue Ann. "Figuring Sexual Difference: Paul Cézanne's Bathers (1870–1906) and Pablo Picasso's Nudes (1906–1907)." PhD diss., University of Pennsylvania, 1997.

"Le professeur Pozzi tombe victime d'un dément." *Le Siècle* (Paris), 15 June 1918, 1–2.

Promey, Sally M. *Painting Religion in Public: John Singer Sargent's "Triumph of Religion" at the Boston Public Library.* Princeton: Princeton University Press, 1999.

Prosser, Jay. "Transsexuals and Transsexologists: Inversion and the Emergence of Transsexual Subjectivity." In *Sexology in Culture: Labelling Bodies and Desires,* ed. Lucy Bland and Laura Doan, 116–31. Chicago: University of Chicago Press, 1998.

Proust, Marcel. *Against Sainte-Beuve and Other Essays.* Trans. John Sturrock. Harmondsworth: Penguin, 1994.

———. *À la recherche du temps perdu.* 1913–27. Ed. Pierre Clarac and André Ferré. 3 vols. Paris: Gallimard, 1961–62.

———. *Contre Sainte-Beuve.* Paris: Gallimard, 1954.

———. *Jean Santeuil.* 3 vols. Paris: Gallimard, 1952.

———. *Jean Santeuil.* Trans. Gerard Hopkins. Harmondsworth: Penguin, 1985.

———. *Marcel Proust on Art and Literature, 1896–1919.* Trans. Sylvia Townsend Warner. New York: Carroll and Graf, 1997.

———. *Remembrance of Things Past.* Trans. C. K. Scott Moncrieff and Terence Kilmartin. 3 vols. Harmondsworth: Penguin, 1981.

———. *Selected Letters.* Vol. 2. Ed. Philip Kolb. Trans. Terence Kilmartin. London: Collins, 1989.

Proust, Robert. *Chirurgie de l'appareil génital de la femme.* Paris: Masson, 1904.

Quétel, Claude. *History of Syphilis*. Trans. Judith Braddock and Brian Pike. Baltimore: Johns Hopkins University Press, 1990.

———. *Le mal de Naples: Histoire de la syphilis*. Paris: Seghers, 1986.

Quillard, Pierre. *La fille aux mains coupées*. In *Deux pièces symbolistes: Pierre Quillard, "La fille aux mains coupées," et Charles van Lerberghe, "Les flaireurs,"* ed. Jeremy Whistle, 1–30. Exeter: University of Exeter, 1976.

Rachilde [Marguerite Eymery]. *La jongleuse*. 1900. Paris: Des Femmes, 1982.

———. *The Juggler*. Trans. Melanie Hawthorne. New Brunswick: Rutgers University Press, 1990.

———. *Madame La Mort and Other Plays*. Ed. and trans. Kiki Gounaridou and Frazer Lively. Baltimore: Johns Hopkins University Press, 1998.

———. *Monsieur Venus*. Trans. Liz Heron. Sawtry, Cambridgeshire: Dedalus, 1992.

———. *Monsieur Vénus*. 1884. Paris: Flammarion, 1977.

———. *Rachilde–Maurice Barrès: Correspondance inédite, 1885–1914*. Ed. Michael R. Finn. Brest: Centre d'Étude des Correspondances et Journaux Intimes des XIXe et XXe Siècles, 2002.

———. "Volupté." *Mercure de France* 9 (September 1893): 26–33.

Rado, Lisa. *The Modern Androgyne Imagination: A Failed Sublime*. Charlottesville: University Press of Virginia, 2000.

Ramírez, Juan Antonio. *The Beehive Metaphor: From Gaudí to Le Corbusier*. Trans. Alexander Tulloch. London: Reaktion, 2000.

Ratcliff, Carter. *John Singer Sargent*. New York: Artabras, 1982.

Reade, Brian, ed. *Sexual Heretics: Male Homosexuality in English Literature from 1850 to 1900*. New York: Coward-McCann, 1971.

Redon, Odilon. *À soi-même: Journal, 1867–1915*. Paris: J. Corti, 1989.

Rees, William, ed. and trans. *The Penguin Book of French Poetry, 1820–1950*. Harmondsworth: Penguin, 1990.

Retrospective Exhibition of Important Works of John Singer Sargent. Exh. cat. New York: Grand Central Art Galleries, 1924.

Reynolds, Gary A. "John Singer Sargent's Portraits: Building a Cosmopolitan Career." *Arts Magazine* (November 1987): 42–46.

Ricci, James V. *The Development of Gynaecological Surgery and Instruments*. San Francisco: Norman, 1990.

Richard, Jean-Pierre. "Deux écrivains épidermiques: Edmond et Jules de Goncourt." In Richard, *Littérature et sensation*, 263–83. Paris: Éditions du Seuil, 1954.

Richardson, Edgar P. "Sophisticates and Innocents Abroad." *ARTnews* 53 (April 1954): 20–23, 61–62.

Richardson, Maurice. "The Psychoanalysis of Ghost Stories." *Twentieth Century* 166 (December 1959): 419–31.

Ridge, Jacqueline, and Joyce Townsend. "John Singer Sargent's Later Portraits: The Artist's Technique and Materials." *Apollo* 148 (September 1998): 23–30.

Rieder, William. *A Charmed Couple: The Art and Life of Walter and Matilda Gay*. New York: Harry N. Abrams, 2000.

Rilke, Rainer Maria. *Letters on Cézanne*. 1907. Trans. Joel Agee. New York: Fromm International, 1985.

———. *Neue Gedichte/New Poems*. 1907–8. Trans. Stephen Cohen. Manchester: Carcanet Press, 1997.

———. *New Poems (1908): The Other Part*. Trans. Edward Snow. San Francisco: North Point Press, 1987.

Rimbaud, Arthur. *Oeuvre-vie*. Ed. Alain Borer and Andrée Montègre. Paris: Arléa, 1991.

———. *A Season in Hell and The Drunken Boat*. Trans. Louise Varèse. New York: New Directions, 1961.

Rivière, Georges. *Renoir et ses amis*. Paris: H. Floury, 1921.

Robb, Graham. *Strangers: Homosexual Love in the Nineteenth Century*. New York: W. W. Norton, 2004.

Robertson, Bruce, ed. *Sargent and Italy*. Exh. cat. Princeton: Princeton University Press and Los Angeles County Museum of Art, 2003.

Robertson, W. Graham. *Letters from Graham Robertson*. Ed. Kerrison Preston. London: Hamish Hamilton, 1953.

———. *Life Was Worth Living*. New York: Harper and Brothers, 1931.

———. *Pinkie and the Fairies*. London: William Heinemann, 1909.

Robinson, Phil. *Under the Punkah*. London: Sampson Low, Marston, Searle, and Rivington, 1881.

Robson, Catherine. *Men in Wonderland: The Lost Girlhood of Victorian Gentlemen*. Princeton: Princeton University Press, 2001.

Rochefort, Helen. "A Portrait of the Greatest Portrait Painter." *Boston Evening Transcript*, 16 September 1916, 4.

Rodenstein, Louis A. "A Case of Enormous Polycystic Ovarian Tumor." *American Journal of Obstetrics and Diseases of Women and Children* 12 (April 1879): 303–6.

Rodin, les mains, les chirurgiens. Exh. cat. Paris: Musée Rodin, 1983.

Roe, Shirley. "John Turberville Needham and the Generation of Living Organisms." *Isis* 74 (June 1983): 158–84.

Rolfe, Julia Rayer, et al. *The Portrait of a Lady: Sargent and Lady Agnew*. Exh. cat. Edinburgh: National Gallery of Scotland, 1997.

Roper, Laura Wood. *FLO: A Biography of Frederick Law Olmsted*. Baltimore: Johns Hopkins University Press, 1983.

Rose, Jacqueline. *The Case of Peter Pan, or The Impossibility of Children's Fiction*. London: Macmillan, 1985.

Rosenblum, Robert. *The Romantic Child from Runge to Sendak*. London: Thames and Hudson, 1988.

———. "Sargent Major." *Artforum* 37 (September 1998): 52–53.

Ross, Robert. "The Wertheimer Sargents." *Art Journal* 73 (January 1911): 1–10.

"The Royal Academy Exhibition." *Art Journal* 26 (June 1887): 248.

"The Royal Academy (First Notice)." *Athenaeum* 3105 (30 April 1887): 580–83.

"The Royal Academy (Third Notice)." *Athenaeum* 3109 (28 May 1887): 708–10.

Rubin, James H. *Manet's Silence and the Poetics of Bouquets.* Cambridge: Harvard University Press, 1994.

Rubin, Stephen D. *John Singer Sargent's Alpine Sketchbooks: A Young Artist's Perspective.* New York: Metropolitan Museum of Art, 1991.

Ruskin, John. *Proserpina: Studies of Wayside Flowers.* 2 vols. Sunnyside, Orpington, Kent: George Allen, 1879, 1886.

———. *The Queen of the Air.* 1869. Kent: George Allen, 1906.

Ryan, Judith. *Rilke, Modernism, and Poetic Tradition.* Cambridge: Cambridge University Press, 1999.

Sade, Donatien Alphonse François, Marquis de. *The 120 Days of Sodom and Other Writings.* Trans. Austryn Wainhouse and Richard Seaver. New York: Grove Press, 1987.

———. *Les 120 journées de Sodome.* 2 vols. Paris: Union Générale d'Éditions, 1975.

———. *Justine, ou Les malheurs de la virtu.* 1791. Paris: Jean-Jacques Pauvert, 1966.

———. *Justine, Philosophy in the Bedroom, and Other Writings.* Trans. Richard Seaver and Austryn Wainhouse. New York: Grove Press, 1990.

Saintine, Xavier-Boniface. *Picciola.* 1836. Chicago: A. C. McClurg, 1890.

"The Salon of the Champs Elysées." *Times* (London), 1 May 1894, 13.

Santos, José. "Réalité et imaginaire des parcs et des jardins dans la deuxième moitié du XIXe siècle." *Nineteenth-Century French Studies* 31 (Spring–Summer 2003): 278–96.

"Sargent at Harvard." Special issue, *Harvard University Art Museums Bulletin* 7 (Fall–Winter 1999–2000).

Sargent's Women. Exh. cat. New York: Adelson Galleries, 2003.

Sartiliot, Claudette. *Herbarium Verbarium: The Discourse of Flowers.* Lincoln: University of Nebraska Press, 1993.

Sato, Tomoko, and Lionel Lambourne, eds. *The Wilde Years.* Exh. cat. London: Barbican Art Galleries and Philip Wilson, 2000.

Savoy, Eric. "Theory *a Tergo* in *The Turn of the Screw.*" In *Curiouser: On the Queerness of Children,* ed. Steven Bruhm and Natasha Hurley, 245–75. Minneapolis: University of Minnesota Press, 2004.

Scanzoni, Friedrich Wilhelm von. *Lehrbuch der Krankheiten der weiblichen Sexualorgane.* Vienna: Wilhelm Braumüller, 1857.

Scarry, Elaine. *Dreaming by the Book.* New York: Farrar, Straus and Giroux, 1999.

———. "Imagining Flowers: Perceptual Mimesis (Particularly Delphinium)." *Representations* 57 (Winter 1997): 90–115.

Schiebinger, Londa. *Nature's Body: Gender in the Making of Modern Science.* Boston: Beacon Press, 1993.

Schlegel, Friedrich. *Kritische Friedrich-Schlegel-Ausgabe.* Vol. 5. Ed. Hans Eichner. Munich: F. Schöningh, 1962.

———. *Lucinde and the Fragments.* 1799. Trans. Peter Firchow. Minneapolis: University of Minnesota Press, 1971.

Schnalke, Thomas. *Diseases in Wax: The History of the Medical Moulage.* Trans. Kathy Spatschek. Chicago: Quintessence, 1995.

Schneider, Laurie. "Butterfly or Scorpion: A Note on the Iconography of Whistler's Signatures." *Source: Notes in the History of Art* 2 (Winter 1983): 26–29.

Schopenhauer, Arthur. *Die Welt als Wille und Vorstellung.* 2 vols. Leipzig: F. A. Brockhaus, 1859.

———. *The World as Will and Idea.* Trans. R. B. Haldane and J. Kemp. 3 vols. London: Routledge and Kegan Paul, 1957.

Schwab, Arnold. "Irish Author and American Critic: George Moore and James Huneker, Part One." *Nineteenth-Century Fiction* 8 (March 1954): 256–71.

Seaton, Beverly. *The Language of Flowers: A History.* Charlottesville: University Press of Virginia, 1995.

Sedgwick, Eve Kosofsky. *Epistemology of the Closet.* Berkeley and Los Angeles: University of California Press, 1990.

Seneca. *Ad lucilium epistulae morales.* Trans. Richard M. Gummere. 3 vols. London: William Heinemann, 1930.

Senelick, Laurence. "The Homosexual as Villain and Victim in Fin-de-Siècle Drama." *Journal of the History of Sexuality* 4 (October 1993): 201–29.

Shank, Barry. *A Token of My Affection: Greeting Cards and American Business Culture.* New York: Columbia University Press, 2004.

Sharp, William. "The Paris Salon." *Art Journal* 46 (June 1884): 179–80.

Shelley, Marjorie. "'Splendid Mountain': A Sketchbook by the Young John Singer Sargent." *Metropolitan Museum Journal* 28 (1993): 185–205.

Shelley, Percy Bysshe. *The Sensitive Plant.* 1820. With an introduction by Edmond Gosse. London: William Heinemann, 1911.

Sheriff, Mary D. *Fragonard: Art and Eroticism.* Chicago: University of Chicago Press, 1990.

Sherman, Claire Richter, ed. *Writing on Hands: Memory and Knowledge in Early Modern Europe.* Exh. cat. Carlisle, Pa.: Trout Gallery, 2000.

Showalter, Elaine. *Sexual Anarchy: Gender and Culture at the Fin de Siècle.* Harmondsworth: Penguin, 1990.

Shteir, Ann B. *Cultivating Women, Cultivating Science: Flora's Daughters and Botany in England, 1760–1860.* Baltimore: Johns Hopkins University Press, 1996.

Shuttleworth, Sally. "The Psychology of Childhood in Victorian Literature and Medicine." In *Literature, Science, Psychoanalysis, 1830–1970: Essays in Honour of Gillian Beer,* ed. Helen Small and Trudi Tate, 86–101. Oxford: Oxford University Press, 2003.

Sickert, Walter. "Sargentolatry." *New Age* 7 (19 May 1910): 56–57.

———. *Walter Sickert: The Complete Writings on Art.* Ed. Anna Gruetzner Robins. Oxford: Oxford University Press, 2000.

Sidlauskas, Susan. *Body, Place, and Self in Nineteenth-Century Painting.* Cambridge: Cambridge University Press, 2000.

———. "Painting Skin: John Singer Sargent's *Madame X.*" *American Art* 15 (Fall 2001): 8–33.

Sidney, Sir Philip. *Selected Poems.* Ed. Catherine Bates. Harmondsworth: Penguin, 1994.

Silver, Carole G. *Strange and Secret Peoples: Fairies and Victorian Consciousness.* Oxford: Oxford University Press, 1999.

Silverman, Debora L. *Art Nouveau in Fin-de-Siècle France: Politics, Psychology, and Style.* Berkeley and Los Angeles: University of California Press, 1989.

Simmons, Edward. *From Seven to Seventy: Memories of a Painter and a Yankee.* New York: Harper and Brothers, 1922.

Simon, John. "Too Many Sargents?" *Arts Magazine* (September 1964): 18–23.

Simpson, Marc. "Reconstructing the Golden Age: American Artists in Broadway, Worcestershire, 1885–1889." PhD diss., Yale University, 1993.

———. "Robert Sterling Clark as a Collector of Sargent." *Antiques* (October 1997): 554–57.

———. "Sargent, Velázquez, and the Critics: 'Velasquez Come to Life Again.'" *Apollo* 148 (September 1998): 3–12.

Simpson, Marc, Richard Ormond, and H. Barbara Weinberg. *Uncanny Spectacle: The Public Career of the Young John Singer Sargent.* New Haven: Yale University Press, 1997.

Sitwell, Osbert. *Left Hand, Right Hand!* London: Macmillan, 1945.

Smith, Timothy d'Arch. *Love in Earnest: Some Notes on the Lives and Writings of English "Uranian" Poets from 1889 to 1930.* London: Routledge and Kegan Paul, 1970.

Spate, Virginia. *Orphism: The Evolution of Non-Figurative Painting in Paris, 1910–1914.* Oxford: Clarendon Press, 1979.

Spiegel, Adriaan van de. *De formato feotu.* Frankfurt: Mattheus Merian, 1631.

Stafford, Barbara Maria. *Artful Science: Enlightenment Entertainment and the Eclipse of Visual Education.* Cambridge: MIT Press, 1999.

———. *Body Criticism: Imaging the Unseen in Enlightenment Art and Medicine.* Cambridge: MIT Press, 1991.

Stahl, P.-J. *Scènes de la vie privée et publique des animaux.* 2 vols. Paris: J. Hetzel et Paulin, 1842.

Stebbins, Theodore E., Jr., et al. *Martin Johnson Heade.* Exh. cat. New Haven: Yale University Press and the Museum of Fine Arts, Boston, 1999.

Steedman, Carolyn. *Strange Dislocations: Childhood and the Idea of Human Interiority, 1780–1930.* London: Virago, 1995.

Stephanson, Raymond. *The Yard of Wit: Male Creativity and Sexuality, 1650–1750.* Philadelphia: University of Pennsylvania Press, 2004.

Stephenson, Andrew. "'A Keen Sight for the Sign of the Races': John Singer Sargent, Whiteness, and the Fashioning of Anglo-Performativity." *Visual Culture in Britain* 6 (Winter 2005): 207–25.

Stevenson, R. A. M. "General Impressions of the Royal Academy of 1893." *Art Journal* 55 (August 1893): 241–44.

———. "J. S. Sargent." *Art Journal* 40 (March 1888): 65–69.

Stevenson, Robert Louis. *The Complete Short Stories.* Centenary ed. Vol. 2. Ed. Ian Bell. Edinburgh: Mainstream, 1993.

———. *The Complete Short Stories of Robert Louis Stevenson.* Ed. Charles Neider. New York: Da Capo Press, 1998.

———. "Forest Notes." *Cornhill Magazine,* May 1876, 545–61.

———. *Selected Letters of Robert Louis Stevenson.* Ed. Ernest Mehew. New Haven: Yale University Press, 1997.

———. *Strange Case of Dr. Jekyll and Mr. Hyde.* London: Longmans, Green, 1886.

Stewart, Miriam, and Kerry Schauber. "Catalogue of Sketchbooks and Albums by John Singer Sargent at the Fogg Art Museum." "Sargent at Harvard," special issue, *Harvard University Art Museums Bulletin* 7 (Fall–Winter 1999–2000): 16–38.

Stewart, Philip. "Representations of Love in the French Eighteenth Century." *Studies in Iconography* 14 (1978): 125–48.

Stoker, Bram. *Dracula.* 1897. Ed. Glennis Byron. Peterborough, Ont.: Broadview Press, 1998.

———. *The Jewel of Seven Stars.* 1904. New York: Tom Doherty, 1999.

Stokes, Adrian. "John Singer Sargent, R.A., R.W.S." *The Old Water-Colour Society's Club: 1925–26* 3 (1926): 51–65.

Stott, Annette. "Floral Femininity: A Pictorial Definition." *American Art* 6 (Spring 1992): 60–77.

Stretch, Bonnie Barrett. "Fragments of a Lost World." *ARTnews* 86 (January 1987): 122–29.

Strickler, Susan. "John Singer Sargent and Worcester." *Worcester Art Museum Journal* 6 (1982–83): 19–39.

Sturgis, Matthew. *Passionate Attitudes: The English Decadence of the 1890s.* London: Macmillan, 1995.

Sutton, Denys. "A Bouquet for Sargent." *Apollo* (May 1964): 395–400.

———. "The Yankee Brio of John Singer Sargent." *Portfolio* 1 (October–November 1979): 46–53.

Sweet, Frederick. *Sargent, Whistler, and Mary Cassatt.* Exh. cat. Chicago: Art Institute of Chicago, 1954.

Swettenham, Sir Frank. *Unaddressed Letters.* 2nd ed. London: John Lane, the Bodley Head, 1900.

Swinburne, Algernon Charles. *The Springtide of Life: Poems of Childhood.* Ed. Edmund Gosse. Philadelphia: J. B. Lippincott, 1918.

Swinglehurst, Edmund. *John Singer Sargent.* San Diego: Thunder Bay Press, 2001.

Symonds, John Addington. *Essays Speculative and Suggestive.* London: Chapman and Hall, 1893.

———. *New and Old: A Volume of Verse.* London: Smith, Elder, 1880.

Symons, Arthur. *The Collected Works of Arthur Symons.* Vol. 1, *Poems.* London: Martin Secker, 1924.

———. *Studies in Seven Arts.* New York: E. P. Dutton, 1907.

Tadié, Jean-Yves. *Marcel Proust.* Trans. Euan Cameron. Harmondsworth: Viking, 2000.

Taine, Hippolyte. *Philosophie de l'art.* 1865. Paris: Hachette, 1893.

Tardieu, Ambroise. *Étude médico-légale sur les attentats aux moeurs.* 3rd ed. Paris: J.-B. Baillière, 1859.

Tennyson, Lord Alfred. *The Works of Tennyson: The Eversley Edition.* Ed. Hallam Tennyson. 9 vols. London: Macmillan, 1907–8.

Terry, R. C., ed. *Robert Louis Stevenson: Interviews and Recollections.* Houndmills, Basingstoke: Macmillan, 1996.

Thiébaut, Philippe, ed. *1900.* Exh. cat. Paris: Réunion des Musées Nationaux, 2000.

———. "*La main aux algues et aux coquillages:* Le testament artistique d'Émile Gallé." *48/14: La Revue du Musée d'Orsay* 18 (Spring 2004): 20–21.

———, ed. *Robert de Montesquiou ou l'art de paraître.* Exh. cat. Paris: Réunion des Musées Nationaux, 1999.

Thompson, D. Dodge. "John Singer Sargent's Javanese Dancers." *Antiques* 138 (July 1990): 124–33.

Thoreau, Henry David. *Walden, or Life in the Woods.* 1854. New York: Dover Publications, 1995.

Thornwell, Emily. *The Lady's Guide to Perfect Gentility, in Manners, Dress, and Conversation.* New York: Derby and Jackson, 1857.

Thwaite, Anne. *Edmund Gosse: A Literary Landscape, 1849–1928.* London: Secker and Warburg, 1984.

Tintner, Adeline R. "Sargent in the Fiction of Henry James." *Apollo* 102 (August 1975): 128–32.

Toal, Catherine. "Free Slaves: The Construction of Cruelty in French and American Literature and Theory." PhD diss., Harvard University, 2001.

Tolini, Michelle. "'Beetle Abominations' and Birds on Bonnets: Zoological Fantasy in Late-Nineteenth-Century Dress." *Nineteenth-Century Art Worldwide* 1 (Spring 2002). http://www.19thc-artworldwide.org/spring_02/articles/toli.shtml.

Tomes, Robert. *Bazar Book of Decorum: The Care of the Person, Manners, Etiquette, and Ceremonials.* New York: Harper and Brothers, 1870.

Townsend, Julie. "Alchemic Visions and Technological Advances: Sexual Morphology in Loie Fuller's Dance." In *Dancing Desires: Choreographing Sexualities On and Off the Stage,* ed. Janet Desmond, 73–96. Madison: University of Wisconsin Press, 2001.

Troyen, Carol. "Watercolors: Sargent's Pictorial Diary." *Magazine Antiques* (November 1986): 668–77.

Twitchell, James B. *The Living Dead: A Study of the Vampire in Romantic Literature.* Durham: Duke University Press, 1981.

Van Capelleveen, Paul. "Butterfly, Abstraction, or Flower? Another Look at Charles Ricketts' Design for *The Picture of Dorian Gray.*" *Private Library* 1 (Winter 1998): 187–92.

Vanderpooten, Claude. *Samuel Pozzi: Chirurgien et ami des femmes.* Paris: Éditions in Fine, 1992.

Van Dyke, John C. *American Painting and Its Tradition.* New York: Charles Scribner's Sons, 1919.

Van Rensselaer, M. G. "John Sargent." *Century Magazine,* March 1892, 798.

Vartanian, Aram. "Trembley's Polyp, La Mettrie, and Eighteenth-Century French Materialism." *Journal of the History of Ideas* 11 (June 1950): 259–86.

Vasari, Giorgio. *The Lives of the Artists.* 1550, 1568. Trans. Julia Conaway Bondanella and Peter Bondanella. Oxford: Oxford University Press, 1998.

Verlaine, Paul. *One Hundred and One Poems by Paul Verlaine.* Trans. Norman R. Shapiro. Chicago: University of Chicago Press, 1999.

Villiers de l'Isle-Adam, Philippe-Auguste, Comte de. *L'Ève future.* Paris: M. de Brunhoff, 1886.

———. *Eve of the Future Eden.* Trans. Marilyn Gaddis Rose. Lawrence, Kans.: Coronado Press, 1981.

Wagner, Anne Middleton. *Three Artists (Three Women): Modernism and the Art of Hesse, Krasner, and O'Keeffe.* Berkeley and Los Angeles: University of California Press, 1996.

Wagner, Richard. *Parsifal.* Trans. H. L. Corder and F. Corder. Philadelphia: Oliver Ditson, 1903.

Walpole, Horace. *The Castle of Otranto.* 1764. Ed. W. S. Lewis. Oxford: Oxford University Press, 1990.

Warhol, Andy. *The Philosophy of Andy Warhol (from A to B and Back Again).* San Diego: Harcourt, Brace, 1975.

Warwick, Alexandra. "Vampires and the Empire: Fears and Fictions of the 1890s." In *Cultural Politics at the Fin de Siècle,* ed. Sally Ledger and Scott McCracken, 202–20. Cambridge: Cambridge University Press, 1995.

Watry, Maureen. *The Vale Press: Charles Ricketts, a Publisher in Earnest.* New Castle, Del.: Oak Knoll Press; London: British Library, 2004.

Watson, Forbes. "Sargent, Boston, and Art." *Arts and Decoration* (February 1917): 194–97.

Weil, Kari. *Androgyny and the Denial of Difference.* Charlottesville: University Press of Virginia, 1992.

Weinberg, H. Barbara, and Stephanie L. Herdrich. "John Singer Sargent in the Metropolitan Museum of Art." *Metropolitan Museum of Art Bulletin* (Spring 2000): 1–64.

Weinshenker, Anne Betty. "'The Lover Crowned' in Eighteenth-Century French Art." *Studies in Eighteenth-Century Culture* 16 (1986): 271–94.

Weir, James. "Viraginity and Effemination." *Medical Record* (16 September 1893): 359–60.

Weisberg, Gabriel P. "Jacques Émile Blanche and the Stylish Portrait, 1880–1905." *Arts Magazine* (June 1985): 97–101.

Wells, H. G. "The Flowering of the Strange Orchid." In *The Short Stories of H. G. Wells,* 198–206. New York: Doubleday, Doran, 1929.

Wells, Samuel R. *How to Behave: A Pocket Manual of Republican Etiquette, and Guide to Correct Personal Habits.* New York: Fowler and Wells, 1856.

"The Westminster Aquarium." *New York Times,* 24 October 1875, 4.

Whistler, James Abbott McNeill. *The Gentle Art of Making Enemies.* Facsimile of 2nd ed., 1892. New York: Dover, 1967.

Whistler, Sargent, and Steer: Impressionists in London from Tate Collections. Exh. cat. Nashville: Frist Center for the Visual Arts, 2002.

White, Chris. *Nineteenth-Century Writings on Homosexuality: A Sourcebook.* London: Routledge, 1999.

White, Edmund. *Marcel Proust.* New York: Viking, 1999.

Whitman, Walt. *The Correspondence.* Ed. Edwin Haviland Miller. 6 vols. New York: New York University Press, 1961–77.

———. *Leaves of Grass and Other Writings.* Ed. Michael Moon. New York: W. W. Norton, 2002.

Wick, Peter. *The Court of Flora: Les Fleurs Animées, the Engraved Illustrations of J. J. Grandville.* New York: George Braziller, 1981.

Wilde, Oscar. *The Complete Letters of Oscar Wilde.* Ed. Merlin Holland and Rupert Hart-Davis. London: Fourth Estate, 2000.

———. *Complete Poetry.* Ed. Isobel Murray. Oxford: Oxford University Press, 1997.

———. *De Profundis and Other Writings.* Harmondsworth: Penguin, 1954.

———. *Intentions.* London: James R. Osgood, McIlvaine, 1891.

———. *The Picture of Dorian Gray.* 1891. Harmondsworth: Penguin, 1985.

Wilde, Oscar, et al. *Teleny, or The Reversal of the Medal.* 1893. Ed. John McRae. London: GMP, 1986.

Wildenstein, Daniel. *Claude Monet: Biographie et catalogue raisonné.* 5 vols. Lausanne: Bibliothèque des Arts, 1974–91.

Wildman, Stephen, and John Christian. *Edward Burne-Jones: Victorian Artist-Dreamer.* Exh. cat. New York: Harry N. Abrams and the Metropolitan Museum of Art, 1998.

Wilenski, Reginald Howard. *The Modern Movement in Art.* 2nd ed. New York: Frederick A. Stokes, 1935.

Wilhelm, C. "Flowers and Fancies: From the Garden to the Stage, Part 1." *Magazine of Art* (November 1898): 172–79.

———. "Flowers and Fancies: From the Garden to the Stage, Part 2." *Magazine of Art* (January 1899): 121–28.

Williams, Benjamin Samuel. *The Orchid-Grower's Manual.* 6th ed., rev. London: Victoria and Paradise Nurseries, 1885.

Williams, Neville. *Powder and Paint: A History of the Englishwoman's Toilet, Elizabeth I–Elizabeth II.* London: Longmans, Green, 1957.

Williams, W. Roger. *Uterine Tumours: Their Pathology and Treatment.* New York: William Wood, 1901.

Willsdon, Clare. *In the Gardens of Impressionism.* New York: Vendome Press, 2004.

Wilson, Gretchen. *With All Her Might: The Life of Gertrude Harding, Militant Suffragette.* New York: Holmes and Meier, 1998.

Winckelmann, Johann Joachim. *History of the Art of Antiquity.* 1764. Trans. Harry Francis Mallgrave. Los Angeles: Getty Research Institute, 2006.

Wise, John Richard de Capel, with illustrations by Walter Crane. *The First of May: A Fairy Masque.* Boston: James R. Osgood, 1881.

Wood, John George, Rev. *Homes Without Hands.* 1865. London: Longmans, Green, 1884.

Wood, Naomi. "Creating the Sensual Child: Paterian Aesthetics, Pederasty, and Oscar Wilde's Fairy Tales." *Marvels and Tales: Journal of Fairy-Tale Studies* 16, no. 2 (2002): 156–70.

Wood, T. Martin. *Sargent.* London: T. C. and E. C. Jack, 1909.

Wordsworth, William. *William Wordsworth.* Ed. Stephen Gill. Oxford: Oxford University Press, 1984.

Wullschlager, Jackie. *Hans Christian Andersen: The Life of a Storyteller.* London: Allen Lane, 2000.

Wynter, Andrew. *Subtle Brains and Lissom Fingers, Being Some of the Chisel-Marks of Our Industrial and Scientific Progress.* London: Robert Hardwicke, 1863.

Zeigler, Lee Woodward. "John Singer Sargent." Introduction to Nathaniel Pousette-Dart, *John Singer Sargent,* vii–x. New York: Frederick A. Stokes, 1924.

Zola, Émile. *The Abbé Mouret's Sin.* Trans. Alec Brown. London: Elek Books, 1957.

———. *La curée.* 1871. In *Les Rougon-Macquart.* Ed. A. Lanoux and H. Mitterand, 5 vols., 1:319–599. Paris: Gallimard, 1963.

———. *La faute de l'abbé Mouret.* 1875. In *Les Rougon-Macquart.* Ed. A. Lanoux and H. Mitterand, 5 vols., 1:1215–1527. Paris: Gallimard, 1963.

———. *The Kill.* Trans. Alexander Teixeira de Mattos. London: Paul Elek, 1958.

———. *The Masterpiece.* Trans. Thomas Walton and Roger Pearson. Oxford: Oxford University Press, 1999.

———. *L'oeuvre.* 1886. Vol. 15 of *Oeuvres complètes.* Ed. Maurice Le Blond. Paris: Bernouard, 1928.

INDEX

Page numbers in *italics* refer to illustrations.

Abbey, Edwin Austin, 158, 186
Abrams, Meyer, 14
absinthe, 134, 219
abstraction, 16, 20, 229–35
Actaeon, 142–43
Adams, Henry, 78
Adonis, 46, 248n112
Aestheticism, 44, 90, *91*, 177, *179*, 187, 189, 208, *209*, 219, 249n129
"Aesthetic lady artist clutching her mahl stick," 208–9, *209*
Agnew, Lady, 82–83, *85*, 254n61
À la recherche du temps perdu. *See* Proust, Marcel
Albert, Alfred, 57, *60*, *61*
  costumes for *Le papillon*, 57, *60*
  costumes for *La source*, 57, *61*
Alcott, Louisa, *Rose in Bloom*, 55, 246n73
Alibert, Jean-Louis, 127, 264n115
Allen, Grant, *Physiological Aesthetics*, 42–44
Alma-Tadema, Lawrence, 286–87n10
ambergris, 31, 115, 219, 260n36
Andersen, Hans Christian
  fairy tales, 8, *9*, 187, 240n42
  inversion, 187, 278n183
  Tommelise (floral procreation), 8, *9*, 240n42
Andersen, Hendrik, 223
androgyne, model for the artist, 10, 169–73, *171*, 240n46, 240n49, 274n101. *See also* hermaphrodite *and* hermaphroditism
androgyny, anatomical, *132*, 257n113, 273n94
android, 91–92, 96
Anstruther-Thomson, Clementina, 80, 144
anthemion. *See* Moore, Albert *and* Simeon, Solomon
anthology, 2, 13, 46, 279n1
anthurium (painter's palette), 128, 150
anus
  architectural slang for, 219
  fistulae around Lorrain's, 263n105
  flowery, 48, 57, 67, *69*, 249n139, 250n141, 250n156
  fruity, 48
  infundibuliform, 243n4
  in Whistler's signature, 67, *69*
apes, chimpanzees, gorillas, and monkeys, 88, 130, *131*, 213, 264–65n128
apian artist. *See* bee
Apollo and Daphne, 38, 212, 213, 236, 252n199, 287n11
arbor vitae (tree of life), 26–27
Argencourt, Louise d', 259n155
Aristophanes, 169, 173
Aristotle, 87, 93
Armory, Carle, *Le monsieur aux chrysanthèmes*, 61
Artemis (Diana), 142–43, 191–93, 216
The Art of Beauty (Haweis), 61, 253n29
artist
  as plant, 4–8, 10–11, 12, 18, 65, 103–6, 177, 180, 239n26, 239n34
  as pollinator, 1, 3–4, 15–16, 20, 63–64, 65, 67, 106, 145, 166, 185, 186, 195, 196–97, 198, 200–201, 217, 219, 229, 233, 238n6, 239n34, 281n34, 281n39
  as pollinator and plant, 14–16, 166, 195, 211, 237
artistic creation
  as childbirth, 8–10, 109, 153, 157–62, 174, 189, 239n40, 239–40n41, 240n45, 270n42, 270n44, 274n103
  as cross-fertilization, 1, 7, 8, 11–16, 18, 67, 147, 155, 156, 157, 160, 187, 196, 229, 236, 243n98
  painting as hand-pollination, 4, 18, 20, 145, 150–51, 156, 157, 174, 185
  as self-fertilization, 10, 160
artists' honey. *See* honey
aster(s), 35, 46, *47*
atavism/regression, 12, 14, 45, 89, 96, 100, 104, 129–31, 133, 134, 213–14, 258n137, 264n123
  in art, 14, 242n74
Athena, 158–59, 240n45, 270n44
attractors, 61
autochthony, 168, 187, 189, 200, 275–76n139

Babil and Bijou (Boucicault), 38
backdoor pollination. *See* pollination
Bain, Francis, *A Digit of the Moon*, 252n5
ballets, 23, 31, 57, 242n84, 278n183
  Flore et Zéphyre, 14
  Le papillon, 60, *60*
  Le roi Carotte, 38, 57, 59
  La source, 57, *61*, 251n177
  Théa, ou la fée aux fleurs, 245n49
Balzac, Honoré de
  and gay horticulture, 13
  Illusions perdues (pollination in), 13
  Lys dans la vallée, 27
  Splendeurs et misères des courtisanes, 44
Baring-Gould, Sabine, 253n40
Barnard, Dolly and Polly, *146*, 156–57, *157*, *158*, *160*, 185, 272n66, 286n170
Barnes, Djuna, 243n98
Barnes, Robert, gynecological text and illustrations, 120, *120*, *138*, 261n58
Barney, Natalie Clifford, 181
Barrie, J. M., *Peter Pan*, 273n80
bat(s), 4, 13, 16, 20, 180–81, *181*, *182*, 196–98, 200, 219, 241n66, 255n76, 275n136, 279n2, 281n34
Bataille, Georges, 98
Bätschman, Oskar, 238–39n19
Baudelaire, Charles
  artist as child, 166

artist of modern life, 208
artist's childhood awareness of flowers, 23
children as plants, 166
on color, 269n8
on Delacroix's palette, 150
flâneur, 1
Les fleurs du mal, 2, 48, 249n135
on the flowers of artists' brains, 158
invoked in discussions of Sargent, 32
phosphorescence, 225
Bauer, George, 13, 250n156
Bazin, Nancy Topping, 240n49
Beckwith, James Carroll, 271n60
bee(s)
artist as, 1, 4, 65, 104, 145, 166, 184, 186, 217, 219, 229, 231, 233, 238n6
as artists' implements, 4
critic as, 177, *179*, 274n123
Egyptian, 219
human romances with, 53, *56*
leaving calling cards, 51, 237
in modern Zeuxis myth, 3
navigating flowers and honeycombs, 51, 220–23
people figured as, *7*, 7–8, 13, 20, 44, 55, 57, *58*, 60–61, *62*, *64*, 104, 106, 177, *179*, 184, 186, 239n37, 271n48
in pollination scenes, 12–13, 48, 51, 53, 57, 145, 197, 203, 220, 251n163
sex of, 273–74n100
squeezed by wasps, 285n138
sting(er)s of, 53, 197, 282n88
tongue of, 203
vampire, 197, 217
Beer, Gillian, 64
Beerbohm, Max, 191
"London in November, and Mr. Henry James in London," 88, *88*
beetle(s), 60, 61, *62*, 63
Bell, Charles, *The Hand: Its Mechanism and Vital Endowments as Evincing Design*, 87
Bell, Clive, 102, 258n143
Belleroche, Albert de, 271n60
Bellow, Juliet, 124
Benjamin, Walter, 110, 135, 189, 278–79n199
Bennett, Paula, 30
Berliocchi, Luigi, 87
Bernhardt, Sarah
in Clairin's mural, 125
cyst of, 137
and Pozzi, 111, 125, 137–38, 267n163
and vampirism, 196, 280n17
Bernheimer, Charles, 264–65n128
Besnard, Paul-Albert, 271n61
Bey, Khalil, 133
Billy Budd (Melville), 46
bimanual exploration. *See* Pozzi, Samuel Jean
bird(s)
and art, 1, 3, *3*, 16, 174, 261n56
artist as, 4, 15–16, 145, 166, 195, 196, 219
colombe poignardée, 211
as ornament, 61
penis symbol, 57, 251n173
people or characters figured as, 15, 22, 53, 57, 60, 61–63, *62*, *64*, 209, *210*, 242n84, 274n120
Sargent and, 4, 16–18, 61–63, *64*, 145, 166, 195, 196, 209, *210*, 219, 258n146
birth pangs, 8, 239n40, 240n45, 270n42
birth topos. *See* artistic creation as human procreation
Blanc, Charles, *Grammaire des arts du dessin*, 260n34, 264n126
Blanc, Louis, 158, *161*
Blanche, Jacques-Émile, 64, 98–102, 211, 258n139, 258n140, 258n143
hands of, 95–96
Little Girl with Hydrangeas, 183, *185*
Marcel Proust, *50*, 51
Self-Portrait with Raphael de Ochoa, 100, *102*, 217, 258n140, 259n155
on Sargent, 258n143, 270n42
Sargent's portrait of, 98–100, *101*, 258n142
and Wilde, 102
blanching, 183–84
Bland-Sutton, gynecological text and illustrations, *114*, 115, 261n47
Blashfield, Edwin, 74, 158, 166, 271n49
Bloch, Gabrielle, 16
bloom (blush), as symbol, 27, 31, 215, 217, 284n135
Blossom Hosts and Insect Guests (Gibson), 51
Boime, Albert, 268n189
Bois de Boulogne, 195
Boit daughters, 31–32, 272n67
Boit, Mrs. Edward Darley, 61–63
Le Bon Marché, 37–38, *41*
Bosie (Alfred Douglas), 48, 249n29
Botticelli, Sandro
Birth of Venus, 123
hands in works of, 72, *72*
The Primavera, 127, *128*, 263n105
Sargent touching, 139
Boucicault, Dionysius, *Babil and Bijou*, 38
Boulanger, M., "L'oeillet," 29, *30*
Bourget, Paul, 32, 78
Brabazon, Hercules Brabazon, 153
breasts, flowery, 27–28, *29*, 42
Bretonne, Raitif de la, 198
Broadway, 65, 155, 159, 186, 187, 273n76, 277n158
Broca, Hôpital. *See* Pozzi, Samuel Jean
Broca, Paul, 111
brothels
heterotopia of, 261n48
male, 46, 48, 195
vegetable associations of, 46, 48, 246n68
Browning, Robert, 286n10
Buckley, Arabella, *The Fairy-Land of Science*, 189, 278n189
Bunker, Dennis Miller, as failed vegetable artist, 18, 243n96
Burke, Edmund, on vegetable longevity, 275n129
Burnand, Francis Cowley, 177, *179*, 274n123
Burne-Jones, Edward
caricatured in Burnand's *The Colonel*, 177
and Connal, 271n48
influence on Sargent, 159, 271n48
Romaunt of the Rose: The Heart of the Rose, 159, *163*, 271n48
Sambourne caricatures of, 177, *178*, *179*
Burns, Sarah, 174
Burton, Richard, and L. C. Smithers, *Priapeia*, 50

Busst, A. J. L., 169, 273n91
buttercup(s), 46
butterfly/ies
and artists, 3–4, *5*, *15*, 15–16, *17*, 20, 51, 67, *69*, 196–97, 198, 200–201, 219, 231, *232*, 233, 239n21, 239n22, 243n98, 281n34, 281n39
collected by Sargent, 18, 201, 281n52, 282–83n91
diseased, 282–83n91
and fairies, 189
images of human/ized, 38, 51–53, *52*, *53*, *54*, 55, 57–61, *57*, *59*, *60*, *62*, *63*, 242n84, 250–51n162, 251n177, 251n189
landing on hands or flowers, 79, 80, *80*, *84*, 239n21, 243n98
and libertinage, 57, 279n2
as ornament, 61
and sadism/violence, 195, 196–97, 219
buttocks, vegetable image of the, 48–49
buttonhole, 7, *7*, 50–51, *50*, 61, 223, *236*, 237, 250n156, 250n157
Byron, Lord, 46, 248n124, 256n104

Caffin, Charles, 164, 165, 186, 191, 272n67, 280n22
calamus, 46, 249n130
calling cards, 79
bees leaving, 51, 237
camellia(s), 31, 35, 38, 50, 79, 245n43, 246n68, 259n155, 264n115
Campbell, Lady Archie, 277n156
canes, *40*, *54*, 57, *59*, *66*, *67*, 80, *83*, *188*, 191, 196, *197*, 198, 199, 203–4, *216*, 217, *218*, 279–80n8, 281n40
Carlier, François, 243n4
carnation(s), 29, *30*, 34, *34*, 48, 145, 151, 160, 250n156, 251–52n197, 252n198. *See also* carnations, green
Carnation, Lily, Lily, Rose. *See* Sargent, John Singer
carnations, green, 4, 46, 249n129, 284–85n136
Carolus-Duran
caricature by Stop, 63, *67*, 78
portrait by Sargent, 63, *66*, 78
Sargent's posing for, 75, 164
Carpenter, Edward, *The Intermediate Sex*, 173, 273–74n100
Carr, J. Comyns, 271n49, 271n53
Castagnary, Jules-Antoine, 133–34
Castiglione, comtesse de, 275n139
Castle, Terry, on birth topos, 8–9, 240n45
Cattleya, 53, 116, 134
Caylus, Comte de, "Rosanie," 53, *56*, 251n167
Cellini, Benvenuto, 12
Cézanne, Paul, 237
Champfleury (Jules Fleury-Husson), *La reine des carottes*, 38
Chard, Cecil, 151
Chardin, Jean-Baptiste-Siméon, 109, 124, 151, 262n85
Charteris, Evan, 272n64, 272n71
Chase, William Meritt, 150–51, *151*, 245n51
Chaucer, *Romaunt of the Rose*, 159, *163*
Chicago, Judy, 274n115
Chicorée trade card, 53, *56*
child(ren)
art as. *See* artistic creation as procreation
development of, 166, 169, 176, 272n74
as figure(s) for the artist, 20, *146*, 157, *157*, 164–76, *167*, *175*, 195, 274n114
as figure(s) of interiority, 165, 272n67, 272n74
hands of, 79, *80*, 87
as plant(s), 31–32, *33*, *146*, 156, *156*, 166, *167*, 168, 176, 183, *185*, 272–73n75
as pollinator(s), 60, *146*, *156*, 156–57, *167*, 168, 175, *175*, 270n38
sexuality of, 166, 169, 170, 272n71, 272n72, 273n91
vision of, 166–68, 176, 273n81
Child, Theodore, 75
childbirth as model for art creation. *See* artistic creation as childbirth
childbirth, vegetable forms of, 8, *9*, *10*, *11*, 180, *180*, 219, *220*, 283n109
childhood awareness of flowers, 16, 23, 119, 145, 166, 176
children's portraiture, 174
Ciboulette, 38, 247n89
cigarettes, 63, *65*, *102*, 103, *104*, 104–6, *105*, 219, *236*, 237, *237*, 259n151, 259n153
cigars, *40*, 104, 201, 259n153, 259n155
Clairin, Georges, *Health Restored to the Sick*, 125–26, *126*, 133, 135
Clairville, *6*, *7*
class, hands as index of, 73–74, 88–90, 253n19
Cleland, John, *Fanny Hill*, 37, 61
clitoris, vegetable images of the, 30, 245n33
Cohen, William, 78
Coleridge, Samuel Taylor, vegetable productions of, 10, 11, 240n52, 256n104
Colette, 243n98
colombe poignardée. *See* Proust, Marcel
columbine(s), 35, 246n67, 279n2
Colvin, Sidney, 283n115
Connal, William, *7*, 7–8, 239n37, 271n48
convolvulus, 13, 34, *34*, 35, *36*, 48, 50
Coombe, 277n156
Cooper, James Fennimore, *The Oak Openings, or the Bee-Hunter*, 55
coral, 117, 122, 123, 261n56, 263n105
corsets, 27, 29, *29*, 260n38, 261n47
cosmetics, 31, 140–43
Courbet, Gustave
apple-tree production, 2
floriferous aesthetics, 3
WORKS
La dame verte, 134, *135*
The Origin of the World, 124, *133*, 133–34, 143, 262n77, 265n142, 265n143, 265–66n144, 266n148
The Source of the Loue, 134, *134*
The Trellis, *2*, 3
Couvreur, André, 127
Crane, Walter, 216, 270n39
The First of May: A Fairy Masque, 189, 278n191
Flora's Feast, a Masque of Flowers, 35, *36*
Crichton-Browne, Sir James, on dreamy mental states, 177, 274n119
crocus(es), 30, 49, *49*
cross-fertilization, 3, 11, 51, 129, 133, 203, 220–23, 241n60, 251n163. *See also* artistic creation
Cullerier, Auguste, *Atlas of Venereal Diseases*, 127, *128*
Culver, Leigh, 164

Curran, Charles Courtney, *Scent of the Rose*, 29
Curry, David Park, 75, 253n32
Curtis, Daniel, 277n167
Cypripedium, *117*, 129, *129*
cyst(s)
Bernhardt's, 137
gigantic ovarian, *114*, 115–16, *116*
jewel-like, 137
Montesquiou's, 266n157
Pozzi popping and collecting, 135, 137–38, 266n152, 266n156
like pregnancy, 116, *116*, 137
vegetal images of, 115

daisy/ies, 4, 16, *25*, 26, 31, 42, 46, 227, 284–85n136
Dame aux camélias, La (Dumas *fils*), 31, 245n43
dame verte, la, 134, *135*
Daphne. *See* Apollo and Daphne
daphnes, 104
Darwin, Charles
on atavism or reversion, 131, 213, 264n123
on bee as brush or pencil, 4, 239n23
on bee navigation, 220–23
on coral, 119
on cross- and self-fertilization, 129, 133, 213, 254n55
on eels, 257n113
on floral morphology, 97, 220–23
on hands, 88, 255n76
on petal's sensitivity to light, 4
popular influence of, 1, 11, 14, 61, 64, 119, 130, 131, 134, 189, 213, 258n137, 276n140, 278n191
WORKS
"Cypripedium," 129, *129*
The Different Forms of Flowers on Plants of the Same Species, 251n163
The Effects of Cross and Self Fertilisation in the Vegetable Kingdom, 129, 133, 254n55
On The Expression of the Emotions in Man and Animals, 111
On the Origin of Species, 88, 255n76
"Orchis Mascula," 97, *97*
The Power of Movement in Plants, 239n30
The Various Contrivances by which Orchids are Fertilised by Insects, 51, 97, *97*, 129, *129*, 189, 239n23, 250n159
Darwin, Erasmus, 26, 46, 189, 248n112, 248n122
Daudet, Alphonse, 127, 213
Daudet, Léon, 131–33, 214, 258n139, 265n138
Daumier, Honoré, 276n148
death drive, 130, 133, 213, 283n115
Decadence, 2–3, 10, 23, 30, 44, 50, 87, 219
Degas, Edgar, 4, 251n177, 270n42
Delacroix, Eugène
dream hand, 92, 96, 122, 267n162
palette, 150
Delafosse, Léon, 78, *78*, 254n47
Delarue-Mardrus, Lucie, *Hortensia dégénéré*, 181, 268n185, 275n138
Delbourgo, James, 92
Della Porta, Giambattista, *26*, 26, *87*, 87
Deleuze, Gilles, 14
Des Esseintes. *See* Huysmans, Joris-Karl
Diana. *See* Artemis
Diaz, Narcisse, 147
Dick, Brigadier Archibald Campbell Douglas, 80–82, *83*, 254n54
Dick, Isabelle (Mrs. Douglas Dick), 254n54
Dickinson, Emily, 22, 30
Didelot, Charles, *Flore et Zéphyre*, 14
Didier, Jules, 4, *5*, 239n22
A Digit of the Moon (Bain), 252n5
Dilsner, Ludwig, *Jasminblüthe*, 249n129, 264n120
Dixon, Marion Hepworth, 282n69
Dr. Warner's Coraline Corsets, 29, *29*, 261n47
Donne, John, 73
Douglas, Alfred (Bosie), 48, 249n129
Doyle, Arthur Conan
and fairies, 189
Sherlock Holmes, 72
Doyle, Charles, 189
Dracula. *See* Stoker, Bram *and* vampirism
dragonfly/ies, 4, 15, 16, 57, 60
dreamy mental states (Crichton-Browne), 177, 274n119
Dreyfus, Alfred, 180
Du Camp, Maxime, 262n77, 266n148
Duez, Ernst-Ange, 181–83, *183*, 276n141
Dumas, Alexadre (fils), *La dame aux camélias*, 31, 245n43
Dumas, Alexandre (père), *La tulipe noire*, 42
Du Maurier, George
"Edison's Anti-Gravitational Under-Clothing," 67, *68*
The Martian, 249n129
Music Hall Moth and Nocturnal Cad-Fly ("Entomological Studies"), 53
Trilby, 30, 199
and Whistler, 199
Duncan, Isadora, 16
Dunham, Helen, 78–79, *79*
Dyce, William, *Titian Preparing to make his First Essay in Colouring*, 145, *147*, 269n9

Edison, Thomas, *68*, 91–92, 96, 259n155, 280n27
"The Eel" (postcard), 94, *94*
eel(s), 93–95, *94*, 97, 256n110, 256–57n111, 257n113, 257n118
Eells, Emily, 90–91
electricity, 91–93, 103, 256n104, 256n105, 286n167
Elkins, James, 117
Ellis, Havelock, 241n56, 248n122, 250n149, 273n94
Empress Eugénie, 57, *58*, 278–79n199
Ephrussi, Charles, 145
Epstein, Julia, 241n59
erotica, botanical, 26–27
Essai de papillonneries humaines (Saint-Aubin), 51, *52*, 250–51n162
etiquette manuals, 71–72, 88, 89, 90, 252n7
L'Ève future. *See* Villiers de l'Isle-Adam, Philippe-Auguste
evolutionary psychology, 166

Fabre, J.-H., *Souvenirs entomologiques*, 203, 281–82n57, 285n138
Fairbrother, Trevor, 216, 225, 243n97
Fairchild, Lucia, diaries, 139, 215, 270n37

fairies
absinthe (*la fée verte*), 134
dame verte, la, 134
difficulty procreating, 187
and evolution, 189, 278n194
flowers, 252n1
and inversion, 48, 187, 237
and science, 189, 278n189, 278n191
and the theater, 57, 64, 216
as pollinators and plants, 189, 278n195
fairyland (fairy-land, fairy land), 119, 187, 189, 278n184, 279n200
The Fairy-Land of Science (Buckley), 189, 278n189
fairy lights, 187, 278n180. *See also* lanterns
fairy tales, 8, *9*, 53, *56*, 187–89, 226, 240n42, 251n167, 271n48, 278n194, 278–79n199
and inversion, 278n183, 278n184
Falchetti, Alberto, 220, *221*, 223
Fanny Hill (Cleland), 37, 61
fée verte, la, 134
femme-fleur, 15, 31–35, 57, 60, 125, 140, 156, 183, 276n145, 282n72
Fialetti, Odoardo, *Petal Venus*, 8, *9*, 42
La fille aux mains coupées (Quillard), 86–87
finger painting, 175, 185, 274n114
fistulae, 134, 138, 263n105, 266n152
Flament, Albert, 34
Flaubert, Gustave
Salammbô, 224
and syphilis, 129, 130
La tentation de Saint-Antoine, 96, 98, 131
Flers, Robert de, 247n89
flesh-pollen, 64–65, 106, 252n198
Fletcher, John, *Faithful Shepherdess*, 277n156
fleur(s) du mal, 32, 48, 211, 249n135
Fleurs du mal, Les (Baudelaire), 2, 48, 249n135
fleur(s) du mâle/man-flowers, 35, 37, 61, 106
Fleurs, fruits et légumes du jour. See Le Petit, Alfred
Flore et Zéphyre (Didelot), 14
flowers, sex lives of, 19, chapter 1. *See names of particular flowers*
flowers (menstruation), 31, 113, 125, 158
flowers of the palette (and paint-pollen), 4, 145–153, *147*, *148*, *149*, *150*, *151*, *152*, 233, 269n8
forcing (botanical), 35, 183, 277n154
Ford, H. J., "The Princess Carried Off by the Bees," 53, *56*
Forest Brush, George de, 4, 174, 274n108
Forster, E. M., pollination in *Maurice*, 13, 241n66
Foucault, Michel, 261n48
Fox, Thomas, 215, 267n176
Fragonard, Jean-Honoré, 27, 251n173
fragrance. *See* scent
Frankenstein (Shelley), 92, 256n104, 286n167
Fraser, Robert, 250n159, 284–85n136, 285n157
Frazer, James
The Golden Bough, 45, 191–93
influences of the sexes on vegetation, 45
King of the Wood, 191–93
and Sargent, 45, 191–93, 248n111
Freeman, Alma, 240n49
Freud, Sigmund
death drive, 213
eels, 93, 257n113
evolution and perversion, 134
fetishises, 137–38, 209, 282n84
and Morelli, 72
narcissism, 248n122
syphilis, 129
Fried, Michael, 3
Friedman, Susan Stanford, 239–40n41, 270n44
Frutex Vulvaria, 26
Fry, Roger, 187
Fuller, Loïe, *15*, 15–16, 242n82, 242n84, 242n86

Gagnier, Regenia, 174
Gaillard, Françoise, 134, 265–66n144
Gallati, Barbara Dayer, 276n150
Gallé, Émile, *Hand with Seaweed and Shells*, 98, *100*, 258n137, 275n132
Gallienne, Richard Le, 284–85n136
Gammell, R. H. Ives, 209
gardening tools, 50, 97, *98*, *167*, 168, 263n105
Gardner, Edward, 189
Gardner, Isabella Stewart, 32, 78, 276–77n153
Garnier, Pierre, 273n94
Gauguin, Paul, *Still Life with Japanese Print*, 158, *162*, 276n149
Gautier, Judith, 267n163, 268n185
Gautier, Théophile, 19, 48, 87, 127
Mademoiselle de Maupin, 78, 106
Gautreau, Virginie, 19, 139–43, *140*, 208, 268n189, 268n208
Genet, Jean, 87, 187
Geoffroy Saint-Hilaire, Étienne, 241n59
Geoffroy Saint-Hilaire, Isidore, 241n59
Gerdts, William, 166–67, 243n96
The Germ, 11
Gérôme, Jean-Léon, 243n96
Gibson, William Hamilton, *Blossom Hosts and Insect Guests*, 51
Gide, André
Les faux-monnayeurs, 283n105
influence of Sand's *Gribouille* on, 45
and polyps, 23
Gilbert and George, 20, 235–37
George the Cunt and Gilbert the Shit, *236*, 237
A Touch of Blossom, *235*, 236–37, 287n11
Gilbert and Sullivan
caricatured in *Illustrated Sporting and Dramatic News* ("Queer Fish at the Westminster Aquarium"), 94–95, *95*
Patience, 44
Gilmore, Paul, 256n104
Goelet, Beatrice, 32, *33*
Goethe, Johann Wolfgang von, *Roman Elegies*, 50
Golden Bough, The. See Frazer, James
Goncourt, Edmond de
La Faustin, 90
and Lorrain, 225, 263n105
and Montesquiou, 198
Goncourt, Edmond and Jules de
on bananas, 30
on Chardin, 151–53
Idées et sensations, 262n80
Manette Salomon, 142
on Mirbeau, 86

on Whistler, 67
Gosse, Edmund, 18, 23, 145, 186, 191, 243n3, 271n49, 273n76
on *Carnation, Lily, Lily, Rose,* 145
"Discipline," 194, 195
Father and Son, 119–20
and Frazer, James, 191
lilac hair, 65
on literary biographies, 145, 269n2
"Manes, The Heretic," 248n122
"Moorland," 27
and polyps, 23, 119–20
and Shelley, 254n63
and Stevenson, 283n109
on Verlaine, 219
and von Gloeden photographs, 286–87n10
"The Whitethroat," 253n39
Gosse, Philip Henry
"Actinia mesembryanthemum," *118,* 119–20
on bees (bottom-licking), 51
and Edward Gosse, 145
natural history of, 51, 243n3
polyps, *118,* 119–20
Gosse, Tessa, 271n49
Gourmont, Remy de
on hands, 88, 258n133
The Physiology of Love, 88, 174
Grandville, J. J.
living vegetable work, 3
Works
"The Adventures of a Butterfly," 51, *55*
"Dried Periwinkle," 37, *38,* 276n145
Les fleurs animées, 23, *24,* 31, 32, 35, 37, *38,* 42, *43,* 48, 53, 140–41, *141,* 243–44n7, 276n145
"Le Louvre des marionettes," 3, *3,* 238–39n19
"An Old Butterfly," 51, *54*
"Pansy," 140–41, *141*
Scènes de la vie privée et publique des animaux, 51, *54, 55,* 61, *63*
"Tuberose and Jonquil," 23, *24,* 31, 48, 243–44n7
"Wallflower," 42, *43*
"Wasp," 61, *63*
Zeuxis parody, 3, *3*
grapes, 3, *113,* 113–15, *202*
Gray, Dorian. *See* Wilde, Oscar
green carnations. *See* carnations, green
The Green Carnation (Hichens), 249n129
greeting cards, 8, *10, 11,* 19, 23–25, *25,* 79, *81,* 147, *220*
Greffulhe, comtesse, 111, 283n104
Greuze, Jean-Baptiste, 244n19
Grew, Nehemiah, *The Anatomy of Plants,* 244n14
Grosvenor Gallery, 177, *179,* 274n124
Guattari, Félix, 14
Gubernatis, Angelo de, *La mythologie des plantes,* 253n21
gynecology, 19–20, 106, 108–43, 193, 271n56

Haddad, Michèle, 133, 265–66n144
Hadley, James, *Aesthetic Teapot,* 90–91, *91*
Haeckel, Ernst
"Actiniae" (*Kunstformen der Natur*), *119,* 120
ontology recapitulates phylogeny, 130, 264n126
Hahn, Reynaldo, 211, 247n89, 258n140
Hale, Mary, 153
Hallward, Basil. *See* Wilde, Oscar
Hammer-Purgstall, Joseph, 48
hands, chapter 2
of artists, 4, 19, 72, *72,* 74–75, 95, 98–100, 103, 104, 121, 139, 213
of doctors, 19, 106, 111–12, 121, 123–24, 135, 137, 138
and evolution, 87–88, 95–96, 98
and flowers, 19, 69, 70–71, 73, 75–77, *76,* 78, 79, *80,* 80–87, *81, 83, 84, 85, 87,* 97, 98–100, 103, *104,* 104–6, *105,* 125, *126,* 217, 219, 229, 252n1
and identity, 19, 63, 71–75, 78–80, 87, 88–90, 95–96, 234
and sea creatures and snakes, 94–96, 98–100, *99, 100, 102, 104,* 109
and sex, 45, 53, 78–79, 80–93, 94, 95, 97, 122, 191
on scrap, greeting, and calling cards, 79, *81*
hand-pollination (botanical), 4
hand-pollination, painting as. *See* artistic creation
handshake, 89–92, 95–96, 255n86
Harrison, Helen (Mrs. Robert Harrison), 206–8, *207,* 209, 282n73
Harrison, Peter, 251n193
Haweis, Mary, *The Art of Beauty,* 61, 253n29
Hawthorne, Nathaniel, 4, 121
The Marble Faun, 42, 130, 247n96, 276n151, 282–83n91
and Sargent, 211, 247n96, 276n151, 282–83n91
The Scarlet Letter, 211
headbirth, 158, 240n45, 270n44
Heade, Martin Johnson, *Roses on a Palette,* 147, *147*
Helleu, Paul
and hydrangeas, 180, *181,* 183, 275n134
Hydrangeas and Bats, 180, *181*
and Montesquiou, 180, 275n134
portraits by Sargent, 102–3, *103,* 104–6, *105,* 211, 219
Helmreich, Anne, 186
Henschel, Sir George, 269n3
hermaphrodite
model for the artist, 16, 166, 168–74
in *Splendor Solis,* 168, *168*
hermaphroditism
and androgyny, 169–73, 273n94, 274n101
biological, 12, 19, 96, 130–32, *132,* 169, 173, 240–41n55, 241n56, 241n59, 254n55, 265n131, 265n134, 273n94
and childhood, 166, 273n91
of floral hands, 19, 77, 78, 79, 82, 87, 96
of flowers, 12, 19, 77, 166, 168, 248n122, 254n55
of Hadley's teapot, 90–91, *91*
implied in Sargent's paintings, 78–83, *79, 80, 83, 85,* 87, 140, *140, 167,* 168–74, *169,* 265n132
and inversion, 12, 78, 130, 240–41n55, 241n59, 265n134
of Pierrot, 276n140
in Pozzi, *Traité de gynécologie,* 130–31, *132*
psychosexual, 12, 241n56, 273n94
Hermaphroditus, 169–70
heterotopia, 116, 134, 142, 261n48
Hichens, Robert, *The Green Carnation,* 249n129

Higginson, Thomas Wentworth, on water lilies, 35, 86
Hoffmann, E. T. A., 199
Holmes, Oliver Wendell, 247n86
Holmes, Sherlock, 72
Homer, Winslow, 187
Homes Without Hands (Wood), 87–88, 255n77
honey, artists', 1, 145, 217–19, 224
honeycomb, 115, 217, 223, 224
honeysuckle, 35, *36*, 71, 239n26
Hôpital Broca. *See* Pozzi, Samuel Jean
Hopkins, Gerard Manley, 86
Hortensia dégénéré (Delarue-Mardrus), 181, 268n185, 275n138
hortensias. *See* hydrangeas
Houssaye, Henry, 142, 268n203
hummingbird(s), 22, 57, 61, 261n56
Hunter sisters, 31
Huysmans, Joris-Karl
  À Rebours, 2–3, 30, 83, 128–29, 180, 219
  artists as horticulturalists, 2–3, 235
  En rade, 49
  hands in, 19, 87
  on Ligeia (Poe), 77
  on syphilis, 129
  syphilitic plants, 128–29, 207
  on Whistler, 198
hyacinth(s), 44, 46
hydrangea(s), *2*, 3, 16, *34*, 34, *175*, *176*, 175–76, 180–84, *181*, *182*, *183*, *184*, *185*, 196, 219, 251–52n197, 274n115, 275n131, 275n132, 275n134, 275n136, 275n138, 276n144

Illustrated Sporting and Dramatic News
  "Aesthetic Lady Artist Clutching her Mahl Stick," 208–9, *209*
  "Queer Fish at the Westminster Aquarium," 94–95, *95*
  "Spring Flowers," 55, *57*
Impressionism, 153, 155, 157, 174, 185, 187, 189, 219
  "La Capucine" (potential name), 4, 239n27
  See also Degas, Monet, Morisot, *and* Renoir
intertwingles, 226, 286n170
inversion, sexual
  and atavism, 12, 129–31, 213
  causes, 133, 134
  definitions and types, 12, 240–41n55, 241n57, 241n59, 243n4
  naturalization of, 12–16, 18–20, 46, 67, 87, 124, 174, 180, 187, 240–41n55, 241n56, 241n59, 241n60, 273–74n100
  and sterility, 16, 133, 174, 184, 193
inversion, bodily
  fantasies of turning inside out, 120–24, 138, 139, 213, 224, 262n85
  uterine, 122, *122*, 138, 261n74
inverts
  and flowers, 12–13, 15–16, 23, 46–51, 61, 64, 180–81, 243n4, 249n129, 249n130, 258n135, 264n120, 275n136
  and pictures, 195, 214, 284n118
Iota (Katherine Mannington), *A Yellow Aster*, 35

Jacobi, Mary, 31
James, Henry
  and Andersen, 223
  and Broadway, 277n158
  caricatured by Beerbohm, *88*, 88
  and children, 166
  dessication of characters, 177
  and electricity, 286n167
  and fairies, 187–89
  floral and insect artists in, 7, 239n34
  image of the blush in, 215
  and inversion, 18
  parturition model of writing, 174, 271n54
  pollination model of writing, 7, 174, 177, 214, 274n103, 275n128
  and Sargent, 20, 104, 206, 272n70
  and Stevenson, 213
  and Symonds, 212
  WORKS
  The Sacred Fount, 60
  Watch and Ward, 35
James, William, 286n167
Le jardin des supplices (Mirbeau), 108, 128
Javanese dancers, 73, 253n15
Jekyll, Gertrude, 283–84n117
jellyfish, 97, 117
Jenkins, David Fraser, 75
Joux, Otto de, 243n4

Kahn, Gustave, 219, 258n146
Keats, John, 119–20
Kent, Abraham, *Hands of Royal Child(ren)*, 79, *80*
Kerner, Anton von Marilaun, 250n160. *See also* backdoor pollination
Kilmurray, Elaine, 208, 272n64
King, Amy, 27
King of the Wood, 191–93
Kingsbury, Martha, 155–56, 166, 215
Kingsley, Charles, polyps in *The Water-Babies*, 122
Knight, Philip, 1–2
Konody, P. G., 279n207
Krafft-Ebing, 240–41n55, 265n134
Kristeva, Julia, 14, 90–91, 256n93
Kupka, František, 20
  The Beginnings of Life, 231
  Roman du Rose study, 231–33, *232*
  Self-Portrait Among Roses, 229–31, *230*
  Tales of Pistils and Stamens, 231

Lacan, Jacques, 122, 143, 254n42, 261n74
Laocoön, 97
ladybird(s), 20, 61, 144, 205, 279n2
Laib, Wolfgang, 20, 229, 233
Lalique, René
  Gynécologie, 134–35, *136*, 193
  pendant, 258n134
La Mottraye, Aubry de, 249n133
lamps. *See* lanterns
Lang, Andrew, *The Green Fairy Book*, *56*, 251n167
Langtry, Lillie, 61, 74, 191, 239n28
language of flowers, 48, 249n133, 274n115, 276n145
  Bataille essay on, 98
lanterns, 142, *146*, 155–60, *156*, *157*, *158*, *160*, 162, 169, 175, 181, *182*, 185, *186*, 270n39, 270n40, 271n52, 271n57, 276n140, 278n180
Leadbeater, Charles, 189
League of the Rose, 111, 259n13
Lear, Edward

"Queeriflora babyoïdes," 180, *180*
"There was an old person so silly," 44, *44*
Lears, T. J. Jackson, 272n71
Lee, Vernon (Violet Paget)
on art and gardens, 144
and Darwin, 189
and Empress Eugénie, 278–79n199
fairy story, 278–79n199
on Fletcher's *Faithful Shepherdess*, 277n156
on handshakes, 255n86
name of, 4, 18
on nationality, 277n171
painted by Sargent, 273n87
on portraits, 200
and Priestley, 209
and Sargent, 142, 187, 206, 247n96, 268n182, 271n48, 271n61, 276n147, 277n156, 279n206, 282n68
and smoking, 259n151
Legh, Mrs. Thomas Wodehouse, 82, *84*
Leighton, Frederick, *Athlete Wrestling with a Python*, 97
Leonardo da Vinci, floral artist, 7
Lepaute, Hortense, 274n115
Le Petit, Alfred
Fleurs, fruits, et légumes du jour, 37, *39*, 158, *161*, 177
"Louis Blanc," 158, *161*
"M. Littré," 37, *39*, 177
"La Rose Impériale," 57, *58*
leprosy, 264n115
lesbian(s), 48, 242n82, 249n135
Liebig cards, 23, *25*, 244n8
Ligeia. *See* Poe, Edgar Allen
lily/ies, 3, 4, 15, 16, 27, 31, 42, 44, 46, 48, 61, 71, 90, 104, *146*, 156, 157, 158, 160, *166*, *167*, 167, 168, 185, 197, 203, 212, 237, 269n23, 270n38, 270n39, 277n154, 284n135
Linnaeus, Carolus
botany of, 12, 26, 69, 244n9
nomenclature of, 27, 248n122
and syphilis, 262n95
Littré, Emile, 37, *39*, 177
Livry, Emma, 57, *60*
Loeffler, Charles, 102, 258n146
Looby, Christopher, 246n69, 248n114, 248n124, 250n157
Lorrain, Jean
on the Bois de Boulogne, 195
and Botticelli, 127, 263n105
on eels, 94
fistulae of (treated by Pozzi), 134, 225, 263n105
on Fuller, 15
on handshakes, 95
and Proust, 239–40n41
syphilis of, 127, 263n105
tubercular symptoms of, 264n115
Works
"Bathyle," 15
Monsieur de Phocas, 95
"Ophelius," 127, 263n105
"Les yeux glauques," 94
Loti, Pierre (Louis Viaud), 4, 239n29, 278n194, 280n17
Lovejoy, Lucretia, *Elegy on the Lamented Death of the Electrical Eel*, 93–94
Lovell, Margaretta, 74
Low, Will, 285n137
Luft, Joanna, 45
Lynes, George Platt, 236

MacColl, D. S., 205, 206
Macfall, Haldane, 196
Madden, Ed, 250n157
Mademoiselle de Maupin (Gautier), 78, 106
Maeterlinck, Maurice
on bees, 203
on Darwin, 189
on flowers, 86, 128, 189
on hands, 71, 86
Works
Le double jardin, 86
L'intelligence des fleurs, 128, 189
Pelléas et Mélisande, 86
Serres chaudes, 71
La vie des abeilles, 203
magnetism, 91–93, 199
magnolias, *frontispiece*, 31, *33*, 73, 75, *76*, 77–78, 245n53, 253n139, 273n96
Maison du Pont-Neuf trade cards, 51, *53*
Mallarmé, Stéphane
faun-like ears, 264n123
flesh-pollen, 64–65, 106
floral poetics, 2, 146–47
and Loïe Fuller, 15
on Poe, Villiers, and Whistler, 280n27
Malvasia, Carlo, 1, 238n4
Manet, Édouard
Déjeuner sur l'herbe, 271n52
Olympia, 141, 264–65n128, 268n204
Mantz, Paul, 75
Mapplethorpe, Robert, 235
The Marble Faun. *See* Hawthorne, Nathaniel
Mardi (Melville), 121, 262n85
Marey, Étienne-Jules, 73
Marlowe, Christopher, 12
The Martian (Du Maurier), 249n129
Marvell, Andrew, 38–42, 45–46
Marx, Roger, 16, 242n86
masturbation/self-fertilization
and art, 160, 271n54
botanical and gardening images of, 31, 37, 45, 245n44, 246n79, 248n122
and hands, 88, 89, 91, 92
health consequences, 254n61, 264n115
suggested in Sargent's paintings, 80–83, *83*, *84*, *85*
Matthews, Charles Skinner, 46
Maturin, Charles, *Melmoth the Wanderer*, 219
Mauclair, Camille, 139, 278n181
Maule's Seeds, *149*, 150, 269n12
Maupertuis, Pierre, 255n76
Mazzinghi, Joseph, "The Wreath," 159, 271n50, 277n169
Melmoth, Sebastian (Wilde), 219
Melmoth the Wanderer (Maturin), 219
Melville, Herman
ambergris, 115, 219
Billy Budd, 46
eels, 95, 257n118
Mardi, 121, 262n85
Moby-Dick, 95, 109, 115, 116, 257n118
Mengs, Anton Rafael, 238n6
menstruation

botanical images of, 31, 113, 125, *126*, 158, 244n19
Pozzi and, 124, 267n162
Merleau-Ponty, Maurice, 92, 109–10, 111, 112, 121, 122, 213, 259n15, 261n61, 267n172
Méry, Joseph, *Monsieur Auguste*, 48, 249n136
mesmerism, 90, 91, 97, 104, 199, 201, 281n34
Metamorphoses. *See* Ovid
Meyer, Mrs. Carl, 32
Meynell, Alice, 211, 272n65
Michaelson, Alexander. *See* Raffalovich, Marc-André
Michelangelo, 8, 12, 16, 250n149
Michelet, Jules
America, 277n168
blue hydrangeas, 275n131
botany as science of love, 14
coral and polyps, 119, 123
emotions of flowers, 23
intestines, 121–22
jellyfish, 97
meadows, 102
possessing both sexes, 124
seawater, 95, 257n122
whales and women, 260n38
Mikschik's instrument, *119*
Miles, Frank, 239n28
Millais, John Everett, *Apple Blossoms*, 3
Millet, Frank (Francis Davis), 187, 277n162
Millet, John Alfred Parsons, 277n162
Millet, Kate, 271n52, 272n66
Millet, Lily, 271n49, 271n53, 277n158
Millet, Lucia, 185, 270n27, 271n49, 271n52
Mirbeau, Octave
Le jardin des supplices, 108, 128
relationship to flowers described by the Goncourts, 86
Moby-Dick. *See* Melville, Herman
modernism, 3, 4, 15, 18, 19, 20, 109, 110, 157, 166, 233, 237
Moll, Albert, 91, 256n95
Monet, Claude
allusions to in Sargent's work, *146*, 157, *158*, 159, 160, *160*, 269n23
garden of, 3
painted by Sargent, *152*, 153, 175, 269n23
painting with Sargent, 155
palette of, 269n22
Works
Caricature of Jules Didier as butterfly artist, 4, *5*, 239n22
Impression, Sunrise, 157, *158*
The playwright François Nicolaïe, called Clairville, *6*, 7
monkeys. *See* apes, chimpanzees, gorillas, and monkeys
Monsieur Auguste (Méry), 48, 249n136
Le monsieur aux chrysanthèmes (Armory), 61
Montagu, Lady Mary Wortley, 249n133
Montesquiou, Robert de
on artists' honey, 217–19
as bat and blue hydrangea, 16, 20, 181, 196, 219, 275n136, 281n34
canes of, *197*, 199, 281n40
as "chasseur de chauves-souris," 181, *182*, 275–76n139, 276n140
floral poetics of, 212
hands of, 95, 275n132
"Hortensiou," 180
inversion of, 18, 254n47
model for Charlus and Des Esseintes, 20, 180
portrait by Whistler, 196–201, *197*
and Pozzi, 111, 125, 137, 266n157, 275n134
and Proust, 18, 180, 286n163
on Sargent, 98, 130
and spiritualism, 281n34
and Whistler, 196–201, *197*
Works
art and furniture by, 275n132
"Ego Imago," 275n132, 275–76n139, 276n140, 281n39
Les hortensias bleus, 180, 219, 275n134, 276n144
Moore, Albert
anthemion signature, 4, *7*, 8, 239n26, 239n37, 252n206
End of the Story, 239n26
and flowers, 64, 251–52n197
Reading Aloud, 239n37
William Connal, Esq. Jr., of Solsgirth, *7*, 7–8
and Robertson, 216
Moore, Kate, 80
Morand, Paul, *La fleur double*, 254n55
Moreau, Gustave, 180
Morelli, Giovanni
and comparative anatomy, 87
Italian Painters, 72, *72*
method of, 72, 75, 252n11
"Typical Hands," 72, *72*
Morisot, Berthe, palette of, 145
Morris, Harrison, 174
moth(s)
artist as, 4
Dracula as, 197
metamorphosis of, 77
proboscis of, 203
sitters imaged as in Sargent's paintings, 20, 63–64, *66*, *67*, *146*, 157, 175, *175*, 203, *204*, 270n38
slang for prostitute, 279n2
Verlaine as, 219
in visual culture, 53, 57, 60, *61*, *63*
Wilde as, 219
writings on, 51, 203
mother(s)
artist as, 8, 16
botanical versions of, 8, *9*, *10*, *11*, 53, *55*
botanist as defiler of, 86
Egyptian, 116, 261n44
male rivalry with artistic, 184–85, 276n149
phallic, 137
Moths. See Ouida
Munhall, Edgar, 275n131, 281n34
mushroom(s), 37, 49, 115, 189, 260n28, 260n36
Musil, Robert, *Confusions of Young Törleß*, 195
Muybridge, Eadweard, 73, 253n16
Myers, Frederick William Henry, 284n118

namby-pamby, 177, *179*
names, floral and vegetal, 4, 31, 46, 177, 246n68
narcissism, 137, 248n122, 262n91
Narcissus, 36–37, 45, 46, 159, 164, 236
Needham, John, 117

Neugebauer's speculum, *139*
New Woman, 35, 282n72
Nicolet, Charles. *See* Pozzi, Samuel Jean
nipple, flowery, 27–29, *28*, 171, *171*
Nordau, Max
  on degeneration, 7, 130
  English translation, 275n130
  on Mallarmé's faun-like ears, 264n123
  on oversensitivity to plants, 45
  on plants/artists, 7, 240n52
Novalis, "Pollen," 10–11, 240n51

O'Brien, Maureen, 181–83
Ochoa, Raphael de, 100, *102*, 217, 258n140, 259n155
octopi, 95, *96*, 100, 117, 122, *123*, 258n134
Offenbach, Jacques, *Le roi Carotte*, 38, 57, *59*
O'Keeffe, Georgia, 233–35
Olmsted, Frederick Law, *188*, 189–91, 269n3, 279n201
Olson, Stanley, 169, 268n208
onions, 73–74, *74*, 253n21
ontogeny recapitulates phylogeny, 130, 169, 264n126
operas
  Ciboulette (chives), 38, 247n89
  Parsifal. *See* Wagner, Richard
  Patience, 44
  Le roi Carotte, 38, 57, *59*
Oppian, *Halieuticks*, 93
orchid(s)
  associations of, 37, 51, 87
  attacked by suffragettes, 35
  and bee scene in Proust, 12–13, 14, 50, 97
  contrivances of, 250n159
  Darwin on, 51, 97, 129, 189, 239n23
  fertilization of, 12, 14, 51, 203, 239n23
  Fuller dance, 16
  human resemblance of, 128–29, 244n13
  images of, *26*, *50*, *87*, *97*, *117*, *129*
  names of, 26, 42, 247n103, 252n1
  portrait sitters described as, 32, 140
  Ruskin on, 42, 247n103
Orchid-Grower's Manual (Williams), *117*
Ormond, Richard, 156, 185, 270n30, 271n61, 276n141
Order of the Knights and Nymphs of the Rose, 27
Ouida, *Moths*, 31, 55, 60, 61, *63*
ovarian cysts. *See* cysts
Ovid, *Metamorphoses*, 38, 45, 64, 127, 142–43, 169, 212, 248n122, 252n199

Paget, Violet. *See* Lee, Vernon
Pailleron, Mme., 32
Pailleron, Marie-Louise, 201, 281n52, 282–83n91
paint and pollen. *See* flowers of the palette
paintbrush
  as pollination instrument, 4, 20, 65, *146*, 150, *151*, 156, 157, *157*, *158*, 212
  as torture instrument, 198–99, 206, 208
paintbrushes as flowers, 147, *147*, 150, *151*, 220, *221*
palette. *See* flowers of the palette
Palmer, Elsie, 208, *208*, 209
Pangbourne Weir, 158, *159*
pansy (*pensée*), 16, *17*, 31, 46, *47*, 67, *69*, 79, *81*, 140–41, *141*, 158, *161*, 166, 217, 237, 281n39
pansy sailors, 46, *47*
pantomimes, 23, 38, 60, 252n200
Le papillon, 60, *60*
parallel brushstroke, 157, 231, 233
Parsifal. *See* Wagner, Richard; Rochegrosse, Georges
Parsons, Alfred, 270n27, 271n49, 277n162
parturition, artistic fantasy of. *See* artistic creation
pastoral fantasies and landscapes, 20, 102, 115–17, 124–26, *125*, *126*, 159, 177, 181, 185–86, 261n47, 277n156, 277n158
Pater, Walter, 7, 16, 18, 215, 259n151, 278n184
Patience (Gilbert and Sullivan), 44
Peakman, Julie, 27
"The Pearl" (postcard), 122, *123*
Pedersen, Vilhelm, "Tommelise," 8, *9*
Péladan, Joséphin, 31–32
penis
  as baby, 122
  as cigarette, 104
  as finger, 97
  as gardening tool, 50, 247n82
  as pollinator or pollinator part, 57, 251n173
  vegetable images of the, 37, 49, 244n14, 246n79, 254n51
perfume. *See* scent
Pernice, Luwig, 113–15
perversion
  and eels, 93–94
  and evolution, 121, 134, 174
  and hands, 86–91, 257n124
  and plants, 13, 23, 45, 249n129
  and syphilis, 129, 131, 264n115
  of children, 166, 272n72
  of plants, 19
Peter Pan (Barrie), 273n80
Petrarch, 1
Philips, Ambrose (namby-pamby), 177
Phillips, Claude, 32
phosphorescence and phosphorus, 224–25
photography, 20, 73, 153, 184–85, 229, 235–37, 253n16, 275n132, 275–76n139
Picciola (Saintine), 42
The Picture of Dorian Gray. *See* Wilde, Oscar
Pierrot, 276n140
pigment. *See* paint
pimp, as flower-fancier, 246n68
Plato
  birth topos in, 8
  myth of androgynes, 169–70, 173
  poet as bee, 1
Platonic love mocked in *Patience*, 44
Poe, Edgar Allen, 72, 201, 256n104, 280n27
  Ligeia, 75–77, 91, 224, 254n43, 256n98, 286n161
  "The Oval Portrait," 198–99, 206
pollen (botanical), 20, 233, 239n23, 241n60, 251n163
"Pollen" (Novalis), 10–11, 240n51
pollen brushes/brushing, 1, 57, 63, 82, 175, 229, 233
pollination, 1, 3, 4, 11, 13, 51, 145, 203, 220
  backdoor, 20, 196, 203, 220–23, 224
  painting and writing as. *See* artistic creation

pollinators, in 19th-century culture, 51–64
Polwhele, Richard, *The Unsex'd Females*, 42
polyps, 23, 115, 117–122, 124, 130, 260n28, 261n50, 261n70
Ponsonailhe, Charles, 75
Pope, Alexander, 8, 240n45
postcards
  "L'anguille," 94, *94*
  with bee-woman, *62*
  with blossom-womb motif, *11*
  "L'oeillet" (Boulanger), 29, *30*, 244n27
  "La perle," 122, *123*
  A Touch of Blossom (Gilbert and George), *235*, 236–37, 287n11
pots, 8, 167–68, 175, 177, 200, 273n84, 274n112, 277n154, 283n109. *See also* tea, teapots, and teacups *and* vases
Pozzi, Samuel Jean, 18, 106, chapter 3, 162, 223
  art collection, 111, 116, 260n36
  bimanual exploration, 19, 111–12, *112*, 121, 122–24, 143, 260n19
  hands of, 111, 135–37, 266n152
  League of the Rose, 111, 259n13
  murder of, 138, 267n167
  Nicolet's illustrations for, 112, *112*, 117, *117*, 121, 122, *122*, 131, *132*, 135, *137*, 260n22
  portrait by Sargent, *110*, 111, 122–24, 262n75, 262n77, 265n132, 282n68
  Traité de gynécologie, 19–20, 111–17, 122, 127, 128, 130–31, 135–37, 260n39, 261n58, 267n162
  translation of Darwin, 111
  ward at Hôpital Broca, 111, 125, *125*, *126*, 133, 134, 135, 265n131
  and women, 111, 124, 262n82, 267n163
  See also cyst(s) *and* vegetations
Pratt, Katherine Chase, 183, *184*
Pre-Raphaelites, the
  The Germ, 11
  Sargent's interest in, 277n167
  subject matter of, 187
Priapeia (Burton and Smithers), 50
Priestley, Flora, 4, 209–11, *210*, 282n82, 282n84, 282–83n91
primrose(s), 13, 270n39, 278n189, 282n63
proboscis/ides, 4, 57, 65, 157, 168, 191, 196, 198, 203, 204, 219, 220, 226
prolapse, uterine, 122, *122*, 138
Promey, Sally, 165, 268n199, 269n23, 272n67, 274n116
Prosser, Jay, 241n57
prostitutes
  male (and flowers), 243n4
  as pollinators, 195, 246n67, 279n2
  vegetable images of, 35, 246n67, 246n68
prostitution, of plants, 42, 46
Proust, Marcel
  as bee, 1
  and Desjardins, 269n2
  and gynecology, 19, 109
  and Hahn, 211, 247n89
  and hawthorns, 23
  inversion of, 239–40n41
  and lavatories, 104, 256n93
  and Montesquiou, 18, 180, 286n163
  and Pozzi, 109
  portrait by Blanche, *50*, 51
  reading of, 13, 42, 247n98, 251n163, 285n157
  and Stevenson, 284n123, 285n157
  Works
  À la recherche du temps perdu: atavism, 130; brothel, 46–48, 195; buttonholes in, 50; calling cards left by bees, 51, 237, 251n163; cattleya, 53; *colombe poignardée*, 211; cross-fertilization, 12–13, 14, 51, 133, 168, 219, 229, 237, 251n163; hands, 19, 72–73, 87, 89, 96, 257n124; handshaking, 89, 257n124; hermaphrodites, 168; inversion, 18, 50, 97, 109, 130, 137, 195, 211, 212, 213, 215, 216, 266–67n161; jellyfish, 97; orchid and bee scene, 12–13, 14, 50, 97, 216; page boy/shrub, 268n185; roots, 227; Saint-Loup, 215, 284n123; self-fertilization, 129, 133, 168, 265n141; sleep, 226; summaries of, 224, 286n163; tea, 256n93; tumbles in the grass, 195; vegetable code words, 46–48
  "Chardin" (art and gynecology), 19, 109, 124, 262n84, 262n85
  "Gustave Moreau," 180
  Jean Santeuil, pollen in, 104, 283n105
Proust, Robert, 109, 259n11, 262n77, 265n131
pubic hair
  in Courbet, *133*, 134
  in Pozzi's illustrations, 117
  vegetable images of, 29, 134
Punch, or the London Charivari
  F. C. Burnand, 177, *179*, 274n123
  Sargent and, 274n121
  Caricatures and Articles in
  "A Bird of Prey," 61
  "The Butterfly Was A Lady," 251n189
  "Edison's Anti-Gravitational Under-Clothing," 67, *68*
  "Entomological Studies," 53
  "Flowers of 'Culture,' or a Swinburne-Jones Cutting," 177, *178*
  "Last Sweet Thing in Toilettes," 31, *32*
  "A Maudle-In Ballad to His Lily," 44
  "Our Guide to the Grosvenor Gallery," 177, *179*, 274n124
  "O.W.," 177, *179*
  "Suffrage for Both Sexes," 251n189
  "Wilde Flowers," 274n125

Queensberry, Marquis of, 49–50, 282n63
Quillard, Pierre, *La fille aux mains coupées*, 86–87

Rachilde (Marguerite Eymery)
  ape/woman image, 130
  leprosy, 264n115
  plants and perversion, 23, 44, 243n4
  Works
  La jongleuse, 130, 264n115
  Monsieur Venus, 243n4
  Volupté, 44
Rado, Lisa, 240n46
Raffalovich, Marc-André
  and Alfred Moore, 64, 251–52n197
  and inversion, 18, 250n157
  floral tropes, 48, 250n157
  portrait by Starr, 51
  "Verglas and Pollen," 251–52n197
Raffele, Ambrogio, 153, *154*, 191–93, *192*
Raile, Arthur Lyon, 249n130

Raphael, 72, 238n4
Raphael Tuck and Sons, *220*
Ray, Man, 235
Redon, Odilon
and Darwin, 4
on flesh-pollen, 252n198
floral model of perception and art, 4–7
The misshapen polyp, 261n70
There Was Perhaps a First Vision Attempted in the Flower, 4–7, *6*
regression. *See* atavism
La reine des carottes (Champfleury), 38
Remembrance of Things Past. *See* Proust, Marcel
Reni, Guido, 238n4
Renoir, Pierre-Auguste, 64
resurrection plants, 191
Ribblesdale, Lord, 204, *205*, *206*
Ricketts, Charles
cover of *The Picture of Dorian Gray*, 16, *18*
cover of *Intentions*, 49, *49*
Ricord, Philippe, 129, 133, 135
Rilke, Rainer Maria, 183, 276n144
Rimbaud, Arthur, 2, 48, 219, 238n10, 250n156
Robertson, Walford Graham, 18, 20, 211, *216*, 216–17, *218*, 219, 223, 284n134, 285n138
Robinson, Edward, 220, *222*, 223, 274n120, 279n206
Robson, Catherine, 273n91
Rochegrosse, Georges, *Knight Among the Flower-Maidens, from Wagner's 'Parsifal,'* 32–34, *34*, 246n62, 262n91
Rococo art, flowers in, 27, 244n19
Rodd, James Rennell, *Rose Leaf and Apple Leaf*, 215
Rodin, Auguste, 15, 31, 245n54
Le roi Carotte, 38, 57, 59
Roman de la rose, 38, 45, 49, 159, *163*
Romanticism, 1, 8, 10–11, 13–14, 169, 240n46, 242n73, 242n84, 256n104, 273n81
Romaunt of the Rose (Chaucer), 159, *163*, 271n48
root(s), botanical, 27, 37, 73–74, *74*, 98, 137, 177, 240n52, 246–47n80, 284–85n136
rootlessness/uprooting, 14, 42, *43*, 137, 174, 187, 195, 275–76n139
roots/root taking/rootedness, 14, 36, 123, 137, 166, 168, 176, 187, 189, 200, 227, 275–76n139. *See also* autochthony
"Rosanie" (Caylus), 53, *56*, 251n167
rose(s). See also *Carnation, Lily, Lily, Rose* and *Roman de la rose* and *Romaunt of the Rose*
artists and, 64, 139, 158–59, 173, 231, 237, 239n34
body parts like/as, 27, 29, 30, 45, 46, 48, 49, 50, 57, 78, 82–83, 87, 96, 111, 142, 158, 197, 244n29, 249–50n139, 250n141, 263n105, 284n135, 284–85n136
humanized, 31, *32*, 35, 38, 46, 50, 51, *55*, 57, *58*, 73, 239n34, 246n73
images of, *3*, *6*, *10*, *25*, *28*, *32*, *54*, *55*, *58*, *60*, *62*, *73*, *80*, *81*, *85*, *146*, *147*, *148*, *160*, *163*, *172*, *230*, *236*
sexual acts like or involving, 15, 27, 38, 45, 48, 50, 57, 111, 237, 244n19, 245n44, 250n141
wreaths and garlands of, 159, 243n4, 244n19, 259n151, 271n49, 271n50, 271n52, 286n10
Rose in Bloom (Alcott), 55, 246n73
Rose, Jacqueline, 168, 272n72
Rose Leaf and Apple Leaf (Rodd), 215
Rosenfeld, Paul, 234
Rosny, Joseph Henry, 189
Ross, Robert, 139
Rubin, James, 145–47
Runge, Otto, 273n81
Ruskin, John
on fairyland, 278n184
flowers, 23, 42, 97, 151, 247n103
onions, 74
orchids, proposed renaming of, 42, 247n103
Sesame and Lilies satirized in Proust, 46–48
snake dream, 96
Ryan, Judith, 276n144
Sackville West, Vita, 31, *32*
Sade, Marquis de
sadistic pollinators in, 197–98
vegetable characters and parts in, 46, 48, 57, 249–50n139
sailor(s), 15, 46, *47*, 263n105
Saint-Aubin, Charles-Germain de, *Essai de papillonneries humaines*, 51, *52*, 250–51n162
Saint-Aubin, Gabriel de, *Comparison of the Rosebud*, 27, *28*, 250–51n162
Saintine, Xavier-Boniface, *Picciola*, 42
Sambourne, Linley, *Punch* caricatures
"Flowers of 'Culture'; or, a Swinburne-Jones Cutting," 177, *178*
"Our Guide to the Grosvenor Gallery," 177, *179*, 274n124
"O.W.," 177, *178*
Sand, George
Delacroix's dream, 92
Gribouille, metamorphosis, 45
Sardou, Victorien, *Le roi Carotte*, 38, 57, *59*
Sargent, Emily, 155, 169, *170*, 248n111
Sargent, FitzWilliam, 16, 187
Sargent, John Singer
allegories of cross-fertilization in works of, 18, 20, 82, 147, 154–56, 160, 187, 203–5
anal penetration/backdoor pollination suggested in paintings of, 20, 217, 219–23, 224
atavistic painting of, 14, 242n74
as bee, 4, 65, 106, 166
as bird, 166. *See also* as wagtail
at Broadway, 65, 155, 159, 186, 187, 273n76, 277n158
and Burne-Jones, 159, 271n48
butterfly collector, 18, 201, 281n52, 282–83n91
and Carolus-Duran, 63, *66*, *67*, 75, 78, 164
childhood of, 16–17, 166, 169, *170*, 187, 278n175
dislike of own image, 164–65, 271n59–60
and fairies, 64, 187
as flower, 65, 104, 166, 237
and Frazer, 45, 191–93, 248n111
hands (his own), 19, 74–75, 95, 139

hands in paintings, 19, 69, 71, 73–87, 98–106, 123–24, 209, 253n17, 265n132, 268n203
and Hawthorne, 247n96, 276n151, 282–83n91
and Impressionism, 153, 155, 187
and James, 20, 104, 206, 272n70
and inversion, 18, 102, 104–6, 243n97, 258n143
medical metaphors for his artistic practice, 139, 142, 185, 189
and modernism, 18, 19, 237
and Monet, *152*, 153, *158*, 160, *160*, 269n21–23
national identity, question of, 20, 155, 186–87
and natural history, 16–18, 95, 102
paintings bursting into bloom, 3, 153
paints of, 151
photographs of, *155*, 155, 164, 169, *170*, *237*, 237
portraits of, 75, 164, 271n59, 271n60
and Pre-Raphaelitism, 159, 187, 271n48, 277n167
and puns, 273n76
sadism of, 142, 205–8, 209–11, *210*, 285n138
self-portraits, 160, *163*, *164*, 164–66, 271–72n62, 272n64
sitters as flowers, 31–32, 35, 64, 65, 103–6, 206, 207–8, 212, 226–27
sitters as pollinators, 61–64, 203, 204–5
and the theater, 64–65
and Wagner, 156, 270n37
as wagtail, 4, 145, 195, 196, 219, 279n2
and Wilde, 18, 215–16, 284n126–28
working method of, 65, 145, 155, 259n159, 269n3, 269n5, 270n29, 274n114
Works
Alberto Falchetti, 220, *221*, 223
Alfred, Son of Asher Wertheimer, 223, *224*, 285n159
An Artist in His Studio, 153, *154*, 191, 193
Beatrice Townsend, 79, *80*, 82
Brigadier Archibald Campbell Douglas Dick, 80–82, *83*, 254n54
Candelabrum, The, 259n151
Carnation, Lily, Lily, Rose, 20, 143, 145, *146*, 155–64, *158*, 166–67, 168, 171, 173, 175, 176, 180, 187–89, 205, 270n30, 270n32, 270n38, 270n40, 270n46, 271n57, 272n67, 277n169, 279n200; caricature of, 160, *163*, 271n53; oil studies for, 157, *157*, 158–62, *160*, 173, 271n56; pencil sketches for, 156, *156*, 162, *163*; photograph of Sargent painting, 155, *155*
Carolus-Duran, 63, *66*, 78; caricature of, 63, *67*
Claude Monet Painting by the Edge of a Wood, *152*, 153, 269n23
Daphne, 252n199
Daughters of Edward Darley Boit, 31–32, 272n67, 273n76, 273n84
Dr. Pozzi at Home, 19, *110*, 111, 122–24, 262n75, 262n77, 265n132, 282n68
Edward Robinson, 220, *222*, 223
Ellen Terry as Lady Macbeth, 63
Elsie Palmer, 208, *208*, 209
Ernest-Ange Duez, 181–83, *183*, 276n141
Flora Priestley, 209–11, *210*, 282n84, 282–83n91
Frederick Law Olmsted, *188*, 189–91, 279n201
Fumée d'ambre gris, 260n36
Garden Study of the Vickers Children, *167*, 167–74
Garden Study with Lucia and Kate Millet, 271n52
Gathering Flowers at Twilight, *172*, 173
Group with Parasols, 226–27, *227*
Helen Dunham, 78, *79*
Helen Sears, 20, 174–79, *175*, *176*, 183–84, 185, 189, 274n112, 274n113, 274n116
Hell, 224–25, *225*
Hercules and the Hydra, 97, *99*
Home Fields, 165, *165*
The Honourable Victoria Stanley, 204–5, 206, *207*, 282n63
Hydrangeas at Lange's Greenhouses, 183
Il Solitario (The Hermit), *190*, 191–93, *192*, 279n206, 279n212
Isabella Stewart Gardiner, 32, 78
Jacques-Émile Blanche, 98–100, *101*, 258n142
Javanese Dancers, *73*
Katharine Chase Pratt, 183, *184*
La Carmencita, 32
Lady Agnew of Lochnaw, 82–83, *85*, 254n61
Lady Sassoon, 63, *64*
Léon Delafosse, 78, *78*, 254n47
Lord Ribblesdale, 204, *205*
Lord Ribblesdale, sketch for, 204, *206*
Madame Édouard Pailleron, 32
Madame X (Virginie Gautreau) 19, 139–43, *140*, 155, 208, 253n32, 268n189, 268n192, 268n204
Man Screaming, 225, *226*
Mauresque, 260n36
Miss Beatrice Goelet, 32, *33*
The Misses Hunter, 31
The Misses Vickers, 73, *73*, 82, 273n87
Mrs. Albert Vickers (Edith Foster), 75–78, *76*, 253n32, 253n38; study for, 75–77, *77*
Mrs. Carl Meyer and Her Children, 32
Mrs. Charles Thursby, 80, *82*
Mrs. Douglas Dick, 254n54
Mrs. Edward Darley Boit, 61–63
Mrs. Kate Moore, 80
Mrs. Robert Harrison, 206–8, *207*, 209
Mrs. Thomas Wodehouse Legh, 82, *84*
Nude Oriental Youth, 170–73, *171*, 273n96, 273n97, 273n98
Padre Sebastiano, 153, *154*, 269–70n24, 270n25, 272n65
Paul Helleu, 104–6, *105*, 211, 219
Paul Helleu Lying in a Field, 102–3, *103*, 211
Paul Helleu Sketching with his Wife, 211
Pressing the Grapes: Florentine Wine Cellar, 201, *202*
Robert Louis Stevenson, 20, 103–4, *104*, 211
Robert Louis Stevenson and his Wife, 20, 212–13, *212*, 220, 223, 283n109; study for, 214, *214*
Rose Branch, 162, *163*
Self-Portrait, *164*, 164–65, 272n64
Sir Frank Swettenham, 203, *204*
Standing Male Nude with Arms Akimbo, 216–17, *217*

Tommies Bathing, 226, *227*
Two Girls on a Lawn, 168–69, *169,* 273n87
Two Men Smoking and Writing(?), 63, *65*
Two Octopi, 95, *96*
Venetian Onion Seller, 73, *74,* 273n88
W. Graham Robertson, 20, 211, *216,* 216–17, 219–20, 223, 284n134, 285n137, 285n138; study for, 217, *218*
William M. Chase, 150–51, *151*
The Wyndham Sisters: Lady Elcho, Mrs. Adeane, and Mrs. Tennant, *frontispiece,* 31, *33,* 73, 245n53
Sargent, Violet, 209, 211, 273n87
Sartiliot, Claudette, 1, 37
Sassoon, Lady, 63, *64*
Savoy, Eric, 215
Saxoléine, *156,* 156
Scarry, Elaine, 38, 145
scent
of ambergris, 31, 260n36
of cigarettes and cigars, 103, 104, 106, 219, 259n155
of flowers, 12, 29, 31, 37, 44, 86, 104, 126, 196, 223, 250n157, 275n138
of fruits, 194
of garlic and onions, 74, 195
and queer community, 37, 46, 195, 223, 249n130, 250n157
of Whistler's portrait faces, 67
Schlegel, Friedrich, 10, 242n73
Schneider, Laurie, 67
Schopenhauer, Arthur, pederast as fly, 12, 241n60
scrap
with flower-palette motif, *148*
with hands, 79, *81*
scrapbooks (Sargent's), 274n121
sea anemone(s), 23, 98, 117, *118,* 119, *119,* 120, 122, *123,* 123, 243n3
Sears, Helen, 20, 174–79, *175, 176,* 180, 183–84, 185, 189, 225, 274n112, 274n113, 274n116
Sears, Sarah Choate, 184–85, 208
Helen Sears with Japanese Lantern, 184–85, *186,* 276n150
Seaton, Beverly, 276n145
Sebastiano, Padre, 153, *154,* 269–70n24, 270n25, 272n65
Sedgwick, Eve Kosofsky, 173
seed trade cards and catalogues, 19, 37, *40,* 46, *47, 149,* 150, 269n12
sélam, 48
Sélavy, Rrose, 235
self-fertilization (botanical), 129, 133, 213, 248n122, 254n55
semen
and hands, 90–91, 96
vegetable images and associations, 37, 42, 244n14, 246n77
Seneca, 1
sensitive plant, 37, 246n79, 254n63
sesame, 46–48
sexual inversion. *See* inversion
Shelley, Percy Bysshe
The Sensitive Plant, 246n79, 254n63
Zoroaster, 201
Shelley, Mary, *Frankenstein,* 92, 256n104, 286n167
Sheriff, Mary, 251n173
Shuttleworth, Sally, 176
Sickert, Walter, 219
Sidlauskas, Susan, 141–42, 165, 272n67, 272n74
Sidney, Sir Philip, 8
Simmons, Edward, 143
Simpson, Marc, 186, 277n158
Sitwell, Edith and Osbert, 272n69
slang, 29–30, 31, 35, 37, 46, 48–49, 57, 91, 94, 97, 104, 126, 156, 195, 219, 244n29, 245n45, 246n68, 246n79, 256n93, 260n28, 273n85, 279n2, 279–80n8, 280n21
Sleeping Beauty, 189, 271n48
Smith, Logan Pearsall, 280n24
Smithers, L. C. and Richard Burton, *Priapeia,* 50
smoking. *See* cigarettes *and* cigars
snake(s), 18, 93, 95, 96, 97, *98, 99, 101,* 102, 258n133
sodomites
codes of, 195, 212
"infâme," 250n145
Proust's chronicle of, 109, 216
vegetable associations of, 46, 49–50, 237, 249–50n139, 250n141
Solomon, Simeon, 239n26, 252n206
La source, 57–60, *61,* 251n177
Spate, Virginia, 286n3
sphincter. *See* anus
Sphincter-Grip Armoured Hose, 97, *98,* 258n133
Splendor Solis (Trismosin), hermaphrodite, 168, *168*
"Spring Flowers," *Illustrated Sporting and Dramatic News,* 55, *57*
Stafford, Barbara, 127
Stanley, Victoria, The Honourable, 204–5, 206, *207,* 282n63
Starr, Sydney, 51
Steedman, Carolyn, 272n67, 272–73n75, 273n91
Steichen, Edward, 235
Stephanson, Raymond, 240n45
Stevenson, Fanny, 212, *212,* 214, 258n147, 283n109
Stevenson, R. A. M., 201
Stevenson, Robert Louis
atavism of, 283n106, 283n115
and Gosse, 283n109,
illness of, 213, 215
influence on Proust, 284n123, 285n157
and inversion, 18, 213
like a plant, 103, 212
portraits by Sargent, 20, 103–4, *104,* 211–13, *212, 214,* 214–15, 220, 283n109
and Sargent, 157–58, 211, 215
Works
"The Rajah's Diamond," 137, 266n159
Strange Case of Dr. Jekyll and Mr. Hyde: aestheticism in, 214, 284n118; hands in, 72; inversion in, 213–15, 223–24, 283–84n117, 285–86n160; regression in, 130, 213–14, 264n124
"The Yellow Paint," 285–86n160
Stoker, Bram
Dracula, 196, 197, 201, 280n21
and syphilis, 282–83n91
Stop (L. P. G. B. Morel-Retz), caricature of Carolus Duran, 63–64, *67*
Stott, Annette, 31, 32, 245n51, 245n53

Strange Case of Dr. Jekyll and Mr. Hyde. See Stevenson, Robert Louis
strawberry/ies, 31, 48, 247n86
sunflower(s), 4, 90, *91*, 177, *179*
superposition, 168–69
Svengali, 199
Swettenham, Sir Frank, 203, *204*, 281n53
Swinburne, Algernon, 82, 87, 255n72
  Sambourne caricature of, 177, *178*
La sylphide, 57
Sylvestre, Armand, 32
Symbolism, 86, 180, 184, 231
Symonds, John Addington
  and dreamy mental states, 274n119
  and inversion, 18, 49, 212, 242n89, 250n149, 258n135
  natural historical art criticism of, 12
  and von Gloeden, 286–87n10
  Works
  "An Allegory," 258n135
  "The Valley of Vain Desires," 49
Symons, Arthur
  poetry of, 15, 60–61, 253n15
  on Sargent, 206
  on Whistler, 74
syphilis
  and atavism, 129–30, 134, 213–14
  and creativity, 131–33, 219, 265n136, 265n138, 282n88
  doctors specializing in, 129, 133, 262n95, 264n115
  and leprosy, 264n115
  metamorphoses induced by, 127–30, 213, 263n105
  suggested in Courbet's *Origin of the World*, *133*, 133–34, 265n143, 265–66n144
  suggested in Lalique's *Gynécologie*, 135, *136*
  suggested in Sargent's paintings, *140*, 141–42, *190*, *192*, 193, 207, *207*, *210*, 282–83n91
  and tuberculosis, 264n115
  and vampirism, 282–83n91
  vegetable associations of, *127*, 127–30, *128*, 262n95, 263n105, 271n56
Taglioni, Marie, 57
Taine, Hippolyte, botanical art criticism, 11–12, 240n53
Talbot, William Henry Fox, 20, 235
Tardieu, Ambroise, 243n4
tea, teapots, and teacups, 90–91, *91*, 256n93, 256n95, 284n134
Teleny. See Wilde, Oscar
Tennyson, Alfred, Lord, 219
Tentation de Saint-Antoine. See Flaubert, Gustave
Terry, Ellen, 63
testicles
  of eels, 93, 257n113
  vegetable images and associations of, 26, *26*, 37, 49, *49*, 80–82, *83*, 244n14
Thea, ou la fée aux fleurs, 245n49
Thiébaut, Philippe, 98, 258n137, 275–76n139
Thomas, Théophile, costumes for *Le roi Carotte*, 57, *59*
Thompson, Henry, 200
Thousand Nights and One Night, 46–48, 181, 199, 281n39
Thursby, Mrs. Charles, 80, *82*
Titian, 72, 145, *147*, 269n9
tobacco, *40*, 104, 219, 259n153. *See also* cigarettes *and* cigars
torpedoes. *See* eels
torture, 60, 142, 195, 201, *202*, 203, 206, 281n50
Townsend, Beatrice, 79, *80*, 82
trade cards
  Le Bon Marché, 37–38, *41*
  Chicorée, 53, *56*
  Dr. Warner's Coraline Corsets, 29, *29*, 261n47
  with flower-palette motif, 147, *148*
  Liebig cards, 23, *25*, 244n8
  Maison du Pont-Neuf, 51, *53*
  with pollinators, 60, *62*
  Saxoléine, 156, *156*
  seed and fertilizer cards, 37, *40*, 259n153
Trilby (Du Maurier), 30, 199
Trismosin, Salomon, *Splendor Solis*, hermaphrodite, 168, *168*
tuberculosis/consumption, 104, 141, 213, 264n115
tuberose, 23, *24*, 243–44n7, 249n129
tulip(s), 8, *9*, *11*, 32, *33*, 35, *36*, 42, 150, 270n39
Tulipe noire, La (Dumas *père*), 42

Ulrichs, Heinrich, 243n4
Uranists/Uranians, 91, 243n4
uterus
  absence of, *137*
  blossom image of, 8, *9*, *10*, *11*
  botany of, 20, 113–15, 116, 138, 244n14, 261n49
  diseases of, *113*, 115
  explored, 112, 259–60n17
  inversion, prolapse, and retroflexion of, 122, *122*, 138, *138*, 261n74
  lining of, 261n44

vagina
  botanical associations, 26, 29–30, 126, *126*, 244n13, 244n29
  Courbet painting of, 133–34, *133*, 265–66n144
  explored bimanually, 111–12, *112*
  geological images of, 126, *127*, 134, *134*, *135*
  gynecological illustrations of, *112*, *117*, *122*, *127*, *132*, *137*, *138*, *139*
  invagination and consciousness in Merleau-Ponty, 110, 121, 122, 213
  medical procedures involving, 111–12, *112*, 116, *117*, 138, *138*, *139*
  other slang for, 156
valentines, 8, *10*, 23–26, *25*, 50, 240n43
Valéry, Paul, 282n88
vampirism, 181, 183, 196–201, 211, 217, 253n40, 280n21, 281n34, 282–83n91, 284n135, 286n167
van Capelleveen, Paul, 16
van Dyke, John, 270n38
The Various Contrivances by which Orchids are Fertilised by Insects. See Darwin, Charles
vase(s), 119–20, 147, *162*, 177, *179*, 258n135, 273n84, 275n134, 276n149
vegetations, anatomical, 20, 113–15, 127, *127*, 133, 264n115
Venice, 49, 73, 74, *74*, 102, 196, 220, 286n161

Verlaine, Paul, 48, 219, 250n156
Vicinus, Martha, 274n104
Vickers, Billy and Dorothy, 167–74, *167*
Vickers, Edith (Mrs. Albert), 75–78, *76*, *77*, 253n32, 253n38
Vickers, Katie, *169*, 273n87
Vickers, Misses, 73, *73*, 82, 273n87
Vick's Magazine, *113*
Vick's seeds, 46, *47*
Villiers de l'Isle-Adam, Philippe-Auguste, *L'Ève future*, 91–92, 96, 256n98, 259n155, 264n115, 267n164, 280n27
violet(s), 15, 16, 30, 31, 35, 42, 46, 104, 246n68
Virgil, 1, 143
von Gloeden, Wilhelm, 20, 235–36, 286–87n10

Wagner, Anne, 233–34
Wagner, Richard, *Parsifal*, 32–34, 127, 156, 246n64, 270n37
wagtail(s), 4, 145, 195, 196, 219, 237, 279n2
Warhol, Andy, 224, 229
Warwick, Alexandra, 282–83n91
wasp(s), 14, 51, 57, 61, *61*, *63*, 281–82n57, 285n138
water lily/ies, *29*, 48, 86, 159, 231, 249n136, 261n47
  Victoria Regia, 35
Watson, Forbes, 237
Weil, Kari, 169–70, 171, 274n101
Weininger, Otto, 241n56, 241n59
Weir, James, 240–41n55
Wertheimer, Alfred, 223, *224*, 285n159
Wertheimer, Asher, 285n159
Westminster Aquarium, 94–95, *95*, 257n117
whales, 109, 115–16, 260n38
whips, 195, 196, 203, 204, 205, *205*, 206, *206*, *207*, 279–80n8
Whistler, James McNeill
  art as pollination, 18
  atavistic painting of, 14
  black portraits by, 199, 281n34
  butterfly signature/identification, 3, 16, *17*, 20, 65, 67, *69*, 196, 219, 252n206, 281n34
  caricature in *Life*, 150, *150*
  fairy-land of, 279n200
  hands of, 74–75
  and inversion, 18, 67, 69
  Mallarmé on, 280n27
  and Montesquiou, 196–201
  nationality, question of, 187
  painting as nature produces plants, 2, 3
  painting in the dark, 198, 280n22
  and Poe, 198–99, 280n27
  sadism of, 196, 199, 211, 280n24
  and Sargent, 14, 18, 65, 123, 187, 196, 201, 205, 206, 211
  sexuality of, 67
  sitters as plants and pollinators, 65–67, 196, 245n51
  stinger, 67, *69*, 196, 198
  as vampire, 196–99
  whips, 196, 279–80n8
  Works
  Arrangement in Black and Gold: Comte Robert de Montesquiou-Fezensac, 196–201, *197*
  Butterfly, 67, *69*
  The Gentle Art of Making Enemies, 67, 279n200
  Large Bowl and Cover, 200, *200*
  Loïe Fuller Dancing, 15, *15*
  Note in Green and Brown: Orlando at Coombe, 277n156
  Three Butterflies, 16, *17*, 196
Whitman, Walt
  albescent honey, 42, 219
  "Bunch Poem," 53
  and calamus, 46, 249n130
  and electricity, 92–93, 256n104
  "Faces," 42
  and flowers, 42
  on hands, 82, 88, 92–93
  "Song of Myself," 88, 92–93
  "Spontaneous Me," 82
  Symonds on, 12, 250n149
Wilde, Oscar
  on art and life, 4
  art for art's sake, 174
  and Blanche, 102
  caricatured in: *The Green Carnation*, 249n129; *Patience*, 44; *Punch*, 177, *179*, 274n125; *Washington Post*, 130, *131*
  fairy tales of, 278n184
  and flowers, 4, 46, 177, *179*, 239n28, 248n115, 249n129, 282n85, 284–85n136
  on hands, 19, 70, 71, 95, 209
  inversion of, 18, 213, 214
  Queensberry's vegetable bouquet for, 49–50
  and Robertson, 215–16
  and Sargent, 18, 215–16, 284n126–28
  as Sebastian Melmoth, 219
  trial of, 180, 284n126
  and von Gloeden, 235–36
  Works
  "Athanasia," 284–85n136
  Intentions, 49, *49*
  The Picture of Dorian Gray: Basil Hallward, 4, 13, 201; bookstand in, 219; cover of, 16, *18*; Dorian as plant, 46, 209, 284n135; Dorian as vampire, 201, 284n135; Dorian's portrait in, 124, 217, 284–85n136; hands in, 70, 71; pollination scene in, 13, 48; preface mocked in *Punch*, 274n125
  Teleny: buttonholes in, 50–51, 223, 250n157; catalogue of hand types in, 88–89, 95, 209, 253n19; flowery genitals and sex acts in, 30, 37, 48, 49, 97, 245n33, 246n77; handshake in, 90, 91; sex in, 97, 122
Wildenstein, Daniel, 239n22
Winckelmann, 1, 215, 238n6
Wise, John Richard de Capel, *The First of May: A Fairy Masque*, 189, 278n191
Wood, John George, *Homes Without Hands*, 87–88, 255n77
Wood, T. Martin, 206
woodbine(s), 26, 73, 104
Woodbines, Will's Wild, 104
Wordsworth, William, 86
wreaths, 159, 229–31, *230*, 244n19, 271n49, 271n50, 271n52, 277n169
Wyndham sisters, *frontispiece*, 31, *33*, 73, 245n53

A Yellow Aster (Iota), 35

Zeus, 158, 240n45, 270n44
Zeuxis myth, vegetable versions, 3, *3*
Zola, Émile
artists in, 8, 139, 239n40
tableaux vivants in, 35–37, 122
vegetation in, 122, 125–26, 128
WORKS
La curée, 35–37, 122, 128
La faute de l'abbé Mouret, 125–26
L'oeuvre, 8, 139, 239n40